ENCYCLOPEDIA OF
BRIDGES AND TUNNELS

ENCYCLOPEDIA OF
BRIDGES AND TUNNELS

STEPHEN JOHNSON
ROBERTO T. LEON, P.E., PH.D.

☑®

Facts On File, Inc.

Encyclopedia of Bridges and Tunnels

Copyright © 2002 by Stephen Johnson

Facts On File, Inc.
132 West 31st Street
New York NY 10001

Library of Congress Cataloging-in-Publication Data

Johnson, Stephen (Stephen, Paul), 1954–
The encyclopedia of bridges and tunnels/Stephen Johnson, Roberto T. Leon.
p. cm.
Includes bibliographical references and index.
ISBN 0-8160-4482-1
1. Bridges—Encyclopedias. 2. Tunnels—Encyclopedias.
I. Leon, Roberto T. II. Title.
TG9 .J64 2002
624'.2'03—dc21   2001053200

Facts On File books are available at special discounts
when purchased in bulk quantities for businesses, associations, institutions,
or sales promotions. Please call our Special Sales Department in New York at
(212) 967-8800 or (800) 322-8755.

You can find Facts On File on the World Wide Web at http://www.factsonfile.com

Text design by Joan M. Toro
Cover design by Cathy Rincon

Printed in the United States of America

VB FOF 10 9 8 7 6 5 4 3 2

This book is printed on acid-free paper.

CONTENTS

For our parents

FOREWORD

The history of civilization is the story of travel and transportation. The building of crossings in the form of bridges and tunnels has always been a measure of technological development.

Every bridge and tunnel contains engineering concepts and structural materials created through years of trial and error. They have also been made possible by engineering mechanics, the study of motion and forces on structures. As deoxyribonucleic acid (DNA) represents the genetic history of a human being, an accumulation of concepts and innovations developed decades, if not centuries, before is revealed by a close examination of any bridge or tunnel.

It is for this reason that the ancient wooden bridges of China and the stone arches built by Roman engineers are no less important than the Brooklyn Bridge or Japan's Akashi Kaikyo Bridge, two of the world's great engineering achievements. The world is filled with magnificent spans, both large and small, that serve not only as routes of travel and commerce but also as monuments and beloved icons.

Although tunneling was originally used to create shelter and later to allow the extraction of salt, coal, and ores, this engineering art was eventually used for the creation of extraordinary routes through great mountains and beneath rivers and seas. When the need to leap across rivers and chasms or to burrow through hard rock or dangerously soft ground became apparent, engineers devised ways of accomplishing these incredible feats. The result has been a steady progression of construction marvels as engineers and workers overcame technical obstacles and hazards previously considered insurmountable.

Humankind's ability to conceive of such great projects, plan them, and eventually build them is a testament to technical skill and ingenuity. The erection of a bridge and the boring of a great tunnel are just as often monuments to the courage, vision, and will of those involved in these projects. The stories behind history's great bridge and tunnel efforts are far more complex than most of us who use them would ever suspect.

The Byzantine political processes that often precede great bridge and tunnel projects are just the beginning. When the plans are made and the effort begun, those involved are in a race with time and nature to complete the project on schedule and within budget.

Financial ruin, scandal, and the death of workers struggling to meet schedules have marked many of these great projects throughout history.

Today, modern structural materials, computer technology, and daring new designs have been coupled with a heightened awareness of worker welfare. Engineers continue not only to pursue the dream of creating a stunning engineering achievement but to do so with the maximal degree of safety, while reducing the risk of cost overruns and delays, issues that still plague some of the world's biggest and most expensive projects.

Modern bridges and tunnels are designed to withstand numerous types of loads, including those due to large-magnitude earthquakes, typhoons, and collisions with ships. As such, modern long span bridges and tunnels are immensely complex structures. In general, such projects represent the culmination of long-held dreams and the work of thousands of individuals. The Eurotunnel beneath the English Channel was something considered impossible for hundreds of years that today carries high-speed passenger and freight trains between Great Britain and France. Bigger and grander bridges and tunnels are on the drawing board as humankind continues its quest to conquer obstacles.

No single book can catalog all these projects nor fully explain each one. Although the scale and impact of the projects were criteria in the selection of the projects described in this encyclopedia, space limitations meant that many interesting and important structures could not be included. *The Encyclopedia of Bridges and Tunnels* introduces the reader to some of the most important and extraordinary bridge and tunnel efforts.

The encyclopedia also reveals the fascinating technology that makes these projects possible while examining the human and political dramas that occurred during their creation. It is a window into a little-known world filled with both exhilarating success and crushing failure. The encyclopedia is written with the lay reader in mind, and thus the terminology is not that of engineering texts.

After reading *The Encyclopedia of Bridges and Tunnels* and gaining a better understanding of the sacrifices and the engineering artistry that went into these structures, you will never be able to look at a bridge or tunnel in quite the same way.

—Roberto T. Leon, P.E., Ph.D.

INTRODUCTION

If you have traveled anywhere in the world by any means, you have crossed a bridge or passed through a tunnel. Even air travel that can loft us across oceans and over mountains does not eliminate the trip to and from the airport. The world is full of chasms, rivers, straits, and mountains. For ancient humans, negotiating such obstacles meant swimming, climbing, or taking the long way around. In many cases, people simply stayed where they were, imprisoned by geographic barriers.

No one should doubt the importance of eliminating these obstacles. It was, after all, the natural emergence of the Bering Land Bridge between northeast Asia and the northwest corner of North America that allowed humans to populate the Americas more than 10,000 years ago.

With tunnels and bridges, we simply cross impediments to mobility as though they did not exist. Although most are free to their users, the only delay we might encounter while crossing rivers and mountains in modern times is a brief stop at a tollbooth.

If humans needed food to eat and clothes for warmth, they also needed bridges and tunnels to allow them to sate the desire to conduct commerce or simply to wander. In the beginning, fallen trees or slabs of stone connecting the banks of a stream allowed the crossing of small obstacles. Simple post and beam bridges of timber or stone emerged next. These were followed by more complex stone arch spans. Eventually, refined, truss-reinforced bridges of timber and then of iron and steel emerged in the 19th century. The ancient compound of concrete reinforced with iron and steel during the latter half of the 1800s would mature into one of the most dependable structural materials for both bridge and tunnel construction. As innovations and materials were developed, suspension bridges haltingly came into use during the 19th century, and by the 20th century these breathtaking structures were appearing around the world. Daring steel arch bridges grew between shores, and dashing cable-stayed bridges spanning considerable distances began to appear by the 1950s.

The engineering art of tunneling is no less ancient than that of bridge-building. As Paleolithic humans were quick to use existing caves for shelter, both the Neanderthal and the Cro-Magnon were the first to tunnel as they expanded the caves they inhabited. The ancient

Babylonians used tunnels to carry water for irrigation, and the Romans used the first hard rock mining method of fire-setting to tunnel 4,920 feet from Pozzuoli to Naples in 36 B.C.

As the transport of goods became essential in a world increasingly connected by trade during the 18th century, tunneling became a science unto itself. Engineers devised better ways of boring through rock to allow the passage of canal, train, and finally automobile traffic. Tunnels assumed even more importance with the advent of railroad technology during the 19th century, which also caused an unprecedented explosion in bridge construction.

Although there was a bridge-building boom during the 19th century, this did not mean the technology of bridge construction was without problems. Bridge disasters were common, particularly among rail bridges and immature suspension bridge designs. The understanding of loads and stresses was still in its infancy. The science of building longer, lighter, and more attractive bridges was a complex task. The 1940 collapse of the spanking new Tacoma Narrows Bridge proved that even relatively modern engineering techniques are no guarantee of success.

The nearly five-mile-long Hoosac Tunnel in western Massachusetts is an example of the great 19th-century tunneling boom in ways good and bad. It was eventually completed but only after 24 years of agonizingly slow and fatal effort. Beset by difficult geologic conditions, tunneling technology that was inadequate for the task, and unreliable funding, the Hoosac Tunnel was an example of everything that could go wrong underground. Despite the accumulation of knowledge and experience in the intervening 140 years, the complexities of underground work saw the Eurotunnel project beneath the English Channel become a nightmare plagued with technical and financial problems.

When properly constructed, bridges and tunnels can become enduring symbols of human ingenuity. Some bridges thousands of years old still stand, and many are now officially recognized as cultural treasures. The stone arch span known as the Alcántara Bridge was built in Spain by the Romans in ca. A.D. 100. Wrecked by war during the intervening centuries and rebuilt, it stands today as a beloved symbol to the people of Spain and as a testament to the prowess of ancient technology.

In ancient times, a bridge spanning a river or a chasm where one never existed before must have been a wonder to behold. Although bridges and tunnels are now ubiquitous, we still cannot help but be mesmerized by these bigger-than-life accomplishments. They are nothing less than symbols of a society's ability to conquer nature. Bridges and tunnels are our silent servants and deserve our collective respect since they are the arterial system of our mobile societies.

The story of bridges is the story of history, and even the smallest and simplest of bridges has played a role. Who is not familiar with a bridge that has not existed for nearly a thousand years, immortalized by the nursery rhyme "London Bridge Is Falling Down"? A close look at the story of this bridge reveals a complex tapestry of events. This bridge did not fall down: It was pulled down by Viking mercenaries fighting on behalf of an English king to rout another group of Vikings who controlled England at the time.

In fact, bridges are so crucial to military operations that both their preservation and their defense have figured heavily during the desperate days of armed conflict. In 1945, the Ludendorff Bridge, known as the Bridge at Remagen, miraculously survived demolition attempts by German troops and remained erect long enough to hasten the Allied advance into the heart of Germany. In 1297, Scotland's rebel leader William Wallace—"Braveheart"—craftily allowed half an English army to cross a narrow wooden span known as Stirling Bridge. Once they had squeezed across the wobbly bridge, Wallace slaughtered the divided force.

Because mobility is so essential in warfare, engineers have devised portable bridges that can be carried to a location and rapidly erected. Ancient military commanders in the Iraq-Persia region were using pontoon bridges nearly 3,000 years ago to move their armies over rivers. During World War II the Bailey bridge was developed, providing a portable span bridge strong enough for tanks that enabled the Allies to liberate Europe. When American peacekeeping troops entered Bosnia in 1995, they did so on a modern pontoon bridge assembled by combat engineers.

Tunneling has also been a tool of war. Armies attacked enemy forts by using tunnels during siege warfare while their adversaries tunneled toward the approaching excavations to head them off. In an attempt to defend itself against attack after World War I, the French government constructed a massive series of tunnels and underground fortifications known as the Maginot Line along its border with Germany. The Swiss, also fearful of Germany, constructed a similar defensive system during the 1940s.

The price paid in the construction of both bridges and tunnels has always been high in human and financial terms. Some underground workers have died in cave-ins and explosions, and others fell victim to the silent death of carbon monoxide asphyxiation. During the 1930s, workers fell to their death from the Golden Gate Bridge or died of heat stroke in the poorly ventilated diversion tunnels of the Hoover Dam. Hundreds of miners, most of whom were African Americans, died of the nefarious lung disease silicosis during the Depression Era excavation of West Virginia's Hawk's Nest Tunnel. This event ignited a workplace health scandal still resonating within the tunneling industry.

The high-wire act of bridge construction also claimed casualties underground. Before the hazards of working in a compressed air environment were understood, workers were crippled and killed by caisson disease, or the bends. When they left the pressurized confines of the caissons during underwater foundation work, the nitrogen in their blood expanded, with terrible consequences. Even Washington Roebling, the chief engineer of the Brooklyn Bridge, found himself disabled for more than a decade after being stricken with caisson disease. When compressed air was subsequently used to prevent flooding in subaqueous tunnels during the latter half of the 19th century, caisson disease also struck tunnel workers in scandalous numbers.

The steep financial and human costs often made high-profile bridge and tunnel projects lightning rods for criticism and scandal. These great structures not only defeated the challenges of physical impediments, but also survived outright ridicule, eventually to become revered symbols. Dogged by criticism and scandal over the deaths of its workers and the high cost of its construction, the brilliantly designed and steadfastly reliable Brooklyn Bridge endured as an example for other large-scale suspension bridge projects.

Without these expensive, frequently controversial, but invariably impressive wonders, life as we know it would cease: All forms of commerce and travel would halt. Trains would be stranded in sidings and tractor–trailer rigs would be idled at truck stops. Passenger vehicles by the millions would be perched at the edge of mountains and riverbeds with their occupants staring at the obstacles that have blocked attempts to travel since the beginning of time.

Although the majesty of crossings such as the massive Verrazano Narrows Bridge between Brooklyn and Staten Island is obvious from miles away, the remarkable technology of tunnels is hidden beneath wide straits and towering mountains. The drama of these engineering and construction feats is mostly buried. Japan's Seikan Tunnel and the Anglo–French Eurotunnel make it possible for rail traffic to travel dozens of miles under water. What the passengers see is a long, gleaming tube whose orderly appearance belies the floods, the explosive rock bursts, and the heartbreak of struggling through the innards of the Earth.

The great tunneling projects of the Alps, including the Simplon Tunnel, Mont Cenis Tunnel, and Saint Gotthard Tunnel, were monumental efforts that cost hundreds of lives, often bankrupting the firms doing the digging and taking years to accomplish. As the tunnels became longer and penetrated more complex geologic barriers, engineers came up with solutions to allow the efforts to proceed. Rock drills were developed, and explosives like black powder, nitroglycerin, and eventually dynamite were used. Finally, the dream of tunnelers was realized in the 1950s with the American creation of tunnel-boring machines that could methodically grind through the Earth and leave a reinforced lining in its wake.

With the growing popularity of the automobile during the first two decades of the 20th century, New York City realized the necessity of building the first tunnel for vehicular traffic and completed the Holland Tunnel in 1927. It was up to engineers to solve a safety problem that threatened the project's success—how to ventilate internal combustion engine exhaust fumes from the 8,500-foot tunnel. The ingenious ventilation system helped everyone breathe easier while driving beneath the Hudson River between New York and New Jersey. The result of this pioneering effort was that vehicular tunnels were subsequently built around the world.

Although these bridges and tunnels are easy to take for granted because they serve in silence, each was a unique project demanded by a world desperate to remove impediments to trade and travel. But America has taken its bridges and tunnels for granted and is only now attempting to make up years of neglect. Nearly 30 percent of the nation's 592,091 bridges are structurally deficient.

This state of affairs is almost unbelievable since bridges and tunnels not only are a means of traveling from one point to another, but have themselves become travel destinations. So beloved are these monuments to ingenuity that the Golden Gate Bridge and the Eurotunnel attract tourists they also help to transport. Who is so world-weary that he or she has not paused to stare in awe at soaring spans like Saint Louis's historic Eads Bridge or even Louisiana's nondescript Pontchartrain Causeway, which is so long it disappears beneath the curve of the horizon?

Like the monolithic structures they designed, the bridge builders have also become symbols, particularly those who pushed the limits of modern bridge design. The innovative accomplishments of 19th-century bridge designers such as America's John Augustus Roebling and Britain's Thomas Telford paved the way for structures such as the Akashi Kaikyo Bridge at Kobe, Japan. This suspension bridge—the world's longest—spans an ocean strait six miles wide. Despite its seemingly impossible dimensions, the bridge is based upon the same suspension bridge principles that Roebling perfected in his design of the Brooklyn Bridge in 1869.

Other projects were born under unlucky stars. The Quebec Bridge in Canada, which collapsed twice during its construction in the early 1900s at a high cost in human life, is an example of engineering failures still mulled by experts. Even Boston's ambitious tunnel and bridge project known as the "Big Dig" has come under federal criticism as technical problems stoked multibillion-dollar cost increases while it crawled toward completion at the beginning of the 21st century.

Although thousands of bridges and tunnels have already been built, advances in technology and techniques have made it possible to tunnel through or bridge across obstacles that would have been considered impassable even a few years ago. The use of refrigeration to solidify the ground beneath buildings is allowing Boston to construct tunnels beneath historic buildings without damaging irreplaceable structures or interfering with surface activities. Modern, and eerily beautiful, new cable-stayed bridges are a visual treat with their skeins of brilliantly painted cables radiating from tall support towers to the road decks below. This visually striking design allows lighter bridges to be built in less time. Modern bridge designers like Santiago Calitrava in Spain have specialized in cable-stayed bridge concepts, applying the design to small footbridges as well as massive spans. Functional bridges in every sense, they are also seen as a perfect marriage between artistry and engineering.

The future of bridge and tunnel construction will be exciting and stamped with the imprint of new technologies. Bridges are now being built out of fiber-reinforced polymer composites—the structural material of advanced aircraft—and tunnel-boring machines are now guided by lasers and computers. It has been recommended by some that future tunnels be suspended between the surface of the water and the seafloor, testing not only the skill of the engineers but the confidence of the public.

Although some bridge and tunnel projects have been flawed, sometimes to the point of catastrophic failure, each has provided lessons for the efforts that followed. Although the World Wide Web has connected the planet electronically, the need and the desire to go from place to place in the physical world remain deeply ingrained. Bridge and tunnel projects both past and future will continue not only to amaze us but to remove barriers so each of us can explore the world firsthand.

ENTRIES A-Z

A

abutment An abutment is the structural component that supports the ends of a bridge. In older bridges, an abutment can be as simple as a rock outcrop located on either side of a river or on the wall of a valley. In modern bridges the abutments are earth-retaining structures that serve as both supports and gradual approaches to the bridge. Because abutments are rigid structures, they are often designed to take many of the horizontal or side-to-side forces.

A classic example of a bridge completely supported by a pair of opposing abutments was the 533-year-old Mostar Bridge in the Bosnia and Herzegovina village that took its name from the bridge. The bridge spanned nearly 90 feet between the shores of the Neretva River, and its single arch gracefully rose 62 feet above the water. Built by the Turks, the arch was handsomely supported by a pair of stone and masonry abutments built into the steep walls of both riverbanks.

Although the bridge collapsed under Croat tank gunfire in 1993, the massive abutments remained essentially intact. The superb construction of the abutments made them fully capable of supporting a rebuilt bridge (using the original stone) in an international reconstruction effort that is expected to be completed by 2004.

See also MOSTAR BRIDGE.

adit An adit is a horizontal passage to a tunnel or mine. It can be used to provide access to the working face of a tunnel heading or to a mine and is different from a shaft, in that a shaft is a vertical entrance to a tunnel.

Intermediate shafts and adits between the end openings or portals of a tunnel can be used to provide additional working faces or headings within a tunnel while also serving as ventilation outlets.

afterdamp and blackdamp Afterdamp is a mixture of carbonic gas and nitrogen produced during a coal mine fire or explosion that can cause death if inhaled. This combination is also known as aftergases. A second gas produced during coal fires and explosions is blackdamp, also known as chokedamp.

Blackdamp is heavier than air and can flow along the floor of a tunnel. Consisting of carbon dioxide and nitrogen, it contains so little oxygen that it can asphyxiate someone who has only blackdamp to breathe. A warning that a concentration of blackdamp is present is the extinguishing of a lighted flame.

Both gas mixtures can be deadly in a cruelly ironic way since those killed by them have usually survived the initial fire or explosion that produced the gases. Although modern emergency respiration equipment can provide protection from these gases, ventilation is a crucial defense against gas buildup.

See also CARBON MONOXIDE ASPHYXIATION; VENTILATION.

airlock *See* caisson.

Akashi Kaikyo Bridge (built 1988–1998) *Japan's amazing suspension bridge* As a collection of

2 Akashi Kaikyo Bridge

islands, Japan is faced with an infrastructure problem that has forced it to spend billions of dollars to set records in bridge and tunnel construction. The 1998 completion of the mammoth Akashi Kaikyo Bridge linking the main island of Honshu with the southern island of Shikoku gave Japan a world record in bridge construction for the longest such bridge.

At the dawn of the 21st century, the Akashi Kaikyo Bridge—named after the *kaikyo*, or the strait that it spans—was the longest suspension bridge in the world by a huge margin at the time of its completion. The center span between its gargantuan towers is 6,532 feet and the total length of the bridge's three suspended spans is 12,831 feet. By comparison, the bridge is twice the total length of the famous Golden Gate Bridge, which has a center span of only 4,200 feet. The Strait of Messina Bridge, which was still in its planning stages as of 2000, will, when built, smash this record with a center span reaching a length of 10,826 feet. Until then, the Akashi Kaikyo Bridge will serve as a mammoth testament to construction and engineering ability.

The Akashi Kaikyo span handily made off with the title of the world's longest suspension bridge held by England's Humber Bridge since 1981. The 4,626-foot center span of the mighty Humber Bridge is 1,906 feet shorter than that of the Akashi Kaikyo. The center span of the Akashi Kaikyo Bridge even bested the length of Denmark's Great Belt (suspension) Bridge completed in the same year as Japan's behemoth. The center span of the Danish bridge is 5,328 feet, or 1,204 feet shorter than that of its Japanese counterpart.

But it is not that Japan had a choice in the matter, or that the government was eager to spend $7.7 billion on the world's most expensive bridge. For decades, the Japanese government's declared goal has been to link all of what are termed its home islands with bridges and tunnels. Until air travel became a viable alternative, commuters between Honshu and Shikoku depended upon relatively slow and costly ferryboat service, which was at the mercy of the weather.

As a testament to the difficulty of the project, the bridge was the subject of official consideration for 33 years before construction actually began in 1988. The idea of a bridge across the strait had grim beginnings: a powerful 1955 storm that swept the strait, sinking two ferryboats and drowning 168 schoolchildren. Despite the desire to create a safer way of crossing the strait, officials naturally wanted time to ponder what sort of bridge they would build, considering the engineering challenges facing the project.

Foremost among the obstacles to construction is that the Akashi Strait is 2.4 miles wide. It separates the southern end of Honshu (Japan's largest island) from the northern tip of Awaji Island, the stepping stone to Shikoku Island's city of Naruto. Any bridge crossing the strait would have to be a suspension structure, and when built it would be the longest such structure. The Ohnaruto Bridge at Awaji Island's southern tip completes what is known as the Kobe-Awaji-Naruto Route by running to Shikoku Island across the mile-wide Naruto Strait, famous for massive tidal whirlpools.

Aside from the difficulties of erecting a huge bridge over a wide strait, Japanese engineers had to concern themselves with the savage earthquake hazards in the Kobe region. Because an unstable seismic fault existed only 90 miles from the proposed site of the Akashi Kaikyo Bridge, it was decided the bridge would have to be able safely to survive an earthquake measuring 8.5 on the Richter scale. And, as fate would have it, the bridge would be tested by nature before it was even completed.

And if the earthquakes were not concern enough, the blue waters of the Akashi Strait are as treacherous as they are beautiful. Part of Japan's Inland Sea, the waters of the strait reach a depth of 350 feet. Powerful ocean tides cause the water to race at speeds of up to 9 knots, or just over 10 mph. Sea conditions in the strait add to the woes of any construction effort since waves there can grow as large as 30 feet when storms, including typhoons, strike the region.

Because of this, the bridge design was subjected to intensive aerodynamic studies. Scale models, including one that measured 131 feet in length, were subjected to a variety of wind conditions at varying speeds. This was done to determine whether the bridge could endure winds of 179 mph without shaking itself to pieces. Engineers designed enough give in the bridge to allow its road deck to sway 100 feet during extremely high winds. When such windy conditions do occur, the remarkably stable bridge is closed to traffic.

This treacherous strait would essentially be the "office" for those building the Akashi Kaikyo Bridge. The length of the bridge, the extreme wave conditions, storm winds, and treacherous currents meant the bridge would have to incorporate new materials and techniques. In 1985, officials with the Honshu–Shikoku Bridge Authority, the governmental agency that finances and builds bridges and tunnels in the region, ended a long period of discussion by eliminating a rail line on the proposed bridge, thus allowing finalization of the design that year.

Nothing would come very easy for the builders of the Akashi Kaikyo Bridge, owned by the Honshu–Shikoku Bridge Authority. Although it was determined that a 4,291-foot-wide channel between the supporting towers of the Akashi Strait would be adequate, the Bridge Authority decided to make the center span even wider for safety reasons. This was because the channel at the time accommodated a huge amount of maritime traffic: an average of 1,400 ships navigating the strait every 24 hours. The bridge's center span would gain nearly 2,300 feet in length to provide ships more room to pass and ideally to prevent collision of a ship with one of the two suspension tower piers.

By 1988, work on the foundations for the bridge's two suspension towers was under way. This involved the lowering of clamshell grab buckets 200 feet into the strait's fast-moving waters. The cranes methodically lowered and raised digging buckets at the two pier sites until they had excavated more than 200,000 cubic meters of rock and soil. By 1989, a pair of huge holes had been dug 45 feet through the seabed to reach a layer of compacted rock and sandstone known as the Kobe stratum. It would be upon this stratum that cylindrical steel caissons would be seated to serve as the framework for concrete foundations. The powerful currents of the strait made the seemingly simple process of excavation difficult since the work barges had to remain in precisely the same position. The current also vibrated the steel cables controlling the digging buckets that had to conduct the excavation with great precision. Since it was too dangerous to place divers below the strait to supervise the work, a remotely operated vehicle equipped with low-light video cameras became the eyes of the workers.

Prior to the excavations, the huge cylindrical steel caissons were fabricated. The caisson for the Kobe side measured 262 feet in diameter and was as tall as a 21-story building—over 213 feet. The caisson off the Awaji shore was 255 feet in diameter with a height of 219 feet. Each was equipped with compartments that would be ballasted to allow it to be sunk gradually into its submerged excavation.

Kobe's steel caisson weighed 15,800 tons, its "lighter" sister weighed in at a svelte 15,200 tons. It was decided the caissons would be cylindrical, a relatively neutral hydrodynamic shape. Because water will not flow smoothly around a square caisson, positioning the caisson during its descent to the floor of the strait would be difficult. In addition, the circular shape of the caisson would reduce the amount of turbulence that could worsen the problem of foundation scour, current action that removes stabilizing rock surrounding the caisson's base.

With the excavation work completed by March 1989, the Kobe caisson was tethered to a fleet of seagoing tugboats and towed from the shipyard where it was built. The tugs were dwarfed by the monolithic cylinder, which was made stable by partially flooding its compartments, much like tossing a half-filled soft drink bottle into a pond, but on a much grander scale. Once over its site, a ballet of placement began, in which high-tech equipment carefully determined the positioning of the caisson as the tugs moved it gingerly to and fro. More controlled flooding was allowed inside the caisson so it could slowly settle to the bottom. As it did so, last-minute positioning adjustments were made. The Kobe caisson finally landed less than one inch from where bridge engineers originally wanted it. During June 1989, the Awaji caisson was installed in a similar fashion.

Once in place, the caissons would be pumped full of various types of concrete not only to provide them with massive weight, but also to turn the empty steel drums into piers capable of withstanding 100,000 tons of pressure transmitted to them by the towers. Because seawater contains salts and other elements that can infiltrate and deteriorate steel reinforcement in concrete, high-performance concrete was used to remedy this problem. The majority of each caisson's internal volume was filled with 270,000 cubic meters of this deterioration-resistant concrete. The concrete-filled cylinders were then topped off with a layer of 89,000 cubic yards of concrete reinforced with steel. All the concrete was pumped into the caissons from a barge serving as a floating concrete factory. Protruding from the top layer of concrete were a pair of 200-ton steel anchors—huge brackets to which the tower legs would be attached.

In January 1990, work began on the Awaji anchorage, and three months later workers began the Kobe anchorage. That March, work began on the second anchorage at Awaji Island. These massive concrete structures have the all-important job of holding the cables that suspend the road deck. The weight of the steel in the truss-reinforced deck and the cables alone is a staggering 147,000 tons. Because land is at a premium in Japan, it was necessary for the Bridge Authority to create two artificial peninsulas on which it would construct the suspension cable anchorages at each end of the bridge. Fill was hauled in and compacted to create a foundation for the anchorages, which were to be the largest and heaviest ever constructed. Since the anchorages would be doing the work of holding up the bridge's road deck, they had to be capable of holding the anticipated live and dead loads of the bridge, the weight of traffic, and the structure, respectively. They also had to be heavy enough to compensate for additional loading brought about by high winds or earthquake activity.

Designed like battleship gun turrets, the anchorages consist of huge steel-reinforced concrete boxes that rise skyward atop massive concrete cylinders concealed in the earth. To reach the granite bedrock on the Kobe side, the cylinder had to protrude 200 feet into the ground. On the Awaji side, the anchorage did not have to extend so deep, although steel pipes were used to reinforce the earth surrounding the anchorage. The cylindrical foundation stems of both were surrounded by a special type of concrete that is poured as an extremely dry mixture and then compacted with heavy rollers to cause it to cure with extraordinary strength. The anchorage on the Kobe side of the strait weighed 350,000 tons; the Naruto anchorage weighed 370,000 tons. The anchorages were then completed with steel eyebars that would accept the 290 strands of wire composing each of the two suspension cables.

With the anchorages and piers completed, it was time to assemble the huge twin-legged towers of the bridge, which would rise 927 feet above the strait. The summits of their supporting uprights would be crowned with saddles for the suspension cables. Construction on the Kobe tower began in April 1992, with erection of the Awaji tower starting two months later. The components of the towers were prefabricated and then carried by barge to the site where they were erected. Each tower consists of 30 segments with each segment containing three cells that stand roughly 30 feet high. This cellular design first appeared in suspension bridge construction with the 1926 Philadelphia–Camden Bridge and was

replicated in the towers of the Golden Gate Bridge because of the massive strength it provides.

A traveling crane supported by its own gantry was erected at each pier to lift the steel tower cells into place. As the tower rose skyward, so did the crane. Unlike the erection of the Golden Gate Bridge towers, in which red-hot rivets were used, high-strength steel bolts tightened by the workers connect the tower sections of the Akashi Kaikyo Bridge.

Because of the earthquake threat facing the Kobe region, special precautions were taken in the design of the towers to prevent excessive movement during a temblor. For added strength, the twin legs of the towers are connected by cruciform cross bracing whose open design allows wind to pass through the structure. To counteract swaying in the towers by high winds and any movement of the Earth, each tower is equipped with a series of 10-ton dampers. These pendulum-like devices absorb the energy of any sudden movement to keep the towers stable. The Kobe tower and its movement dampening devices were completed in January 1993, and workers completed the Awaji tower in April.

The towers are now ready to be draped with the steel suspension cables. Although the massive weight of the bridge originally required four cables, the wire manufacturer, Nippon Steel Corporation, managed to strengthen the galvanized steel wire just enough to reduce the number of required cables from four to two. This design also slashed the cost of the monumentally expensive bridge by 20 percent. It was decided that the cables would not be spun by using a moving carriage as they were on the Brooklyn or Golden Gate Bridge. Instead, each of the 290 strands composing a cable would be prefabricated from 127 wires (each 1/20 inch in diameter) and then pulled across the strait. This would reduce the risk to workers who otherwise would have to accompany the wires back and forth across the bridge in a spinning operation.

In a first for suspension bridge construction, Japanese engineers decided to pull the strands not with another length of heavy steel cable, as was common practice, but with high-strength synthetic fiber ropes. Ordinarily, a workboat would pull heavy steel "pilot cables" across the waterway that would be used for hauling across the remaining strands. It was decided that the swift current of the strait might play havoc with a boat unwinding a cable in an operation that might also interfere with heavy maritime traffic. The only other option was to fly a line over the strait by helicopter, but engineers immediately knew no helicopter had the ability to drag tons of steel cable across a 2.4-mile-wide strait. However, engineers figured a helicopter could carry the necessary length of lightweight rope made of superstrong synthetic aramid fibers, the same type used in bullet-resistant vests. The weight of the aramid rope was one fifth that of a steel cable with comparable strength. In November 1993, the quarter-inch-thick rope, referred to as a "pilot rope," was flown across the strait without incident.

The woven steel strands were prefabricated in 2.5-mile lengths before being shipped to the work site. Each cable required 290 of these strands, which were systematically dragged across the strait, placed in the tower saddles, and secured at both anchorages. The work of erecting all 580 strands began in June 1994 and was completed seven months later. Workers then used a hydraulic device to squeeze the bundled strands into a circular cable 44 inches in diameter. To prolong the life of the wire exposed to the corrosive effects of rain and the briny environment of the strait, blowers were installed to ventilate the insides of the cables with dehumidified air. Steel bands were then clamped around the compressed strands. By the end of 1994 all the cables were in place, and it appeared that the task of completing the world's biggest bridge was turning out to be little more than business as usual.

What the engineers and workers of the Akashi Kaikyo Bridge project feared most finally came to pass. At 5:46 A.M. on January 17, 1995, one of the largest earthquakes recorded in Japan's earthquake-plagued history shook the Kobe area as well as Awaji Island. The quake measured a powerful 7.2 on the Richter scale. The Great Hanshin Earthquake made headlines around the world, smashing reinforced concrete structures so massive that their destruction caused disbelief among engineers. Wet soil was vibrated so heavily it suffered liquefaction, meaning the grains of sand vibrated to the point of losing the friction that held them together. The results were the sinking or the collapse of buildings large and small and general destruction on a mass scale. At least 5,000 died in Kobe. There, the temblor and its aftershock ruptured gas lines, causing huge fires. It also split water lines, preventing frustrated firefighters from extinguishing the blazes. The devastation left by the quake was extraordinary, and scientists are still studying its effects in order to protect existing structures and those to be built in the future.

Another aspect of the earthquake is that it drastically reduced maritime traffic to Kobe's bustling port facilities. Since the destructive quake badly disrupted the port, cargo ships went elsewhere, taking hundreds if not thousands of jobs with them. One of the main reasons for placing the piers of the Akashi Kaikyo Bridge so far apart—the presence of heavy sea traffic—had disappeared in a matter of minutes.

Incredibly, the one structure undamaged by the quake was the partially completed Akashi Kaikyo Bridge, with its towers and cables still awaiting its truss-reinforced road deck. Worried engineers and safety experts gave the towers, foundations, cables, and anchorages a close inspection to determine whether the quake had caused any structural damage. To the relief of the experts there was none, but they discovered the quake had moved the towers apart by nearly 40 inches and had pulled both anchorages back from the towers another 12 inches. Although this widening of the strait by the quake was testimony of its power, work on the bridge nonetheless resumed after a delay of only 30 days.

By June 1995, it was time to put the final pieces of the bridge puzzle together with the assembly of the highly refined truss-stiffened road deck. The reinforcing truss skeleton below the roadway is not only designed to make the road deck stiff but engineered with a slew of aerodynamic tricks to counteract the effects of strong wind. Even minor winds can lift and slam down a poorly designed suspension bridge deck with destructive consequences. Beginning in June 1995, the truss deck sections were fabricated and carried by barge to the work site. Cranes then lifted them so workers could connect the sections to vertical steel rope suspenders attached to the suspension cables by hangers. Each of these massive sections is 116 feet wide to accommodate the six-lane roadway of the bridge. The reinforcing truss girder below the road deck is massive, measuring 46 feet in depth.

The importance of aerodynamics precautions in the design of a suspension bridge as large as the Akashi Kaikyo span cannot be overstated. Stone and concrete bridges are rigid and comparatively heavy and therefore rarely prone to wind-induced movement. But suspension bridges are relatively light for the distance they span, and they are designed to flex, but only within limits. A time-tested way of stiffening the flexible roadway is with a truss girder. This provides extraordinary stiffness while adding as little wind resistance as possible as a result of its open, skeletal construction. Vertical steel plates within the truss girder's framework channel wind upward toward a ventilation slot running down the roadway's center to equalize air pressure between the deck's top and bottom. When air pressure beneath a horizontal surface is greater than that above, a lifting effect occurs, causing the bridge to act as the wing of an airplane does. Because a suspension bridge is just that—a roadway suspended from wire cables—aerodynamic flutter can cause the road deck to twist violently. This effect has torn down numerous suspension bridges throughout history; the most recent American collapse was that of the Tacoma Narrows Bridge on November 7, 1940. The longer and wider the road deck, the greater the potential for "flutter," or the flexing of the bridge by wind and the vortices formed on the lee side of a structure. The Akashi Kaikyo Bridge's reinforcing truss is also designed to break up the vortices or curlicues of wind that can generate destructive vibrations.

The truss girder sections bearing the roadway were completed by September 1997, triggering the paving of the steel road deck. On April 5, 1998, a ribbon-cutting officially ended a decade of work. Japanese drivers now had the longest if not the most advanced suspension bridge in the world.

Unfortunately, Japan's 30-year spending binge on bridge and tunnel construction had as its price the stiffest tolls in the world. Taking the family car on a drive across the Akashi Strait cost $48 in 2000. Virtually all of Japan's bridge and tunnel construction is paid for through loans granted to various bridge and tunnel agencies, which are then repaid with tolls.

Perhaps the most amazing fact about the Akashi Kaikyo Bridge is that no one died during its construction. A handful of accidents resulted in an amazing total of only six injuries, a safety record that is virtually unmatched on a massive, long-term construction project.

Despite the bridge's impressive center span length, longer suspension bridges are still being planned. The length of the Akashi Kaikyo Bridge will stand as the record until construction of the proposed bridge across Italy's Strait of Messina, an even larger bridge project proposed during the late 1990s.

See also CAISSON; SUSPENSION BRIDGE.

Akeley, Carl Ethan (1864–1926) *Taxidermist savior of miners and gorillas* Carl Ethan Akeley was not a miner but a taxidermist and naturalist, who was maniacal about accurately representing animals to museum visitors. However, a device he invented for spraying mortar into forms to help him create breathtakingly realistic wild animal exhibits inadvertently made him one of tunnel engineering's great innovators.

Akeley traveled widely in Africa, killing species of animals whose hides he needed for his dioramas. Dissatisfied with the unrealistic forms over which animal hides had been previously stretched, Akeley invented a compressed air technique for spraying mortar into wire frames of animal shapes. The mortar was smoothed and the resulting sculptures were lifelike shapes for the animal hides to cover. Akeley's magnificent African animal dioramas can still be seen at the Chicago Field Museum of Natural History and at New York's American Museum of Natural History.

Akeley's 1907 invention of what became known as the cement gun was patented by Akeley and developed further by the Allentown Cement Gun Company. By the 1950s, an Austrian tunnel engineer realized that concrete sprayed with Akeley's invention could rapidly and effectively reinforce tunnel walls, providing protection against cave-ins. The use of sprayed concrete, or "shotcrete," formed the foundation of an entirely new method of safer soft earth tunnel construction. This tunneling technique, now known as the new Austrian tunneling method, has become widely used around the world.

In soft earth tunneling the ground can cave in or simply squeeze together unless it is immediately reinforced. The time between excavation and reinforcement is a dicey period for miners. Using a modified version of Akeley's cement gun to apply shotcrete speeds the reinforcement process while providing an extremely strong inner shell of high-strength concrete.

As it turned out, Akeley would become a savior not only of miners working in dangerously unstable tunnels but also of Africa's endangered mountain gorilla. During his African travels, Akeley spent much time in the bush and once had to kill a leopard with his bare hands. Although he shot animals of all types, including the mountain gorillas, for his exhibits, Akeley was in his time a lover of wildlife because of his passion for realistically displaying Africa's animals to the public. However, even Akeley knew nature could absorb only so much encroachment. After encountering the once-mythical mountain gorillas in the mid-1920s in Belgium's colonial possession of Zaire (at the time the Belgian Congo), Akeley pressed King Albert of Belgium to create the Prince Albert National Park. This preserve now lies in Zaire, Rwanda, and Uganda as a sanctuary for the apes, albeit one threatened by poaching. The preserve and its gorillas, which live exclusively in the Virunga Mountains, were made even more famous by the movie *Gorillas in the Mist,* which portrayed the life of the gorilla researcher Dian Fossey.

Akeley, who subsequently returned to Africa to collect additional specimens for his exhibits, became ill with a fever and died in Uganda in 1926.

See also SHOTCRETE.

Alamillo Bridge   *See* Calatrava, Santiago.

Albert Bridge   *See* cable-stayed bridge.

Alcántara Bridge (built A.D. 103–106)   It is a bit redundant to refer to the magnificent Roman-built structure over Spain's Tagus River as the Alcántara Bridge since its name is a contraction of the Arabic *al-kántara,* meaning "the bridge." To both civil engineers and ordinary bridge enthusiasts, the Roman-built span is one of the world's great examples of bridge design and construction.

Its beauty and its 2,000-year history have made the bridge an integral part of Spain's culture. For generations the important bridge has made life far easier for Roman troops, various invaders, and ordinary folk who had to cross the rugged gorge that it spans. So revered is the bridge that its partial destruction in 1809 during warfare with Napoleon created a serious rift between local Spaniards and the British and Portuguese troops that demolished one of its arches.

The Tagus River runs roughly east and west from Spain into Portugal's Estremadura region and lies along a route that is historically important for both trade and military purposes. Its architect was a Roman engineer, Gaius Julius Lacer, who was ordered to construct the bridge during the reign of the Roman emperor Trajan, famous for his enthusiasm for bridge projects during his reign. After damage, alterations, and partial rebuilding during the previous centuries, an inscription dedicating the bridge to Trajan remains on the structure, as does an inscription listing the names of all the towns that contributed to its construction.

Built between A.D. 103 and 106, the bridge is composed of three large central arches and three smaller arches. There are no records detailing how the bridge was built, leaving engineers mystified as to how the arches were erected above the waters of the Tagus, which is frequently swollen by flooding. The construction of these arches must have taken place with great speed since the required wooden falseworks would have been swept away by spring flooding. The massive strength of the bridge has been proved time and again, as the Tagus has often risen from its normal level of 37 feet to nearly 180 feet without sweeping it away.

The Alcántara, like other exquisite examples of Roman bridge construction, survived because its foundations are set upon solid granite and because an arch is an extremely stable structural form. Under Lacer's direction, cofferdams may have been set in the middle of the Tagus, during low-water conditions, to allow the construction of the two middle piers down to bedrock. These foundations are 30 feet square with cutwaters pointing upstream to deflect the current or, if need be, debris and ice.

The bridge is constructed of square blocks of granite so finely shaped and fitted that no cement was needed to bond them. The stonework is held together by the compression of its own weight, as are the Alcántara's six arches. The two largest arches each support spans measuring 98 feet. The 636-foot-long bridge has a roadway 26 feet wide that stands 170 feet above the Tagus River, and in what was once a custom in ancient times, the nearby town bears the name of the bridge. In its original form, the bridge included extensive walls and fortifications at both ends for defense. These were removed in past centuries, leaving only a granite gateway at the midpoint of the bridge that towers 60 feet above the roadway.

It is not surprising that Trajan ordered the construction of the Alcántara Bridge since he was an enthusiastic supporter of numerous bridge projects during his reign, including the construction of a 3,000-foot bridge across the lower portion of the Danube River. The bridges were far from monuments to Trajan, whose own ego seems to have been healthily contained; they were essential to the Roman Empire, which reached its greatest level of territorial expansion under his leadership. He was

The Alcántara Bridge crossing Spain's Tagus River stands as one of the finest surviving examples of imperial Roman bridge construction. Built during the reign of Emperor Trajan and completed in A.D. 106, the bridge has survived nearly 2,000 years of warfare and flooding. (Robert Cortright, Bridge Ink)

undoubtedly familiar with the need for bridges in Spain, where he was born.

When bridges were built on the orders of Rome, funds for their construction were sometimes solicited from adjacent towns, and highly skilled engineers attached to the Roman army were assigned to direct the work. The construction force might have consisted of slaves, local artisans, and Roman troops. Lacer was the highly capable engineer assigned to the Alcántara Bridge and apparently took the project very seriously. The project was such a wonder at the time that Lacer himself apparently considered the bridge his legacy to civil engineering. He designed

his own tomb to stand in the vicinity of the bridge with the prophetic inscription "Pontem perpetui mansuram in saecula mundi," (I leave a bridge forever in the centuries of the world.)

However, this almost did not happen; wars raged across the Iberian Peninsula for hundreds of years, threatening to destroy a bridge that nature's floods lacked the power to topple.

Muslim invaders, who ruled much of the Iberian Peninsula between A.D. 800 and 1492, managed to smash one of the arches around 1214, though it was later rebuilt. Nearly 600 years later, when the French armies of Napoléon Bonaparte campaigned in Spain in 1809 during the Peninsular War, the bridge would face its greatest disaster. To amputate Napoléon's supply line, the future duke of Wellington dispatched a strike force into western Spain from Portugal to seize the Alcántara Bridge. The north side of the bridge was fortified by 2,000 Portuguese troops supported by artillery, all under British command. One of the large arches of the bridge was mined with gunpowder in case the French appeared to be about to capture the span.

On May 14, 1809, the British-led Portuguese saw a force of nearly 12,000 French troops, including 1,500 cavalrymen with a large number of cannon, approach the bridge. The French and the Portuguese began shelling one another and 1,200 Portuguese militiamen fled, leaving fewer than 800 troops to face 12 times their own number of French. The British had no choice but to retreat and ordered the charge on the bridge detonated, knocking only half the width of the roadway into the river below. This allowed the French cavalry to charge across the bridge and chase away the fleeing defenders. Although the Roman-built arch held, it crumbled into the Tagus within a year. The Spaniards in the vicinity were incensed that first Portuguese troops mined their bridge (under British orders) and then their cowardice forced its destruction.

In the spring of 1812, the duke of Wellington ordered the bridge repaired for the convenience of the British troops. British Army engineers built a suspension span that was then pulled by ropes across the 100-foot gap. The suspension span was then tightened with ropes and cables, making it capable of withstanding virtually any load. Wellington was not as interested in the transportation needs of the local populace as he was in his desire to shorten his own supply lines, as the Peninsular War continued until 1813. When Wellington invaded France, his need for the bridge once again reinforced its strategic importance, highlighting perhaps the same reasons why the Romans were so eager to build a bridge there. Spanish officials had the bridge repaired by the 1860s, and a model of the successful British effort to patch the bridge can be seen today at Madrid's Spanish Army Museum.

The skills of the stonemasons and stonecutters repairing the bridge 17 centuries after it was built were not adequate to replicate exactly the original work on the bridge. Although the Roman design allowed the bridge to stand without mortar, the repair work was done using mortar to bond the granite blocks of the repaired arch. The repairs to the bridge were carefully carried out using granite identical to that of the original structure, and visitors are hard-pressed to find any signs of the destruction. The bridge remains fully functional, carrying automobile traffic across the Tagus River.

See also ARCH BRIDGE; PONT DU GARD.

Allegheny Aqueduct (built 1844–1845) *John Augustus Roebling's first suspension bridge*   Sick of farming and a failure at raising singing parakeets, the immigrant engineer John Augustus Roebling burned with the desire to build suspension bridges. By 1844, the manufacturer of wire rope had also determined the best way to create wire suspension cables and the anchorages for holding them.

The one thing that the man who would eventually design the Cincinnati-Covington Bridge and the Brooklyn Bridge had not done by 1844 was build his first bridge, but that was soon to change.

Roebling's opportunity arrived in May 1844, when he learned that a fire had destroyed a Pittsburgh aqueduct designed to carry barges over the Allegheny River. Better yet for Roebling, the Pennsylvania Canal Company sought a design that could be built cheaply and quickly. One of the exquisite attributes of a suspension bridge is that it can sustain greater loads while weighing less than conventional beam or arch bridges. The reason for this, Roebling knew, was that the loading is not transmitted solely to the span but to anchorages through the suspension cables.

When he learned that a new bridge was sought, the indefatigable Roebling worked around the clock to design a span, prepare models, and conduct tests. Although suspension bridges were nothing new, Roebling's suspension cable system was unique. Roebling's cables were not merely wires bundled together on shore and draped over their suspension towers but wires that were individually spun to a precise curvature so each uniformly carried its portion of loading.

To finish off this wire suspension cable method, Roebling had also devised a technique for giving the cable a tight external wrapping of wire that would be painted to provide a weatherproof covering. The suspension cables would be connected to a patented anchorage in which a massive iron plate would be secured aboveground in a masonry foundation. The wires of the cables would be attached to the ends of articulated iron eyebar chains seated in the anchorage.

Once in place, this gravity anchorage could sustain twice as much loading as the cables could be expected to receive from the combined live and dead loads of the bridge. All these innovations would be employed for the first time in the Allegheny Aqueduct and would be adapted to virtually every major suspension span built in the next 150 years.

In August 1844, Roebling's proposal of a 1,092-foot bridge with seven spans and six intermediate piers was accepted and Roebling was given nine months to complete the work. Of the 43 bids submitted, Roebling's was the lowest, amounting to $62,000. His bid was a bargain for the Pittsburgh Canal Company since it included his services as designer and contractor. This meant the aqueduct would cost the canal company a mere $56 per foot to build, an unbelievably low figure even for 1840s America: The previous bridge had cost $104,000 to construct. Roebling was courting financial ruin with his low bid, but the precision-minded entrepreneur had calculated his expenses shrewdly and viewed the fiscally anemic contract as the only sure way to ignite his career as a bridge builder.

The bridge was an early tour de force of what was to be Roebling's trademarks—innovation, attention to detail, and scientific knowledge. The aqueduct's seven flume spans, or channels, would be rigid, self-supporting U-shaped wooden beams, each weighing 120 tons. So rigid were the beams that they were able to support their own weight independently of the cables. The wire suspension cables would easily assume the load of the water contained in the flume, which amounted to roughly 2,100 tons. In this way Roebling saved both weight and money by eliminating tons of lumber.

Confused Pennsylvania Canal company officials had quizzed Roebling as to why he failed to add the weight of the canal boats to the weight of water in the flume. Roebling explained that a boat would displace its own weight in water that would never rise above a set level, a basic bit of physics understood by the designers of ships. Roebling's engineering was of the best type: It involved the use of the most simple and effective means of solving a problem at the lowest cost with the greatest margin of safety.

The effort to build the bridge was a race against time and a harsh winter. This process was complicated since Roebling's innovations required detailed explanations to workers unfamiliar with his concepts. It was necessary for Roebling to be everywhere at once, issuing directions, making decisions, and supervising work. Despite the winter months and the almost experimental nature of the design, Roebling completed the job on schedule in May 1845.

A victorious Roebling later wrote an article about the project and predicted that suspension bridges could carry heavy railroad trains across rivers. Roebling would begin construction on the Niagara Falls Bridge, the first suspension span for a rail line, six years later.

The Allegheny Aqueduct carried the canal for the next 16 years and resisted any serious damage from the river ice that had smashed its predecessor. The aqueduct went out of service only because the canal it served ceased operation in 1861 as a result of railroad competition. The success of the bridge launched Roebling's career as one of America's greatest bridge builders.

See also BROOKLYN BRIDGE; CINCINNATI-COVINGTON BRIDGE; NIAGARA FALLS BRIDGE; ROEBLING, JOHN AUGUSTUS; SUSPENSION BRIDGE.

Ammann, Othmar Hermann (1879–1965) *Designer of the magnificent George Washington Bridge, among others* The Swiss-born Othmar Hermann Ammann was planted in the 19th century to grow artistically and technically into the 20th century. He was in many ways the last of a long line of pre-20th-century engineers who immigrated to America from Europe to create greater bridges than they could have built at home.

He became a protégé and confidant of the famed Austro-Hungarian engineer Gustav Lindenthal, chief engineer of New York's Hell Gate and Queensboro Bridges, only to experience an acrimonious break when the older man's designs outgrew the bounds of reality. The rift would derive from Ammann's success in competing with Lindenthal to design the first major span across the Hudson River. The span was the mighty and elegant George Washington Bridge, a project that would propel Ammann to the forefront of American bridge engineering.

Ammann's success did not occur overnight but came about slowly, thanks to his legendary attention to engineering detail, his dogged ambition, and his dedication to the exquisite promise of the suspension bridge: the ability to span great distance with an economical use of materials.

The visionary Lindenthal once considered Ammann his most capable assistant, because of Ammann's dedication to detail. The Swiss immigrant, who was highly regarded for his superb management of bridge-building budgets and schedules, also became a high priest of 20th-century architectural form with his strong, spare, modernistic suspension bridge designs.

Raised in Feuerthalen on Switzerland's northern border with Germany, the artistically inclined Ammann was the son of a hatmaker. Schooled at the Swiss Federal Institute of Technology, Ammann was instructed by the engineer Wilhelm Ritter, who also influenced the brilliant Swiss engineer Robert Maillart, the designer of capable and visually daring concrete bridges that remain engineering and aesthetic wonders. Ammann's suspension bridges, including the Bronx-Whitestone and the monumental Verrazano Narrows, would possess a reserved and functional beauty in which form is the handmaiden to function. Not surprisingly, this is a trait shared by Maillart's concrete arch bridges.

After a brief period as a draftsman in Switzerland and later as an assistant engineer in Germany, Ammann hungered to build great bridges. Heeding the advice of his engineering professor Karl Emil Hilgard, who had worked in America, Ammann decided to go to America, where young European engineers had gone for more than a century to spread their professional wings. The master-apprentice mode of European engineering was still much in vogue, and those wearing the clothes of older men were the ones who oversaw great projects. America, in its race for growth, was willing to bank on any man, including a young one, if he could perform the task.

Ammann arrived in America in 1904 and found work at a bridge engineering firm before moving on to work as an engineer at the Pennsylvania Steel Company, then building New York's Queensboro Bridge, designed by Lindenthal. During his employment with the Pennsylvania Steel Company, Ammann assisted his superior, Frederic C. Kunz, who took part in the investigation of the 1907 Quebec Bridge collapse. By 1912, Ammann had been named chief assistant engineer to Gustav Lindenthal, who was then working on the largest steel arch span in the world, Hell Gate Bridge.

Another engineer working for Lindenthal at the time was David Barnard Steinman. The self-promoting Steinman, who carried the title of special assistant engineer, would later acquire a level of fame approaching that of Ammann. Steinman, who would design the stunning Mackinac Bridge in Michigan during the 1950s, would strike disagreeable sparks with Ammann. When the 1914 outbreak of World War I pitted Germany against much of Europe, Ammann returned to Switzerland, where he was an officer in Switzerland's famed citizen army. During Ammann's absence, Steinman was promoted into his position. Within four months Steinman was demoted to make room for Ammann, whose return was made possible by Germany's failure to invade Switzerland.

Ammann's family would later recall that Steinman's name was not to be spoken in the Ammann household. Although the intimate details of this acrimony are not fully known, the differences between the two men are apparent. Steinman was the energetic self-promoter authoring books to guide the public through the art of bridge design while brassily pushing his own projects. Ammann was the deliberate professional quietly pursuing his goals while methodically offering sensible engineering solutions. Ammann was the solid company man, and Steinman was the eager public speaker penning poems to bridges. Both men were brilliant. But, in terms of personality and approach, they were worlds apart.

In 1918, Ammann distinguished himself by publishing a highly regarded 150-page report on the construction of the Hell Gate Bridge in the *Transactions of the American Society of Civil Engineers*. Although Lindenthal was the chief engineer of the project, Ammann wrote the paper with Lindenthal's blessing, clearly demonstrating the extent of Ammann's involvement in what was the longest steel arch bridge of its day. Lindenthal published his own paper in 1921 and in it mentioned Ammann and Steinman for their contributions in designing the Hell Gate Bridge.

Like Steinman, Ammann became a capable writer, although his target audience was the officialdom of public works and his fellow engineering professionals. Compared with 21st-century America, where a debate rages about requiring immigrant children to learn English, Ammann became proficient enough in a foreign language to inform a discriminating audience of his engineering prowess and to propel his career. Both Steinman and Ammann understood that writing was a powerful sales tool in the competitive business of bridge engineering. The two eventually became rivals in the small world of large bridge design.

Ammann's professional trajectory deviated from bridge design in 1920, when a slowdown in bridge construction forced Lindenthal to offer Ammann employment at a New Jersey clay mine. The business was one Lindenthal co-owned with a New York businessman and politician, George S. Silzer. Lindenthal, like any businessman who wants to hang on to a valued employee, sought to keep Ammann in his stable. The Swiss immigrant managed the initially unprofitable mine for nearly three years. With his habitual attention to detail Ammann was able to make the operation profitable. While Ammann worked at the clay mine, he was planning the design and construction of a bridge across the Hudson River.

Such a bridge had become a great dream of Lindenthal's, but the egotistical bridge designer's vision had become too large, too expensive, and too unrealistic. Ammann, over the years, had tried to talk Lindenthal into moderating his design to make it less heavy, less expensive, and less impossible. A frustrated Ammann wrote to his mother in 1923 of his frustrations with Lindenthal, "In vain, I as well as others, have been fighting against the unlimited ambition of a genius that is obsessed with illusions of grandeur."

Ammann, comprehending that Lindenthal's great dream would be nothing more than that, communicated with Silzer, who was by 1923 the governor of New Jersey. His former clay mine employer was a shrewd and realistic politician who understood the need for a bridge across the Hudson. Despite the start of construction of the competing Holland Tunnel in 1920, Silzer threw his support behind Ammann's design, which was nearly 10 times less expensive than Lindenthal's.

Silzer wanted a bridge but knew the one proposed by his friend Lindenthal was not affordable. After asking Lindenthal to review Ammann's plans, the 73-year-old engineer wrote a reply that soiled Ammann's character for proposing a design that would be in competition with his own while claiming Ammann's cost estimate for such a bridge was unrealistically low. Lindenthal added, untruthfully and pompously, "He [Ammann] never was necessary or indispensible [sic] to me."

The Port Authority of New York and New Jersey had rejected Lindenthal's monstrosity but was interested in Ammann's design. Although Ammann's bridge was still massive—twice as long as any previous suspension span—it was still far less expensive than that of his former mentor. Ammann was named chief engineer of the Port Authority in 1925 and oversaw the final design and construction of his bridge, which was completed in 1931. Ammann had intended to sheath the steel latticework of his innovative suspension towers in granite, but the Great Depression forced an abandonment of this façade, resulting in a structure that was even more unique and beautiful.

At the same time, Ammann had served as chief designer on a bridge over New York's Kill van Kull waterway between Staten Island and New Jersey. The span was to be a steel arch similar to Lindenthal's brilliant Hell Gate Bridge. Ammann's experience on Hell Gate prepared him well for his design of what was eventually dedicated as the Bayonne Bridge. The resulting bridge presented a sleek and graceful arch to the public at its completion in 1931. Bowing to both economy and style, Ammann decided against employing traditional stone towers that rose above the level of the road deck. Concrete piers sunk to bedrock on each end secured the springings of the arch, unhidden by the nonfunctional towers.

For the next 14 years Ammann threw himself into the role of an engineer in government service overseeing the construction and design of bridges as well as tunnels. In 1933, the newly appointed Triborough Bridge Authority chief, Robert Moses, asked Ammann to serve simultaneously as the chief engineer of his agency. Ammann's task was to oversee construction of the trio of bridges and a viaduct forming the Triborough Bridge system. Ammann redesigned all but one of the spans, including the famed suspension span commonly known as the Triborough Bridge, which was completed in 1936.

Ammann's next notable accomplishment, and the closest he came to possible failure in his otherwise impeccable career, was the design of the Bronx-Whitestone Bridge. With the assistance of the noted consulting engineer Leon Moisseiff, the guru of thin bridge design, Ammann created a bridge with a wafer-thin deck and sleek steel towers. But the wind-induced collapse of Moisseiff's Tacoma Narrows Bridge in 1940, six months after the Bronx-Whitestone Bridge became operational, reawakened engineers to the aerodynamic hazards that had smashed Charles Ellet's Wheeling Bridge 86 years earlier.

Moisseiff had designed the Tacoma Narrows Bridge with an exceedingly thin and light road deck, and his career was smashed by its failure. So great was Ammann's reputation by 1940 that he was one of three experts asked to review the destruction of the Tacoma Narrows Bridge. The final report indicated that Moisseiff had followed accepted practice and principles of the period, and no one was quite sure exactly how the bridge reacted with the wind. Nonetheless, Ammann went away realizing that the finest theories of thin bridge construction could be undone by an invisible wind.

Nonetheless, Ammann believed the Bronx-Whitestone Bridge was heavy enough to endure aerodynamic forces without damage, although the bridge exhibited appreciable movement that was worrisome to its toll-paying users. In 1946, Moses ordered the bridge made more rigid, and Ammann, who argued that reinforcement was unnecessary, took on the task of adding a substantial truss railing that stiffened the deck while adding stabilizing weight. The two catenary suspension cables that curved almost to the roadway were accommodated in scalloped recesses designed in the truss railings on each side. Although the thin, economical design had lost some of its sleekness, Ammann's alterations tamed much of the movement that so unnerved drivers.

In 1939, Ammann left both the Port Authority and Triborough to start his own engineering firm with the landscape architect C. C. Combs. However, Ammann had taken outside engineering work in the private sector during his tenure in public service, most notably as one of three consulting engineers on the Golden Gate project, the others were Moisseiff and the University of California at Berkeley engineering professor Charles A. Derleth. The Golden Gate project was perhaps the most stunning bridge-building accomplishment of the first half of the 20th century.

Although World War II stilled much large infrastructure work, Ammann was kept busy in 1941 reviewing the structural integrity of John Augustus Roebling's great Brooklyn Bridge, which Ammann found in surprisingly good shape. Ammann, born a decade after Roebling's 1869 death in the course of surveying the site of the great East River suspension bridge, made some strengthening recommendations to Roebling's great bridge. In 1946, Ammann and Combs merged with the Charles S. Whitney firm to form Ammann & Whitney, Consulting Engineers.

Soon after this, Ammann undertook the design of the Delaware Memorial Bridge between New Castle, Delaware, and Pennsville, New Jersey, which was completed in 1951.

Moses and Triborough turned to Ammann & Whitney to design New York's Throgs Neck Bridge, and Ammann met this latest challenge with an effort to create a beautifully slender but nonetheless stable bridge. To reduce deck movement, endemic to suspension bridges but irksome to motorists, Ammann employed a hefty road deck reinforced with a substantial truss on its underside. He also designed transverse steel reinforcement to provide weight while making the road deck even more rigid. This East River bridge was completed in 1961.

Ammann, who was involved in engineering virtually until the end of his life, had one last great design left. In a cooperative arrangement between the Port Authority and what was by the late 1950s the Triborough Bridge and Tunnel Authority, Ammann was asked to design the Verrazano Narrows Bridge across the entrance to New York Bay. It was to be a distillation of his previous suspension designs and his greatest bridge in terms of size.

The Verrazano Narrows Bridge would have a two-level road deck joined by massive reinforcing trusses. The two-story road deck would be supported beneath two pairs of suspension cables running down each side of the bridge, whose main span would measure 4,260 feet. Its two steel supporting towers would rise 690 feet toward the sky. Despite its massive girth, from a distance the lines of the bridge are sleek. It was an exercise in both strength and gracefulness.

The Verrazano Narrows Bridge was completed in 1964 and Ammann died the following year at the age of 85. His three greatest bridges are perhaps the George Washington, the Bayonne, and the Verrazano Narrows. All three are examined in detail with wonderful accompanying illustrations in Darl Rastorfer's thoughtful book *Six Bridges: The Legacy of Othmar H. Ammann*.

See also BAYONNE BRIDGE; BRIDGE AERODYNAMICS; BROOKLYN BRIDGE; BRONX-WHITESTONE BRIDGE; GEORGE WASHINGTON BRIDGE; GOLDEN GATE BRIDGE; HELL GATE BRIDGE; LINDENTHAL, GUSTAV; MOISSEIFF, LEON; MOSES, ROBERT; ROEBLING, JOHN AUGUSTUS; STEINMAN, DAVID BARNARD; TACOMA NARROWS BRIDGE; THROGS NECK BRIDGE; TRIBOROUGH BRIDGE; VERRAZANO NARROWS BRIDGE.

anchorage *Gravity's role in suspension bridges* Suspension bridges are perhaps the more admired bridge type since they enable a roadway to cross bodies of water using cables for support rather than arches or massively thick road decks. Ironically, although the fundamental enemy of any elevated roadway is gravity, suspension bridges fight fire with fire by employing gravity through the use of anchorage blocks to help bear the weight of a bridge's roadway.

Although early vine or rope suspension bridges were far less complex than San Francisco's Golden Gate Bridge, even they relied on crude anchorages to enable them to remain erect. In early suspension bridges, vines or ropes were often tied off to the trunks of large trees or held fast with deep stakes. Since the towers over which suspension cables are draped are not sufficient to support a suspension span by themselves, it was imperative that the cabling be attached to anchorages to resist both horizontal and vertical forces.

This complementary task means the cables transmit a vertical force—the weight of the road

Anchorage

deck—into the towers and ultimately the foundations of the towers through compression. The anchorages play a role in resisting both the horizontal and vertical forces transmitted through the cables so the cables remain taut and in position in their tower saddles.

For example, if a person were to attach a weight to a rope and throw it over a limb, the tree would accept the vertical force of the weight only as long as the rope was tightly held. If the rope were released, the weight would fall because the tension allowing the limb to play a role would be absent.

As suspension bridges became longer and heavier, the anchorages too had to become heavier since gravity was in competition with itself. Early anchorages were composed of large blocks of stone set by masons with the wire cables sometimes running deep underground to buried masonry foundations. This enabled the mass of the anchorage to combine with the resistance of the Earth to provide a fantastic amount of dead weight resistance.

Nineteenth-century French engineers employed anchorages that were completely open inside to allow the inspection of the iron components that might be subject to corrosion. The German–American engineer and Brooklyn Bridge designer John Augustus Roebling improved upon this system during the 1840s by using an anchorage that secured articulated iron eyebars to a base plate embedded in masonry. The suspension cabling was then attached to the ends of these eyebars, allowing an easy means of inspection without creating large and expensive subterranean chambers.

Subsequently, anchorages were refined into massive monuments to gravity with stupendously heavy foundations built in deep excavations using reinforced concrete. The structures were then built high above ground using poured concrete until their aboveground segments appeared as great blockhouses. The interiors of modern wire suspension cable anchorages are huge cathedrals where massive bundles of wire splay out from the cables to be attached to eyebars angling outward from deep beneath the ground.

See also AKASHI-KAIKYO BRIDGE; BROOKLYN BRIDGE; SUSPENSION BRIDGE.

Aqualine Tunnel *See* combination bridges.

aqueduct *See* viaduct.

arch bridge   In arch bridge construction a structural arch or series of arches of stone, steel, or concrete are used to support the road deck of the bridge. Arch bridges possibly represent the oldest form of short- to medium-span bridge construction and were used by the ancients, including the Romans, to carry bridges across wide obstacles.

In a properly designed arch, most of the forces are transferred by compression from the roadway to the arch and into its abutments. Under adequate compression, the stones, or other structural elements that make up the arch, are uniformly pushed against one another, creating a self-stabilizing effect responsible for the longevity of arch bridges. This arch action can be visualized by bending a flexible plastic ruler or even a credit card into a semicircular curve. If the ends of the plastic arch are tightly secured and a load is placed at the top or the crown of the arch, the flexible ruler assumes a rigid and secure shape.

As the efficiency of the arch is easily demonstrated by this example, three problems facing bridge arches are also exposed. The first problem is that the arch needs to be supported at its ends by some horizontal force in order to maintain its curved shape; otherwise, it will flatten and collapse. This horizontal force is provided by the construction of solid foundations or abutments that are required for the arch to work properly. Before excavation machinery and rock-blasting became common, arch bridges could not be erected economically unless soil and rock conditions provided adequate support with little digging.

The second problem is that the arch needs to be completed before any load is applied to it in the form of roadway construction. This means that for many older arch bridges, which often consisted of masonry blocks, extensive falsework erected from the floor of the valley or in the moving water of a river was needed to support the structure until the arch was completed. Modern steel arch bridges are constructed by techniques in which the arch is held with cables until the arch is completed, eliminating expensive and complex falseworks.

A third and potentially devastating problem facing arch bridge construction is to avoid the application of a load along the side of the arch, rather than at its crown, from which it is uniformly carried down to its end points. Pushing one's finger downward on the top of the plastic model arch transmits the forces evenly to the position where both ends of the arch are supported. If force is applied along the

This 15th-century stone bridge crosses Scotland's Forth River, where Scottish patriot William Wallace led a 13th-century attack on English troops attempting to rush across a small timber bridge. The earlier bridge eventually collapsed beneath the confused English troops during what became known as the Battle of Stirling Bridge. *(Author's Collection)*

side of the arch between the crown and its abutments, the arch will tend to bend since its components will no longer be compressed in a uniform fashion. Although older stone arch bridges tend to resist this type of bending because of the rigidity of the materials used, steel arch bridges must be carefully designed to prevent asymmetrical loading. The slenderness of the ductile steel arch depends on uniform compression in order to retain its shape and therefore its strength.

The most traditional examples of this type of arch bridge construction using stone are Roman arch bridges such as the Alcántara Bridge in Spain and the Pont du Gard aqueduct in France. Some Roman arch bridges are so superbly designed that mortar was not needed to bind the stones together, held as they were by the compression of the arch stones.

In traditional stone arch construction, falseworks are usually built to support the arch, which cannot support its own weight until its final component, known as a keystone, is installed. When under compression, the arch holds not only its shape but massive loads as well. The arch, or series of arches, is then ready to have a roadway built above it. During the construction of steel or iron arches, notably those of the Eads Bridge in Saint Louis, cables on shore held the incomplete arches erect until they were completed and capable of supporting themselves.

The introduction of new bridge materials, particularly iron, steel, and various forms of concrete, meant the arch bridge concept—one of the oldest—would continue with a new look. By the late 19th century, reinforced concrete that could be formed into any desired shape became a material of choice for some arch bridge designers. The reinforced concrete arch bridge reached a high state of development by the 1930s with the bridges designed by the Swiss engineer Robert Maillart, who eliminated unnecessary bulk to produce modernistic, skeletal arches.

When iron and later steel were used, metal beams or tubes would replace stonework by forming the chords of arches. Often these metal arches were reinforced by a latticework of iron or steel trusses with supporting struts connecting the arch to the roadway. The use of metal for the structure of a bridge meant not only a lighter structure but also one that even appeared lighter because of the ventilated look of the metal arch.

These advances in bridge design, materials, and construction techniques meant engineers could create shallower arches that weighed less. As a result, the familiar semicircle design of Roman and medieval arches gave way to shallower, flatter curves. Whereas traditional arch bridges usually placed the arch beneath a roadway, engineers began developing steel arch bridges with the arch rising well above the road deck. Gustav Lindenthal's Hell Gate Bridge in New York (1916) is one such design. The truss arch is reinforced by a latticework of steel struts to support the steel beams forming the arch that are known as chords. The inside radius of the Hell Gate arch is connected to the railway deck of the bridge by vertical steel beams.

Some arch bridges, with their supporting arch rising above the roadway, utilize cables running the length of the underside of the roadway to connect the ends of their arches. In this way, the cables transfer the weight of the elevated roadway to the arch while locking the entire structure together as a tight, cohesive unit. Known as a tied-arch bridge, this type of span minimizes the need for large foundations, since the horizontal forces that would tend to bend the arch are locked into the structure.

See also ALCÁNTARA BRIDGE; CONCRETE; HELL GATE BRIDGE; MAILLART, ROBERT; PONT DU GARD.

armored vehicle launched bridge   Driving across a bridge is exceedingly convenient, but having your own personal bridge driving around with you for almost instantaneous use, an innovation developed by the United States Army during World War II and now used by armies around the world, is even better.

The result was a class of military bridge known as the armored vehicle launched bridge, or AVLB. These were also known as mobile assault bridges, or MABs. By the 1950s these had become bridge sections that unfolded scissors-style to span a distance of 60 feet and were mounted on the turretless tank chassis of the period. The spans would be driven to a water obstacle and hydraulically moved across a stream or chasm to allow the passage of troops and vehicles. A huge spade on the lip of the bridge would dig into the ground to hold it firm. Once the troops had crossed, the bridge's two hinged, aluminum alloy sections were folded and the portable bridge was driven off to its next assignment.

The latest version of the U.S. Army's MAB is known as the "Wolverine" heavy assault bridge and is mounted on a modern M1A2 tank chassis to allow it to keep pace with high-speed tank units. Rather than being hinged like a folding door, the Wolverine system's two bridge deck sections retract, one beneath the other. At an obstacle one section is emplaced and the second extended beyond it and locked to complete the bridge. It can span streams or chasms up to 85 feet and can be erected in five minutes and reloaded for transport in 10 minutes. Placement of this new bridge system on the modern M1A2 tank chassis was necessary since the older versions of AVLBs were unable to keep pace with the faster M1A2 tanks during the Gulf War.

See also BAILEY BRIDGE; PONTOON BRIDGE.

Ashtabula Bridge disaster   *See* Stone, Amasa.

B

Bailey bridge (invented 1940) *Prefabricated bridge that helped win World War II* Donald Bailey, a civil servant, was a little known worker in the British War Office in 1940 when he came up with the idea of building prefabricated steel components for portable combat bridges capable of bearing the weight of modern military vehicles.

Recognizing that such a bridge system was essential if Europe were to be retaken from Germany's occupying forces, Bailey presented the idea to his superiors. They were impressed with the concept and ordered Bailey to refine his invention. By July 1941 what became known as the Bailey bridge was in production. The Bailey bridge, with some modifications, was then adopted by the United States Army and was soon in use in the European and Asian theaters of the war.

Bailey's invention was a masterpiece of both simplicity and innovative thinking. By combining modular components including a truss railing and decking, Bailey designed a bridge that could be carried in trucks, assembled, and then extended in a cantilever fashion across a river by cables and jacks. Combat engineers could add more railings side by side or make higher truss railings by placing them atop each other for additional strength. Although far from lightweight, the components could be maneuvered into place by portable jacks and hand tools. As sections were assembled, a span could be counterbalanced on one shore and shoved to the other side. Although the inherent rigidity of a Bailey bridge could sustain its own weight and that of traffic, existing piers, suspension cable systems, and even floating pontoons could also support the spans.

The Bailey bridge was an absolute necessity in Europe, crisscrossed as it is by an untold number of streams and rivers. After D-Day, Allied engineers erected Bailey bridges by the thousands to speed the advance into Germany. By the end of the war Bailey bridging components totaled 200 miles of rigid spans and another 40 miles of floating bridging material. The aggregate weight of the steel bridging produced amounted to 490,000 tons.

General of the Army Dwight D. Eisenhower considered Bailey's invention one of the most important innovations of World War II, with radar and the long-range bomber. Bailey was knighted for his contributions to the war effort in 1946 and died in relative obscurity in 1985.

At the end of World War II, many of the wartime bridges remained in place for some time as permanent bridges were rebuilt. Elsewhere, use of the Bailey bridge by civilian road and bridge departments became common around the world. In the United States, Bailey bridges were kept on hand by public works departments and used in case a bridge failed. Updated versions of the Bailey bridges remain in use in the United States.

The bridging system has also become a weapon in the arsenal of relief workers who must transport supplies to stricken areas. When a massive tsunami swept over Papua New Guinea in 1998, it was decided that a Bailey bridge would be transported and erected to replace an existing Bailey bridge

U.S. Army combat engineers erect a Bailey bridge over a chasm using suspension cables. The brilliant concept developed in Great Britain during World War II allowed the rapid construction of prefabricated bridges necessary for Allied troop movements in every theater of the war. These bridges remain in civilian and military service. *(National Archives)*

swept away by the seismically triggered wave that killed thousands.

Bailey bridges are still manufactured in Great Britain and the United States by Mabey and Johnson Ltd., who began manufacturing the bridge system soon after World War II. Another firm, Bailey Bridges, Inc., also produces the bridge components. Websites for both firms can be found at http://www.mabey.com/Bailey_bridge.htm and http://www.baileybridge.com/index.html.

Barlow shield *See* GREATHEAD, JAMES HENRY.

bascule bridge Though less common than they once were, bascule bridges remain in the United States and the rest of the world an inexpensive way

of allowing land and maritime traffic to coexist. More commonly known as "drawbridges," these spans have existed in one form or another for thousands of years. Perhaps the most enduring image of a bascule bridge is that of a rope-operated drawbridge spanning a protective moat before a castle.

Although bascule bridges are among the world's oldest types, they enjoyed a renaissance during the last decades of the 19th century and the first half of the 20th century. The invention of steam and electric power made it possible to operate heavier and stronger bascule bridges to enable river and canal traffic to pass. The alternative would have been the construction of tall, expensive bridges in developed urban areas where land for bridge approaches was pricey. River and canal traffic, it should be noted, once carried most of the goods transported between cities throughout the world before the emergence of railroads, and later, comprehensive highway systems.

Bascule bridges are divided into two main types, those with single-leaf spans and those with double-leaf spans. The former simply possesses a single horizontal road deck that elevates to allow passage of maritime traffic, and the latter possesses a double span that opens in the middle to allow watercraft to pass. These bridges were essentially a combination of bridge types. Not only were they drawbridges, but they were also constructed in a cantilever fashion since bascule spans had to be stable when secured at only one end. They were often constructed as through-truss bridges since this design provided additional reinforcement to stiffen the road deck. Although these bridges open and close rather slowly, they are incredibly strong.

Although ferries often carried land traffic from shore to shore in the absence of bridges, that means of crossing waterways is time-consuming, expensive, and sometimes dangerous. As America grew during the Industrial Age, bridges and not ferryboats were the waterway crossings of choice. With the popularization of the automobile during the early 20th century, bascule bridges were among the hottest items in the public works budgets of cities like Chicago with numerous water crossings and sizable waterway traffic.

One of the most prolific American bascule bridge builders was Robert Strauss, who lobbied for and obtained the commission to design and build the Golden Gate Bridge. Strauss's firm designed and erected hundreds of bascule bridges during the first decades of the 20th century, having developed a bascule bridge design that was simpler, cheaper, and more reliable. His Chicago-based Strauss Bascule Bridge Co. designed and built bascule bridges on an international scale, shipping their components and then erecting them wherever necessary.

Bascule bridges are unusual not only because they have moving sections to allow the passage of traffic on waterways, but also because, as a rule, they are large, powered machines staffed 24 hours a day. It is necessary for an operator to raise and lower the span or spans of a bascule bridge when waterway traffic approaches. Notable bascule bridges include London's famed Tower Bridge over the Thames River. Built in 1894, it has a double-leaf design and is possibly the most famous bascule bridge of all. Its classic Tudor era design makes it

appear far older than it is. The Tower Bridge is often mistakenly identified by Americans as the "London Bridge," which is a stone arch bridge that was disassembled and moved to the United States during the late 1960s.

The largest bascule bridge, built in 1914 and rebuilt in 1941, is one constructed by the Canadian Pacific Railroad over a canal between Sault Sainte Marie, Michigan, and Ontario. Its movable spans bridge a gap measuring a total of 336 feet.

The actual design and operation of a bascule bridge are relatively simple, as the base of the movable span usually ends at a trunnion, or a horizontal axle with its ends held in hubs. A trunnion was also provided for the counterweight that was elevated above and to the shore side of a bascule span. When a motor raises the bridge, the counterweight reduces the energy needed to lift it. This trunnion arrangement is similar to the mounting of a cannon barrel at its center of gravity on a gun carriage so its muzzle can be raised and lowered. The bascule bridge tender, working controls in the tender's house overlooking the bridge, sets in motion the machinery that raises the bascule span. During the 19th century, bascule bridges were often operated by using steam, but later models were equipped with electric motors. The counterweights were often made of expensive iron until Strauss initiated the use of concrete counterweights to reduce costs.

A Strauss bascule bridge of note is Boston's Congress Street Bascule Bridge, one of the last electrically operated bridges of Strauss design in the United States. The bridge, rehabilitated during the early 1990s, is near the Boston Tea Party Museum, where its tender house is on display. The bridge was built in 1930 in close proximity to four other bascule bridges built during the first half of the 20th century.

See also MOVABLE BRIDGES; STRAUSS, JOSEPH BAERMANN.

bat use of bridges *Concrete shelters for flying mammals* Although a tad ugly to some and downright repellent to others, bats are misunderstood creatures whose shrinking natural habitat has forced them to seek refuge beneath certain types of bridges, much to the surprise of maintenance personnel. This discovery has prompted investigations into the compatibility of bats and bridges while spurring efforts to convert bridges into habitats in hopes of preserving the flying mammals.

This phenomenon seemed to happen without anyone's noticing until the people of Austin, Texas, realized in the early 1980s that hundreds of thousands of Mexican freetail bats were inhabiting the underside of their city's Congress Avenue Bridge. The citizens as well as the Texas Department of Transportation officials were alarmed to find that 1.5 million bats had turned the underside of one of the city's major bridges into their home.

Studies were conducted to find out whether the bats and their guano would have a corrosive impact on the steel reinforcement of the bridge's concrete piers and beams. There was also a health concern since Texas, which reports a sizable number of confirmed rabies cases each year, routinely warns its citizens to avoid handling bats or any other wild animal because of the threat of the virus. Bats are blamed for nearly 50 percent of human rabies infections nationwide.

Studies conducted in Texas and elsewhere indicated that the bats and their droppings were not harmful to the bridge or the public, although spectators who foolishly handle grounded bats are urged to have precautionary rabies vaccinations. The Austin-based group known as Bat Conservation International that assisted in the Texas bat study began promoting the idea that the city treat its topsy-turvy guests with respect. Austinites soon warmed up to the creatures, whose growing fame turned the bridge into one of the city's most important tourist attractions.

Not only do the bats serve as superb natural exterminators by devouring millions of pounds of insects nightly in the Texas Hill Country, but their sunset forays also provide spectators an extraordinary sight. Their evening emergences, which occur during the summer months, send a swirl of bats into the darkening sky to the delight of tourists.

As a result of the publicity surrounding the Congress Avenue Bridge bat colony, it has been discovered that other bridges around the United States are also providing habitats for bats or have the potential to do so. The flying mammals seem to find the spaces between tightly spaced concrete support beams perfect for napping and for sheltering their young from the elements. Bats, which are susceptible to cold, are able to huddle between the beams to share their body heat.

A study conducted by Bat Conservation International and the Texas Department of Transportation has determined interesting facts about the relationship between bats and bridges, including the finding that bats prefer concrete structures to those made of steel. More specifically, the bats prefer hollow-box concrete beam bridges, perhaps because the dead air space in the beams provides insulation from cold and sound.

Since the bats also prefer crevices between the concrete beams that measure no more than 1 1/4 inches, plywood bat abodes with such crevices were placed between concrete beams spaced farther apart to provide additional roosting locations. As many as 85 percent of the concrete bridges studied in West Texas showed signs of bat habitation.

Bat Conservation International points out that half of America's bat species are endangered because their cave habitats have been destroyed or made inaccessible by human development.

A 1998 study by Oregon's Bureau of Land Management found that 94 percent of the bridges suitable for bat roosting were used by some species of bat. The study also found that numerous other concrete bridges and culverts along the southwest coastal region of the state, known as the Coos Bay District, could be converted into prime bat real estate with the addition of plywood bat houses. The Oregon study can be viewed online at http://www.or.blm.gov/coosbay/Bat/bci_rprt.htm#Abstract.

Bayonne Bridge (built 1928–1931) *New Jersey's magnificent steel arch bridge* The Bayonne Bridge is a timeless work of structural architecture that deftly accomplishes the double-edged mission of a bridge by being both beautiful and strong in equal proportions. Seen from a distance the graceful lines of the bridge belie the thought and effort that went into a span that seems to fit so comfortably over the Kill van Kull channel. The bridge arcs so effortlessly over the waterway (whose Dutch name means "Van Kull Channel") that it appears as if it had always been there connecting peninsular New Jersey with Staten Island.

Othmar Hermann Ammann, who also oversaw its construction in his position as engineer for the Port Authority of New York and New Jersey, designed the bridge. Although famous for his suspension bridges such as the George Washington Bridge and the Verrazano Narrows Bridge, Ammann was no stranger to the steel arch design. As the chief assistant engineer to Gustav Lindenthal during design and construction of the Hell Gate Bridge,

The steel arch of Othmar Hermann Ammann's Bayonne Bridge yawns across the Kill van Kull between New Jersey and Staten Island. In a fortunate twist of fate, a cost-cutting measure that eliminated stone facades that were to cover the section where the arches met their abutments actually added to the spare, sleek beauty of the bridge. *(Historic American Engineering Record, National Park Service)*

completed in 1916, Ammann played a role in the building of what was at the time the longest-span steel arch bridge.

The Bayonne Bridge was needed for vehicular traffic traveling southward from New Jersey to the northern edge of New York's Staten Island across the Kill van Kull channel. Although a suspension span, a project well within Ammann's capabilities, was considered, it was decided to utilize the far stiffer arch bridge design. The goal was to make the bridge capable of supporting a commuter rail line if necessary. The flexible nature of a suspension bridge made it less than ideal for the movement and vibrations generated by a train.

Ammann, assisted by the engineer Allston Dana, who also helped Ammann design the Bronx–Whitestone and Triborough Bridges, developed a design that would incorporate a spandrel-braced, parabolic arch from which the road deck was suspended by steel cables.

The vertical spaces between the arches were reinforced with the zigzag of steel latticework trusses. The material of the lower chords would be of manganese steel; the upper chords would be built of heavier but weaker silicon steel. The chords were

all of hollow box construction, with the bottom chords carrying the load of the bridge while the top chords served as a stiffening member to the arrangement. Altogether, the arch itself was something of a skeletal, hollow-box design held together by rivets. On each side, the road deck was suspended by 27 vertical steel cable arrangements composed of four massive cables each.

During its erection, the arch growing from its abutments was supported both by a falsework of temporary piers and also by steel cables from the banks of the channel in cantilever fashion. Full falsework supporting the entirety of the arch was out of the question since it would have blocked the busy channel. Solid rock ran very close to the surface of the ground, and concrete abutments were placed atop the rock after excavation. The two lower chords that met the abutments were connected by hinges consisting of 16-inch steel pins. This allowed movement of the arch due to thermal expansion and contraction without damaging either the abutment or the steel chords. The upper chords merely ended at a vertical steel framework in line with the terminus of the lower chords of the arch.

The steel towers erected to carry the roadway from the approach viaducts were designed to accept granite facing, but Depression era hardships made politicians extremely sensitive to unnecessary expenses. Ammann's requests for funds to complete the bridge fell on deaf ears as politicians boasted that the bridge came in at $13 million, or 14 percent below the original predictions. (Suffice to say the Depression had squeezed the contractors so badly nationwide that other Depression era bridge projects, including the Golden Gate Bridge, became unexpected bargains.)

Although Ammann wanted to see the stone in place, it never arrived. However, the nakedness of the steel framework at each abutment is arguably more attractive and in keeping with the overall architectural minimalism of the span. Lindenthal had placed stone towers at each end of his Hell Gate Bridge's arch to provide the appearance of strength, although they did nothing at all structurally. Ammann was deterred from his cosmetic touch by political caution.

The record-breaking 1,675-foot distance spanned by Ammann's Bayonne Bridge is another curious feature of the span since it managed to best the length of Australia's Sidney Harbour Bridge by 25 feet. The bridge in New South Wales began construction two years before Ammann's span. One cannot help wonder whether Ammann and his colleagues participated in a bit of one-upmanship by ensuring that their bridge exceeded the length of the bridge in the land down under. The Bayonne Bridge held the record as the world's longest steel arch bridge until the 1978 completion of the New River Gorge Bridge in West Virginia. This new bridge at Fayetteville was only 24 feet longer than the Bayonne Bridge, proving that bridge-building is not only an art form but a competition.

Although the structure was built to endure the loading of a commuter rail line, no rail line was ever placed upon the bridge. The span retains a pedestrian walkway cantilevered out from one side of the roadway. The bridge, which was to help connect New Jersey, Staten Island, and Manhattan, was not as heavily used as planners expected. Although it is capable of accepting additional lanes, the bridge traffic capacity has not been augmented since its construction.

See also AMMANN, OTHMAR HERMANN; ARCH BRIDGE; HELL GATE BRIDGE; LINDENTHAL, GUSTAV.

Beach, Alfred Ely (1826–1896) *Renaissance man who built New York's first subway* Born to a life of privilege and prestige, Alfred Ely Beach became a patron saint of sorts to American inventors, publisher of *Scientific American,* and the first man to tunnel beneath New York City for the purpose of building a subway.

His father was the publisher of the *New York Sun,* who made his son a partner in the enterprise at the age of 19 in 1845. At 20, Beach purchased *Scientific American* magazine to make it one of the nation's most unique publications and became partners in a patent firm. Beach not only published stories about inventiveness, but was also an inventor in his own right, obtaining a patent on a typewriter when he was 21. At 27, he unveiled a typewriter that produced embossed type that could be read by the blind.

Like many technically inclined men of the 19th century, Beach was fascinated with transportation issues. From his office, Beach watched New York City become more crowded and polluted and pondered ways of solving the traffic problems choking the city. Horse-drawn wagons and carriages were competing dangerously, and sometimes fatally, with throngs of pedestrians on streets littered with dung. The city's horse-drawn trolleys were carrying 100 million passengers yearly and had reached their capacity.

Beach knew the only unobtrusive place for New York's traffic to go was underground. By 1867 Beach had designed a model of a pneumatic underground subway and demonstrated it at New York City's Exhibition of the American Institute. Beach was in favor of air pressure as the method of moving his subway cars underground since he had always abhorred steam locomotives for their noise and pollution. The miniature pneumatic subway car that carried 10 people amazed visitors to the exhibit, and 100,000 boarded it for a 100-foot aboveground ride during the exhibition.

Propelled by air pressure generated by a fan, the pneumatic tube concept was fascinating but frowned upon by Tammany Hall's William Marcy Tweed, who held a virtual stranglehold on the public works programs and the politics of the city. Tweed liked aboveground transportation that needed bridges and elevated rail lines because there was more graft in such projects—although it is assured that Tweed would have undoubtedly found a way to make a dollar off a subway project if one had gotten under way during his tenure as a power broker.

Amid all the attention given to Beach's pneumatic subway concept, Beach only inflamed Tweed further by publishing a pamphlet touting the advantages of an underground pneumatic transit system.

Recognizing the influence of "Boss Tweed" but nonetheless committed to building a subway, Beach engaged in subterfuge to accomplish his goal. He applied for a charter to construct a small pneumatic tube for delivering mail. With Tweed temporarily mollified by this ruse, Beach secretly began construction of a 312-foot subway tube from the basement of Devlin's Clothing Store.

Beach employed a tunneling shield equipped with hydraulic rams to push it through the soft ground as miners excavated the spoil at the open front. As the eight-foot-diameter bore progressed, masons laid a brick lining to the rear of the shield. In a curved section of the tunnel, workers erected a section of cast iron support rings. Dirt was hauled away under the cover of darkness in wagons with wheels padded to prevent the clatter of their iron rims. Although the actual excavation took only 58 overnight shifts to complete, Beach would take two years to line the tunnel, install the air blowing equipment, and produce a tube-shaped rail car.

Concerned that the public would shy away from entering a dark, underground tunnel, Beach ordered its interior painted a bright white and lined with gaslights to dispel the gloom of traveling underground.

Beach also financed a finely appointed 22-person tube car as well as beautifully decorated stations beneath Warren Street and Murray Street. The car itself was a small work of art with a separate section for women traveling alone. The cat was finally out of the bag when a *New York Tribune* reporter disguised as a workman infiltrated the workings and reported the details of the subway. With his project exposed, Beach threw it open to the public on February 28, 1870.

It was a resounding success, as the public flocked to pay a 25-cent fare to ride what Beach proclaimed the Pneumatic Transit. A steam engine near the Warren Street station powered a fan to blow and then suck the car back along its route. The car was actually nothing more than a rail-mounted piston moved along like a large mail tube. Although

The small, 312-foot tunnel and the single cylindrical car constituted the Pneumatic Transit of the publisher and inventor Alfred Ely Beach. Secretly constructed in 1870, the car was powered by pressurized air like a mail tube and was New York's first subway. Political opposition and funding problems precluded its expansion. *(Library of Congress)*

the speed was kept to 6 mph, the conveyance was probably capable of faster speeds though safety dictated moderation.

Tweed fought Beach's Pneumatic Transit in court, claiming Beach had overstepped the bounds of his charter—which was true—but ultimately failed in his legal challenge as the court threw the matter to the New York Legislature. Beach struggled to obtain a bill allowing the extension of his subway to Central Park, but Governor John T. Hoffman eventually vetoed the legislation. Despite clamorous public support, Beach was stymied when a competing bill proposing a massive and hideous elevated railway through the heart of Manhattan appeared in 1871.

This Tweed-backed project could have lined Tweed's pockets with millions in graft, but Tweed's days in power were numbered. Eventually Tweed was prosecuted for his graft while Beach fought for his subway. On April 9, 1873, the reform governor John A. Dix signed into law a bill granting Beach a charter to enlarge his Pneumatic Transit. Unfortunately, one of the 19th century's numerous financial crises erupted the following September, dooming any hopes of financing for the project. Beach's superb little tunnel later became a target range and ultimately a massive cellar for wine storage. Eventually, Beach ordered the tunnel sealed.

Contractors revisited Beach's marvelous pneumatic subway in February 1912 as they prepared to excavate a passage for the Brooklyn-Manhattan Transfer. Although the myth persists that workers came upon the tunnel like archaeologists discovering a pharaoh's tomb, the location of the pneumatic subway line was well known. Entering the ventilation grate used by Beach's subway at City Hall Park, workers found the tunnel remarkably intact, although the rails had corroded to nothing. Beach's tunneling shield remained at the terminus of the excavation, it was recovered and given to Beach's son, Frederick, who had helped build the tunnel, and the younger Beach gave the shield to Cornell University, where it was placed on exhibit. The present-day whereabouts of the shield are not known.

See also TWEED, WILLIAM MARCY.

beam bridge One of the simplest concepts in bridge design is the beam bridge, which consists of a horizontal structure sometimes supported by nothing more than the sides of a riverbank or by posts or piers sunk into the ground. Although prehistoric bridges of this type have been found, they usually span extremely narrow obstacles. This is because the ability of the beam to bear weight depends upon its own ability to resist deformation or deflection between its end supports.

On a beam bridge, the bending of the main beams or girders carries the forces. Bending force can be visualized by thinking of a yardstick supported at its two extremes and loaded with a weight at its center. Under this loading the ruler deforms downward in a parabolic shape, as the fibers at the top contract (or take compression forces) and the bottom ones expand (or take tension forces).

Although the earliest beam bridges were simple ones consisting of a stone slab or a log crossing a brook, stone and wood lacked the ability to resist destructive bending in the form of what engineers call flexural strength. Stone deformed only slightly before cracking, and wood, although it would bend a great deal more, would also eventually snap. Unreinforced concrete was later used, but this brittle material lacked the capacity to endure tension without cracking like stone. Beam bridges require materials that can withstand both tension and compression. Steel can generally endure both types of forces. Steel reinforcement and prestressed, internal tendons added to concrete later made it capable of enduring tension forces that had bedeviled it.

Concrete's ability to withstand the bending force of tension was so greatly increased by adding steel reinforcement and prestressed tendons that concrete beam bridges are now among the most common. They are inexpensive and capable spans often used for highway overpasses or for crossings of small waterways. Numerous other short bridges in America use composite steel–concrete beams. In these bridges a concrete slab on top takes all the compression and the steel in the bottom takes all the tension, combining the properties of the two materials.

Because the maximal forces are concentrated at the extreme top and bottom areas of a beam, designers determined that the most efficient structural form is one that concentrates most of the materials in these locations. This has given rise to the I-shape beams (often called wide flange beams) that are often used today in steel and prestressed concrete bridges.

Some of the most interesting examples of ancient beam bridges are found in Britain. Usually consisting of a flat stone set upon other stones on each side of a narrow stream, these bridges are known as "clapper bridges" by the locals. An untold number of small wooden beam bridges exist, many of them in the timber-rich United States. Simple in concept and relatively easy to build, beam bridges remain a good solution for spanning relatively short distances.

The length of a beam bridge segment is self-limiting since it cannot be expected to reach very far without being supporting by vertical columns. This is best explained with the example of a two-by-four piece of lumber 12 inches long that is terrifically rigid. A piece of lumber of the same dimensions but 100 feet long will sag (or deflect, in the lexicon of bridge engineers). When a beam is excessively long between supporting piers, its deflection, or bending, caused by its own weight, makes it fail.

Theoretically, a beam bridge can have an indefinite length as long as intermediate supporting piers are installed to carry beam segments. Not surprisingly, the longest bridge in the world is a beam bridge, the Lake Pontchartrain Causeway. The bridge runs 24 miles across the widest part of Lake

Pontchartrain between New Orleans and Covington, Louisiana. A pair of two-lane bridges was driven across the lake by methodically installing supporting piers. Prestressed concrete beams were then fitted to the supporting columns until the lake was spanned.

See also LAKE PONTCHARTRAIN BRIDGE; PRESTRESSED CONCRETE; REINFORCED CONCRETE; STEEL.

bearings   Engineering terms are like medical terms in that a word usually describes the precise function of a component. In the case of bridge *bearings,* the word is succinctly accurate since the purpose of a bearing is to provide a load-transfer mechanism for all the vertical and some of the horizontal loads on the bridge.

Bearings provide an attachment interface between a supporting pier and a steel or concrete beam that allows for the expansion and contraction of bridge components due to temperature changes. Other forces that can cause a bridge's components to shift include movements of the Earth such as potentially destructive earthquake activity. For these reasons, bearings can play an important role in bridge construction since some have the ability to absorb differential movement between supporting piers and bridge deck beams.

A crude bearing might consist of little more than a stack of steel plates between a vertical column and a horizontal beam. More complicated bearings can consist of large steel rockers (rocker bearings), large neoprene pads mounted between steel plates, or sliding bearing assemblies.

In consideration of the contraction and expansion of bridge components due to temperature changes and support movements, bearings are designed to allow for movement of the bridge components without causing a failure of the span. A breakthrough in this type of bearing is the elastomeric bearing, which consists of a rubberized or plastic material between thin steel plates, a concept akin to a rubberized motor mount on an automobile. Any horizontal movement of the bridge components is transmitted to the elastomeric bearing, whose flexibility can absorb it. The steel plates, on the other hand, help provide vertical stiffness so that the elastomeric bearing does not deform excessively under vertical loads. When the movement stops, the bridge components are drawn back into position by the elasticity of the bearing. This return of the components to their original location by a bearing is known as self-centering.

As civil engineers have become more concerned with the survivability of buildings and bridges in earthquake-prone areas, additional research has gone into improving the capabilities of these bearings, which can sustain tens of thousands of tons of weight. These bearings play an important role in efforts to retrofit bridges with components to resist damage caused by seismic movement. The technology of elastomeric bearings has also been applied to the foundations of buildings in areas with seismic activity. Modern bearings are designed to be inspected and replaced if damaged by excessive loading, corrosion, or other environmental factors.

Bearings took on new significance after the Loma Prieta earthquake in northern California on October 17, 1989, and again after the earthquake that devastated Kobe, Japan, on January 17, 1995. In these events, many concrete structures and their steel components failed because they lacked bearings that could deform sufficiently to allow the superstructure of the bridge to survive. Improved bearings are being installed in bridges worldwide since adding deformation capacity into a bridge can mean the difference between destruction and survival under severe circumstances. Engineers also consider high-tech bearings a relatively cheap investment that can not only save lives but ensure the structural well-being of an expensive bridge as well.

See also SEISMIC RETROFIT.

Beaumont, Frederick *See* Beaumont tunneling machine.

Beaumont tunneling machine (patented 1880) *Forerunner of today's tunnel boring machines* Known as the Beaumont tunneling machine, the precursor for modern tunnel-boring machines was actually invented by one man and perfected by another. Frederick Edward Blackett Beaumont (1833–1899) devised a boring machine in 1875, and Thomas English (1843–1935) obtained a patent on critical improvements to the machine in 1880.

Although tunneling machines had been proposed, and at least one other had been built, by the 1870s, the Beaumont machine was not only capable of making considerable progress in soft rock but also provided a way to remove the spoil while supplying fresh air to the tunnelers. Before the introduction of this tunneling machine, 19th-century miners were forced to use explosives packed into drill holes or undertake the backbreaking work of excavating by hand.

The original machine was the brainchild of Beaumont, a British army officer who entered the Royal Engineers as a second lieutenant in 1853. During his military service, Beaumont helped plan and supervise construction of military fortifications at Dover, whose famed white chalk cliffs overlook the English Channel. Beaumont, like anyone involved in construction, undoubtedly became intimately familiar with the properties of the ground on which he was working. As it turned out, this experience would stand Beaumont in good stead since his machine would eventually be used in the first major effort to dig a Channel tunnel between Dover and France.

Beaumont's original design was not a truly practical boring machine. It might never have even been built because of this. Necessary improvements were made to the design by his fellow Royal Engineering officer Thomas English, who was apparently equipped with a considerable amount of metallurgical and engineering knowledge. English eliminated a conveyor belt for removing debris and improved the cutting heads of the machine.

The final machine looked something like a jumble of machinery on a sled, which is precisely what it was. A compressed air engine, supplied by hoses running from outside the tunnel, powered a series of gears that rotated a cutting head equipped with steel chisels. The gearing allowed the large auger to turn 1 1/2 revolutions per minute.

Before starting the machine, workers would anchor it to the floor of the tunnel and activate the cutting head. A hydraulic piston then pushed the cutting head forward on an upper frame until its reach was exhausted. The same hydraulic pump then raised the upper frame of the machine so that the lower frame resting on the tunnel floor could be elevated and advanced. The upper frame was lowered and the drilling began anew. Although it sounds ungainly, the Beaumont–English machine produced stunning results under certain conditions.

The clever Beaumont–English machine also incorporated a bucket chain that would travel the length of the machine, carrying spoil to the rear, where it would be poured into small rail cars and removed. And because it was powered by compressed air, the exhaust air expelled by the machine would provide miners a supply of fresh air in tunnels where the atmosphere could be stale if not downright dangerous.

Because of the limitations of metallurgy of the period, no drilling or cutting machine could be truly effective until the development of harder cutting materials such as tungsten carbide, which occurred during the 1920s. However, the Beaumont–English boring machine was well-suited for relatively soft chalk.

The first major test of its capabilities would garner Beaumont and English an unparalleled amount of publicity. Their machine would be used by French and English promoters in attempting to drill a tunnel beneath the English Channel.

See also TUNNEL-BORING MACHINE.

Benjamin Franklin Bridge   *See* Modjeski, Ralph.

bentonite   *See* slurry shield machine.

Bienvenue, Fulgence   *See* cut-and-cover tunneling.

blackdamp   *See* afterdamp.

black lung   *See* coal dust.

black powder   *See* dynamite.

blowout   For subaqueous tunnel and caisson workers laboring in compressed air, the one hazard feared more than any other is the sudden loss of the pressure holding back tons of water. Known as a blowout, this accidental decompression can eject workers out of the rupture and cause catastrophic flooding in their once-dry work space.

When a section of a soft ground, subaqueous tunnel is excavated, there is a brief interim between when it is dug and when its interior is reinforced and sealed. It is during this period that a weak section of the tunnel may give way and allow compressed air to blast out of the excavated space. The penetration through which the air escaped then becomes the entryway through which water can fill the void.

During compressed air caisson work, the potential for blowout also exists, but blowout has generally been more likely to result from a loss of pressure in a watertight door or another mechanical fixture on the roof of the caisson.

At one point during the excavation of the pneumatic caisson on the Brooklyn side of the Brooklyn Bridge, a blowout occurred when the water level in an open-ended excavation shaft decreased so much that the pressurized air was able to expel the column of water. The incident took place on a Sunday morning, when the site was fortuitously unoccupied. When air pressure spewed from inside the massive caisson, a geyser rocketed 500 feet above the East River, pelting surrounding buildings with rocks and an unpleasant coating of reeking riverbed muck. Water rushed into the caisson but no one was injured.

When the Hudson Tunnel Railroad Company began working on a rail tunnel beneath the Hudson River between New Jersey and Manhattan in 1874, work was done in such a way as to invite a blowout. The railroad engineer Dewitt Clinton Haskin decided to use compressed air without a tunneling shield that would support the soft earth as workers dug and lined the tunnel. Haskin would instead rely only on compressed air to prevent the tunnel from collapsing as workers raced to reinforce it with iron plates and brick masonry.

When the soft ground of the tunnel failed on July 21, 1880, it allowed a jet of compressed air to shoot from inside the excavation. A falling timber killed one man and 20 others were drowned in the ensuing flood.

See also HUDSON AND MANHATTAN RAILROAD TUNNELS; MABEY, MARSHALL; TUNNELING SHIELD.

boomer The nickname for any itinerant worker who travels seeking work, often during a boom in economic activity. In the United States, the word *boomer* has come to be a nickname given to bridge construction workers because of the temporary nature of their work.

The term probably originated during the 19th century during widespread expansion of railroad bridge construction and other building booms across the United States. The term has nothing to do with the builder Lucius Bolles Boomer, who built a large number of railroad bridges during the mid- and late 19th century.

Boomer, Lucius Bolles *See* Eads Bridge.

box tunnel *See* Brunel, Isambard Kingdom.

Brandt, Alfred (1846–1899) *Hydraulic rock drill inventor and Simplon Tunnel engineer* The engineering career of the German-born Alfred Brandt reached a high point with the invention of his innovative water-powered rock drill and fell to its low at the time of his ill-fated construction of Switzerland's Simplon Tunnel.

Brandt began his career as a railroad engineer and was employed by Switzerland's Gotthard Railway during construction of the Saint Gotthard Tunnel in the Swiss Alps. This project was one of the great Swiss tunnel projects, though one plagued by hundreds of worker deaths, disease, and financial ruin for the contractors. The project began in 1872 and concluded in 1882.

Brandt took careful note of the Saint Gotthard debacle and would go on to become a partner in his own engineering firm. Though he would later gain a contract to build Switzerland's Simplon Tunnel, he would first gain fame by developing a new type of rock drill whose beginnings lie in one of the many problems encountered on the Saint Gotthard Tunnel effort.

During that hellish project, Brandt was dispatched to the Swiss village of Airolo to assess problems with the compressor system that provided pressurized air to the pneumatic rock-drilling machines. Brandt became fascinated with the inefficiency of compressing air through the use of waterpower since a by-product of the process was the generation of useless heat.

Like all innovators, Brandt cut through traditional thinking and decided that a direct use of water pressure would be far more efficient. At the time, trapping water at a higher elevation and releasing it through a pipe to a lower elevation was the method used to create water pressure. Brandt wanted to use this water pressure directly to power a rock drill and in the process eliminate the mechanical production of compressed air produced by water-driven turbines.

His first drill attempted to use the water pressure to power a piston in the same way an airpowered percussive rock drill worked, but this was a failure. He then used pressurized water to make a hollow drill steel revolve as it was locked against a rock face with 12 tons of force. By 1876, Brandt had developed a hydraulic drill with hardened teeth that could penetrate rock. The rock core would emerge from the rear of the hollow drill steel to be broken up. Water vented from the system would then flush the drill hole clear during the process. Brandt's drills were attached to a carriage that was rolled by rail to the tunnel face to perform its task.

Brandt's innovative drill was then put to the test on the Sonnstein Tunnel in Austria, where it managed to drill a 12-foot shot hole in 2 1/2 hours. This was an excellent result, and two drills mounted on carriages working night and day allowed the tunnel to be driven six feet in 24 hours.

This water-powered drill did not operate at high speeds and could turn at a maximum rate of only 10 revolutions per minute. When combined with the force with which it was held against the rock, this was adequate for penetration. A pair of the drill carriages was subsequently used on the Arlberg Tunnel between Austria and Switzerland during the 1880s.

By 1893, Brandt was a partner in his own engineering firm, Brandt, Brandau & Company, and had obtained what at the time seemed a plum contract to help construct the Simplon Tunnel. The geological plan on which the tunnel design was based anticipated solid rock throughout the proposed path of the 12-mile tunnel, whose construction was to use Brandt's hydraulic rock drill.

Work began on the Simplon Tunnel's northern heading August 1, 1898. The group of other engineering firms with which Brandt's company was allied was known as the Simplon Consortium. Brandt threw himself into the job. After a stretch of working night and day without sleep, Brandt died suddenly of a stroke on November 29, 1899, at the age of 53. The Simplon Tunnel's rock conditions were horrific and labor unrest stalled the project.

Although the project started out well, crews working both headings ran into massive amounts of flooding, and miners were forced to build what amounted to a protective steel cage through a seam of decomposed rock. The contractors involved were saved from financial devastation only by the nationalization of Switzerland's railways, which prompted the renegotiation of favorable terms for the contractors.

As for Brandt's rock drill, it had seen its final use on the Simplon Tunnel since a new compressed air rock drill manufactured by Ingersoll had captured much of the market after the American drill was introduced in 1871. A Brandt drill remains on display at the Bergbau Museum in Bochum, Germany.

See also ROCK DRILL; SAINT GOTTHARD TUNNEL; SIMPLON TUNNEL.

break step (1831) *Order for soldiers to march out of step*   Perhaps one of the longest-lived myths surrounding suspension bridges is the belief that a column of troops marching in step will create a swaying effect that will invariably collapse the flexible spans.

The myth is drawn from a pair of catastrophic European bridge failures that occurred nearly 30 years apart in the 19th century. True, the Broughton Suspension Bridge collapsed in 1831 as a column of infantry marched across it near Manchester, England. Also true is that on April 16, 1850—in what remains the deadliest recorded bridge disaster—226 French soldiers died when the Basse-Chaine (Low Chain) suspension bridge they marched across collapsed in Angers, France.

In the case of the Broughton structure, it may not have been the rhythmic marching of the troops alone but the failure of a weakened bridge component that allowed the collapse. As for the bridge at Angers, its iron suspension cables were badly corroded and the bridge was generally in need of serious repairs. A perfectly good bridge will endure the marching of soldiers within its designed load limit.

Bridges of all types were collapsing in the 19th century, and not infrequently. Most of the collapses occurred because bridges were poorly designed and maintained or were loaded beyond their limits. But the myth about suspension bridges may have caught on because of the airy and frail appearance of such spans, which did not appear as solid as traditional stone bridges and did sway side to side, something they were designed to do.

In view of the high failure rate of suspension bridges during the 1800s, it is not surprising that so few were willing to take unnecessary risks. In no time at all, military commanders in many nations were instructed never to march a unit in step across a suspension bridge. For any officer who forgot, bridges were often marked with large signs reminding them not to do so. It was common for marching troops in Great Britain to receive the command "Break step!" as they approached any type of bridge and to resume marching in step after the crossing. In the United States the command to walk normally is "Route step."

Although little more than an old wives' tale, the myth persisted until the 1930s and undoubtedly beyond. When the massive Golden Gate Bridge was opened to the public for the first time May 27, 1937, only pedestrians were allowed on the bridge for the day. Although 20,000 flooded across the bridge in the first hour it opened, bridge officials were horrified to see a relatively small U.S. Army unit marching in step toward the bridge. Still believing the myth that marching troops would collapse a suspension bridge, officials chastised the proud commander of the well-drilled unit. The officer reluctantly ordered his men to walk across like tourists. A newspaper account later compared the appearance of the Depression era soldiers ambling across the massive span to that of unemployed men trudging through the line of a soup kitchen. When the bridge opened to auto traffic the following day, heavy autos of all types poured across without causing the slightest problem.

Although it probably remains mainly as a historical curiosity, a sign warning of this "danger" still adorns England's Albert Bridge, a cable-supported span. The placard posted by the Royal Borough of Kensington & Chelsea bluntly warns:

ALL TROOPS
MUST BREAK STEP
WHEN MARCHING
OVER THIS BRIDGE

See also SUSPENSION BRIDGE.

breakthrough   *See* holed through.

bridge aerodynamics   As bridges evolved from heavy stone arch structures into thinner, lighter spans of greater length, engineers were unwittingly experimenting with the unseen forces of aerodynamics, which could turn a solid, heavy bridge into a writhing serpent alive with dangerous movement. With the advent of lighter and more flexible structures, such as long-span suspension bridges, during the late 18th and early 19th centuries, it became apparent that the wind could not only make such bridges sway and oscillate but could tear them to pieces as well.

The introduction of stronger and lighter building materials such as iron and later steel allowed construction of thinner, less weighty bridges. The suspension spans using iron chains and eventually steel cables allowed engineers to span considerably longer distances with a bridge that weighed a fraction of their stone arch predecessors.

Gone are the good old days of stone arch bridges when designers built structures that were inherently so heavy and of such great mass that wind was scarcely a factor. Engineers once were more concerned with the force of a river on a bridge's foundations than with the effects of wind on elevated superstructures.

Before the advent of aviation, aerodynamics in past centuries was left to the designers of sailing ships and cannonballs, items that had a more intimate relationship with the air. Not until the November 7, 1940, wind-induced collapse of the Tacoma Narrows Bridge were bridge designers compelled to wed the steel and concrete world of bridge engineering with the comparatively ethereal science of aerodynamics. Engineers, who understood for nearly two centuries that the wind could have an adverse effect on suspension bridges, decided to ignore this knowledge. Instead they flirted with disaster during the 1930s in the far-too-light design of the Tacoma Narrows Bridge and a number of other notable suspension bridges, including the Golden Gate Bridge.

Modern long-span-bridge designers would relearn that they were building not only elevated roadways but also aerodynamic structures that could be devastated by a force as seemingly innocuous as the wind. To their chagrin, the long road decks of their bridges were also wings with a terrific amount of surface area that would attempt to "fly" if aerodynamic countermeasures were not included in the design.

The wind presents dizzyingly complex aerodynamic problems to engineers and a wide array of unwanted movements to bridges. In one type of destructive movement or oscillation, called flutter, a bridge deck is rhythmically lifted and dropped. Sometimes a strong wind can shake the components of a bridge, creating buffeting, an effect that may

Just before its 1940 collapse, the deck of the lightly built Tacoma Narrows Bridge was twisted by aerodynamic forces that revealed the complex impact of wind on long, lightweight bridges. This event sparked intensive study into aerodynamics and ways of controlling effects on long suspension bridges. *(MSCUA, University of Washington Libraries, Farquharson #4)*

not destroy a bridge but can accelerate the deterioration of the components. Although engineers had long calculated the drag, or the "static" horizontal wind load that pushed against bridges, these other potentially damaging effects would not be systematically studied in America until the Tacoma Narrows Bridge collapse. Torsional and lateral movements that occurred on the Tacoma Narrows Bridge resulted in its flexible road deck's swaying and twisting.

Bridges, rising to challenge the elements over river valleys and waterways, are strategically placed in the path of the strongest winds. Growing taller, longer, and lighter, bridges became ripe for damage by wind by the beginning of the 19th century, something that Brooklyn Bridge designer John Augustus Roebling recognized as early as 1841. Suspension bridges in particular, with their lighter weight, collapsed on a regular basis during the 1800s. Some early suspension structures could not handle the weight of traffic placed upon them; many, however, succumbed to the wind. Though lacking a detailed understanding of aerodynamics, Roebling instinctively understood that making bridges heavy and stiff would dampen the effects of the strongest winds.

Perhaps the earliest notable example of a catastrophic American suspension bridge failure brought about by wind is the 1854 collapse of the Ohio River's Wheeling Bridge. Designed by the maverick engineer Charles Ellet, who was supremely confident of its structural integrity, the bridge was completed in 1849 and boasted a 1,010-foot suspended span. Although the bridge seemed substantial enough, a strong wind caused its 460-ton road deck to collapse thunderously into the Ohio River after only five years of service. Even before the Wheeling mishap, the 1836 destruction of the Royal Suspension Chain Pier in Brighton, England, by strong winds provided another catastrophic example of the force of wind. Despite these and many other examples of wind-related disasters, 20th-century bridge designers would drift away from the lessons that Roebling—who never lost a bridge to wind damage—had taken seriously and countered.

Bridges that stood against the wind were designed to be heavy and rigid enough to counteract not only the lifting effect of the wind and unnecessary vibration, but also the compressive effect of wind's pushing horizontally against the structure. Although the destruction of bridges by the compressive force of winds is infrequent, it may have played a role in the collapse of Scotland's Tay Bridge on December 28, 1879. This rail bridge, the longest in the world at the time, collapsed during a gale, killing all 75 aboard a passing train. It had been designed to endure a "pushing" wind load of 10 pounds per square foot, but poor materials and bad maintenance conspired with the wind to destroy the iron truss bridge 19 months after its completion.

However, as suspension bridge design flourished in 1930s America with the magnificent George Washington, Golden Gate, and the Bronx-Whitestone Bridges, the influential designer Leon Moisseiff was pushing for thinner, lighter designs that were both aesthetically pleasing and less expensive. Although the deflection theory applied by Moisseiff to suspension bridges supported the notion that these slender designs were adequately strong, Moisseiff and others failed to appreciate that the complex interaction of massive bridge structures and wind invalidated these computations. Moisseiff had the unhappy distinction of designing the extremely slender but ill-fated Tacoma Narrows Bridge, the third-longest suspension bridge of its day. The Bronx–Whitestone Bridge, on which Moisseiff contributed design input, was given additional reinforcement after proving to be too light and flexible, as was the famous Golden Gate Bridge, on which he was also a consulting engineer.

When the Tacoma Narrows Bridge did collapse on November 7, 1940—after four months of frighteningly wobbly and undulating service—its destruction was investigated by a three-man team appointed by the Federal Works Administration. This investigative triumvirate was composed of not only the esteemed bridge engineers Othmar H. Ammann and Glenn B. Woodruff (a designer of the Oakland Bay Bridge) but Theodore von Kármán, a gregarious and ostentatious bachelor who was considered one of the fathers of modern aerodynamic theory. Although the report was less than conclusive, the Tacoma Narrows collapse and the work of the investigative committee would inextricably fuse aerodynamics and bridge design. The results of the investigation would be debated for decades, but what was obvious was that the bridge was unable to endure a relatively modest 42-mph wind.

Even von Kármán's inclusion on the Tacoma Narrows Bridge investigative committee was not a foregone conclusion, reflecting a desire of the engineering community to delve into the science of aerodynamics. Instead, the dauntless aerodynamicist played a major role in making himself part of the investigation of an event that was a national sensation at the time. Stunned upon hearing the Washington state governor's self-serving announcement that the bridge was adequately designed, von Kármán subjected a rubber model of the bridge to wind generated by a small fan to replicate the destructive motion. The personable scientist then notified state authorities of his findings and found himself on a committee with two of America's most staid and conservative engineers.

What von Kármán claimed at the time was that vortices, or swirling eddies created by the wind on the lee side of the bridge, generated turbulence that caused it to flounce up and down or, more scientifically, to oscillate. This uniform, turbulent flow pattern created in the wake of an object was named the von Kármán vortex street after its discoverer. Thus was triggered a long-term debate over the issue of "resonance," in which some believed the frequency of the wind matched the natural frequency of the structure to trigger massive oscillations. This unfortunate match of the frequency of an external force to that of an affected object is best explained by the movement of a pendulum pushed into motion by someone's hand. If the pendulum is pushed again as it returns with the same force in a uniformly periodic way, its motion increases because the frequency of the push matches its own frequency of movement.

Although von Kármán's theory about the bridge disaster remains a topic of debate, if not outright confusion, it was apparent that bridges had to be viewed not only as traffic-bearing roadways but also as structures whose aerodynamic attributes had to be neutralized or counteracted for the sake of stability. Subsequent wind tunnel tests have indicated it was unlikely that the 42-mph wind that tore the Tacoma Narrows Bridge road deck to pieces created the condition of resonance, because the calculated frequency of the wind was incorrect and because the wind's velocity would be less than uniform. However, it was determined that von Kármán was partially correct in assuming that turbulence due to the shedding of vortices played a major role, although the vortices were not the symmetrical and uniform type he proposed.

Subsequent analysis of the Tacoma Narrows collapse revealed that the movement of the bridge deck created a variety of vortices, which resulted in an increasingly powerful flutter that finally exceeded the load limits of the steel suspenders of the bridge.

This flutter caused the bridge to twist along its axis during the last 45 minutes before the collapse, a startling sight captured on film by a University of Washington engineering professor, Frederick B. Farquharson.

It was also realized that suspension bridge decks in particular, as they became thinner and lighter, could behave in the same way as the wing of an airplane does when, under the right circumstances, it generates high pressure underneath and lower pressure above to create the lifting effect. A bridge with these aerodynamic characteristics could be lifted in high winds and dropped with great force when the wind velocity decreased. If not enough to destroy a bridge, such reactions would at least provide an uncomfortable crossing for motorists. Aerodynamic forces like these present serious problems to suspension bridges since such bridges are designed to move up, down, and sideways as a result of the contraction and expansion of the steel suspension cables and the sideways force of the wind.

That it took so long for 20th-century bridge engineers to understand this is surprising to the Duke University engineering professor and bridge design historian Henry Petroski, who has pointed out that American bridge engineers after 1900 displayed an incredible myopia by ignoring the historical record of wind-induced bridge collapses.

However, Professor Frederick B. Farquharson, who filmed the collapse of the Tacoma Narrows Bridge, knew something was drastically wrong and studied the instability before, during, and after the official inquiry. As a result of the collaboration of Ammann, Woodruff, and von Kármán and with the painstaking research of Farquharson, wind tunnel testing was subsequently recognized as an essential element of the process of establishing stable aerodynamic qualities for bridge designs.

What researchers recognized as a result of the Tacoma Narrows collapse was that suspension bridge roadways could be too lightweight, too narrow, and too thin to dampen the effects of wind. Their structures could also present to the wind a nonaerodynamic shape that would increase turbulence. During the second year of operation of the Golden Gate Bridge, a powerful wind caused massive oscillations, the first sign that this landmark bridge required additional stiffening reinforcement, which was added 14 years after its completion. Moisseiff, the guru of "thin is in" bridge design, played a central role in formulating the design of the

Golden Gate Bridge, which also reflected his passion for thinner decks.

Engineers have also learned that besides returning to beefier bridge designs, other ways had to be found to make bridges aerodynamically stable. The Tacoma Narrows Bridge used a solid plate girder beneath the road deck for stiffening to save money, but wider and heavier truss-reinforced decks with an open latticework of steel became preferable since they provided little resistance to the wind while stiffening the deck and adding stabilizing weight. Such trusses would eliminate the flat, unventilated shapes of plate girders that were known in aerodynamics as bluff bodies. Such flat, square-edged shapes generate far more turbulence and stronger vortices as wind passes over them. In addition, many modern long-span bridges now utilize ventilated grating on the roadway to equalize pressure above and below the deck, thereby robbing it of the ability to act as a wing. This innovation was included in the bridge that replaced the collapsed Tacoma Narrows span in 1950 and is in place on the world's longest suspension bridge, Japan's Akashi Kaikyo Bridge completed in 1998.

Scale models of bridges are now routinely subjected to extensive wind tunnel testing to determine whether the structure contains hidden aerodynamic flaws. Wind tunnel research flourished before and during World War II in military aviation research and has been heavily applied to construction applications since the end of the 1940s. Before the Tacoma Narrows Bridge collapse, horizontal wind forces were estimated and a bridge was made rigid and heavy enough to survive them. Even after the vagaries of dynamic wind forces were recognized and studied, it became apparent that the complexity of aerodynamics meant that not all the problems could be foreseen.

Just when bridge designers thought it was safe to go out in the wind again, a new aerodynamic problem that was afflicting some of America's modern bridges emerged during the 1990s. With the advent of cable-stayed bridges in Germany that followed World War II, a whole new era of bridge design was ushered in and, with it, a whole new era of serious aerodynamic problems.

Easy to erect, strong, and strikingly beautiful, cable-stayed bridges became increasingly popular in the United States during the 1980s. Cables that run directly down to the roadway from towers support this type of bridge. As these bridges have become more popular, engineers determined that they are also subject to a complex set of aerodynamic conditions. When raindrops cling to the sides of the cable stays, they alter the aerodynamic profile of the cables and their ability to shed the wind in a smooth fashion. The result is that the taut, massive cables vibrate excessively, degrading the components of the bridge faster than anticipated. By 2001, at least 14 cable-stayed bridges in the United States were identified as having this previously unrecognized and potentially serious problem.

This problem has generated studies on bridges across the United States to determine ways of reducing the vibration and increasing the structural longevity of the bridge. The 2,475-foot Fred Hartman Bridge, which crosses an estuary of Galveston Bay just southeast of Houston, is a magnificent-looking span that was one of the first to be seriously afflicted with this vibration problem. Completed in 1995 at a cost of nearly $118 million, the bridge will have a severely curtailed lifespan as a result of vibration if the problem is not mitigated. University and private sector engineers have been studying the issue in an effort to cure the problem. Some experts have recommended that cable-stayed bridges receive dampers—devices akin to shock absorbers—between the cables and their connections to the roadways they support.

Curiously, it is only rainfall accompanied by mild wind that creates the damaging vibration problem. It seems that high winds blow the raindrops away from the plastic-encased cables and restore their smooth, aerodynamic profile. Surprisingly enough, this could mean cable-stayed bridges are less prone to aerodynamic vibration in a hurricane packing a 90-mph wind than in a light shower with a mild 20-mph wind. Oddly, one cable may vibrate while the one next to it does not. It should be pointed out that this vibration is no small matter. The phenomenon is so noticeable on Texas's Fred Hartman Bridge that it has been reported by passing motorists who spot the cables oscillating up to five feet side to side. Engineers have used dimples and other types of textured surfaces on the polyethylene covers of cable stays to remedy the problem.

See also AKASHI KAIKYO BRIDGE; AMMANN, OTHMAR HERMANN; BRONX-WHITESTONE BRIDGE; BROWN, SAMUEL; CABLE-STAYED BRIDGE; ELLET, CHARLES, JR.; GOLDEN GATE BRIDGE; MOISSEIFF,

LEON SALOMON; ROEBLING, JOHN AUGUSTUS; TA-
COMA NARROWS BRIDGE; TAY BRIDGE.

Bridge of Sighs *See* Rialto Bridge.

bridge protection *Collision mitigation* Despite
the reassuring bulk and seemingly invincible grandeur of bridges crossing wide expanses of water, any
bridge—no matter its size—is vulnerable to collision
with a ship if its piers extend from a navigable
waterway. Unfortunately for bridges, which are
among the world's largest stationary structures,
modern tankers and cargo ships are the largest moving structures, and the hazard of collision with one
of these floating behemoths is a real and a potentially deadly possibility, especially for those on the
bridge. The possibility of collision between a ship
and a bridge pier must be reckoned with by engineers but received little attention until the 1990s.

This sedan narrowly escaped a plunge into Tampa Bay after a
ship knocked down a span of the Sunshine Skyway Bridge in
1980. Unfortunately, 35 others were less fortunate and dropped to
their deaths. This horrifying event forced officials to recognize
the necessity of protecting bridges from errant ships. *(Florida
State Archives)*

Although the first line of defense against a collision is the navigational and ship-handling skill of a
maritime crew, accidents do occur on a regular basis.
One of the most serious was the May 9, 1980, collision between the 609-foot cargo ship *Summit Venture* and a pier of Florida's Sunshine Skyway Bridge
during a blinding rainstorm. Lacking fenders or protective islands, the pier took the brunt of the collision, causing one of its spans to collapse into Tampa
Bay. As a result, 35 people died when the vehicles
they occupied plunged over the edge of the damaged
bridge.

Another horrific collision occurred in the
predawn hours of September 22, 1993, when six
barges being pushed by a towboat struck a rail
bridge's pier in a remote section of Alabama. A span
and the rails it supported were displaced minutes
before an Amtrak passenger train plowed into the
damaged bridge segment. The engine derailed and
plunged into the 15-foot-deep water below, pulling
with it a chain of passenger cars. The fuel tanks of
the engine ruptured and fed a massive fire that lit the
disaster with a deadly and hellish light. Of those
aboard the train, 47 died, including five Amtrak
employees.

To protect bridge piers, a variety of engineering
solutions can be constructed to absorb the impact
of a wayward ship. By the 20th century such systems became necessary as technology allowed construction of gigantic bridges over waterways used
by seagoing ships that were growing ever larger in
size. During the 1930s a massive 300-foot-long concrete fender was constructed to protect the south
pier of the Golden Gate Bridge, mainly to shield the
pier's foundation from erosion by wave and tidal
action. Although costly and sometimes difficult,
construction of fenders around bridge piers is considered by most engineers to be a worthwhile safety
feature. By precluding damage to the bridge and
loss of life, such improvements can perhaps repay
the investment.

Until 1970, however, few bridges were built with
any type of protective system for their piers, this,
despite the fact that merchant ship traffic has
increased by 33 percent in American ports since
1970, with much of that traffic passing beneath
bridges having vulnerable foundations and piers.

A variety of bridge pier protective techniques
exist. Concrete fender walls like that employed by
the Golden Gate Bridge's south pier is one. Another
type of fender is a wall of protective steel or timber

pilings driven into the water in front of a bridge foundation. One form of protective structure also uses pilings driven into the water to form a cylindrical barrier that is known as a dolphin. Advanced dolphin designs include hollow cylinders of concrete or cylindrical piles of heavy rock encircled by a ring of pilings. When the cable-stayed Sunshine Skyway Bridge was constructed in 1987, three-dozen impact-absorbing dolphins were built to protect its piers. An extremely effective and relatively simple method is construction of protective islands out of rock of such substantial size that they cannot be washed away by normal wave action.

The primary goal of any bridge protection system is to absorb the energy of a ship collision or to deflect a ship away from a bridge pier. It is unrealistic to think a 900-foot ship or a chain of barges weighing thousands of tons can be stopped instantly. Some cargo ships weigh more than 100,000 tons. Barges and ships, though usually traveling slower than 15 knots when entering a harbor, nonetheless pack tremendous energy. Modern fender and dolphin designs are built to absorb this energy in several ways.

Some structures can compress to absorb the energy, such as the piled stone of the artificial islands or the shearing or crumbling of a hollow concrete dolphin. Steel pilings serving as fenders can bend and flex to absorb the energy of a moving ship. Florida's Department of Transportation now provides some bridges with a fender system that deflects a ship away from the pier without absorbing the full impact of the collision.

Engineers recommend the cost-efficient method of building protection systems with replaceable components that are much like the impact-absorbing crash barriers on freeways. Should a ship collide with such a protective device, its modular components could be quickly replaced to await the next befuddled sailor.

However, retrofitting bridges with such protective structures can reach a cost as high as that of the bridge. Designing and building protective structures for the foundations and piers during initial construction of the bridge can amount to only 50 percent of the cost of the bridge since the equipment and labor are already at hand, according to some studies.

Aside from the costs, public officials recognize the problem of ship-bridge collision as a potentially serious threat to the public. The American Association of State Highway and Transportation Officials has issued a set of design guidelines for protecting the supporting piers of highway bridges. As a result of the 1993 Amtrak derailment, the federal government has ordered that towboat operators demonstrate a high proficiency with radar in order to retain their operating licenses. It is hoped that the use of advanced electronic navigational and positioning equipment, including the Global Positioning Satellite System, or GPS, will reduce navigational confusion.

Britannia Bridge (built 1846–1850) *Tubular bridge that gave the world the hollow-box beam* If any waterway in the world has inspired innovative bridge design, it is the Menai Strait between Wales and the island of Anglesey. Thomas Telford's Menai Bridge, the world's first long-span suspension bridge, crossed the strait in 1826. Within 20 years, the expansion of railways prompted the need for a rail bridge across the Menai Strait, and a government decree that 100 feet of overhead clearance be provided prompted the development of one of the world's most unique bridges.

The man who was going to build this bridge was no less than George Stephenson, the son of the self-taught engineer and locomotive inventor Robert Stephenson. The younger Stephenson wanted to complete a rail line between London and Holyhead on the western coast of Anglesey, at a point closest to the Irish coast. Incredibly, Stephenson contracted to build the bridge on a tight schedule despite the fact that he had not yet developed a design with which to accomplish the task.

A crossing point north of Telford's Menai Bridge was selected. A rock outcropping in midstream known as the Britannia Rock could serve as a location to anchor one of the necessary piers and eventually provide a name for the span. In 1845, Stephenson began his planning and first proposed an iron arch bridge. Since the lower chords of an arch curving down to their springings would have violated the 100-foot clearance requirement, Stephenson had to find a new design strong enough to support a pair of railroad tracks but capable of providing the necessary clearance.

Stephenson heard of the launching of an iron-hulled ship that had gone awry, leaving part of the vessel suspended in midair without collapsing. Stephenson hit upon the idea of creating a hollow, enclosed span that could be placed atop the piers. The structure would, like the hull of a ship, be able

Robert Stephenson's 19th-century Britannia Bridge over the Menai Straits as it originally appeared, before it was rebuilt after a fire. One of the world's most unusual bridge designs, it carried rail traffic through a series of hollow-box beams, a structural concept originally developed for the bridge. *(Godden Collection, Earthquake Engineering Library, University of California, Berkeley)*

to support its own weight as well as any live load placed on it over a considerable distance.

The novel idea was easy to conceive but hard to put into practice, and Stephenson's father recommended that his son enlist the help of William Fairbairn. A self-educated engineer, Fairbairn had pioneered the use of iron in ships, was a toolmaker, and was a man capable of both theoretical and practical thinking. Fairbairn suggested recruiting a mathematician, Eaton Hodgkinson, to help them understand the performance of a round tubular structure, the first type of hollow beam Stephenson intended to try.

The initial tubular models tended to crack along the top and bottom edges when loading was applied. Fairbairn deduced that tubes of uniform thickness or strength were not going to work and recommended adding reinforcing at the roof and floor of the tube. It had also become apparent that a rectangular "tube" would work better. When the rectangular

shape was tested, they found that the iron plates on the sidewalls distorted under loading. This was corrected by the addition of iron stiffening plates to the walls.

On February 6, 1846, Stephenson reported to the directors of the Northern and Eastern Railway that he had developed a method of bridging the Menai and estimated that he would complete the bridge by 1849. Despite the daring nature of his plan, Stephenson was nonetheless cautious about his design and planned to incorporate suspension chains for additional support.

Stephenson's plan was to construct three masonry piers on bedrock and to connect these with four rectangular tubes. Of these, two central segments would measure 479 feet each and the two tubes approaching from shore would measure 229 feet apiece. The total length of the tunnel-like spans would measure 1,417 feet. Since the bridge was to handle two lanes of rail traffic, the rectangular tubes would be erected in pairs. Since 1,750,000 rivets would be required for the assembly of the tubular spans, workers used a hydraulic hammer tool developed by Fairbairn to increase their efficiency.

Work proceeded smoothly and rapidly, as Stephenson carefully planned the operation with military precision. The tubes would be constructed on shore, transferred to barges, and then lifted with hydraulic jacks from the surface of the Menai Strait at high tide. Stephenson decreed that the raising of the spans would be done slowly and deliberately with supports placed beneath the tubes to prevent a catastrophe should a hydraulic jack fail. This precaution proved wise when one jack failed, dropping a span. Although some damage to the tubular section occurred, it was repaired and eventually placed into service.

The sight of the long central sections being floated at high tide below the piers was breathtaking. A system of cables pulled by capstans was employed to haul the floating sections into place and then hold them against the current during lifting. At one point, the famed engineer Isambard Kingdom Brunel, who used a similar system for lifting sections of his Royal Albert Bridge at Saltash five years later, was on hand to help direct the positioning of a tubular span. Brunel and Stephenson were both friends and rivals, and Brunel's brilliant bridge at Saltash is often compared to the Britannia span.

Once the spans were in place and secured, it became apparent that the tubular sections alone were enough to support their own weight and that of any rail traffic, and the expense of the suspension chains could be eliminated. The portals at the summits of the 200-foot towers still stand open today, devoid of the chains. The two longest iron boxes were considered susceptible to thermal expansion and contraction, so their ends were rested on heavy roller bearings on the two shoreward towers. In this way, movement of the iron tubes would not create a destructive push-pull effect where they met the piers.

The first train crossed the completed bridge on March 5, 1850, and a second test of the span supported a train weighing more than 700 tons. Although the design exceeded expectations and was a curious sight to those accustomed to conventional bridges, it was not a beautiful bridge.

David Billington, in *The Tower and the Bridge*, points out that the willingness of the railway to pay any price for the bridge meant Stephenson could employ a heavier and more expensive structure that was, invariably, far less appealing aesthetically. Although the bridge proved the box beam concept, this type of rail span would not catch on since other types of spans, including cantilever bridges, provided comparable strength with less weight and cost.

Nonetheless, Stephenson's extraordinary bridge served on for 120 years, until an unlikely tragedy destroyed the magnificent tubes. In 1970, a pair of teenage boys were hunting for birds inside the superstructure of the bridge using lighted torches. When 17-year-old Peter Owen ignited a strip of tarred burlap with his rolled newspaper torch, it traveled to the timber roof covering the two center tubes. The fire burned for nine hours, softening the iron tubes until they slumped from their own weight. Owen was fined five British pounds for trespassing.

The bridge underwent a reconstruction process from 1970 until 1980 that included the installation of a pair of truss arches between the three piers to support two rail lines, a vehicular road deck, and pedestrian walkways.

During the five years of the original construction of the bridge, 15 men were killed in accidents. A memorial was erected to them at a churchyard near a North Wales village with a characteristically unpronounceable Welsh name believed to be the longest in the world: *Llanfairpwllgwyngyllgogerychwyrndrobwllllantysiliogogogoch*. The name means "the Church of Saint Mary in the hollow of white hazel, near the rapid whirlpool of Saint Tysilio and a red cave." Sadly, during the reconstruction of the bridge two more workers were killed. Their names were also placed on the memorial.

Despite the fire, the demolition of the original tubes crossing the Menai, and the passage of more than 150 years, a cross section of the Stephenson-Fairbairn tube was preserved and can be seen on display near the present bridge.

See also BRUNEL, ISAMBARD KINGDOM; FAIRBAIRN, WILLIAM; HOLLOW-BOX GIRDER; MENAI BRIDGE; STEPHENSON, GEORGE; STEPHENSON, ROBERT; TELFORD, THOMAS.

Bronx-Whitestone Bridge (built 1937–1939) *New York's wiggly suspension span* Like most bridge projects, the Bronx-Whitestone Bridge was discussed for decades before the first cubic yard of concrete was poured or a single rivet driven. The political impetus that propelled the graceful bridge across the

East River was the 1939–1940 New York World's Fair.

The New York City infrastructure czar Robert Moses, who was in charge of, among other things, developing parks, applied his full powers in pushing development of the site. Since the improvements were to revert to the ownership of the city's Parks Commission, as well was the initial $2 million in revenue, Moses had a bureaucratic stake in its profitability. Although the project was privately funded, Moses oversaw it as a military commander, or more precisely, a warlord far removed from the oversight of authority. To site the park, contractors reclaimed a former garbage dump with a night-and-day landscaping effort while using muck from local tunneling projects as fill.

Moses, who viewed his projects as monuments to his own vision, decided that road access to the World's Fair was essential. Moses loved building bridges, but he also wanted to make it easy for New Englanders to drive south to his Flushing Meadows extravaganza. And, since he was chairman of the Triborough Bridge Authority, Moses could put a bridge up just about anywhere as long as tolls could generate enough income.

As planning for the bridge began, Moses received resistance from the Regional Planning Association, a citizen group founded in 1922 to impose order on the chaos of New York area development. The group would support a Bronx-Whitestone bridge only if Moses agreed to make it capable of supporting the subsequent addition of mass transit rail lines. Moses, who loved the automobile but who never drove one, was loath to comply with this request, which would increase not only the cost of the bridge, but also the time necessary to construct the span, which Moses wanted ready for the spring opening of the fair. Fortunately for Moses, the Regional Planning Association could count upon no other allies in the fight with the redoubtable parks commissioner, who built the bridge as soon as he obtained state permission to do so.

Moses had ordered the Triborough Bridge Authority chief engineer, Othmar Hermann Ammann, to produce a bridge for a minimal amount of money in record time. Ammann complied with his usual efficiency and attention to detail. The result was a design that lent itself to rapid construction and low cost. One design feature providing savings was the first use of suspension towers devoid of cross bracing between their legs. The bridge had a spare, modern look that would set the style for suspension bridges built for the next 60 years. The concrete foundations for the two 377-foot towers were rapidly sunk to bedrock, and erection of each tower required less than a month. Ammann, who was assisted in the design by Allston Dana, also depended heavily on the consulting engineer Leon Moisseiff, who had worked with Ammann on the Golden Gate Bridge.

Moisseiff had pushed for thinner and lighter bridge designs and had found an eager ally in Ammann, who was committed to building a long-span suspension bridge with an uncluttered appearance. Moisseiff advised Ammann on the design of the bridge at the same time he was adding the final touches to his design of Washington state's exceedingly slender Tacoma Narrows Bridge.

The Tacoma Narrows Bridge did not use a stiffening truss but rather a narrow plate girder along the underside of its deck, a feature also included in the design of the Bronx-Whitestone Bridge. A plate girder was lighter, less expensive, and easier to erect. Moisseiff's march toward lighter and thinner bridges had been given a considerable boost during the impecunious depression years when infrastructure money was scarce, and the Bronx-Whitestone Bridge was a depression-era structure. Moisseiff's ideas were not only aesthetically beautiful but less expensive.

The Ammann-Dana-Moisseiff team nonetheless produced a graceful bridge exhibiting an artistic minimalism. The plate girder that was a mere 11 feet deep and the slender, unobtrusive towers gave the bridge a stiletto appearance. The building of the crossing set something of a record for a long-span suspension bridge: With its main span of 2,300 feet and side spans of 735 feet, the bridge was built in an amazing 23 months at a cost of $19 million. The work began on the bridge on June 1, 1937, and the span was opened April 29, 1939, the day before the official opening of the World's Fair.

During construction the highly flexible and light bridge had demonstrated a tendency to oscillate in prolonged strong winds. Ammann knew this, as did the contractors, but the calculations showed the bridge structurally capable of standing firm against the winds blowing down the East River. Six months after the Bronx-Whitestone Bridge was built, its nearly identical twin, the Tacoma Narrows Bridge, was flouncing wildly in a relatively mild gust. The bridge rose and fell until one of its suspenders con-

The truss railing installed to tame the unwanted movement of the deck of the Bronx-Whitestone Bridge is apparent in this view. Designed by Othmar H. Ammann and Salomon Moisseiff—the latter an influential proponent of lighter and thinner suspension bridge designs—it was notoriously unstable. Although it did not collapse, Ammann (who always maintained the bridge was structurally sound) was nonetheless ordered to install the truss railing. *(Historic American Engineering Record. Photograph by Jet Lowe)*

necting the suspension cable to the deck broke. Soon after, the deck collapsed into the water.

Ammann, one third of a federal team appointed to investigate the Tacoma Narrows collapse, concluded that Moisseiff had followed existing bridge engineering practices—practices that Moisseiff had helped to establish and practices to which Ammann also subscribed. There is still uncertainty over the precise cause of the Tacoma Narrows Bridge collapse, but engineers realized that lighter, long-span bridges were susceptible to aerodynamic damage even though engineers had come to believe that they were not.

Although calculations showed that the Bronx-Whitestone Bridge was probably far more stable than the Tacoma Narrows crossing, motorists were unnerved by the sizable motion they felt when crossing the bridge in a stiff wind. Ammann hotly defended the stability of the bridge, and most believed the span was far more stable than Moisseiff's Tacoma Narrows Bridge. By comparison, the depth to span ratio of the Tacoma Narrows Bridge was one foot of deck thickness for each 350 feet of its central span's length, compared to a 1/209 ratio for the New York bridge, and, although 500 feet longer than the Bronx-Whitestone span, the bridge over the Tacoma Narrows weighed only half as much.

In 1940, the Bronx-Whitestone Bridge received alterations to counteract the effects of the wind, including the installation of cable stays running diagonally from the towers to the deck. This innovation, originally utilized by John Roebling on his 19th-century suspension bridges including the Brooklyn Bridge, was designed to stabilize lateral movement due to horizontal wind forces.

Despite this modification, which detracted from the clean lines of Ammann's creation, the Bronx-Whitestone Bridge still oscillated enough to perturb pedestrians and drivers. In 1943, movement of the bridge's deck became so pronounced that the crossing was closed until the winds died.

After the war, Moses ordered Ammann to modify the bridge to arrest the wind-induced oscillations despite protests from Ammann that the movement was both harmless and not unanticipated. Ammann could not dissuade Moses, who was concerned that public distrust of the span might reduce traffic on the bridge and thus reduce the all-important toll revenue.

In 1946, Ammann installed a 14-foot truss down each side of the road deck's surface in addition to the existing plate girder beneath. Ammann's thin whisper of a bridge had gained weight and a heftier silhouette, but it had also gained more rigidity. It was hoped that it had also gained the driving public's trust, even if those using the bridge had never been in danger. At the same time, two pedestrian walkways were eliminated from the bridge and the space allocated to widen the existing six traffic lanes.

The Bronx-Whitestone, in spite of the diagonal cable stays and the addition of truss railing, remains New York's most lively bridge. On November 12, 1968, a 70-mph wind produced 10-inch vertical movements of the bridge deck, causing it to be closed for nearly five hours. Triborough Bridge and Tunnel Authority officials reassured the public at the time that the bridge was prone to movement that was harmless to both drivers and the structure.

See also AMMANN, OTHMAR HERMANN; BRIDGE AERODYNAMICS; DEPTH TO SPAN RATIO; MOISSEIFF, LEON SALOMON; MOSES, ROBERT M.; TACOMA NARROWS BRIDGE.

Brooklyn-Battery Tunnel (built 1940–1950) *Moses vs. the Tunnel* Although tunnels often have unusual histories if not deeply tragic ones—particularly when it comes to casualties and cost overruns—New York's Brooklyn-Battery Tunnel not only involved a struggle between engineers and nature, but a collision between two of America's most powerful men.

The fate of the tunnel was decided in a titanic political struggle between New York's commanding public works czar Robert Moses and President Franklin Delano Roosevelt, two men who hated each other years before their clash over Moses's desire to build an obtrusive bridge rather than an unseen tunnel. What resulted, after more than 20 years of planning, the Moses-Roosevelt run-in, and a three-year delay caused by World War II, is the longest vehicular tunnel in the United States.

The drama began when it became apparent by the late 1920s that a new vehicular artery was needed between Manhattan and Brooklyn. On January 15, 1929, New York City officials decided upon a two-mile tunnel in lieu of a bridge, a decision that predated Moses's rise to full power. The subsequent descent of the country into economic depression brought infrastructure construction to a virtual standstill, including plans for the Brooklyn-Battery Tunnel. The proposed project languished until January 9, 1936, when the New York City Tunnel Authority was granted approval to build the Brooklyn-Battery Tunnel by the New York State Legislature. Two years before, Moses had been appointed one of three commissioners to the Triborough Bridge Authority.

Moses, who by some accounts wanted to control all intracity transportation projects in New York, found himself excluded from the Tunnel Authority when it was conceived by the New York mayor Fiorello La Guardia in 1936. It was La Guardia, who had a complex and sometimes difficult relationship with Moses, who ordered the legislation establishing the Tunnel Authority to exclude Moses. The mayor and Moses had often clashed, but La Guardia was also being practical in his exclusion of Moses, whose acrimonious relationship with Franklin Roosevelt dated back many years. In the seventh year of the Depression, the first thing La Guardia needed was federal money and the last thing was alienating the president.

Moses, piqued at being excluded from control over the proposed tunnel, expressed disgust with the tunnel project, calling it overly expensive and incapable of carrying an adequate amount of traffic. Some, including Moses's biographer Robert Caro in *The Power Broker,* claim that Moses's opposition to the tunnel was based upon his desire to acquire complete power over all transportation projects rather than any real objection to a tunnel.

The Tunnel Authority, with the help of Mayor Fiorello La Guardia, took on the considerable task of acquiring funding for not only the tunnel but the roadways that would connect it to existing and planned traffic arteries. Despite his opposition to a tunnel, Moses saw a Brooklyn-Battery crossing as a way of completing his ambitious Belt Parkway project, which would curve around Brooklyn and then to the northeast into Long Island by 1940.

The total cost of the tunnel and the attendant roadways was $105 million. La Guardia knew the city was incapable of financing the project so he dunned the Reconstruction Finance Corporation chief Jesse Jones, a Houston millionaire and Roosevelt confidant, for the funding. La Guardia's hopes of obtaining approval for government assistance were probably based in no small measure on his appointment of Jones's cousin, Alfred B. Jones, to the Tunnel Authority board.

Himself an expert practitioner of political manipulation, Jones turned La Guardia away, saying New York City had already received more than its share of federal loan assistance. La Guardia then went hat in hand to Secretary of the Interior Harold Ickes, who ran the Public Works Administration (PWA). Again, La Guardia was told that his city had already received more than its fair share of federal grants. The PWA, said Ickes, would provide no more than $5 million in funds for any project. La Guardia did, however, receive permission to transfer $7 million from federal school and hospital construction funds for the tunnel project.

La Guardia, despite all his pleading and political guile, could muster only $12 million with which to build the tunnel, but this amount eventually persuaded Jones's Reconstruction Finance Corporation to agree to purchase as much as $43 million in self-liquidating bonds in the project. Despite having $55 million in funding, the entire project was still $50 million short of what it needed, and La Guardia hungrily eyed the fat coffers of Moses's Triborough Bridge Authority, a surplus that La Guardia could not legally require Moses to contribute.

If the Brooklyn-Battery Tunnel would open a new doorway between Battery Park in Manhattan and the rest of New York City, it would also open a door of opportunity for Moses. A renowned expert at close-in political combat, Moses did agree to contribute $22 million in Triborough funds to build the tunnel, which was initially estimated to cost nearly $65 million. Moses's gesture was hardly magnanimous since it had a price: La Guardia had to agree to spend $12 million in city funds to build traffic links desired by Moses to both ends of the proposed tunnel. La Guardia also said he would seek state legislative approval to allow Triborough eventually to acquire not only the Tunnel Authority's Queens-Midtown Tunnel project, which began construction in 1936, but also the Brooklyn-Battery Tunnel.

Moses and La Guardia, despite their complicated relationship so often filled with friction, were both addicted to the power of their respective offices and their mutual love of grand infrastructure projects. As a result, one needed the other to accomplish his goals. However, the stripping of $7 million from school and hospital construction funds to build a tunnel and portions of a roadway project—later named the Belt Parkway—outraged the New York City comptroller, Joseph D. McGoldrick. As planning on the Brooklyn-Battery Tunnel progressed during 1938, the comptroller argued that the project was siphoning money away from children's health care and the maintenance of aging subway facilities.

A vote of the city's Board of Estimate (of which McGoldrick was a member) was necessary to approve the project. When the Board of Estimate convened, its meeting erupted into a raucous event. Moses's supporters heckled McGoldrick and two other board members opposed to diverting school and hospital funds. The board adjourned to a closed session, in which La Guardia pressured the trio to approve the project. After the four-hour private session, McGoldrick and his allies, fearful of fracturing the city's fragile political coalition, emerged to vote unanimously for the Brooklyn-Battery Tunnel. Such are the unhealthy birth pangs of great infrastructure projects.

The following year, Moses, who now held sway over the project thanks to his agency's financial contribution, announced that he intended to build a suspension bridge rather than a tunnel. The "Brooklyn-Battery Bridge" would be a more sensible option during the troubled economic times, argued Moses, who despised tunnels.

The bridge—a pair of suspension spans placed end to end with a common monolithic anchorage at Governors Island—would have huge elevated approaches and a pair of blockhouselike anchorages on each shore. Needless to say, this would have been a terrific eyesore, and many recognized it as such straightaway. On the Battery Park side, construction of the bridge and its approaches would have required the demolition of historic Fort Clinton and much of the park itself.

A bridge would have also marred the seaward view of the lower Manhattan skyline, an issue that would eventually upset many of the city's wealthy and powerful, including the New York native First Lady Eleanor Roosevelt.

The issue of Moses's decision to convert the tunnel to a bridge simmered with flattering drawings of the proposed bridge originating from Moses and frighteningly gothic depictions of the same bridge disseminated by the tunnel engineer Ole Singstad. Singstad was a tough, unyielding opponent of the bridge who helped design the Holland Tunnel and was the chief designer of the Brooklyn-Battery Tunnel.

Moses's drawing revealed the bridge from an overhead perspective as a slight, unobtrusive, and narrow line running almost invisibly across New York Harbor. Singstad, Moses's equal in making his points with the public, came up with a drawing showing the bridge from a sea level perspective as a gigantic eyesore clawing its way across New York's skyline. This more realistic view of the bridge revealed 10-story concrete anchorages and massive elevated approaches smothering the view of Brooklyn and Battery Park.

The bridge versus tunnel debate continued, accusations flew, as Moses claimed Singstad and other tunnel supporters were providing unrealistically low figures for the cost of building the Brooklyn-Battery Tunnel. Moses also claimed he was telling the truth about the financial debacle the tunnel would become. Moses was ruthless and did his best to tar the tunnel project and anyone who supported it. One elderly opponent of the bridge was referred to publicly by the ruthless Moses as an "exhumed mummy"; another was labeled a "communist."

Moses was eventually caught lying about the cost of his massive bridge project when he falsely attributed an astonishingly low cost estimate of his proposed bridge to the austere and highly regarded engineer Othmar Ammann, designer of the George Washington Bridge and later the Verrazano Narrows Bridge. The honest and rigidly correct Ammann was caught in a terrible political situation and quietly directed the tunnel supporters to the correct (and far higher) cost of the bridge Moses preferred.

Moses, famed for his brutish political behavior and his willingness to push through projects no matter their fiscal or social cost, was fighting the fight of his career in public service against seemingly insurmountable odds. By 1939, the unions and reform-minded citizens were in favor of the tunnel, and the no-nonsense Singstad proclaimed a tunnel a cheaper alternative to an unsightly bridge that would consume the priceless land that held Battery Park. Singstad believed a tunnel could be built for just over $65 million. Moses claimed the tunnel would cost nearly $20 million more than Singstad's figure. (The tunnel was eventually built for $59 million.)

Steeled against the opposition he faced, Moses bullied his way through the bureaucracy and obtained the approval he needed to build the bridge from state and local officials. By 1939, those opposed to Moses on the matter of converting the tunnel to a bridge had been strewn across the political landscape. The bridge Moses considered a more fitting symbol of his power and influence needed only the approval of the U.S. Army Corps of Engineers. That approval seemed all but assured until First Lady Eleanor Roosevelt weighed in with a one-paragraph mention of the proposed bridge in her newspaper column "My Day."

Mrs. Roosevelt's column, carried in newspapers across the nation, appeared on April 5, 1939, and devoted a single paragraph to Moses's bridge:

> I have a plea from a man who is deeply interested in Manhattan Island, particularly in the beauty of the approach from the Ocean at Battery Park. He tells me that a New York official, who is without a doubt always efficient, is proposing a bridge one hundred feet high across the river, which will go across to the Whitehall Building over Battery Park. This, he says, will mean a screen of elevated roadways, pillars, etc., at that particular point. I haven't a question that this will be done in the name of progress and something undoubtedly needs to be done. But isn't there room for some consideration of the preservation of the few beautiful spots that still remain to us on an overcrowded island?

Moses, who had acquired so much power that he had literally become a power unto himself, had unwittingly aroused the opposition of the one man in the United States who could successfully thwart his plan, the husband of Eleanor Roosevelt, who happened to be the 32nd president of the United States.

There is little doubt that Mrs. Roosevelt mentioned the issue to her husband, whom their relative Alice Roosevelt Longworth once described as "two-thirds mush and one-third Eleanor." The president was also to hear of the tunnel controversy from an influential New York attorney, Charles C. Burlingham, a bridge opponent and friend of the Roosevelts. The 81-year-old lawyer made it clear to the president that the U.S. Army Corps of Engineers should kill the bridge project by denying approval of its construction at the mouth of the East River, a suggestion that Roosevelt apparently took.

Soon after Roosevelt became aware of the controversy, Moses realized that an army report that would have provided the needed federal approval for his bridge had been delayed. Even more perplexing was the fact that his inquiries into the delayed report were not answered. This occurrence puzzled Moses since the Corps of Engineers and other military experts had previously assured him that a suspension bridge posed no hazard to maritime traffic if bombed or damaged. Nonetheless, the report and its endorsement of a Brooklyn-Battery Bridge never materialized.

This was because Roosevelt's secretary of war, Harry H. Woodring, had inserted himself into the process soon after the appearance of Mrs. Roosevelt's critical column, obviously at Franklin Roosevelt's request. As Moses continued to dun Washington for approval of his bridge project, Woodring secretly ruled that a Brooklyn-Battery bridge could block access to the United States Navy Yard (also known as the Brooklyn Navy Yard) if damaged during wartime. When Moses was finally given this news publicly on July 17, 1939 (two months after Woodring made his decision), an outraged Moses pointed out that the Brooklyn Bridge and the Manhattan Bridge already stood between the Navy facility and the sea. Woodring stood firm. The commanding Moses was handed the worst defeat of his bureaucratic career as he was ready to savor one of his greatest victories.

Soon after the obliteration of Moses's plan to construct his bridge, Jones's Reconstruction Finance Corporation, which had originally agreed to purchase only $39 million in bonds for the tunnel project, changed its course and announced it would provide $54 million in financing. With the Triborough Bridge Authority's funds no longer so necessary for the construction of the Brooklyn–Battery crossing, Moses lost his ability to dictate terms for the tunnel project to La Guardia. An immediate result was La Guardia's decision to delay the construction of traffic links between the tunnel and Moses's beloved Belt Parkway.

After the well-camouflaged intervention by President Franklin Roosevelt, final plans were made and tunneling efforts begun. During the October 28, 1940, groundbreaking for the tunnel, Roosevelt was in attendance and boldly recounted the falsehood that was necessary to save the magnificent view of New York City while positioning himself even further from Moses's defeat.

"I feel like saying today—'at last!' because for some time there was some dispute as to whether we would cross . . . under the water or over the water," said Roosevelt. "That was a question that mere laymen couldn't decide. We had to call in the people who were primarily responsible for the national defense, and they told us . . . it would be far safer for America to have a tunnel than a bridge. That is why essentially, it is a tunnel and not a bridge." With Roosevelt's disingenuous and untruthful statement, the tunnel work began.

Compared to the politics that swirled around the Brooklyn-Battery Tunnel, the actual feat of constructing the crossing was relatively straightforward. The tunnel was actually a pair of parallel tubes excavated side by side from Battery Park to Brooklyn. Construction of the tubes began at both ends, using the pneumatic shield method that would drive both tunnels 9,117 feet between the portals on the Battery Park and Brooklyn shores.

Pneumatic shields were started at each end of both tunnels. These massive steel tubes were slightly larger than the finished diameter of the tunnel and protected the sandhogs from a collapse of the soft earth during excavation. The working area containing the shield would be filled with pressurized air to restrain flooding. As muck was hauled away, the shield was pushed forward by hydraulic rams, and the sandhogs used gigantic wrenches to bolt massive semicircular iron rings together behind the shield. Once in place, these iron rings formed a strong lining for the tunnels' interiors, which were 31 feet in diameter.

Working at the face of the tunnel, sandhogs were encased in a compressed air atmosphere of 37.5 pounds per square inch to forestall seepage. Workers entered airlocks where the air was slowly pressurized so they could then enter the already-pressurized working area. The precaution was necessary since opening the airtight doorway between the working area and the unpressurized portion of the tunnel would have caused an explosive escape of pressurized air and the resulting infiltration of water.

Workers exiting the pressurized work area would reenter the airlocks, which would then serve as decompression chambers. After being slowly returned to normal air pressure that allowed the dangerous nitrogen absorbed in their blood to dissipate, workers could then depart for home. Despite highly refined precautions against what was known as "the bends," or caisson disease, workers on the project nonetheless wore badges identifying them as tunnel workers. If they were found unconscious on the street after their work shift, the badges directed bystanders to have them transported to a nearby hospital, where they could undergo additional decompression.

In clay or mud the workers were able literally to dig the tunnel, but when they encountered rock, drilling machines bored holes for dynamite charges. These carefully placed charges artfully blasted through the ground, leaving an opening of the precise dimensions. The opening would then be "scaled," or smoothed, and reinforced by iron rings.

Each time the shield advanced 32 inches, or the width of a ring, an erector arm at the rear of the shield lifted the iron sections into place so mud-covered workers could bolt together a complete ring. Once completed, the shields were shoved forward with potent hydraulic jacks pushing against the forward edge of a completed ring.

The depression-era project, designed to provide employment as much as to serve transportation purposes, gave work to New Yorkers of all colors. Workers of African descent labored on the tunnel, as did Irish Americans, who traditionally constituted the majority of New York sandhogs, as tunnel workers were known.

However, if the pestilence of a bad economy and the plague of municipal politics were not enough to delay the Brooklyn-Battery Tunnel, the involvement of the United States in World War II was. As workers began to burrow beneath New York Harbor, the December 7, 1941, Japanese attack on another harbor, in the Hawaiian Islands, catapulted America into world war.

After a year of work, Ole Singstad's efforts to complete the tunnel were halted and nearly destroyed. The war provided the vindictive Moses another chance to destroy the tunnel he hated. Soon after the outbreak of war, Moses recommended that the 28,000 tons of segmented iron lining rings already placed inside the tunnel be melted down for the war effort. Removal of the reinforcement rings would have allowed the tunnel to flood while permitting the soft earth and rock to squeeze and deform, in effect collapsing the tunnel. The suggestion was ignored and the tunnel was unmolested throughout the war.

By mid-1942, the project was halted, to remain inactive for three years. Japan's August 14, 1945, capitulation triggered the sandhogs' return to the tunnel in the fall of that year. As work progressed, Moses's Triborough Bridge Authority began to absorb the Tunnel Authority. Not one to forget his opponents, Moses fired Singstad as the chief engineer of the project, replacing him with Ralph Smillie. Though Singstad had been instrumental in designing the tunnel and its massive ventilation system, he had also committed the unpardonable sin of opposing Moses. When the tunnel was completed, Singstad's name was excluded in an opening ceremony brochure, in which 21 other engineers were mentioned.

As the tunneling progressed, work began on two ventilation buildings in Battery Park, a third on Governors Island, and a fourth rising above the tunnel's toll plaza in Brooklyn. Vehicular tunnels, because of the potentially deadly carbon monoxide fumes and other gases released by internal combustion vehicles, had the technical challenge of providing fresh air within the tunnel. The answer was to place a fresh air duct beneath the roadway of the tunnel and an exhaust air duct running the length of the tunnel's roof. Massive fans in the ventilation buildings made it possible for more than 6 million cubic feet of fresh air to be circulated through the tunnel every minute.

On September 16, 1948, workers driving toward each other from opposite directions met inside one of the tunnels to shake hands. Although completing the connection beneath the East River was a grand symbol, both subterranean tubes had to be given a concrete lining and a roadway and then be lined with 799,000 square feet of ceramic tiles.

The cut-and-cover tunneling method was used to provide the dry-land subterranean approaches to the tunnel's pair of two-lane tubes. Essentially this meant the digging of a pair of massive trenches that were then provided a steel-reinforced concrete roof and walls to contain the lanes leading into the submerged tunnel.

By the spring of 1950 the Brooklyn-Battery Tunnel, which had taken more than two decades to move from an idea to a reality, had been completed. It was officially opened on May 25, 1950, with hundreds of dignitaries, including the tunnel's new boss, Robert Moses, the man who unsuccessfully attempted to metamorphose the project into a bridge. Moses's agency now "owned" the tunnel in the bureaucratic sense, but the tunnel was nonetheless a bitter political defeat for him. Despite the smile of those surrounding him during the opening day ribbon-cutting ceremony, Moses is pictured looking slightly dour. The man who outmaneuvered him on the project, Franklin Roosevelt, had died five years and one month before.

The Brooklyn-Battery Tunnel was a resounding success. During its first full year of operation, the tunnel carried 13.9 million vehicles between Manhattan and Brooklyn. During the late 1990s the number of vehicles traveling through the tunnel increased to nearly 20 million yearly.

See also AMMANN, OTHMAR HERMANN; CAISSON DISEASE; CUT-AND-COVER TUNNELING; HOLLAND TUNNEL; MOSES, ROBERT M.; QUEENS-MIDTOWN TUNNEL; SINGSTAD, OLE; TUNNELING SHIELD.

Brooklyn Bridge (built 1869–1883) *The controversial miracle*   It has become so synonymous with New York City that it is almost impossible to imagine the East River devoid of the gracefully magnificent Brooklyn Bridge, a span connecting Manhattan with the borough that is its namesake. Considered a transportation necessity long before it was built, the Brooklyn Bridge is more than a mere span over a waterway by virtue of its innovative design, the hazards of its construction, its functionality, and its powerful beauty.

Half again longer than any other previous bridge at the time of its construction, the Brooklyn Bridge was designed not only to carry the commerce and commuters of the world's greatest city, but also to serve as a visually inspiring landmark. It has succeeded admirably on both counts.

Designed by the leading bridge builder of his day—the Prussian-American engineer John Augustus Roebling—the Brooklyn Bridge had a gestation period that would span 30 years from concept to completion. The project would cost the life of the legendary Roebling and subsequently cripple his son, Washington Augustus Roebling, who succeeded his father as chief engineer on the project.

The elder Roebling initially applied his engineering genius to the perfection of woven wire rope, which he later recognized as ideal for use on suspension bridges where cabling with a high strength to weight ratio was needed. Roebling then embarked on a stunning career of bridge design and construction throughout the United States. As early as 1852, the successful entrepreneur and bridge designer—said to have been frustrated by a painfully slow ferry crossing of the heavily iced East River—decided to design and promote a series of bridge concepts capable of spanning the waterway.

With his plans completed, Roebling mounted a campaign in 1857 to promote his vision for a suspension bridge linking Brooklyn with Manhattan. Although the famed newspaper publisher Horace Greeley published the plans in the *New York Tribune,* opposition to what was then considered an impossibly long bridge prevailed. The influential Greeley also weighed in as a doubter.

In 1866, the indefatigable Roebling touted a second suspension bridge design, which seemed destined to meet a second successful wave of opposition. Thanks to the capriciousness of the weather, however, Roebling's plan for a suspension bridge was taken out of the deep freeze of public opinion and given a warm reception when a harsh winter badly obstructed East River ferry traffic. On April 16, 1867, the New York State Legislature created the New York Bridge Company to finance, construct, and maintain a bridge between New York and Brooklyn. Roebling was named chief engineer on the project and set about organizing the project with his son, Washington Augustus Roebling.

The dream became a nightmare for the Roebling family near the end of June 1869, when the elder Roebling's toes were crushed between pilings struck by an East River Ferry as he scouted for a suitable location to place the bridge. Roebling's toes were amputated, but tetanus apparently developed, killing the engineer a month later. The 32-year-old Washington Roebling, a civil engineer and a former Union Army engineering officer in the Civil War, was picked to oversee the construction of the bridge. As it turned out, the bridge would also take an awesome toll on his health and prove to be one of the most controversial civil engineering projects in American history.

This stunning image of the Brooklyn Bridge at night looking toward Manhattan reveals the majesty of the bridge, which took 14 years to complete. The brainchild of the civil engineer John Augustus Roebling, the project was one of the nation's most controversial construction efforts and one of its most successful and enduring. *(Historic American Engineering Record, National Parks Service)*

The proposed suspension bridge would measure a total of 5,989 feet. Of this distance, 1,595.5 feet comprised the central span erected over the river between a pair of masonry towers—a feat considered impossible by many. These towers would be emplaced upon foundations beneath the East River and constructed to rise 276.5 feet above the water. Massive stone anchors would be built on each side of the East River, where the steel suspension cables passing through the towers would be connected. These anchors would essentially bear the weight of the bridge's roadway hanging from the cables by steel rope suspenders. As large as a city block, the anchors would form the approaches to the suspension bridge.

Once the steel cables were in place, they would curve up to the towers from their anchors and down again toward the center of the center span. The bridge roadway and steel cables were designed to weigh 15,000 tons. Although this sum seems heavy,

a stone bridge built by using older stone arch technology would have been far heavier and would have required substantially more foundation work to support a series of arches across the river. Such a design would have been an obstacle to maritime traffic in one of the world's busiest waterways. The Brooklyn Bridge's central span would stand 135 feet above the water at its center, allowing ships to pass easily beneath it without requiring a drawbridge. (It should be pointed out that although it is referred to as a river, the East River is actually a saltwater strait with powerful currents generated by tidal action.)

The sight of the bridge's slender silhouette raised eyebrows among some who believed the bridge to be too spindly to stand up to the strong winds that traversed the length of the East River. However, Roebling's brilliant design had taken the aerodynamic effects into account, and he designed the bridge to be hefty enough to counteract any destructive oscillations caused by wind (although 1,000 tons of additional steel was used in the latticework of trusses to be on the safe side). Roebling also overengineered the bridge to such an extent that it was capable of handling the heavier styles of rail cars that were being used by the time the bridge was completed in 1883. The bridge was designed to support 18,700 tons, or nearly 4,000 tons more than its own weight.

Perhaps the greatest engineering and safety challenges involved construction of the foundations for the bridge's two towers just offshore of the New York and Brooklyn sides of the river. These massive towers would require huge foundations sunk deep into stable soil or bedrock at a depth of 78.5 feet on the New York side and 44.5 feet near Brooklyn's shore. To do this job, massive pneumatic caissons—essentially 168-foot by 102-foot boxes—were ordered constructed at a shipyard. These were the largest such caissons then constructed. These huge boxes contained airlocks and fittings to allow pressurized air to be pumped into the lower compartments, where men would dig to allow the caissons' iron edges to cut through the muck to bedrock. On the New York side, workers encountered a layer of foul-smelling muck created by years of dumping garbage and the carcasses of dead animals. This added to the bouquet of sewage that poured into the river in the days preceding sewage treatment systems.

Before entering the lowest level of the caisson, men would enter airlocks, where they would be "blown down" (in modern diving parlance) so they could enter the pressurized working area without allowing water to rush in. The air in the chamber would be pressurized to match the pressure level of the air inside the caisson. After the airlock pressure was equal to the caisson pressure, the men would open the doorway leading to the working area at the base of the caisson. Simply opening a doorway between the caisson and the unpressurized air outside would have allowed the internal pressure to escape, flooding the caisson.

The pressurized air in the six massive work chambers at the bottom of the caisson was needed to prevent water from infiltrating into the enclosure through the muck. The work not only took place in miserable conditions, but was horribly dangerous as a result of the hazards of caisson disease, or the "bends." Little understood at the time, this condition occurs when nitrogen is forced into the bloodstream under pressure. When the worker leaves the

high-pressure confines of the excavation chamber without proper decompression in the airlock, the nitrogen bubbles out of the tissues, sometimes causing irreparable harm or death. During construction of the tower foundations, workers entered the hazy, dark, pressurized atmosphere of the caissons, where they would remove muck and rock to allow the caisson's iron cutting edges to settle into the mud. Washington Roebling, a considerate and humane boss, employed a physician to oversee the health of his workers, reduced the working shifts to roughly three hours, and required rest periods. But the limits of 19th-century medical science did not allow a complete understanding of a problem that was not fully comprehended until the middle of the 20th century. As a result, caisson workers died horrible, painful deaths when they failed to decompress properly to allow the safe dissipation of the nitrogen bubbles pushed into their tissues. At least 20 men died during construction of the bridge, but only a handful fell victim to caisson disease as nitrogen bubbles expanded uncontrollably inside their bodies.

When it became apparent that digging alone would not submerge the caissons quickly enough, Roebling—a former Union army combat engineer known to everyone as Colonel Roebling—experimented with black powder charges to remove rock and soil. At great risk to himself he detonated charges of increasingly greater power until he determined that the caisson (and the workers) could endure the blasts. Although the explosive charges were safe if precautions were taken, the compressed air inside the caisson's working area was heavy with oxygen, making fire a tremendous threat. One such fire would nearly doom the pneumatic caisson on the Brooklyn side.

On December 2, 1870, a carpenter's candle apparently set fire to oakum, or caulking, that waterproofed the 15 layers of oak timbers reinforcing the roof of the Brooklyn caisson. Although it was a small fire, the pressurized, oxygen-rich atmosphere caused it to burn deep within the thick timber ceiling. The fire infiltrated upward into the wood and drew Washington Roebling into the excavation chamber to direct fire-fighting efforts throughout the night. When this failed, Roebling ordered the caisson flooded for 24 hours. The fire was extinguished, but the departing Roebling, who apparently failed to decompress adequately in the airlock, collapsed as expanding nitrogen gas bubbles attacked his body. Roughly two years later, Roebling collapsed again

1.—ENTRANCE TO THE SUPPLY-SHAFT OF THE CAISSON.

2.—MOUTH OF SUPPLY-SHAFT OF THE CAISSON.

3.—DRILLING THE ROCK AT THE SHOE OF THE CAISSON.

4.—DOOR THROUGH PARTITION, SHOWING DIFFERENT APARTMENTS IN THE CAISSON.

5.—FILLING THE BUCKET OF THE WATER-SHAFT IN THE CAISSON.

6.—WORKMEN SAWING TIMBER FOR WEDGES.

INSIDE VIEWS OF THE EAST RIVER BRIDGE CAISSON, BROOKLYN, N. Y.—FROM SKETCHES BY OUR SPECIAL ARTIST.

This series of illustrations reveals the activities within the Brooklyn caisson of the Brooklyn Bridge. Working in the dangerous compressed air environment, workers labored to excavate mud and rock to sink it to a solid footing beneath the soft silt of the East River. *(Library of Congress)* while in one of the caissons and became an almost-blind invalid. He would direct the remainder of the work through his wife, Emily Warren Roebling, who became well versed enough in the intricacies of bridge-building technology to be his eyes and ears on the project. Eloquent and knowledgeable, she successfully convinced officials to retain Roebling as the project's chief engineer when he was on the verge of being fired in 1882. Washington Roebling had come under criticism since he had not been physically able to visit the bridge during the previous 10 years because of his health.

It had been Roebling's desire to place both caissons on the incredibly dense and hard gneiss, a rock that supported much of the New York area. Even ancient bridge builders preferred a natural bedrock base as the foundation for the piers of their bridges, and Roebling knew that sinking the caissons to rest on rock millions if not billions of years old would enhance the longevity of his own bridge. The caisson on the Brooklyn side struck bedrock at a depth of 45 feet in March 1871. Unfortunately, on the New York side of the East River, the caisson would never find bedrock, and as the caisson settled to nearly 75 feet beneath the East River, workers contracted the bends, and some died.

For every two feet beneath the water that the caisson descended, air pressure was increased by one pound. By the time the caisson reached 68 feet, the pressure was at 34 pounds per square inch. At a depth of 71 feet, a German immigrant, John Myers, completed his first day on the job and made it home only to die near his own doorstep. Just over a week later, on April 30, 1872, the Irish American Patrick McKay died at the age of 50; however, an autopsy showed that he had a fatal kidney disease. As more suffered the ill effects of the bends, the caisson workers went on strike, demanding that their daily wage be increased from $2 to $3. Officials of the Bridge Company offered them $2.75 a day, and when that offer was rejected, the company broke the back of the strike by threatening to fire all the workers.

Faced with the prospect of having to dig for as long as another year and the estimated loss of dozens of more lives, Roebling decided to gamble on the incredibly resilient, packed-sand bottom encountered by his mostly immigrant laborers. The day after a third caisson death on May 17, 1872, Roebling made up his mind to halt the excavation at 78 feet, six inches. Roebling pronounced the compacted sand and gravel bottom as good as bedrock, a decision that not only saved lives but proved valid: The New York tower has stood firm for more than 100 years. Although 110 men were officially listed as victims of the bends while working within the pressurized confines of the New York caisson, it is believed that many more suffered from decompression sickness without reporting it.

As the caissons slowly sank to the bottom of the East River, stone blocks were set upon their timber roofs until the caissons reached their respective footings below the muck. When the caissons were finally sunk to their respective depths by the summer of 1872, their empty excavation chambers were filled with concrete, and the construction of the New York and Brooklyn towers that would hold the suspension cables aloft was begun. Stonemasons slowly built the towers, finishing the Brooklyn tower in May 1875 and the New York side's tower in July 1876.

The completion of the towers allowed the laying of the four suspension cables. These were arranged with two on the outside edges of the towers and two more at the center of the towers. A carrier unreeled the steel wire back and forth until 286 lengths of wire were accumulated and bundled into each of 19 strands. Of these, seven inner strands were bundled into a nine-inch cable, which formed a core around which the remaining 12 strands were bundled. To compress and protect these bundles, softer iron wire was then spun around the length of the cables. Each of the final four cables was composed of 5,439 wires. The diameter of the 19 bundled and finished cables was 15.75 inches.

Ironically, the innovative firm founded by the elder Roebling that had perfected steel cable and rope could offer a bid on providing cabling for the project only after Washington Roebling sold his stock in the Trenton, New Jersey, firm. A rule prohibiting contractors from supplying the project was the reason. Although John A. Roebling's Sons Company did eventually offer the lowest bid, a firm owned by J. Lloyd Haigh offered the second-lowest bid for crucible steel wire, which was then considered superior to Bessemer process steel offered by the Roebling firm. It would later be learned that Haigh was actually providing the cheaper Bessemer steel wire while Abram Stevens Hewitt, the official who recommended acceptance of the Haigh bid, not only held the title to Haigh's home but would receive a 10 percent kickback of the steel payments. (Accusations of graft had flown frequently and hotly since the inception of the project, even targeting the younger Roebling, who, by all accounts, conducted himself in a manner that was beyond reproach.)

Although Roebling might have endured the graft of others making an unholy buck off his beloved Brooklyn Bridge project, he took action when he found that greed might have compromised the structural integrity of the bridge. Learning that coils of Haigh's steel wire already rejected by quality control inspectors were nonetheless being placed into the bridge, Roebling assigned trusted men to observe this subterfuge. He then ordered additional wire woven into the bridge's cabling to compensate for any potential weaknesses. The "wire scandal" once again fanned fears about the bridge's reliability. Such allegations—whether cast about for political reasons or as the result of shoddy journalism of the time—plagued the bridge throughout its construction and for a few years after its completion.

To ensure that the steel cabling would remain corrosion-free for as long as possible, Roebling instituted a revolutionary process of galvanizing the delivered wire on the job site. The wire was then given multiple coats of thick oil until each strand was encased in a heavy, greasy coating. This was the first instance in which galvanized wire—steel wire coated with molten zinc to inhibit rusting—was used in bridge construction.

With the cables in place and linked to four rows of massive iron anchor bars set deep within the limestone anchor foundations at each side of the East River, the Brooklyn Bridge began to assume its final form.

The private corporation established to finance the bridge had been taken over by the cities of New York and Brooklyn in 1874, and although the original investors recouped their monies, the city of New York began to balk at the cost overruns of the bridge project. In 1878, New York filed a lawsuit to enable it to halt payments. This lack of funds shut down the project temporarily and threatened Roebling with the loss of the talented pool of engineers assisting him. New York eventually lost its legal battle and was forced to pull out its checkbook once again. The workers were called back and construction on the long-suffering project resumed.

By December 10, 1881, the 85-foot-wide floor of the bridge was completed and was hung from the cables by suspenders running to the truss system. As an added measure of strength and stability, diagonal stays were run from the towers connecting with alternating suspenders attached to the road deck. It

A family strolls across the elevated walkway of the Brooklyn Bridge in this 19th-century image showing the four suspension cables of the bridge as well as one of its two masonry towers. *(Author's Collection)*

was estimated at the time that the diagonal stays alone would have held the bridge's roadway in place. What resulted was the supreme roadway of its day. It provided room for a pair of railroad tracks and two trolley car tracks along with two one-way roadways or carriageways on each side of the trolley tracks. An innovative 15-foot-wide elevated walkway ran the length of the bridge for pedestrian traffic.

John Augustus Roebling's genius and long experience with building innovative bridges, such as the Niagara Gorge Bridge, taught him the necessity for countering the force of the wind. His original design did so by providing an extremely thick road deck that was reinforced with open trusses to allow the wind to pass through. Working in conjunction with the diagonal stays connected to the suspenders, the Brooklyn Bridge's deck stood solidly against all winds, even when subsequent bridges collapsed as a result of wind forces. One such failure involved the brand new Tacoma Narrows Bridge, which collapsed in a relatively mild 40-mph wind on November 7, 1940.

The original cost of the Brooklyn Bridge had been estimated at $7 million, but caisson fires, construction delays, and reinforcing efforts conspired to drive the cost to $15.1 million, an overrun that can hardly be considered unusual for a pioneering engineering project that took 14 years to complete. Although heavily criticized before and after construction began, the bridge is today considered a visual and engineering feat with few parallels in history.

The Brooklyn Bridge was opened for traffic on May 24, 1883, to an outpouring of community pride on both sides of the East River. Tragically, the public's infatuation with the bridge led to a fatal incident in which 12 people died. With the opening of the bridge, thousands of people streamed to the span for a chance to walk across its elevated pedestrian walkway. So many people thronged to the bridge that the two carriageways were also made available to pedestrians. By late afternoon an estimated 20,000 people were on the bridge, causing a bottleneck of humanity as pedestrians from both directions collided at the stairs leading to the New York approach. The press of people did not cause an immediate problem, but when one woman began screaming at the sight of another's losing her footing, a panic ensued. Adults and children were trampled and crushed on the narrow stairway. As is the case in such confusing events, no final answer as to what caused the panic was ever ascertained, although it was said that on seeing people falling down the stairs, others shouted that the bridge itself was collapsing, adding to the hysteria. Lawsuits followed, and although negligence was not cited nor damages granted, the Bridge Company blamed the panic on newspaper allegations about the bridge's lack of reliability. Bridge officials said such inaccurate stories created an unfounded "insecurity" in those on the bridge and contributed to the tragedy.

The bridge's fame was so great that it became a lightning rod for stunts performed by those seeking fame or fortune. Sadly, it was also a magnet for those interested in doing away with themselves. The first known incident of someone's leaping from the Brooklyn Bridge was the fatal dive of Robert E.

A view of the Brooklyn Bridge across the East River from the Brooklyn side. Arguably one of the world's most famous bridges, it was a stunning engineering achievement of its time and opened the door to even more extraordinary bridge projects, particularly those involving suspension bridge construction. Completed in 1883, it was built in spite of corrupt city politics, workplace fatalities, and seemingly insurmountable technical challenges. *(Historic American Engineering Record, National Park Service)*

Odlum, a swimming instructor from Washington, D.C. His considerable swimming skills were of little use to him since he was killed on impact with the water on May 19, 1885, after falling 135 feet. To prevent being thwarted in his stunt by police, Odlum arranged for a friend to go to another section of the bridge and pretend to jump as a diversion for his own ill-fated attempt. Steve Brodie, the most famous "jumper" of them all, may never have leapt from the bridge. He became a celebrity after he (and witnesses who happened to be his friends) claimed he jumped from the bridge and lived. Brodie's highly suspect feat occurred on July 23, 1886, and led to his acquiring considerable fame. Brodie even became the star of a play about his feat, *On the Bowery*. The play toured the country and Brodie eventually became a saloonkeeper who regaled his customers with tales of his much-doubted feat. The first intentional suicide may have been that of Francis McCarey, who jumped to his death in 1892. As the bridge became an accepted part of the landscape, the stunt jumping seemed to trail off before the dawn of the 20th century.

With ordinary maintenance the bridge served its first 61 years of duty without incident. When it was given an intensive engineering inspection in 1944, the verdict was a resounding compliment to the elder Roebling and his son: The bridge needed little more than a spot of paint. However, in that same year passenger trains ceased running on the bridge and it was given over to automobile and pedestrian traffic. On the 100th anniversary of its completion in 1983, more than 750,000 turned out to celebrate the elegant bridge's centennial. The Brooklyn Bridge still stands astride the East River as one of America's best known and most loved symbols as well as a marvel of farsighted engineering.

See also ROEBLING, EMILY WARREN; ROEBLING, JOHN AUGUSTUS; ROEBLING, WASHINGTON AUGUSTUS; SUSPENSION BRIDGE; TWEED, WILLIAM MARCY.

Brown, Samuel (1776–1852) *Builder of shaky suspension bridges* The retired Royal Navy captain Samuel Brown made quite a splash during the era of his bridge building in the early 1800s. In fact, Brown made several splashes because many of his bridges collapsed.

Brown is famous for several reasons, however. First, as an inventor he developed the wrought iron link, a flat bar with a hole at both ends, known as an eyebar, that was used for decades as the funda-

mental component of early suspension bridge cables. Brown's iron links were utilized in the Menai Bridge and in many other suspension bridges, including those of Brown's own design. Brown patented the iron link in 1817.

His other claim to fame is his role in designing and building the Union Bridge across the Tweed River near Berwick on the English-Scottish border. Completed in 1820, the bridge lasted only six months before it collapsed, requiring it to be rebuilt. Despite that small glitch, the bridge continues to stand today as the oldest example of an iron chain suspension bridge in Great Britain.

Suspension bridges were all the rage in the early 1800s, and even though many fell down as soon as they were built, designers like Brown were in great demand. A suspension bridge built by Brown in Broughton, England, collapsed in 1831 while a group of soldiers marched across it. One of his most famous bridges was actually a pier, the Royal Suspension Chain Pier in Brighton. The curiosity served as a boat landing and a tourist attraction with up to 4,000 people daily paying for the opportunity to stroll its length. Brown's home, which was located near the Chain Pier, was known as the Chain Pier House. Sadly, the house has outlived the structure for which it was named.

Unlike a number of Brown's bridges, the pier actually was quite long-lived. The 350-yard pier, which was held erect by iron chains suspended between four iron towers, was completed in 1823, during the heyday of Brown's bridge-building. It stood, although a bit worse for wear after storm damage, until it was completely destroyed by wind and wave action in December 1896. The entry portals, which were all that remained of the bridge, remain on display in Brighton.

See also BREAK STEP; MENAI BRIDGE; SUSPENSION BRIDGE.

Brunel, Isambard Kingdom (1806–1859) *Builder of tunnels, bridges, and ships* Big things come in small packages, and the slightly built English engineer Isambard Kingdom Brunel was nothing less than an engineering dynamo. His brilliance and tenacity allowed him to excel in no fewer than three engineering fields, each playing a major role in promoting transportation aspects of the Industrial Revolution.

At four, Brunel demonstrated uncanny artistic abilities, and at 14, he returned to his father's native country of France to study at the College of Caen and then at the Lycée Henri-Quatre, a school renowned for its mathematical instruction. Young Brunel was then apprenticed to the Paris shops of the Swiss-born craftsman Abraham Louis Breguet, who designed watches and scientific instruments.

At 18, Brunel was struggling shoulder to shoulder with miners as his father's chief assistant during the problem-plagued construction of the Thames Tunnel, the world's first subaqueous tunnel. By 25, he was overseeing improvements to canals and designing improvements to various dock facilities so elaborate they were sometimes ignored because of their complexity and cost. In 1831, he was appointed chief engineer at Bristol Docks.

In 1829, Brunel decided to enter his own bridge design to compete with others being considered as potential spans across the Avon Gorge in southwest England. Brunel's plan was eventually selected, though it would not be completed until 35 years later in 1864. Known as the Clifton Bridge, it still stands as a stunning example of Victorian era engineering.

At 27, Brunel was appointed chief engineer of England's Great Western Railway, for which he designed dozens of innovative timber railroad bridges and oversaw the construction of 1,046 miles of rail lines. Brunel's rail bridges were interesting because they were made of rot-resistant Baltic pine and designed in a modular fashion so that components could be fabricated elsewhere and then shipped to the site of the bridge.

Although criticized for using timbers vulnerable to rot and fire, Brunel was in charge of expanding rail service, which was growing during the 1800s as rapidly as the Internet did during the end of the 20th century. He needed bridges that would take weeks instead of years to build and knew timber bridges would serve reliably with proper maintenance. Although the subsequent scarcity of Baltic yellow pine hampered maintenance on the bridges by the end of the 19th century, three of Brunel's timber railway bridges survived until 1931 after more than 80 years of service.

Brunel also constructed bridges that were more permanent, including the magnificent Royal Albert Bridge at Saltash, completed in the year of his death. Like the Clifton Bridge, the Royal Albert span continues in service today. It also serves as a memorial to its designer by bearing the name *I. K. Brunel* on its towers. The unique two-span bridge is supported by a pair of lenticular iron tube arches and is easily one of the world's most unique bridges.

Although his experience with the Thames Tunnel was behind him, Brunel was not through with tunneling. One of his most impressive tunneling efforts as a railway engineer was the Box Tunnel bored through Box Hill between Chippenham and the old Roman city of Bath. For 4 1/2 years 4,000 men aided by 300 horses labored to claw through the hill. The work was relentless and every seven days a ton of gunpowder was expended in blasting rock. As crews encountered flooding, Brunel found himself once again overseeing the problem of draining water and sealing potentially deadly leaks, as he had in the Thames Tunnel.

When the Box Tunnel opened for service in 1841, it measured 1 3/4 miles in length. The grade of the tunnel's rail line attracted criticism as too steep, with its slope of one foot for every 100 feet of length. One critic predicted a train that lost the ability to brake could accelerate on a downhill run to 120 mph, a speed—it was then believed—that would suffocate the passengers. Brunel shrugged off the claims and completed the tunnel. Although 32 men died building Box Tunnel, no one suffocated while riding a train through it.

His single-minded devotion to engineering accomplishment caused him to despise anything that would inhibit engineering developments. In particular, Brunel loathed governmental bureaucracy because of its cautious and lethargic nature and opposition to innovation. He was so disdainful of impediments to technical development that he refused to patent his own engineering creations. An example is Brunel's development of a polygonal gun barrel, a smallish project undertaken when he was not building graceful bridges and difficult tunnels. He declined to patent the design and expressed amazement that the inventor Robert Whitworth did so after seeing it in Brunel's workshop. Brunel simply thought it absurd that anyone could patent a geometric shape!

Scornful of class differences in 19th-century English society, Brunel would pick up hand tools and work beside laborers, something he advised highly educated engineers in his employ to do in order to boost morale and productivity.

When pushing a particularly urgent military project during the Crimean War, he advised a manager to motivate his workers by genuinely working alongside them. Brunel's advice penned in 1855 suggested,

"It is a good thing occasionally to put your hand to a tool yourself and blow the bellows or any other inferior work, not as a display but on some occasion when it is wanted and thus set an example. I have always found it [an] answer." The project to which Brunel referred was his design of a modular field hospital for British troops predating the modern transportable hospitals that became widespread during World War II. Asked to develop the hospital on February 16, 1855, Brunel designed, built, and delivered the innovative facility within six months.

At the time, the government was writhing under the spotlight of scandalously bad medical care and unclean hospitals for its Crimean combat casualties, an embarrassing circumstance brought to light by the famous British nurse and gadfly Florence Nightingale. Nightingale's direct appeal to Queen Victoria to improve conditions led to Brunel's being assigned the task of developing a prefabricated hospital. Brunel personally supervised its design to ensure that its sanitation and ventilation were optimized. During its use in the war, 1,500 patients passed through the hospital, of whom only 50 died, a remarkably good record compared to that of the vile makeshift hospitals where a high percentage of British soldiers had been allowed to die in filth.

Brunel did have his engineering bloopers. His worst was to commit his employer, the Great Western Railway, to rails spaced seven feet apart when every other rail line in Great Britain used narrower, standard-gauge tracks. Although this eventually hurt Great Western financially, Brunel's intransigence was based on his correct belief that wide-gauge trains could travel more safely at higher speeds, something borne out by the railway's safety record. As testament to his stubbornness, the last wide-gauge lines were not converted to standard gauge until 13 years after his death.

Another goof by Brunel was his foray into the design of massive steamships. His first two were financially sound projects; one, the 322-foot S.S. *Great Britain,* was the first to cross the Atlantic using a propeller instead of a paddlewheel. The third and largest ship designed by Brunel, the 622-foot S.S. *Great Eastern,* was both his greatest failure and greatest success. It was never profitable and was sold in 1865 for 25,000 pounds—36 times less than the 900,000 pounds it cost to build.

The gargantuan ship, however, was perfect for the American entrepreneur Cyrus Field, who used it to lay the first transatlantic cable between Britain and North America. It continued laying telegraph cables at various points around the world until 1886, when it became a sightseeing attraction. No larger ship was built until the luxury liner *Lusitania* in 1906.

Brunel's physical courage was boundless, and he was seriously injured twice during his career after throwing himself into dangerous situations to avert disaster related to his projects. His first brush with accidental death occurred during uncontrolled flooding in the Thames Tunnel in 1828, and 10 years later he came close while fighting a fire aboard the S.S. *Great Western* another of his ships.

In January 1828, Brunel was helping two workmen inside the Thames Tunnel during its construction when uncontrolled flooding filled the tunnel with water. A total of six miners were killed, including two with Brunel, who almost shared their fate when the 22-year-old's leg was pinned by a timber as the water rose around him. He managed to free himself and was flushed out of the tunnel's access shaft by rising water. His fellow engineer Richard Beamish pulled the unconscious Brunel to safety.

When Brunel was revived, his leg was so badly injured that he was unable to walk. Undeterred by his injuries, Brunel lay on a mattress thrown down on the deck of a work barge. There he directed efforts to fill the collapsed segment of the tunnel with bags of clay.

During another flooding incident in the Thames Tunnel, Brunel instantly responded to the cries of an elderly pump mechanic trapped inside the dark, inundated tunnel. Brunel, who had left the tunnel just prior to the flooding, grabbed a rope and entered the flooded shaft, where he managed to find and rescue the man.

Brunel's third brush with death occurred aboard the S.S. *Great Western* in 1838 as the ship prepared to race another across the Atlantic. When a flue of one of the ship's boilers caught fire, Brunel rushed to the scene to take charge of the fire-fighting efforts and fell 18 feet when a charred ladder rung gave way. Fortunately for Brunel, the ship's captain, who was below fighting the blaze with a hose, broke his fall. The captain, stunned from absorbing the blow of Brunel's plummeting body, recovered in time to pull the unconscious Brunel from beneath the water that had accumulated during the fire-fighting efforts.

Whereas the carefully schooled engineers of today might wed themselves exclusively to aerospace, automotive, or construction fields for their entire careers, Brunel's restlessness made him apply his considerable talents to a series of endeavors whose only connection was that each usually related to transportation.

See also BRUNEL, MARC ISAMBARD; CLIFTON BRIDGE; ROYAL ALBERT BRIDGE AT SALTASH; THAMES TUNNEL.

Brunel, Marc Isambard (1769–1849) *Inventor of the subaqueous tunneling shield* Sir Marc Isambard Brunel's genius and maniacal devotion to work led him through one of the most extraordinary engineering careers in history. Brunel was a self-taught engineer whose technical accomplishments are as remarkable for their variety as for their brilliance.

Born in Hacqueville in France's Normandy region, Brunel was selected by his Catholic parents as the family's contribution to the Catholic church and was sent to study for the priesthood. Brunel abandoned this project to stay with a family friend, François Carpentier, who was the American consul in Rouen, France. There he studied navigation and soon after joined the French navy as an officer.

Brunel returned to France from sea duty in 1792 in the fourth year of the French Revolution and naively blurted out his support of France's besieged royal families. Realizing he was now guillotine bait, Brunel sought refuge with Carpentier in Rouen, where Brunel met Sophia Kingdom, the orphaned daughter of an English naval contractor, who had been sent to live with the Carpentiers to learn French.

Brunel, on the run but still loyal to France's royalty, began taking part in violent street battles with French republicans determined to exterminate the royalty and all its supporters. In between clashes with French republicans, Brunel managed to court Sophia Kingdom and become engaged to the Englishwoman.

With no other options, Brunel decided to flee to America and gained passage aboard a ship appropriately named *Liberty* that was bound for New York. He arrived in that city September 6, 1793, intending to return to France as soon as possible. Brunel initially found work as a surveyor and decided to become an engineer. In New York he began to gain a reputation as a civil engineer and submitted a design for a new federal government capitol to be built in Washington, D.C. Although the design was considered too expensive, the concept was recycled into New York's Palace Theater, a structure consumed by

fire in 1821. Brunel's brilliance and hard work made a good impression among his adopted countrymen. Once he obtained American citizenship, the transplanted Frenchman managed to be appointed New York's chief engineer.

One of Brunel's most important chores in New York was the establishment of a foundry for the production of cannon. In conjunction with the manufacture of artillery, the former French naval officer also provided input on how to strengthen the shore defenses of Staten Island and Long Island against attack from the sea.

A chance meeting with another expatriate Frenchman in Washington, D.C., led to a discussion of the painfully slow and expensive method of manufacturing ship's blocks then desperately needed by Britain's Royal Navy. These assemblies of pulleys were essential components of military and civilian ships. Brunel designed a mechanized process that reduced by 90 percent the number of men needed to build ship's blocks for Britain's massive navy.

Smelling an opportunity, Brunel soon emigrated to England to build pulleys for the enemy of his original and adopted countries. Romance undoubtedly played a role in his decision since he longed to be reunited with Sophia Kingdom, whom he had not seen in six years and who had returned to England after being held captive for a while in France. Brunel married his beloved Sophia in 1799. Perfecting the block and pulley machinery took six agonizing years.

When the system was completed, it became a showcase of technology, and world leaders such as Russia's czar, Alexander I, marveled at the accomplishment. Brunel's mechanical system for constructing blocks and pulleys for the Royal Navy was a huge success but he was nonetheless treated with suspicion by British officials. The Frenchman who became an American and who now sought to become an Englishman was not even allowed to visit his own block and pulley works without a special permit to allow him inside the Portsmouth naval facilities. In 1806 Sophia gave birth to Isambard Kingdom Brunel.

Despite his successes and growing fame in engineering circles, Brunel literally found himself in the poorhouse by 1821. Brunel agreed to operate a sawmill to provide lumber to the Royal Navy and also designed a system for the mass production of badly needed boots during Britain's war with Napoléon Bonaparte's forces. When the sawmill failed financially and the Battle of Waterloo canceled his boot orders, Brunel was financially ruined. Accompanied by Sophia, Brunel entered debtor's prison and remained there until he threatened to provide his engineering expertise to the Russians.

An 1805 attempt to tunnel beneath the Thames had failed as a result of flooding, and Brunel, on the lookout for a new engineering challenge, believed subaqueous tunneling was possible if a shield protected the miners from a cave-in and the resulting flooding. As earth was removed from the face of the tunnel, the shield would be pushed forward by jacks. Brick masons working near the miners in the excavated area to the rear of the shield would reinforce and seal the tunnel with bricks.

This brainstorm was based on Brunel's observation of the wood-eating mollusk species *Teredo,* commonly known as the shipworm. Brunel's naval experience undoubtedly figured in this revelation since wood-gobbling shipworms could ruin the hull of a ship. Like Brunel's shield concept, the shipworm leaves behind a lining as it bores through wood. Brunel patented his shield in 1818 and put it to use until 1825, after he was made chief engineer of the Thames Archway Company's effort to tunnel beneath the river.

After the massive flooding of January 1828 that filled the unfinished tunnel and temporarily disabled Brunel's son, Isambard, work on the tunnel came to a standstill for seven years, in large part because of England's miserable economy, which stunted the tunnel company's fund-raising efforts. On March 1, 1836, work began again, and although the public and the press had begun referring to the Thames Tunnel as the "Great Bore," it finally reached the access shaft on the north bank of the Thames River on December 15, 1842.

The year before, the refugee naval officer had been knighted by Queen Victoria for his efforts to construct the tunnel with his innovative shield. Brunel, who struggled to complete the tunnel in the absence of his son, Isambard (who was pursuing his own career as an engineer and inventor), suffered a stroke in November 1842 that partially paralyzed him. Despite his infirmity, Brunel was able to attend the opening ceremonies for the Thames Tunnel on March 25, 1843. The 80-year-old self-made engineer received a standing ovation from the crowd.

Brunel's bad luck in financial matters persisted, and he was never granted the remaining 5,000 pounds he was owed by the Thames Tunnel Com-

pany for use of his revolutionary tunneling shield, making him financially dependent upon his son Isambard. A second stroke in 1845 further paralyzed the elder Brunel, who could no longer walk. He died December 12, 1849, and was buried in London's Kensal Green Cemetery. His wife, Sophia, joined him in death five years later.

The engineer Richard Beamish, who was blinded in one eye during his efforts to construct the Thames Tunnel and who replaced Isambard as Brunel's resident engineer in the tunnel, wrote in his own biography that Marc Brunel's name would be "cherished as long as mechanical science shall be honored."

See also BRUNEL, ISAMBARD KINGDOM; THAMES TUNNEL; TUNNELING SHIELD.

Brunton tunneling machine *Early English Channel tunneling machine* Machines for drilling tunnels began appearing in the 1840s, but the first to show a hint of real promise was the Brunton tunneling machine, invented by the English inventor J. D. Brunton. Although a good start, the machine was not the miracle machine that some hoped it would be. Nonetheless, Brunton's invention contained technical innovations that paved the way for more advanced machines.

Brunton and his colleague F. Trier developed a stone-dressing machine during the 1860s that possessed elements Brunton would later design into his tunneling machine. The stone-dressing machine used a series of revolving cutters to grind a smooth face on quarried stone. One of the most important elements of the stone cutter, and later the tunneling machine, was the rotation of the circular cutters fitted to a plate or chuck. Thanks to the rotation of the cutting edge, the portion touching the stone alternated, thereby reducing cutter degradation.

Trier and Brunton had figured out a very important characteristic of cutting through rock: Although rock contained particles that were hard enough to grind down iron and steel cutters eventually, the elements composing rock were not bonded together as cohesively as those in hard metals. On their stone dresser, the stone itself was held on a platform and pushed against the cutters. By keeping the stone pressed against the rotating cutter, the force itself would "chip off" bits of stone.

When Brunton applied these elements to the tunneling machine he developed in the late 1860s he realized the device would have to be mounted on a platform that would itself move against the face of a tunnel. Brunton also knew that the machine had to be large, rigid, and stable. Any flexibility of the machine's frame or cutter arrangement would reduce the pressure against the rock that was so essential in helping the cutters crumble its surface. Keeping the machine shoved against the tunnel face was a large screw jack that was turned to maintain pressure until it reached the end of its length. At that point the jack was repositioned.

Brunton's tunneling machine was essentially a large frame with four grooved wheels that rode on rails on the tunnel floor. To turn the chucks containing the cutters, a compressed air engine turned a large gear-driven shaft. Although some accounts claim Sir Edward Watkin used the machine in his quashed attempt to excavate the first English Channel tunnel during 1880–1882, others say this not the case. If not, the machine may well have been used on another Channel tunneling venture under way at the time.

What is known is that Brunton's machine did not measure up in the demanding world of tunneling. The machine was put to work in an English chalk quarry and performed relatively well, since chalk, which is far softer than other types of rock, did not take a great toll on the cutters. When Brunton's machine was tested against rock harder than chalk, its cutters suffered so much wear that sharpening and replacing them made use of the machine inefficient.

Although Brunton's idea was a solid concept, the metallurgy of the day lagged behind the needs of true hard rock-tunneling machines.

Had Brunton been able to employ superhard materials for his cutters such as tungsten carbide, instead of only iron and steel, his device would have operated far more efficiently. Tungsten carbide would not be developed in usable form until more than 40 years after Brunton's device was judged unfit as a serious tunneling machine. However, his innovation of using rotating cutters is still employed on modern tunnel-boring machines.

See also EUROTUNNEL; TUNGSTEN CARBIDE; TUNNEL-BORING MACHINE; WATKIN, EDWARD.

Buck, Leffert Lefferts (1837–1909) *Chief engineer of New York's Williamsburg Bridge* Although the bridges of Leffert Lefferts Buck may not be considered the most beautiful or technically elegant spans, the engineer gained a well-deserved reputation for building extremely solid bridges. His late entry into

the engineering field (Buck was 31 when he completed his civil engineering degree at Rensselaer Polytechnic Institute) and his work as an assistant engineer on the Brooklyn Bridge contributed to his being a mature and thoughtful designer of bridges.

Buck was born in Canton, New York, and was apprenticed to a machinist before enrolling in the original class of Saint Lawrence University. While still a student, Buck enlisted as a private at the outbreak of the Civil War and served with the Union army's North County Regiment, in which he saw a considerable amount of combat in four years of active duty. He was wounded twice and survived the war, having risen to the rank of major. While he was in the army, Saint Lawrence University awarded him bachelor of science and master's degrees. On mustering out of the army, Buck enrolled in the prestigious Rensselaer Polytechnic Institute in Troy, New York, which in 1868 conferred his degree in civil engineering.

With his degree in hand, Buck found employment as an assistant engineer on the magnificent Brooklyn Bridge project, which began construction in 1869. His superior was Washington Roebling, also a graduate of Rensselaer and son of the bridge's primary designer, John Augustus Roebling. Buck parlayed his affiliation with the Brooklyn Bridge, still considered to be one of the most important infrastructure projects of the 19th century, into a career as a railroad bridge builder at a time when railroad construction had reached a crescendo.

Buck's reputation for building reliable railroad bridges in North and South America and his experience in carrying out the elder Roebling's plans in the construction of the Brooklyn Bridge gave Buck substantial credibility. He seemed the logical choice when it came time to select an engineer to rebuild John Roebling's Niagara Falls Bridge, a railway suspension span across Niagara Gorge. Built of timbers, Roebling's bridge could never be considered a permanent span, but it served until wear and increasingly heavy trains took their inevitable toll.

In 1877, Buck began a 20-year affiliation with Roebling's bridge at Niagara that would culminate in its replacement with an arch bridge of his own design. Buck initially supervised the strengthening of Niagara Falls Bridge anchorages, and in 1880 he replaced the wooden stiffening trusses of the span with trusses of iron, providing greater strength while reducing weight. When it was determined that Roebling's stone towers had suffered from the vibration of trains crossing the span, Buck designed new steel towers in 1886. With careful planning, Buck managed to conduct all the initial repairs without taking the bridge out of service.

Within a decade of Buck's last repairs to the bridge it was decided the entire span had to be demolished, and Buck was chosen to design and build a replacement.

Buck, mindful that Roebling's suspension railway bridge had suffered under the vibrations and weight of the trains, opted for a steel arch design. Buck began construction of the bridge in 1896 and completed it the following year. Its main span measured just over 547 feet, and the entire bridge was built beneath Roebling's original suspension span without interrupting traffic. When Buck's bridge was completed, Roebling's span was demolished. The bridge, originally named the Lower Falls Bridge, was renamed the Whirlpool Rapids Bridge in 1939. Its double deck carries rail traffic above and vehicular and pedestrian traffic below.

Even before starting construction of what is now known as the Whirlpool Rapids Bridge, Buck was selected as chief engineer for the design and construction of his greatest project, New York's Williamsburg Bridge across the East River. Buck proposed a monumentally rigid bridge that was to combine a substantial truss deck and suspension cables. The Williamsburg Bridge was begun in 1896 and completed in 1903.

Although the biggest bridge built by Buck, it has been criticized by many as less than elegant. Its massive steel towers are of an open latticework design, and its hefty deck and side-span reinforcing trusses give the bridge an ornate and visually busy appearance. Nonetheless, the bridge continues to carry traffic between Manhattan and Brooklyn's Williamsburg section and is one of the most unique bridges in the world in terms of its visual impact.

See also ELLET, CHARLES, JR.; NIAGARA GORGE SUSPENSION BRIDGE; ROEBLING, JOHN AUGUSTUS; WILLIAMSBURG BRIDGE.

Burleigh, Charles *See* Hoosac Tunnel, rock drill.

C

cable-stayed bridge Although its origins can be traced back more than 400 years, the cable-stayed bridge type began to flower only during the 1950s, when German engineers utilized the design to expedite reconstruction of their bridges after World War II.

The concept of a cable-stayed bridge is deceptively simple. Unlike suspension bridges, whose decks are supported by suspenders running to suspension cables that are then tied to massive gravity anchorages, cable stays support the road deck directly. Radiating from a supporting pylon to the deck, the cable stays directly transmit the live and dead load of the span into the tower and ultimately to its foundation. This design eliminates the massive cost of anchorages but requires towers with adequate compressive strength—the ability to endure top-to-bottom stress.

Longer cable-stayed bridges merely require additional towers for supporting additional segments of a bridge. Although the concept is simple, designing a cable-stayed bridge properly requires attention to details, including the resolution of unique aerodynamic problems, selection of the optimal cable arrangement, and dampening of damaging motion where the cables connect to the deck and the towers. Since the 1980s, cable-stayed bridges have become increasingly popular with engineers and the public as a means of crossing moderate distances with highly attractive structures.

The earliest known example of the cable-stayed concept appears in a famous 1595 book, *Machinae novae* (New machines), published in Venice by the Croatian intellectual Faustus Verantius. Verantius illustrated the collection of devices and engineering concepts with copper plate engravings. Although not all the inventions detailed in his book were his, he played an important role in explaining these advances. The widely read and highly regarded book undoubtedly disseminated the cable-stayed concept that may have predated his description of it.

It was up to the Swiss carpenter C. J. Löscher in 1784 to construct what may have been the first

The diagonal cable stays of Washington's Pasco-Kennewick Bridge transfer the weight of the bridge deck to the towers rather than to anchorages as on a suspension bridge. Completed in 1978, this was the first American cable-stayed bridge to use grout-filled sheathing around its cables. *(Godden Collection, Earthquake Engineering Library, University of California, Berkeley)*

bridge whose roadway was suspended by inclined stays. Although the stays radiating from his tower were of timber and not iron or steel cable, they employed the same concept. However, this form of construction became highly suspect after the 1818 collapse of a pedestrian cable-stayed bridge over England's Tweed River when wind snapped the chain stays from the supporting tower. A second cable-stayed bridge collapse occurred in 1824 over Germany's Saale River.

Although 18th- and 19th-century bridges of virtually all types, particularly early suspension spans, were prone to collapse, the cable-stayed bridge was not to see its heyday before the mid-20th century. There are a few notable exceptions, one of which is London's Albert Bridge, which was completed in 1873. The 400-foot span across the Thames River is a hybrid design using both suspension cables and stays that descend from its four towers. The bridge still stands near Battersea Park and is considered by many to be one of London's most interesting bridges.

Although a proponent of suspension bridge designs, the German-American engineer John Augustus Roebling applied diagonal stays to control wind-induced movement in his suspension bridges, such as the Niagara Gorge Bridge (1855) and the Cincinnati-Covington Bridge (1867). Even the use of cable stays in his Brooklyn Bridge did not signal a revival of true cable-stayed bridge design, but it showed an appreciation of the concept.

Cable-stayed bridges popped up occasionally around the world during the 20th century, when some were built in rural areas of Washington and Louisiana, although on a small scale and often employing timber for the towers and decks.

The advanced technology of destruction used during World War II was to play an indirect role in rejuvenating interest in large cable-stayed bridges. Before World War II, bridges were usually destroyed only by hand-placed explosives. Although hundreds of bridges were destroyed in France and Germany by retreating Nazi forces in World War II, Allied air power smashed thousands of bridges throughout Germany. The result was that an estimated 15,000 German bridges lay in ruin.

As Germany struggled to rebuild its bridges using less steel and other materials, engineers turned to the cable-stayed type. Since the foundations of wrecked bridges were often intact and could sustain new towers, some cable-stayed bridges were built atop the remnants of old spans.

It should be noted that although the first modern cable-stayed bridge was built in Sweden in 1955, it was done so by a German contractor under the supervision of a German engineer, Franz Dischinger, clearly indicating the strides made in Germany to develop this type of bridge.

The Strömsund Bridge was built in 1955 and had many of the features of today's modern cable-stayed bridges: a pair of frame towers to accept the weight of the span, steel cable stays, and a system of connecting the stays to the sides of the road deck.

The pylon legs of the frame towers are inclined outward at the bottom and come closer to one another at the top, where they are connected by a horizontal beam. Pylon heads were placed atop each of the four pylons to accept the cable stays. The modified A-frame type of tower, which is considered a "portal frame" since the roadway passes through it, was constructed so its pylons could flex slightly at its base in a longitudinal direction, or in line with the route of the bridge.

The 1,085-foot bridge was composed of two outer spans of 242.8 feet and a center span of 600 feet. The cable stays were connected to the sides of steel anchorage beams set into the road deck. During installation the stays are tightened to the appropriate tension. The bridge is considered a remarkably advanced design for its time and remains in use in the town of the same name in central Sweden.

Cable-stayed bridges caught on slowly around the world. Walter Podolny, Jr., and John B. Scalzi in *Construction and Design of Cable-Stayed Bridges* point out that between 1955 and 1977 a total of only 62 cable-stayed bridges had been built in 19 countries. Germany, which had turned to the cable-stayed design after World War II, possessed 19, or more than 30 percent of the cable-stayed bridges. The United States had only six such bridges. Japan, another nation that had to rebuild after the war, claimed eight.

During the 1980s, cable-stayed bridge construction gained in popularity, not only because the design was sound but also because the bridges were marvelous visual experiences both for far-off observers as well as for those traveling beneath their cathedrals of cable. Combined with lightweight decks, the bridges could sometimes be built far lighter than other designs and often more rapidly.

Cable-stayed bridges began to take on an even more dramatic appearance during the 1960s, when designers began to employ a single, inclined support pylon instead of two. This required the addition of anchorages so the pylons could be secured with backstays. Australia's Batman Bridge (1968) uses a pylon inclined toward the center of the bridge, whereas most other single-pylon bridges have their pylons leaning back toward shore.

The Spanish engineer-architect Santiago Calatrava's Alamillo Bridge in Seville, Spain, utilizes a single, inclined pylon leaning away from the bridge toward shore. This exciting bridge is unique because it is not back-anchored. Instead, it depends on additional mass and rigidity in the pylon to counteract the weight of the bridge deck.

Although American engineers, and the public officials who must authorize funding for new bridges, were at first reticent about employing cable-stayed designs, it seemed the cable-stayed bridge craze was catching on by the 1980s. After the deadly collision of a ship with Florida's cantilever truss Sunshine Skyway Bridge in 1980, a soaring cable-stayed bridge bearing the same name was built as a replacement in 1987.

The look of the Sunshine Skyway Bridge is dramatic if not thrilling. The taut cables radiating downward from its pylons are a far cry from the seemingly relaxed catenary curve of suspension bridge cables. The dashing look of the cable-stayed bridge has been accentuated further in recent years by wrapping of the cable stays in boldly colored polyethylene sheaths selected by designers to complement the lighting schemes of the bridges.

Nonetheless, modern cable-stayed bridges are still in their infancy, and engineers are learning more of their nuances with each new one. The construction of the Fred Hartman Bridge near Baytown, Texas, (1995) revealed that raindrops clinging to the polyethylene sheathing altered the aerodynamic profile of the cable stays, resulting in serious oscillations in relatively mild winds. To solve these aerodynamic problems, some sheathing is being manufactured with various textures on the surface. Unfortunately, roughly half of 28 cable-stayed bridges in the United States are affected by this problem.

Although cable-stayed spans are usually restricted to crossings of moderate length, some, like Venezuela's Lake Maracaibo Bridge, can span great distances. The Maracaibo Bridge, completed in 1962, uses a series of concrete towers to support a massive concrete deck structure. The road deck is supported from below in cantilever fashion by an X-shaped frame and from above by cable stays strung from parallel A-shaped towers.

The Maracaibo Bridge has a total length of 5.4 miles and has five cable-stayed spans, each measuring 771 feet. What is noteworthy about the Maracaibo Bridge is that until it was built, steel decks were most commonly used for cable-stayed bridges. The success of the concrete deck on the Maracaibo Bridge encouraged the use of less expensive concrete on cable-stayed spans. Unfortunately, the zinc-coated cables used on the bridge showed significant corrosion by 1980, forcing their replacement after only 18 years of service.

The longest cable-stayed bridge built by 2001 was Japan's Tatara Bridge, a three-span structure with a total length of 4,855 feet. Its center span is a record 2,919 feet and is supported by cables radiating from a pair of inverted Y-shaped towers. Completed in 1999 after six years of work, the bridge connects the islands of Honshu and Shikoku across the Seto Inland Sea.

See also BRIDGE AERODYNAMICS; CALATRAVA, SANTIAGO; TATARA BRIDGE.

caisson Caissons of various types have been used for nearly 3,500 years for underwater and underground foundation construction. The word *caisson* is derived from the French word for "box," which is essentially what a caisson is. Caissons were developed to allow the construction of submerged foundations for various structures, including bridges. Developments later allowed caissons to become something akin to stationary submarines in which workers could excavate to bedrock deep underwater in a pressurized, air-filled chamber.

Some of the earliest caissons, known as floating caissons, were simple box-shaped structures with a floor that would be towed to where a foundation was to be built. It was then filled with rock until it sank in place. Additional rock would be added until the foundation was substantial enough to support the pier of a bridge. This was usable only if the ground were solid enough to preclude excavating to an acceptably solid stratum.

Another type of caisson is the open caisson, which is essentially four heavy walls without a top or bottom. When it was placed in shallow water or on land, workers excavated soil to lower the caisson, whose walls were then built higher as it descended.

Once a solid footing was reached, the caisson would be filled with rock or concrete. Such caissons could be pumped free of infiltrating water as they descended.

In order to build a stable and long-lasting bridge, a foundation must rest on solid rock or an extremely stable base of cohesive soil. If the bridge is to cross water, engineers have to find a means of making it possible for workers to excavate on the bed of a river or a strait to reach bedrock beneath the muck and gravel. The answer to this problem was the pneumatic caisson developed in 1851 by England's William Cubitt and John Wright. It was essentially a box with its open side pointed downward; its working chamber could be pressurized to prevent water from infiltrating where laborers excavated soil to sink the caisson.

The invention of the airlock in 1830 by the British Royal Navy admiral Thomas Cochrane was the innovation that allowed development of Cubitt and Wright's pneumatic caisson. The airlock provided an intermediate chamber with one airtight doorway leading to the surface and another to the interior of the caisson's pressurized working chamber. Arriving workers would enter the chamber, where the pressure would be increased by pumps to match that in the excavation chamber. When departing, workers would reenter the chamber and decompress to avoid the effect of "the bends," or caisson disease, a condition that results from the absorption of nitrogen under pressure. The deeper a caisson is sunk, the greater the requirement for air pressure to prevent flooding. The result is an increased risk of caisson disease. Without an airlock, opening the doorway to the pressurized work area of the caisson would cause an explosive loss of protective air pressure.

Because of the problem of caisson disease, which killed a sizable number of workers during the 19th century on such projects as the Eads Bridge in Saint Louis and the Brooklyn Bridge in New York, engineers moved away from sending men into deep pneumatic caissons by the 20th century.

By the 1900s, increasingly massive bridges placed in deeper and deeper water often required foundations or footings in conditions that presented unacceptable risks. By the 1930s, open-bottom caissons were being sunk to the bedrock with the excavation conducted by clamshell buckets lowered into individual shafts through the caisson's roof. During the construction of the massive Mackinac Bridge across Michigan's Mackinac Strait during the

A steel caisson assembled from the modular sections in the foreground is ready to be towed into position in the Straits of Mackinac to be sunk to the bottom of the strait to serve as a foundation for one of the towers of the Mackinac Bridge. This type of caisson gained immense popularity during the 20th century since it can be sunk to bedrock by using clamshell shovels lowered from above, eliminating the need to send workers into a dangerous compressed air environment. *(Michigan Department of Transportation/Mackinac Bridge Authority. Photograph by Herman D. Ellis)*

late 1950s, high-pressure water jets were emplaced along the steel cutting edges of the unmanned caisson to blast rock and muck away from the cutting edges. This assisted the effort to seat the caisson while huge clamshell cranes removed the majority of the loose earth.

Once in position these modern caissons are filled with specially formulated concrete, essentially making the bridge foundation not only one with the rock below but far too heavy to be budged by any imaginable force.

In Japan, engineers have developed remotely controlled machinery to enter pressurized caissons and perform the excavation work without jeopardizing the health and safety of human workers. Operators hundreds of feet above send commands to the machinery by remote control while watching the work on television monitors.

See also CAISSON DISEASE; COFFERDAM.

caisson disease, the bends   If there ever was a medical problem brought about by technology, caisson disease is certainly that. The potentially fatal condition, which is also known as "the bends" or more scientifically as decompression sickness, was unheard of until 1667, when the British chemist Robert Boyle began pressurization experiments on animals.

While depressurizing a snake, Boyle saw the creature grow distressed and then observed a bubble of gas that was expanding inside the viper's eye. Boyle was at a loss to explain what was happening to the snake, but he was able eventually to formulate Boyle's law, which recognizes that pressurizing gas reduces its volume. What Boyle did not know was that when animals are placed under increased pressure, such as that encountered under water, the nitrogen in their bodies is compressed and driven into the bloodstream. This happens to humans when they are returned too rapidly to a reduced atmospheric pressure: The nitrogen violently expands like the carbonated bubbles of a shaken soft drink, with painful and destructive effects on nerves and tissue.

The consequences can be ghastly. At a minimum, a person may suffer a temporary loss of balance and excruciating joint aches. The joints are particularly susceptible to nitrogen expansion and the resulting pain is unbearable. The stooped posture of those enduring this condition inspired its label "the bends." In severe cases, paralysis and a horribly painful death can occur. The problem of decompression sickness became apparent in the first half of the 1800s as men began working under extraordinary sea pressure with newly developed helmet diving equipment.

Unfortunately for laborers inside pressurized caissons, the problem that afflicted the hapless snake and that would soon afflict many of them, sometimes fatally, would not be fully understood until the 1950s. Scientists and victims both came to understand that the length of time working under pressure and the magnitude of the pressure were significant factors. They also understood that gradual decompression was important, but for decades no one agreed on how gradual.

During caisson work on the Eads Bridge in the late 1860s in Saint Louis, workers were so mystified about the causes of caisson disease that many resorted to medicinal bracelets made of zinc and silver. Eventually, 14 workers would die of decompression sickness during foundation work beneath the Mississippi River. Caisson work on the Brooklyn Bridge, built between 1869 and 1883, was no less hazardous. Although no one will ever know the true number of deaths due to caisson disease during con-

struction of John Augustus Roebling's famous bridge, it is generally acknowledged that at least 20 died as a result of their work in its pneumatic caissons.

Ironically, Cochrane's airlock invention not only allowed the cause of caisson disease—the invention of the pneumatic caisson—but was also the cure for the condition. Researchers later determined that a person suffering from decompression sickness could gain a full recovery by reentering a pressurized environment and decompressing more slowly. Even though slow decompression was considered a solution to the problem, many rough-and-tumble workers during the 19th century wanted to get off the job site as quickly as possible and complained about increased decompression periods. Eads's workers, who often spent only five minutes decompressing,

A trio of airlocks is suspended in the roof of the Lincoln Tunnel during its construction in the 1930s. Workers entered the doors at one end of the airlocks, where the air pressure was increased to equal that in the working area. Departing workers would decompress in the airlocks to allow the dissipation of dangerous nitrogen in an effort to prevent caisson disease. *(National Archives)*

should have spent more than two hours gradually readjusting to lower pressure to allow the nitrogen to dissipate. Some compressed air workers would later discover that segments of their bone had literally died when expanding nitrogen bubbles cut off the blood flow. This condition is known as osteonecrosis, or "death of bone."

A systematic approach to solving the problem of decompression sickness was not formulated until 25 years after completion of the Brooklyn Bridge. At the request of the British government, concerned about the threat of decompression sickness, a team led by the Scottish physiologist John Scott Haldane developed tables establishing decompression periods in 1908. Both the Royal Navy and the U.S. Navy adopted and later augmented Haldane's conclusions.

See also BROOKLYN BRIDGE; CAISSON; EADS BRIDGE; RECOMPRESSION, SANDHOGS.

Calatrava, Santiago (1951–) *Blender of style and strength to create astounding spans* Surprising, confounding, thought-provoking, and striking only begin to describe the bridges designed by Spanish-born architect and engineer Santiago Calatrava. And although his spans are also controversial and sometimes pricier than others, Calatrava has the ability to combine form and function while creating something more akin to inspirational sculpture than to a simple bridge.

Calatrava's expertise covers the gamut of three-dimensional structures as a sculptor, civil engineer, and designer of bridges and buildings. As an art student at Spain's Escuela Técnica Superior de Arquitectura in Valencia, Calatrava earned a degree in architecture before attending the Swiss Federal Institute of Technology in Zurich, where he obtained a degree in civil engineering. At the institute the famed Swiss-American engineer and bridge designer Othmar Hermann Ammann also did his civil engineering study.

Intending to become an artist, Calatrava instead decided on architecture as his field, but on realizing that architects were dependent on civil or structural engineers, he obtained his civil engineering degree followed by a Ph.D. in technical science. Calatrava had decided to create not only the skin of a structure but the skeleton as well.

After opening his own design practice in Zurich in 1981, Calatrava took on the prosaic assignments that come with starting out in architecture and civil

engineering but soon demonstrated his creativity and quickly garnered a reputation as an innovator. His 1983 design of Zurich's Stadelhofen Railway Station displayed his ability to create fixed and even subterranean structures that looked as if they were in motion.

In 1992, Calatrava designed the Alamillo Bridge in Seville. Employing a modified cable-stayed design, Calatrava provided the bridge with a single pylon at one end inclined dramatically toward shore. The pylon did not require any backstays—cables to hold the weight of the bridge—since he had designed it with adequate weight and strength to endure the loading of its 820-foot deck. The 531-foot pylon inclines 58 degrees away from the roadway of the bridge.

His smaller but no less dramatic design of the Campo Volantín Footbridge, completed in Bilbao, Spain, in 1997, is a study in Calatrava's mixture of visual artistry and structural engineering. The 246-foot curved bridge leads to the Guggenheim Bilbao Museum and is made unique by a parabolic tubular arch that runs from concrete arms set jutting from one side of the bridge. This tubular arch is then connected by skeins of cables radiating downward to each side of the bridge deck for support. Another daring detail is the deck of the bridge, which contains large glass panels.

Like the German engineer Fritz Leonhardt, Calatrava has proved himself a designer of a variety of structures, including large-scale public structures as well as towers. His 446-foot Montjuic Communications Tower in Barcelona, which was built for the 1992 Olympics, is a stunning example of this versatility. The blinding white steel tower has a pair of horizontal tines angled outward that impale a curved metal arc. This visual effect resembles a model of the starship *Enterprise* on a display stand. Finalizing this exquisitely interesting design is a massive, javelin-shaped antenna jutting upward through the open space between the tower and the arc.

Calatrava's designs tweak ordinary concepts of how structures like bridges are supposed to look and perform. His Alameda Bridge, built in 1995 in Valencia, Spain, uses a hollow-box steel arch located asymmetrically along one side of the bridge. The arch is tied to the roadway with steel hollow-box struts. At first glance the road deck seems to be supporting the arch, instead of the other way around. Sheltered beneath the Alameda Bridge is an underground railway station.

Calatrava's brilliance and vision push the edge of what has been attempted before and the architect-engineer relishes the use of high-tech materials. This sort of daring can be costly, and Calatrava's breath-taking design for the Milwaukee Art Museum is an example. The project features a soaring sunshade reminiscent of a soaring bird seemingly taking flight over Lake Michigan as well as a striking footbridge supported by a tubular pylon. By 2001, the project's total cost had crept upward to nearly $100 million, or nearly three times the original estimate.

Only the adoration of engineering students and other design professionals inspired by his work rivals his popularity with the public. Calatrava is one of the few architects to tackle so fully all the aspects of a structure's design and do so in such an artistic manner. Not surprisingly, Calatrava's structural designs have been displayed at museums around the world, providing the architect-engineer with a superb marketing tool.

As of 2001 the youthful Calatrava had offices in Switzerland, Paris, and Spain. His career promises to be a long and productive one, situating him in a league with other bridge design geniuses including Robert Maillart and Leonhardt. Calatrava's own web site contains not only his biography but also a photo gallery of the numerous projects he has completed or is contemplating. It is at http://www.calatrava.com/indexflash.html.

See also AMMANN, OTHMAR HERMANN; ARCH BRIDGE; CABLE-STAYED BRIDGE; FIBER-REINFORCED POLYMER COMPOSITES; LEONHARDT, FRITZ; MAILLART, ROBERT.

canary *Sacrificial sensor of mine gases* Although animal rights activists would today be appalled at the practice of using canaries for the detection of mine gases, it should be remembered that these birds were often carried by miners and rescuers who were themselves unprotected from poison gases.

The birds fell victim to the ill effects of carbon monoxide or methane far sooner than a human would, making them perfect living sensors. Until the late 1800s there were no efficient breathing devices for use in mines, so the miners and the birds were both taking their chances underground. Since female canaries sang less and were cheaper than the more marketable males, they were invariably the ones put to work in mines.

The use of canaries dates back at least to the 18th century, when the rapid pulse and respiratory

A pair of coal miners check a canary for the ill effects of dangerous gases in a photograph taken during the early 20th century. Canaries, because of their rapid heart rate and metabolism, reveal the presence of hazardous atmospheric conditions long before humans are affected. *(National Archives)*

rate of the tiny birds made them a logical choice as mine gas detectors. Despite the development of 19th-century safety lamps that would go out or burn with a different color when exposed to gases, and the 20th-century development of mechanical detectors, canaries were used for gas detection in American mines until the 1940s.

The canaries provided a warning of toxic conditions by their weak or wobbly behavior. Sometimes the birds would fall off a perch unconscious. Miners who lacked breathing devices would make a hasty retreat to fresh air, where the canary could sometimes be revived to serve again in her dangerous role.

To help preserve the lives of the canaries, miners would sometimes place them into an airtight container with glass or clear plastic walls. A small doorway would be opened in this container, known as a resuscitation cage, to allow ambient air to enter. When the bird exhibited ill effects, the miners, who by the first decades of the 1900s were equipped with self-contained breathing apparatus, could inject fresh air into the container to revive the canary.

See also AFTERDAMP AND BLACKDAMP; CARBON MONOXIDE ASPHYXIATION; SAFETY LAMP.

cantilever bridge The very nature of bridging a gap across a river or gorge cried out for the cantilever method of bridge construction. This method provides a means of significantly increasing span length and thus reducing expensive foundation work because a cantilever span can support itself since it is built out from a single foundation.

There are several variations of this method of construction. In the balanced cantilever construction method, bridge erection starts from each pier and moves outward in both directions. Eventually one side of the cantilevers reaches shore or approaches another cantilevered span. Large cantilever bridges such as Canada's Quebec Bridge and Scotland's Firth of Forth Bridge had planned gaps between the cantilever spans that were bridged with spans raised into place (Quebec) or built to meet in the middle of the gap (Firth of Forth). The concept of a cantilever bridge is centuries old but was not applied on a grand scale with any success until the latter part of the 19th century.

Another variation is the simple cantilever. The concept of the simple cantilever span has been admirably demonstrated by silent film comedies in which one person serves as an anchor for an impromptu cantilever by standing on the back of a plank extended outward from the edge of a roof. An unsuspecting coworker walks out on the cantilever pinned securely at one end by the weight of his

A vintage photograph with the ubiquitous Highlanders in kilts beside Scotland's Firth of Forth Bridge reveals the overwhelming profile of this cantilever span. *(Author's Collection)*

"friend." When the man anchoring the plank absent-mindedly steps away, the cantilever span is no longer anchored, and the board and man descend in a quintessential pratfall. In real bridges, the simple cantilever method often uses cables as the anchors until the bridge is completed and can stand on its own.

The reason cantilever bridges were virtually nonexistent before the late 19th century was that materials and techniques had not been developed to make long cantilever spans feasible. Stone and masonry building materials lent themselves to arch construction, a technique that required numerous arches and accompanying piers. However, the development of iron and steel, as well as truss reinforcement schemes, meant spans of adequate strength could be built. Some of the most successful cantilever bridges were constructed by truss reinforcement in which steel or iron components reinforce a span to provide extra rigidity. A strong pier anchors the cantilever itself.

When it was decided to cross Scotland's Firth of Forth River with a railway bridge, the designers John Fowler and Benjamin Baker opted for a cantilever design made of tubular steel. More specifically, the design used was a cantilever through-truss, which provided triangular steel supports in the form of trusses on both sides and the top of the bridge's structure. This reinforced tunnel of latticed steel, along with its reinforced railway deck, made it possible for the bridge to boast a pair of 1,710-foot cantilever spans. This massively strong bridge was completed in 1890. Only Canada's Quebec Bridge has longer cantilever truss spans.

See also FALSEWORK; FIRTH OF FORTH BRIDGE; QUEBEC BRIDGE; TRUSS BRIDGE.

carbon monoxide asphyxiation   The incomplete combustion of any carbon matter including coal produces the deadly gas carbon monoxide, which is a stealthy killer because of its lack of color and odor. Carbon monoxide asphyxiation is a serious hazard in coal mines, where a fire of virtually any size can generate a high volume of the gas.

Carbon monoxide is particularly hazardous in the way it causes death because it is absorbed into hemoglobin 240 times more efficiently than is oxygen. The displacement of oxygen within the hemoglobin, and not a lack of ambient oxygen, causes asphyxiation. For this reason, only a very tiny amount of carbon monoxide must be present to cause illness or death.

Wives of miners trapped within a coal mine rocked by an explosion wait with their children to find out whether their loved ones are among the living or the dead. This scene has been played out hundreds of times during the history of American coal mining. This hard existence offered few rewards and extreme danger, including carbon monoxide asphyxiation and explosions due to excessive methane or airborne coal dust. *(Bureau of Mines/National Archives and Records Administration)*

An early symptom of carbon monoxide asphyxiation is a tightening feeling in the head that might also be accompanied by a headache. Those with chronic carbon monoxide asphyxiation often complain of symptoms approximating those of flu. An amount of carbon monoxide that is only 0.2 percent of the air can cause death within 30 minutes.

See also AFTERDAMP AND BLACKDAMP; CANARY; HELMET MEN; SELF-RESCUER; VENTILATION.

cathodic protection *High-tech anticorrosion weapon* Excluding stone arch bridges, virtually every bridge built within the past three centuries has had iron or, more recently, steel components. Tunnels are often lined with iron rings and then strengthened with an additional layer of steel-reinforced concrete. By their very use and location, bridges and tunnels are exposed to salt in the air, groundwater, or rainwater laced with deicing salts that can corrode their iron-based components.

Even concrete beam and arch bridges that appear monolithically gray and indestructible contain reinforcing steel that can be corroded by seawater or road salts. Salt water can corrode steel and iron components no matter how thick the concrete. Corrosion causes the steel to expand, and the resulting swelling cracks and spalls the concrete, affecting the strength and structural integrity of the span. In tunnels, which are usually surrounded by earth saturated by water, the same corrosive effects can affect the iron lining rings of tunnels as well as the steel reinforcement of their concrete linings. If allowed to go unchecked, this corrosion can require expensive tunnel and bridge repairs if not outright replacement of the entire structure, a costly proposition for municipalities and agencies.

Because corrosion is an electrochemical reaction, researchers have made use of electrical principles since the 1940s to prevent this type of deterioration. Known as cathodic protection, or "CP," among corrosion engineers, the technology uses an electrical current to counteract the loss of electrons in steel and iron that is part of the corrosion process.

Although iron-based materials appear hard, solid, and enduring, they can actually deteriorate at an atomic level since the escape of electrons alters the properties of the iron atom's nucleus. Corrosion is simply a return of the iron (the elemental component of steel) to its original or lowest energy state of iron oxide.

Cathodic protection uses an anode, a type of electrode, to send electrons to the endangered metal to counteract the electron loss and therefore halt or reduce corrosion. Although some anodes provide electrons in a passive fashion based on their composition, some are energized with electricity to increase their capability. This type of protection is also often applied to steel tanks placed underground.

The traditional method of protecting metals in contact with moisture is painting, which insulates them not only from the electrical currents of the surrounding water or ground but from oxygen as well. If the paint adheres and insulates the metal from contact with surrounding electrical activity, then it will work, but a breach in the coating invites corrosion. Some coatings that will provide cathodic protection themselves are now being developed and are being tested by various highway departments around the nation.

The seriousness of the threat of corrosion is exemplified by the fate of the Alsea Bridge, one of the longest concrete arch bridges ever built. It was also considered a historical treasure, designed as it was by the famed Oregon State Highway Department bridge designer Conde B. McCullough. Completed in 1936 for $5.6 million, the steel-reinforced concrete span contained three concrete tied arches and measured 3,028 feet in length. A decade after its completion, maintenance workers noticed that concrete was being spalled by swelling-induced corrosion of the embedded steel reinforcement. In 1966, inspectors discovered cracks running longitudinally along the concrete floor beams of the bridge, another result of the corroding and swelling of steel reinforcement.

To counteract these problems and prolong the life of the bridge—originally designed to last 75 years—a cathodic protection system was installed to halt the loss of electrons in the steel reinforcement electrically. Unfortunately, the cathodic protection applied in 1976 was installed too late to save the bridge; in 1982 it was decided the bridge would have to be demolished. This decision came about partly because the spalling of concrete from the underside of the bridge's deck was so severe that it posed a hazard to boaters passing beneath the structure. After the 1991 completion of a new bridge, two arches of the original were demolished with explosives and a third dismantled by workers and machinery.

To prolong its life and ensure its structural soundness, the famed Brooklyn-Battery Tunnel was the recipient of a $17 million cathodic protection project in 1993 designed to halt corrosion of its iron and steel components, including its massive 31-foot-diameter iron ring lining and its steel-reinforced concrete walls.

Although cathodic protection is effective, some maintenance engineers are astonished that bridges with exposed steel components are not washed with fresh water yearly to remove road salts and grime that can accelerate corrosion. It is recommended that smaller communities with limited funds dispatch a fire truck to accomplish this simple task.

See also ARCH BRIDGE; BRIDGE MAINTENANCE; BROOKLYN-BATTERY TUNNEL; ELECTROCHEMICAL CHLORIDE EXTRACTION; MCCULLOUGH, CONDE BALCOM; REINFORCED CONCRETE.

cement   *See* concrete.

Central Artery/Tunnel Project (begun 1991–2004?)
Boston's controversial tunnel and bridge project   One of the world's most ambitious megaprojects, Boston's grand bridge and tunnel scheme to alleviate its horrific traffic problems has also created a firestorm of controversy through its massive cost overruns and questionable management. Considering the history of massive tunnel projects, it seems Boston's—also known as the Big Dig—is not much different from other famously expensive tunnel projects of the last 175 years.

By the 1980s, Boston was choking on its own traffic. Interstate 93, a major highway running north and south through Boston, was carried through the center of the city by an elevated freeway that was obsolete before its 1959 completion. Known as the Central Artery, the freeway was an ugly and divisive concrete obstacle between Boston's North End and its downtown business district. In addition, a pair of tunnels built between downtown Boston and Logan Airport were also clogged with bumper-to-bumper traffic for up to 16 hours a day. It was estimated that traffic jams were costing drivers up to $500 million yearly in fuel and other expenses.

For decades plans were proposed, discussed, and shelved as transportation officials sought ways to remedy traffic congestion. During the 1980s, a plan in which a 10-lane tunnel would replace the Central Artery was developed. The decrepit six-lane bridge carrying Interstate 93 traffic over the Charles River was often nothing more than an elevated parking lot. It would also be replaced with two bridges carrying a total of 14 lanes.

Interstate 93, running through Boston, would be extended to intersect Interstate 90 and cross Fort Point Channel with the help of an immersed tube tunnel. Interstate 90 would be extended beneath Boston Harbor toward Logan Airport with an even longer immersed tube tunnel. This would augment the existing Callahan and Summer Tunnels to the north.

By 1987, the federal government agreed to finance part of the $2.6 billion project through the Department of Transportation. The fact that the then Speaker of the House Tip O'Neill represented Boston undoubtedly helped the controversial project. Although the $2.6 billion price tag was considered a massive amount of money at the time, subsequent cost overruns would make the original sum seem like a bargain. The Federal Highway Administration, wary of federal involvement in the project from the start, would pay 90 percent of the cost. The Massachusetts Turnpike Authority would pay the remainder. Unfortunately for taxpayers, the original estimate was ridiculously and perhaps intentionally too low, and, for a variety of reasons that no one seems able to explain fully, the costs would eventually spiral out of control.

The man in charge of the project when construction began in 1992 was James Kerasiotes, a political insider who was appointed secretary of transportation and construction by the then governor, William F. Weld. Tough, outspoken, and considered something of a political bully, Kerasiotes, many felt, had the qualities necessary to shove the massive, complicated, and expensive project through to completion.

As the project's costs increased, Federal Highway Administration officials and members of Congress grew increasingly uneasy. Kerasiotes in 1995 pledged to hold the project's cost to no more than $10.8 billion, soothing the anxiety of officials. However, matters had gone terribly wrong as Kerasiotes's staff approved expensive engineering changes in the complex project as delays and technical problems mounted. Despite Kerasiotes's promise to stay within the budget, costs continued to rise, although Kerasiotes denied the project was exceeding his self-imposed $10.8-billion limit.

An Annual Finance Plan required by the Federal Highway Administration was received three months late, in October 1999. Although a number of policy groups had forewarned that the project was massively over budget, federal highway officials gave tentative approval to the plan after Kerasiotes assured them overruns would be no more than $500 million. The federal approval of the plan was granted on February 1, 2000. Incredibly, on the very same day, Kerasiotes ignited a firestorm by admitting to the press that the project had exceeded his $10.8 billion cap by $1.4 billion.

Federal officials were flabbergasted that they had not been given accurate information about the project, which would now cost $12.4 billion. In April 2000, the Massachusetts governor, Paul Celluci, fired Kerasiotes. By 2001, estimates indicated that the final cost of the project had moved beyond $14 billion, only $4 billion less than the final cost of the 24-mile Eurotunnel beneath the English Channel. Originally, the $2.6 billion project was to be completed by 1998, but it was only 70 percent finished by 2001. By then it was a $14-billion project with a completion date of 2004.

Kerasiotes found himself not only fired but also under investigation by the Massachusetts Attorney General's Office. A number of audits were conducted on the project while critics, including the Federal Highway Administration, claimed that hundreds of millions of dollars were wasted by poor management.

A onetime Big Dig opponent, Andrew Natsios, took over Kerasiotes's position in 2000 and whittled down the fee of the management consultant Bechtel/Parsons Brinckerhoff from 11.5 percent of the total cost of the project to 7 percent. Natsios also offered the contractor financial incentives if the project were brought in for less than $14.05 billion. Unfortunately for Massachusetts's taxpayers, the federal contribution to the Central Artery money pit has been capped at $8.5 billion, or roughly 60 percent of the total amount.

As for the project itself, the gargantuan effort of building the 7.5-mile link of tunnels and bridges was problematical and expensive. The reason was that the work would have to be carried out without halting traffic on the Central Artery and rail lines above its route.

Instead of boring through the soft ground below the Central Artery, it was decided to use a cut-and-cover type of construction in which walls are constructed by the slurry wall method, followed by an excavation of a trench between the walls from the top down. A concrete roadway was built as well as a reinforced concrete roof. During this phase, the massive walls that were originally built served as a foundation for the elevated Central Artery, which carried traffic through the tunnel-building process.

The slurry wall process allowed workers to dig an excavation where the walls of the tunnel would stand. A mixture of water and fill would provide enough resistance to prevent the earthen sides from caving in. A steel-reinforced concrete wall would be poured and the opposing wall would be built in the same way. With the two walls holding back the earth, the ground between them would be excavated.

Another underground segment of the roadway would be carried beneath nine active rail lines. Because tunnel boring might cause dangerous settling of the ground, it was decided to first use ground-freezing to stabilize the soil. A brine solution at below-freezing temperatures was pumped through pipes to freeze the damp soil, making it nearly as solid as concrete.

The method of building the tunnel through the rail yard would be tunnel-jacking, a system that shoves reinforced concrete boxes through the ground by hydraulic rams. Each of these tunnel sections measured 80 feet wide by 40 feet high. A road header, a machine armed with a rotating cutter head, would work at the open front of the tunnel, grinding away earth, which was then removed. Then 50 hydraulic rams capable of exerting 10,000 pounds of pressure per square inch would slowly move the concrete sections forward. The eastbound section of this tunnel was completed in 2000 and the westbound section was finished in February 2001.

One of the earliest successes of the project was completion of the Ted Williams Tunnel extending I-90 beneath Boston Harbor in 1995. The immersed tube tunnel method was employed: A dozen sections containing twin traffic tubes were assembled out of steel and then sunk into a 50-foot-deep trench on the floor of the harbor. Once connected, the 300-foot sections then provided a watertight pair of vehicular crossings beneath the harbor. Although the immersed tube tunnel is just shy of 4,000 feet in length, the entire tunnel measures 1.6 miles, including underground approaches on both sides of the harbor.

To accept the tunnel's approach in South Boston, one of the world's largest cofferdams was built, measuring 250 feet in diameter and 80 feet deep. Once

the immersed tube tunnel segments were connected with underground approaches, a massive ventilation building was erected atop the cofferdam to provide fresh air to the tunnel. A second ventilation building was constructed on the opposite shore.

The second immersed tube tunnel of the project crosses the Fort Point Channel, where I-90 formerly terminated. This project is unique since it involves the first large-scale use of concrete immersed tube tunnel sections in the United States. A total of six sections, the longest of which is 414 feet and weighs 50,000 tons, will be submerged and joined to form a 1,100-foot crossing of the channel. The sections were built in a massive, temporary dry dock adjacent to the channel. When the sections were completed, the dry dock was flooded and the completed sections floated out like barges. Unlike the Ted Williams Tunnel sections, these have a rectangular shape with three segments making up the east and west lanes, respectively.

Once built, the sections were submerged with great precision atop 110 six-foot-diameter concrete piers drilled into the bottom of the waterway. These piers were necessary to straddle an existing tunnel beneath the channel. The piers, seated on bedrock, were needed to prevent the weight of the immersed tube tunnel from impinging on Metropolitan Boston Transit Authority's Red Line subway. As each segment was guided into place and slowly submerged, it was aligned with its concrete footings. By early 2001, four of the six tunnel segments were in place. The remaining two segments were to be positioned later that year. The immersed tube tunnel, known as the Fort Point Channel Crossing, will accommodate six eastbound lanes and up to five westbound lanes. It will be completed in September 2002.

A pair of new bridges carrying I-93's traffic across the Charles River will also constitute the Central Artery/Tunnel Project. One will be a conventional pier and beam bridge composed of nine box girder sections. The 830-foot bridge is composed of two spans of 225 feet and a third, center span of 380 feet. The Leverett Circle Connector Bridge was opened to the public on October 7, 1999.

Standing 100 feet east of the Leverett Circle Connector Bridge will be an exciting cable-stayed bridge designed by the Swiss engineer Christian Menn. With its two towers radiating cable stays downward to support its sleek road deck, the $87 million span will be far less obtrusive than another bridge design that was almost built. An outcry over the ugliness of the other design, which some considered as ungainly as the one it was to replace, resulted in the selection of Menn's design.

The span has an unusual asymmetrical design since two of its 10 lanes will be carried in cantilever fashion on the outside of its east-facing deck. The remaining eight lanes will run between its two towers, which are shaped like an upside down Y. Two rows of cable stays in a fan arrangement will run from the towers to the center span, and a single row of fan arrangement stays will angle downward to the approach spans.

The design is also unusual in that it is America's first hybrid cable-stayed bridge: Cable stays provide support to the road deck, steel box girders and floor beams provide rigidity to the 744-foot center span of the bridge, and outside spans are of post-tensioned concrete decks. The bridge measures a total of 1,457 feet and is also notable because its 185-foot width makes it the widest cable-stayed bridge in the world.

The towers of the bridge will accept the loading of the bridge deck by transmitting it in the form of compression to their foundations. Menn designed the inverted Y-shaped concrete pylons to resemble the granite obelisk that stands as monument to the Battle of Bunker Hill in Charleston. The cable-stayed span, known as the Leonard P. Zakim Bunker Hill Bridge, is expected to be completed in November 2003.

The 10 lanes of the cable-stayed bridge and four lanes of its smaller neighbor will provide 14 vehicular traffic lanes. The badly deteriorated, double-deck truss bridge built in 1959 and commonly known as the "High Bridge" will eventually be demolished.

See also CABLE-STAYED BRIDGE; GROUND-FREEZING; IMMERSED TUBE TUNNEL; MENN, CHRISTIAN; HOLLOW-BOX GIRDER; PRESTRESSED AND POST-TENSIONED CONCRETE.

Channel Tunnel   *See* Eurotunnel.

Channel Tunnel fire   A raging train fire was a greater worry to those who designed and operated the English Channel Tunnel than any other hazard. Aside from flooding and collapse, a fire is the most feared event in a tunnel, where air and escape options may be fatally limited. And although many crossed their fingers and hoped such a fire would not occur, this dreaded event came to pass on November 18, 1996, at 9:45 P.M. The hellish inferno that resulted damaged the tunnel and literally melted much of a train, although serious casualties were miraculously avoided.

A heavy goods vehicle train, whose rail cars were designed to carry 44-ton cargo trucks beneath the Channel, was loaded with 29 such trucks at 9:40 P.M. and began to accelerate toward the tunnel. Because of a strike by workers of Eurotunnel, the Anglo-French consortium that built and operates the tunnel, some of the trucks had been delayed in a parking area from between 57 and 112 minutes before departure.

Just before the train entered the tunnel, a pair of Eurotunnel security guards noticed one of the trucks emitting flame from its underside and notified railway personnel at 9:48 P.M. Precisely one minute later a fire detection device inside the tunnel also sensed the blaze and triggered an alarm. The driver of the train was warned by radio at 9:51 P.M. that he might have a fire on his train and was instructed to continue through the tunnel to the British side of the Channel. This move was part of previously established safety procedures to prevent the buildup of deadly fumes while also preventing heat damage to the tunnel's reinforced concrete lining.

By this time the train was traveling at 87 mph and the rail yard in Folkestone, England, was less than 20 minutes away. Simultaneously with notification of the fire, the train's fire alarm system signaled a blaze in the last car on the train. Aboard the train besides the driver, chief of the train, and steward were 32 drivers and passengers of the cargo trucks. Either independently or as a result of fanning by the passing air, the burning cargo truck had burst into a roaring, superhot fire.

Although the Channel Tunnel's two main rail tunnels run east and west, they are referred to as the Running Tunnel North and Running Tunnel South because of their relationship to one another; and the burning train was in Running Tunnel South. A third, smaller service tunnel runs between the two main tunnels and is accessible from both main tunnels by 270 cross passages spaced every 1,230 feet.

The layout of the Channel Tunnel is unique in that it incorporates piston relief ducts in the wall separating the two rail tunnels. To relieve air pressure ahead of the trains, which run at nearly 90 mph, piston relief ducts were bored every 820 feet between the tunnels. Each 10.8-foot-diameter piston relief duct is fitted with a damper to allow it to be closed. Both the steel doors of the cross passages and the dampers can be closed automatically.

To keep smoke out of the Running Tunnel North, the piston relief ducts were ordered closed, but one damper failed to work, allowing smoke to pour in. Smoke also poured into the Running Tunnel North through a huge crossover on the French end of the tunnel, which allows trains to turn around beneath the Channel.

As the train continued its dash through the tunnel with its tail literally on fire, an extremely worrisome warning light informed the engineer that a stabilizing bar on one of the HGV wagons was in the down position. Since this could cause a derailment, the engineer brought the train to a stop 11.5 miles inside the tunnel. As he did so, the chief of the train opened a door of the club car and was greeted by a blast of dense smoke and intense heat. Although he slammed the door immediately, the passengers had to fall to the floor to obtain breathable air.

Within minutes, French members of the First Line of Response team were on their way through the service tunnel and had begun helping the passengers and crew members into one of the cross passages, where they could seek safety in the service tunnel. By this time other truck-loaded rail wagons had caught fire, creating an even larger fire.

French fire fighters were the first to arrive on the scene since the train was still in the French portion of the tunnel, and they immediately called for assistance from British fire fighters already alerted to wait for the burning train to arrive in Folkestone. However, when the burning train did not arrive, the British fire fighters took it upon themselves to head into the service tunnel to determine what had happened. They sized up the situation and began to attack the blaze at 11 P.M., but the 36-inch ledge next to the train and the excruciating heat from the fire hampered their efforts.

Although those aboard the train were vomiting and choking from the fumes and smoke, the poisonous air overcame only seven, one of them a pregnant woman. These incapacitated passengers were treated at the scene while the others were placed aboard a passenger train waiting in the Running Tunnel North to serve as an evacuation vehicle.

The fire, out of control, triggered explosions as the truck cargoes, including one loaded with highly flammable lard, began to detonate. Working in shifts because of the hellish conditions, teams of fire fighters breathing compressed air could endure only eight minutes in the tunnel before replacements had to relieve them. Eventually, 442 fire fighters were working to save the tunnel.

Fighting the fire in shifts, the fire fighters initially had little effect on the inferno since the confined space next to the train restricted the number of emergency personnel who could approach the burning cars. There was also a delay in obtaining adequate water pressure since the heat had warped and caused leakage in the tunnel's water mains. The subsequent arrival of a Eurotunnel engineer, who rerouted the water system, significantly helped relieve the water pressure problem. Another delay in battling the blaze occurred because no one notified a second wave of badly needed British fire-fighting personnel until more than an hour after the initial alarm went out.

The service tunnel, something rarely seen in railway tunnels, was a haven not only for those evacuated from the train but also for the fire fighters recuperating until it was again their turn to struggle against the conflagration. Technicians increased the air pressure, a technique known as overpressure, within the smaller service tunnel so it was greater than that of Running Tunnel South, to block smoke infiltration. This even presented a small hazard since opening the door to the cross passage released such a rush of air that it threatened to knock any unsuspecting rescue workers down or at least blow their equipment out of their hands.

The amount of air circulating through the Running Tunnel South in the direction of the French side was increased to give fire fighters better visibility, remove heat, and detoxify the air. As the explosions and heat took their toll on the concrete lining overhead, the fire fighters soon had to contend with chunks of concrete falling on them from above and cluttering the narrow walkway next to the train.

As more fire fighters arrived to join the struggle, the blaze seemed to have a life of its own. The truck carrying tons of lard burned so hotly it was first thought to be the 25 tons of highly flammable plastic polystyrene. It was later deduced that the heat in the vicinity of that truck rose to more than 2,300 degrees Fahrenheit. The battle to control the blaze continued until it was finally extinguished at 11 A.M. the next day, just over 14 hours after it began.

Damage to the tunnel included serious spalling to 787 feet of concrete lining, of which 557 feet was on the French half of the south tunnel. In the 15-inch lining segments directly above the trucks that burned, only eight tenths of an inch of concrete remained. Another 1,246 feet of tunnel lining suffered light pitting. Lighting fixtures, the chilled water-cooling system, electrical components, and piston relief duct doors were badly damaged, requiring replacement. The damage to the tunnel was estimated at around $90 million and took six months to repair, although full service was resumed in both tunnels by January 1997.

In the aftermath of the fire it was noted that the open design of the HGV wagons did nothing to contain the fire, and criticisms were leveled at the French and British rail companies for not using enclosed rail cars. An inquiry by the Health and Safety Executive, Britain's version of America's Occupational Safety and Health Administration, made 36 recommendations for safety improvements in the tunnel, including the modernization of fire sensors on the train to obtain earlier warnings. The systems in place were unable to "read" the infrared or ultraviolet signature of the fire before the train entered the tunnel. Eurotunnel also began experimenting with infrared fire detection systems as well as a sprinkler system that could be carried aboard the train to extinguish fires automatically.

French authorities revealed two years after the event that the fire was likely caused by arson. No one was ever charged in the blaze.

See also EUROTUNNEL; VENTILATION.

Châtellerault Bridge (built 1899–1900) Attributed to a design by the reinforced concrete pioneer François Hennebique, the Châtellerault Bridge over France's Vienne River is one of the first long-span, reinforced concrete bridges and certainly the longest built, or at least started, during the 19th century.

Although some bridges in Europe had been built of concrete, most were constructed of imitation stone cast from concrete. Concrete reinforced with iron and later steel allowed Hennebique and others to design structurally sound bridges using homogenous arches.

Work on the three-arch bridge was begun in August 1899 in the city of Châtellerault in eastern France, with the first job the sinking of two submerged piers in the Vienne River. After their completion, falseworks of timber were erected to support construction of the three arches that would span the river and end at abutments on each riverbank. By May 1900, the bridge's reinforced concrete arches were completed and the bridge was subjected to load testing. Its roadway was a fairly thin design for the

time and hinted at how thin and light concrete construction could be, as well as how attractive.

The total length of the bridge is 472 feet with the center span measuring a respectable 164 feet and the two arch-supported side spans measuring 131 feet apiece. The approach spans make up the remaining 46 feet of the bridge's length. The 26-foot-wide roadway of the Châtellerault Bridge is connected to the arch by vertical concrete supports.

The bridge is attributed to Hennebique, as are a number of bridges built in Europe and elsewhere around the world. Hennebique, originally a stonemason and later a general contractor, developed his own system of reinforced concrete design and licensed the use of his renowned system. He maintained a design bureau in Paris that churned out thousands of designs for all types of structures, but many designs may have originated elsewhere. A pair of other notable bridges attributed to the creativity of Hennebique includes the Liège Bridge in Belgium (targeted for destruction by the U.S. Air Force in World War II) and the Risorgimento Bridge in Rome.

The civil engineering professor and author David P. Billington has pointed out that no two of the three bridges share a common design element except that all are constructed of reinforced concrete, indicating that all may have been designed by different engineering teams. However, Châtellerault Bridge seems to have been Hennebique's design and is one of the most important early examples of reinforced concrete bridges.

A major flaw in the bridge is the absence of hinges at its abutments and the crown of the arch. Hennebique ridiculed the use of hinges, but other designers, most notably Robert Maillart, considered them essential. The hinges, which allow for movements in the bridge's concrete components, prevent damage by cracking. The Châtellerault Bridge suffered from this type of damage but continues to stand over the Vienne River.

See also CONCRETE; HENNEBIQUE, FRANÇOIS; MAILLART, ROBERT; MONIER, JOSEPH; REINFORCED CONCRETE.

check-in/check-out procedure The terrible fact about tunneling and mining is that not every worker who checks in checks out. A longtime tradition in mining, particularly in coal mining, the check-in/check-out procedure is designed to account for all miners who enter and depart an underground excavation. Although a relatively routine procedure, it becomes extremely important in case of an explosion or a collapse since rescuers must know the identities and the number of persons still below. The traditional method for this procedure has been for miners to pick up their own tag bearing a code number on entering a mine and then return it after their shift or, in the case of an emergency, evacuation.

Federal safety regulations now mandate this procedure for all underground excavations that are in such a partial state of completion that the hazard of explosion or other mishap still exists. Since coal and other mines generally have an explosion or collapse hazard, the check-in/check-out procedure always remains in place. Underground facilities such as subway or traffic tunnels that have been completed to such a degree that collapse or fire has been eliminated as a hazard can halt the procedure.

Chesapeake Bay Bridge–Tunnel *See* combination bridges.

Chester Bridge collapse *See* Stephenson, Robert.

Cheyenne Mountain Complex *See* rock bolt.

Cincinnati-Covington Bridge (built 1857–1867) *Predecessor of the Brooklyn Bridge* John Augustus Roebling's next great success after his construction of the Niagara Gorge Suspension Bridge would be the Covington & Cincinnati Bridge, commonly referred to as the Cincinnati-Covington Bridge. This span across the Ohio River would take nearly 10 years to complete, interrupted as it was by technical difficulties, money woes, and the Civil War.

With his reputation buffed to a high shine by the completion of his Niagara Gorge Suspension Bridge in 1855, Roebling was no longer a hat-in-hand engineer scratching about for projects against better known rivals such as Charles Ellet.

Ellet's reputation as a bridge builder, tarnished by the Wheeling collapse, waned as he moved on to other types of projects. When the Covington & Cincinnati Bridge Company decided in 1856 that it wanted to build a span across the Ohio River between Kentucky and Ohio, Roebling was invited to submit a proposal. Roebling's design was soon selected, and by September 1856, men were digging the foundations for the bridge piers.

The span was Roebling's most mature design up to that time and would possess the longest unsup-

ported main span of the day with a length of 1,057 feet. Digging the foundations using open cofferdams just off the Kentucky shore went smoothly, but workers began encountering a heavy inflow of groundwater in the excavation on the Cincinnati side.

The water was thick with silt and created problems for the steam-driven pumping equipment of the day. Roebling, facing possible defeat on the project, sketched plans for four massive pumps of his own design, and in two days, carpenters constructed the box-shaped pumps out of timbers. The resourceful engineer then decided to power the pumps by using a chain turned by the engines of a river tugboat. The makeshift pumps accomplished their task, and the Cincinnati foundation work was completed, allowing stonemasons to begin constructing the piers.

The harsh winter of 1856–1857 and heavy spring rains raised the river level so high it was impossible to resume work quickly. A bad economy also stunted the payment of funds pledged by the stockholders for bridge construction. Attempts to sell additional stock in New York went nowhere, and in 1857 financial panic forced the closure of banks.

During 1857, work on the bridge slowed and then stopped as a result of bad weather and a lack of funds. Poor financial conditions meant only minimal work was accomplished in 1858. In 1859, no work at all was done on the piers. Although new funds were raised in 1860, the election that year of Abraham Lincoln ushered in the eventual secession of 11 Confederate states. By the following year war had broken out between the North and the South, halting work on the bridge for another three years. The year the war started, Roebling's son and assistant on the project, the engineer Washington Roebling, joined the Union army.

Although a military pontoon bridge was assembled between Covington and Cincinnati near Roebling's uncompleted suspension span, wartime transportation needs made a permanent suspension bridge even more attractive to investors. For this reason, the enterprise once again became solvent. Work began anew on the bridge in 1863, but a wartime shortage of skilled workers coupled with labor unrest made the going difficult. The tough-as-nails Roebling, keenly aware of the sacrifices of those fighting in the war, greeted a strike by bridge workers for higher wages with the proclamation that he would prefer hard-working German immigrants.

Washington's return from military duty in the spring of 1865 provided Roebling with an able engineering assistant on the Cincinnati-Covington Bridge work, which was once again in full swing. When completed in 1867 it boasted the longest continuous span of any suspension bridge in the world. It was a marvel of strength; with its deck possessing a hefty truss railing. Diagonal cable stays to reduce movement of the deck caused by winds gave the bridge additional stability.

The bridge was officially opened on January 1, 1867, an event marked by a parade and a massive party. More than 50,000 people crossed the span on the day of its opening. By July 1867 all the fine details on the bridge were completed. Perhaps the most remarkable fact related to the Cincinnati-Covington Bridge is that during the 10 years it took to build the bridge, only two workers had perished as a result of accidents.

John Roebling's next great challenge after the completion of the Cincinnati-Covington Bridge was construction of the Brooklyn Bridge. The experience Roebling gained building the record-breaking suspension span over the Ohio River was undoubtedly instrumental in helping him plan the bridge across New York's East River.

Ohio and Kentucky did not forget Roebling and the masterful Cincinnati-Covington Bridge that he designed. In 1984, in recognition of his genius, the span, which remains in use today, was officially renamed the Roebling Suspension Bridge, 115 years after the engineer's death.

See also NIAGARA GORGE SUSPENSION BRIDGE; ROEBLING, JOHN AUGUSTUS; ROEBLING, WASHINGTON AUGUSTUS.

City Water Tunnel No. 3 (begun 1970–2020?) *New York City's greatest tunnel* Although most think of tunnels as conduits for traffic, other tunnels large and small perform yeoman service transporting drinking water into cities or carrying away rainwater and sewage. Of all the tunnels running beneath New York City, the water tunnel project known as City Water Tunnel No. 3 is the grandest. It is a construction effort of such magnitude that in many ways it dwarfs what was required to build the Eurotunnel beneath the English Channel. The project began in 1970 and was expected to take 50 years.

It is not because New York lacks tunnels for carrying fresh water to the city that the 60-mile tunnel is being built, but because two other water tunnels completed in 1917 and 1937 cannot be closed for maintenance until a third exists. When the newest tunnel is finally completed, Water Tunnel No. 3 will be the costliest infrastructure project ever undertaken by New York City, expected to total $6 billion.

By 1830, New York's population had ballooned to 242,000 and the city was in desperate need of a pure and reliable water supply. To slake this thirst for fresh water, the City of New York in 1837 began construction of the first of a series of reservoirs north of Manhattan. Eventually these reservoirs would be connected to the city with a system of tunnels. City engineers have kept their fingers crossed for decades because they know the two original tunnels, known as City Water Tunnel No. 1 and City Water Tunnel No. 2, are in dire need of maintenance. Because a sudden collapse of one of the tunnels could hamper city water supplies, a third tunnel must be built and put into service before the other two can be taken out of service.

Although the necessity of a third city water tunnel seemed obvious to engineers since the 1950s, politics, contracting disputes, and New York City's fiscal crisis in the late 1970s not only delayed the project but nearly doomed it. Although city engineers urged that the work begin during the early 1960s, digging did not start until 1970. New York's sandhogs began blasting their way from Hillview Reservoir toward the city through the extremely hard gneiss, a granitelike material that lies deep beneath New York.

City Water Tunnel No. 3 would be built in four stages, with the tunnel's segments running as deep as 800 feet beneath the surface. Projecting southward, the tunnel passes beneath the Bronx, jogs west, and then turns south again under Manhattan. After passing beneath Central Park the tunnel takes an eastward turn through solid rock to pass beneath the East River and into Astoria, Queens. During construction of Stage One of the work, sandhogs used the traditional means of blasting. Although the interior of the water tunnel begins at 24 feet, it eventually necks down to 20 feet.

A total of four valve chambers have been constructed along the route of the tunnel with the largest beneath Van Cortlandt Park in the Bronx. The chambers can be accessed from above so the water tunnel will not have to be taken out of service for inspection and maintenance. The valve chambers will allow operators to control the flow of water along various sections of City Water Tunnel No. 3. A series of vertical shafts will run from the tunnel to the surface so water can be distributed to homes and businesses. Each of these shafts is a major tunneling project requiring excavation, mucking out of spoil, and reinforcement with a concrete lining.

Although the drill and blast method was used in the original two water tunnels leading into New York City, a tunnel-boring machine (TBM) is being utilized on the project to double the rate of tunneling to 50 feet a day on the remaining four stages of the project. The tunnel-boring machine creates a precise and smooth bore that requires less grouting behind the concrete lining. The 610-ton TBM has a circular cutting head, allowing it to grind a 23-foot-diameter bore.

By the time the tunnel is completed it will meander for 60 miles beneath the various sections of New York City. The Eurotunnel, by comparison, is a twin-bore tunnel with a smaller service tunnel that stretches only 24 miles. Even if the Eurotunnel's two main 25-foot-diameter rail tunnels were placed end to end, they would measure only 48 miles. In addition, City Water Tunnel No. 3 has taken longer to excavate because the rock beneath the city is far harder than the chalk through which the Eurotunnel was dug.

Since great tunneling projects are fraught with danger and massive expense, City Water Tunnel No. 3 was to be surrounded by controversy, financial crisis, and danger as a matter of course. A variety of accidents between 1970 and 2000 claimed the lives of 24 workers, including those of the sandhogs of New York's famed Local 147, which traces its roots to the Irish and African-American workers who excavated the foundations of the Brooklyn Bridge.

The death of the sandhog Anthony Oddo is typical of the sudden and horrible mishaps that can claim lives and maim workers during tunneling. On the day before Thanksgiving in 1993, Oddo and his fellow workers were at the bottom of a 525-foot vertical shaft leading to the water tunnel when a falling winch banged along the sides of the shaft on its way to the bottom. The winch struck Oddo, killing him while seriously injuring three of his fellow workers.

Another sandhog, Joe Shubar, was in a similar access shaft in 1980 tethered to a platform with a safety harness when the platform fell more than 500 feet before it halted 40 feet from the bottom. Amazed to be alive, Shubar then felt the scaffolding shudder before it fell the remaining 40 feet to the floor of the main water tunnel. He survived the fall virtually unscathed but told *People* magazine in 1984, "To this day I wake up with cold sweats."

Along with physical hazards, City Water Tunnel No. 3 has its share of controversy, much of it related to money. After it had been under construction for nearly four years, New York City's fiscal crisis threatened to throw the metropolis into bankruptcy, pitting the tunneling contractors and the city against one another. The contractors responsible for ensuring the structural soundness of the tunnel bitterly complained that the city did not plan for enough massive and expensive iron reinforcing rings in the tunnel's design. The city steadfastly refused to increase payments so the reinforcing rings could be purchased, and work halted.

The debate grew so hot that it stalled work on the tunnel for nearly two years. When work did resume in 1977, money was in such short supply that it was not even a full effort, as some politicians fought to close down the tunneling project entirely. Full-scale work on the massive project did not resume until 1983. The sandhogs of tunnel workers union Local 147 actually played a role in forcing the tunnel project back on track by suing the city and the contractors for $10 million in damages because they were thrown out of work by the dispute.

On August 13, 1998, New York mayor, Rudolph W. Giuliani, officially activated the first 13-mile section of City Water Tunnel No. 3, known as Stage One. Stage Two, to be burrowed from Brooklyn to Queens, will cost an additional $1.5 billion. The Kensico Reservoir and valve chamber beneath Van Cortlandt Park will be connected in Stage Three, costing $1.5 billion. The final segment, Stage Four, will carry the water tunnel from the Bronx to Flushing, Queens, at an estimated cost of $1.5 billion. Although portions of the tunnel will become operational as they are completed, the entire project is not slated for completion until 2020.

On October 31, 2000, officials announced plans to build a memorial to the 24 sandhogs, city employees, and other workers who died during construction of the tunnel. The memorial will eventually stand near Van Cortlandt Park and bear the names of those who died working on the tunnel, and those who will undoubtedly die before its completion will be added. Anthony Oddo's name will be on that memorial.

See also EUROTUNNEL; SANDHOGS; TUNNEL-BORING MACHINE.

Clifton Bridge (built 1836–1864) *Daring design across England's Avon Gorge* Illness and a work-related leg injury that forced a young engineer into a long recuperation gave him time to design one of the most beautiful and interesting bridges in the world across southwest England's Avon Gorge. Isambard Kingdom Brunel, the brilliant 19th-century engineer, had just spent nearly three years working on history's first underwater tunnel—the Thames Tunnel—but flooding in the tunnel that killed six workers and nearly drowned him had taken a toll on his health.

He was sent from his family's home in Brighton to Clifton near Bristol on England's southwestern coast. In the days when medical science was virtually neither medical nor science, it was often believed that climate played a major role in health and recuperation. Packed off by his father, Marc Brunel, designer of the Thames Tunnel, Isambard found himself on an accidental rendezvous with the engineering art of bridge design.

The recuperating Brunel decided to explore the area and prowl the banks of the Avon River, which had cut a gorge through the limestone cliffs on each side of the Avon Gorge. The river fed into the wide estuary of the River Severn, which connected Bristol with the Atlantic, making it a port city. Though a spectacular sight, the deep gorge was an immense obstacle to travel.

More than 75 years before Brunel arrived in Clifton to scramble about the Leigh Woods near Avon Gorge, a Bristol wine merchant, William Vick, took the first step to ensure that a bridge would be built there. In 1753, Vick bequeathed an

Clifton Bridge, as it stands today above the Avon Gorge, was the first bridge designed by the mercurial British engineer Isambard Kingdom Brunel and one that was built after his death. Brunel designed the eyebar chain suspension bridge while recuperating from injuries suffered while working in the Thames Tunnel. It was not built until nearly 40 years after it was designed. *(Godden Collection, Earthquake Engineering Library, University of California, Berkeley)*

endowment of 1,000 English pounds that was to finance a bridge across the Avon Gorge when it had grown to 10,000 pounds.

In 1765, an engineer with the ironic name of William Bridges proposed to span the Avon Gorge with the grandest stone arch bridge ever seen. Its supporting towers would be so massive they would contain dwellings, shops, businesses, schools, and a library. Vick's endowment was little more than its original amount at the time and Bridges's grand concept was doomed.

By 1829, Vick's endowment had swelled to a respectable 8,000 pounds, and land was purchased on each side of the gorge for the approaches to the anticipated bridge. A call was issued for designs, and the vacationing Brunel, who had also assisted his father in the design of the Ile de Bourbon bridges for France, decided he could create a bridge capable of spanning Avon Gorge. The center span of the bridge proposed by Brunel would measure 700 feet, a dizzying distance in 1829 but one the 23-year-old Brunel believed was reasonable. The great distance meant the bridge would have to be of the suspension type. Its center span would be suspended by massive iron links anchored into stone on the walls of the gorge.

Brunel, an accomplished artist and draftsman, produced precise engineering drawings of the bridge along with beautiful paintings of how the completed span would look. To help him design his bridge, Brunel traveled to the chain suspension bridge over the Menai Straits. Known as the Menai Bridge, this magnificent and innovative span crosses 579 feet of water, linking the Isle of Anglesey and North Wales. Its success ignited serious interest in suspension bridges; it was designed by the brilliant Scottish engineer Thomas Telford and built between 1818–1826. Brunel had no way of foreseeing that the septuagenarian engineer would nearly choke the life out of his proposal.

Brunel cranked out four bridge designs, with the proposed central spans of his candidates ranging from 870 feet to 916 feet. Expensive and tall suspension towers over which the iron chains would be draped were part of two designs, and in a third, tunnels would be required to funnel traffic to approaches on both sides of the bridge. His fourth design and most sensible in terms of strength, economy, and appearance anchored two short towers on each wall of Avon Gorge. The iron suspension chains draped over these towers would curve down-

ward toward the road deck's center and then up again. The chain cables would be connected to the road deck by iron suspender rods.

Brunel's economical design would have a more modest profile, gained by using relatively squat towers built into the opposing faces of the gorge. Although the length of the span between the towers would eventually be 702 feet, the towers would need to be only 85 feet tall. These towers would be far cheaper to build and were only one third the height of Telford's.

Although the towers could be placed closer together on piers in the Avon River to reduce the center span of the suspension bridge, doing so would require them to rise nearly 250 feet to meet the roadway. Brunel's plan for the longest suspension span up to that time was too bold for Telford, who declared that no suspension span could safely exceed 600 feet. Telford was obviously influenced by the fact that his Menai Bridge's center span was just shy of 580 feet. It was a daring design by the youthful Brunel, who, in his mid-20s, was promoting the world's longest suspension bridge at a time when confidence in suspension bridges had been shaken by their frequent collapses.

Nonetheless, Brunel was completely confident of his design and worked hard to build faith in it among doubters. Brunel believed his short tower design would be strong enough to endure, even though the greater length of the suspended road deck would be more vulnerable to destructive wind forces.

When Brunel and the other prospective bridge designers submitted their plans for review, they eventually found that one of the competitors was the august Telford, who was also screening the proposals for the Bridge Committee! To no one's surprise, Telford found hopeless flaws in all the proposals except his own, which included a pair of towers hundreds of feet tall anchored into the bedrock of the riverbanks. These massive towers reduced the length of the central span and conformed to his conservative belief that wind would destroy any suspension span exceeding 600 feet. Telford's design, though, was considered an engineering and aesthetic aberration. It would be ridiculously expensive, unnecessarily ugly, and tough to build.

Cowed by Telford's reputation and the success of his stunning Menai Bridge, the Bridge Committee asked engineers to submit a second round of proposals to be reviewed by John Seward, a builder of ship engines, along with Davies Gilbert, past president of the Royal Society of London. Designs were resubmitted and Telford once again turned in his original proposal, which remained popular only with him.

Brunel, bowing to the nervousness of the committee, who believed a larger abutment was needed to shorten the length of the bridge, caved in to the committee's worries. He extended one of the abutments so the bridge's main span would measure only 630 feet. (However, when completed 35 years later, the span would measure 702 feet.) Of the 12 designs submitted, only four were selected as finalists, and Telford's monstrosity was not among them. Brunel's design was selected in 1831, but his joy was short-lived when the Bridge Committee was able to raise only 31,000 pounds of the 45,000 needed for its construction.

What put the project in the deep freeze for four years was nothing short of the worst civil unrest in England since the Middle Ages, and, as luck would have it, massive rioting broke out in Bristol near where Brunel was attempting to interest investors in the bridge. The horrific event would be one of the bloodiest chapters in England's modern history, and Brunel, who wanted only to build his bridge, would be thrown into the center of this maelstrom.

While Brunel was in the Bristol area to encourage investment in the bridge, rioting flared on October 29, 1831, when the House of Lords refused to expand voting rights in England. Brunel found himself with Bristol's mayor, Charles Pinney, as the official tried to restore order. Public buildings and 100 homes were burned during a three-day rampage. When cavalry arrived, they hacked hundreds to death in a bloody putdown of the riot. Brunel later testified on behalf of Bristol's mayor, who was accused of failing to do his duty during the riot. Brunel's bridge was temporarily forgotten as Bristol went about the task of rebuilding its city and burying its dead.

It was not until 1836 that interest in the bridge was revived and enough money raised to begin construction. On August 27 of that year, work began on one of the abutments, during which Brunel pluckily made the first crossing of the Avon Gorge by rolling across in a basket suspended by a 1,000-foot iron rod. Work on the bridge continued intermittently until money problems again stalled its progress. By that time Brunel had moved on to other projects and other bridges. For a long period, the completed abutments stood with no span between

them. At the time of Brunel's death on September 15, 1859, the troubled Clifton Bridge stood unfinished. A year after his death, the Institution of Civil Engineers established a corporation that raised additional funding and pushed through the completion of the bridge.

Unfortunately, economic realities dictated that Brunel's creation be modified to make it less expensive. The roadway was narrowed five feet, and Brunel's beautiful Egyptian-motif suspension towers—minus guardian Sphinxes he envisioned on their peaks—were deleted along with cast-iron bas-reliefs that were to depict scenes from the bridge's construction. As a result of these changes, Brunel's first bridge would not completely match its creator's vision, but the result is beautiful nonetheless. To keep the costs down, recycled parts scrounged from a demolished bridge were used to complete the Clifton Bridge. The dismantling of the Hungerford Bridge provided secondhand cast iron chains to replace the ones designed by Brunel. The original Clifton Bridge chains had been sold for the construction of another Brunel-designed span, the Royal Albert Bridge at Saltash. A final irony is that the Hungerford Bridge, whose remnant chains were destined for the Clifton Bridge, was also designed by Brunel.

Brunel's first bridge—which took nearly 35 years to complete—was opened December 18, 1864. The Clifton Bridge still stands over the imposing Avon Gorge and fits in well with the terrain. It has been refurbished and strengthened several times. Originally designed for pedestrians and carriages, it still supports automobile traffic. Its basic design proved to be nearly flawless, and its 702-foot span has never presented serious problems despite the criticism of Thomas Telford. Telford's famous Menai Bridge, on the other hand—a far shorter span—was smashed twice by high winds during the 1800s.

See also BRUNEL, ISAMBARD KINGDOM; BRUNEL, MARC; SUSPENSION BRIDGE; TELFORD, THOMAS.

coal dust   Coal is a carbonaceous and flammable rock formed from decayed plant matter that was compressed and heated by eons of geologic force. The blasting and digging in mines have always raised the dust, but the problem was made far worse with the 20th-century appearance of digging machines with pick-equipped, rotating drums.

Those who mine coal will produce coal dust, and if this dust is not controlled, it will kill. Coal

The bodies of coal miners killed in a coal dust explosion during the early 1900s are drawn by wagon to a cemetery. Outrageously poor safety conditions caused thousands of deaths in American coal mines during the 19th and 20th centuries. Miners not killed outright in mine explosions often fell victim to gases created by the blast and subsequent fires. *(Bureau of Mines/National Archives and Records Administration)*

miners will die of the lingering effects of black lung disease or in an explosion should the coal dust be ignited. Water can be used to prevent the coal dust from becoming airborne, and ventilation systems that provide fresh air can reduce coal dust particles.

Before miners and mine owners paid much attention to the respiratory dangers posed by coal dust, miners recognized the explosion hazard posed by the dust. If allowed to remain dry, the dust becomes airborne and mixes with the air, creating an oxygen and coal-particle mixture. This mixture allows coal dust to become an explosive powder capable of releasing a stupendous amount of energy when ignited.

Although the link between coal dust and black lung disease seems obvious today, it was not fully accepted until the mid-1960s. In 1965, miners began protesting their exposure to coal dust and the resul-

tant sickness it caused. In 1969, the federal government finally ordered the regulation of airborne coal dust levels in mines. However, this regulation has not been a complete solution to the problem.

Although procedures have greatly reduced the likelihood of a coal dust explosion, black lung continues to be a health problem among coal miners. During a series of federally sponsored meetings in 1995, some miners told government officials that they had never seen an honest coal dust measurement taken. Their claims were given validity by the fact that coal miners continue to fall ill with black lung disease, or pneumoconiosis. In addition, 150 mine operators and contractors pleaded or were found guilty of submitting fraudulent coal dust samples to the Mine Safety Health Administration.

Black lung disease is caused by exposure to coal dust, one component of which is silica, which scars the lungs, leading not only to pneumoconiosis and emphysema but also to a host of other serious lung diseases. As explosions and collapses have been reduced in the mines, black lung disease has actually replaced accidents as a leading cause of coal miner deaths. In 1997, only 30 miners died in American coal mine accidents, whereas in 1994 the number of former miners who died of black lung disease totaled 1,478.

See also HAWK'S NEST TUNNEL; VENTILATION.

cofferdam One of the simplest tools in bridge and tunnel construction, and yet one of the most valuable, is the cofferdam. This is an enclosure that is built in place or lowered into the water and pumped dry to allow the construction of bridge foundations or anything else beneath the surface. Its use goes back thousands of years. Roman engineers often used them when constructing foundations for bridges.

Like many construction terms, *cofferdam* describes not only its form, but also its function, as *coffer* means "box." Cofferdams can be constructed of timber pilings or heavy sheets of steel pounded into the bed of a lake or bay, or they can be prefabricated on land and lowered into position by a crane. Once in place they can be pumped free of water. Some cofferdams, particularly those made of wood, sometimes need caulking to make them watertight. Often, when they have served their purpose, cofferdams are hoisted away or disassembled.

Although caissons and cofferdams in their simplest forms are similar, modern cofferdams often are built in place by the driving of sheet metal pilings, whereas caissons are traditionally assembled elsewhere and floated into position.

Cofferdams are fantastic sights to see since they provide an inverted island of open air in the center of a body of water. In this dry enclosure surrounded by millions of tons of water, workers can prepare foundations for bridges. Although cofferdams can be square, like the prefabricated ones used during construction of piers for the Florida East Coast Railway extension through the Florida Keys, they are often circular because of the inherent strength of a cylinder to resist pressure.

However, cofferdams have their limitations. Although some excavation work can be performed inside them, the fact that their walls are usually driven into the muck means that any large-scale excavation will undermine them and drown everyone inside. Also, the depth at which cofferdams can be utilized is limited because it would be prohibitively costly to built one strong enough to reach more than a few dozen feet into the water.

Cofferdams continue to play a crucial role in major construction projects. In the huge Boston Central Artery/Tunnel Project, what may be the world's deepest cofferdam was built of reinforced concrete. The 250-foot diameter structure was 80 feet deep and was designed to house a ventilation building for the Ted Williams Tunnel running 90 feet beneath Boston Harbor. Unlike most cofferdams, the one that held back the waters of Boston Harbor is permanent. It formed a housing for a ventilation building for the Ted Williams Tunnel that emerges in South Boston. Its cost, size, and depth make it highly unusual.

See also CAISSON; CENTRAL ARTERY/TUNNEL PROJECT.

Colossus (built 1812) *America's miraculous wooden arch bridge* Considered an amazing engineering feat, the Upper Ferry Bridge that was popularly known as "Colossus" was believed to be the longest timber bridge in the United States at the time of its 1812 construction, with a clear span of 340 feet. The bridge was the work of the German immigrant Louis Wernwag, a brilliant timber bridge designer who engineered his bridges by rule of

Louis Wernwag's amazing timber bridge soaring across Pennsylvania's Schuylkill River. At the time of its construction in 1812, it was believed to be the longest timber arch bridge in America, boasting a 340-foot span. So famous was the bridge that it was something of a tourist attraction until destroyed by fire in 1838. *(Author's Collection)*

thumb. The bridge was essential for conducting traffic over the Schuylkill River at Philadelphia.

The bridge was a wooden arch bridge and was a wonder to behold. The graceful bridge rose 20 feet above the Schuylkill and was equipped with a roof and sidewalls to protect it from the elements. So confident was Wernwag of his newly completed span that he tested its structural soundness by having 16 horses pull a wagon loaded with 16 tons of stone across the bridge. Like most 19th-century bridge builders working in a time when large spans generated large doubts in a wary public, Wernwag understood the importance of advertising the strength of his bridge.

The bridge burned during the summer of 1838, when it fell victim to a fate common to many wooden bridges. After the disastrous fire the owners of the "Colossus" decided against rebuilding the span. When others raised money for the task, a suspension bridge design by the American engineer Charles Ellet, Jr., was selected and built in 1839.

See also ELLET, CHARLES, JR.; INSPIRED CARPENTERS; TIMBER BRIDGES.

combination bridges   Although it at first seems an odd thing to do, the construction of water crossings sometimes requires engineers to design either a combination of bridge types to span the distance or to combine bridge and tunnel projects to complete the crossing.

Although combination bridges are not unusual, one of the first great combination bridge projects of the 20th century was the San Francisco–Oakland Bay Bridge across San Francisco Bay completed in 1936. This required four types of bridges: a pier-supported viaduct, a series of through-truss spans, a cantilever truss bridge, and a pair of end-to-end suspension spans. Miners tunneled through Yerba Buena Island in the middle of San Francisco Bay to provide a link between the cantilever span and the suspension bridges.

Another grand project which crosses a total of 17.6 miles of water, is the Chesapeake Bay Bridge-Tunnel, which connects the Virginia Beach/Norfolk, Virginia, area with the state's Eastern Shore. The waterway is sailed by some of America's mightiest warships entering and leaving the historic Norfolk Navy Base.

It was decided that a pair of subaqueous tunnels that would carry traffic beneath the navigation channel should link the bridge system. Cars cross the bay by bridge, descend beneath the water for a mile, and emerge again to resume their journey by bridge.

The majority of the crossing is accomplished by 12 miles of viaduct and two miles of causeway. The complex is also composed of four artificial islands, two of which serve as portals for the tunnels. Built between 1960 and 1964, the effort was a monumental construction project that was fully funded by tolls. In 1999, an additional bridge was constructed parallel to the original system of spans at a cost of $199 million. No additional tunnels were built.

One of the world's newest and most impressive combination bridge and tunnel crossings is Japan's Aqualine Tunnel, which provides commuters a handy but expensive escape route from Tokyo. The system is composed of a 5.9-mile tunnel and a bridge stretching 2.7 miles. Running from Kawasaki to Kisarazu under Tokyo Bay, the system opened December 18, 1997. The bridge is a viaduct running from Kisarazu to the tunnel portal on a man-made island, and the tunnel continues to Kawasaki as deep as 197 feet beneath the seabed. The tunnel, driven by a massive 46-foot-diameter tunneling machine, took six years to dig.

Like the Chesapeake Bay Bridge–Tunnel, the Aqualine Tunnel system depends upon artificial islands in the middle of the bay where cars descend beneath the water for the subaqueous portion of their trip. The $12 billion Aqualine system is a marvel, but its $32 toll has held its daily use to around 14,000 cars per day; the tunnel needs about 25,000 vehicle crossings to break even.

Official web sites for the Bay Bridge, the Chesapeake crossing, and the Aqualine project can be found at http://www.dot.ca.gov/dist4/eastspans/index.html, http://www.cbbt.com/, and http://www.aqua-line.com/eng/index.html.

See also SAN FRANCISCO–OAKLAND BAY BRIDGE.

compression and tension   *See* prestressed and post-tensioned concrete.

concrete   *Ancient and future building material* Concrete, with modern refinements, has become one of the premier bridge and tunnel building materials on Earth. Not only is it the most widely used building material for large structures, but it is also one of the oldest materials, used for at least 2,300 years.

Although its gray presence in everything from sidewalks to lawn birdbaths tends to make concrete appear simple, cheap, and common (as it is), its compressive strength—an ability to stand up to large forces—has made it essential to modern civil engineering. The material has made modern building construction possible, allowed safer tunneling techniques, and provided a means of building beautiful bridges.

Concrete is an amalgam of materials usually consisting of aggregate (small rock and sand) and the component that binds everything together, cement. The manufactured cement used in concrete is itself the result of a combination of materials transformed by extremely high temperatures. The first cement used was found in natural deposits of calcerous material (limestone or chalk). This natural cement, when mixed with water, begins forming millions of microscopic crystals in a process known as hydration that causes it to set within 20 minutes to an hour and continue to gain strength for several weeks. The bonding action not only connects the cement to itself, but also locks together any aggregate materials such as sand and gravel that have been mixed with it.

The history of concrete stretches back thousands of years, as the ancient Greeks and Egyptians discovered and used it, but one of the most important beneficiaries of the substance was the Roman Empire. By 300 B.C., the Romans, the greatest builders in the ancient world, were using pozzolanic concrete, which consisted of crumbled lime mixed with volcanic ash from the village of Pozzuoli near Naples, Italy. Roman concrete was used not only in portions of the Colosseum (A.D. 82) but in bridge piers and concrete pipe.

In one of the great engineering oddities of all time, the world's most successful construction material suddenly disappeared from history for more than 1,000 years. The disintegration of the Roman Empire by the end of A.D. 400 also meant the disappearance of the art of producing and using concrete. Although stonemasons used a weaker mortar for binding bricks for the next 1,300 years, high-quality concrete was just another mystery of the ancient world.

The discovery of naturally occurring cement on England's Isle of Sheppey in 1796 generated a new enthusiasm for the material and spurred efforts to produce it artificially. By 1824, an English bricklayer, Joseph Aspdin, had patented his own manufactured version of cement. He crushed limestone and clay and then heated it to about 1,600 degrees.

The resulting clinker, lumps of fused clay and lime, was then crushed into a soft, fine powder. Gypsum was added to moderate the curing process. The addition of water catalyzed crystalline growth to create a rock-hard material. Aspdin's product was named Portland cement (after the isle of Portland) and continues to be the most widely used type of cement. With the addition of gravel and sand, the cement mixture is given volume. It is as hard as rock and has a virtually limitless life span. Some concrete structures, such as the Roman architectural wonder the domed Pantheon, have lasted nearly 2,000 years.

There are numerous types of cements used today. Some give strength at an early age, an advantage in the production of precast concrete components. Others reduce the heat produced as a by-product of the hydration or hardening process. This is important in large dams, where the heat generated can be so great that it turns the water in the concrete to steam and can cause cracking. Other types of cement provide protection against deterioration by sulfates and similar chemicals. Some cement has been engineered to compensate for the shrinkage that concrete undergoes as it cures.

The period since the 1980s has seen the development of so-called high-performance concrete able to overcome some limitations of traditional concrete. In some cases the term *high-performance* is used to describe concrete modified with mineral admixtures, called superplasticizers, which allow the concrete to remain in a flowable condition for long periods before setting. This property allows concrete to be shipped from distant mixing plants to construction sites, where it may need to be pumped vertically and horizontally for hundreds of feet. The addition of superplasticizers also helps the concrete to fill the small spaces when large amounts of steel reinforcement are present. At other times the term *high-performance* is used to describe concrete that has terrific compressive strength. Ordinary concrete has compressive strengths rated at between 3,000 and 5,000 pounds per square inch. High-performance mixtures can sustain pressures between 15,000 and 20,000 pounds per square inch. This additional strength is gained by adding silica fume (microsilica) and fly ash, fine powders that improve hydration of the cement paste. These additives also result in denser concrete, which is less permeable and more resistant to chemicals that can cause corrosion of steel reinforcement.

From the structural standpoint the main draw-back to concrete is its weakness when subjected to tension, a force that tends to elongate the concrete, as opposed to compression, which tends to push the material together. Concrete has roughly one tenth the resistance in tension than it has in compression, so that it cracks easily. This problem was initially solved early in the 20th century by introducing steel reinforcement in the form of smooth square bars. When tensile forces were applied to steel-reinforced concrete some cracking occurred, but the loading was transmitted to the steel. During the mid-1930s, the smooth, square bars were replaced by circular bars that adhered better to the concrete. In the late 1940s and early 1950s the techniques of prestressing and post-tensioning concrete structures became common. By the 1950s steel-reinforced concrete became essential in reinforcing major tunnel projects.

Concrete's Achilles heel lies in its porosity and permeability. About 1 to 5 percent of the total concrete volume is actually air, which allows the infiltration of water that can expand in freezing temperatures and damage the concrete by chipping or cracking. However, that same porosity can allow the freezing water to expand, negating its ability to cause damage. If these bubbles are not present, the forces arising from the freezing of the trapped water break the concrete.

This porosity, however, also allows chemical compounds such as the deicing salts used in roadways to penetrate the concrete and cause it to deteriorate from the inside, particularly if steel, which can be corroded by these same compounds, is present. The resulting rust occupies roughly 10 times the space of original steel and generates enough force to split the concrete from the inside. These problems can be controlled by either compacting the concrete mixture tightly or adding a finely ground filler or modified cements that make the concrete less porous and therefore less susceptible to corrosive infiltration. In some cases, a sealer can be applied to concrete to prevent chemical infiltration.

The construction techniques used must prevent water or corrosive elements from having contact with the steel components. When a walkway collapsed May 20, 2000, at Lowe's Motor Speedway in Concord, North Carolina, 107 pedestrians were injured. Engineers noted that the relatively new bridge's reinforcing steel showed signs of corrosion. Bridges and tunnels built or reinforced with concrete cannot simply be constructed and forgotten.

Although concrete is a strong material, it also can be subjected to gradual but catastrophic deterioration. Inspections of bridges are essential to ensure their structural integrity.

The beauty of concrete for bridge and tunnel engineers is that it cures, or hardens, under water, an attribute that is reflected in the term *hydraulic cement*. Submerged foundations for bridges can be poured into caissons and expected to harden since the water only activates the solidification process of the cement and does not necessarily retard it, although a drier mix generally yields stronger concrete. Reinforced linings sprayed on walls of recently dug tunnels using a form of concrete called shotcrete can be applied rapidly without fear that moisture or leakage will unduly weaken the concrete as it cures.

Portland cement, the primary component of concrete, is in such demand that an estimated 772 million tons was being used yearly in the United States by the end of the 20th century.

See also ELECTROCHEMICAL CHLORIDE EXTRACTION; FREYSSINET, EUGÈNE; LOWE'S MOTOR SPEEDWAY WALKWAY COLLAPSE; MAILLART, ROBERT; MONIER, JOSEPH; PRESTRESSED CONCRETE; REINFORCED CONCRETE.

Contino, Antonio *See* Rialto Bridge.

Conway Bridge *See* Stephenson, Robert.

Cooper, Theodore (1839–1919) *American bridge engineer discredited by the first Quebec Bridge collapse* Theodore Cooper may not have been a creative genius like the Brooklyn Bridge designer John Augustus Roebling, but a high level of competency displayed during his long career elevated him to an inner circle of American bridge engineers in the last half of the 19th century. Cooper's reputation as a bridge builder, which began with his work on Saint Louis's historic Eads Bridge, followed him into the 20th century to be tarnished by the deadly 1907 collapse of the Quebec Bridge on which he was the consulting engineer.

An 1858 graduate of the Rensselaer Institute in Troy, New York, Cooper was first employed as an engineer on the Hoosac Tunnel project, then America's most famous tunneling effort. At the start of the Civil War, Cooper began a 12-year career with the U.S. Navy, spending the war years aboard the Union gunboat *Chocura*. Three years as an engineering instructor at the United States Naval Academy and another two years aboard ship in the Pacific followed this duty. He capped his naval service with two more years as a Naval Academy engineering instructor.

In 1872, James Buchanan Eads, the designer of the famed Saint Louis bridge that would bear his name, hired Cooper for the crucial job of inspecting the all-important steel components for the world's first bridge to contain the metal. Cooper later became a construction inspector on the bridge and survived a 90-foot fall into the Mississippi River while clambering about the bridge.

Cooper developed a reputation as a hands-on engineer who inspected the bridge's progress daily, a habit he would discard in later years with tragic results. During his inspection work, Cooper discovered that two of the steel tubes forming an unfinished arch on the Eads Bridge were cracked and immediately wired Eads, who recommended loosening cabling that held both ends of the uncompleted arch until they could meet. The solution worked and construction continued. Cooper's involvement in this particular engineering episode would later puzzle historians when Cooper failed to act quickly enough to forestall the disastrous 1907 collapse of the Quebec Bridge.

With the successful conclusion of the Eads Bridge, Cooper launched a career designing numerous rail spans. He also authored technical papers for the American Society of Civil Engineers on rail bridge construction. Henry Petroski noted in *Engineers of Dreams* that Cooper twice received the society's Norman Medal for these writings and endeared himself to rail bridge designers by calculating a new method of understanding how rail bridges respond to loading. Cooper's calculations allowed for more efficient design of steel rail bridges and also for a convenient way of understanding how to strengthen existing bridges. Cooper would eventually serve as a director of the American Society of Civil Engineers.

Cooper's writings also focused on the issue of safety and warned that although each bridge should possess a surplus of structural strength, its design should be based upon a well-rounded approach to engineering and not brute strength alone.

Highly regarded and having a track record for designing bridges that held up to the complex rigors of rail traffic, Cooper was an engineering institution by the late 1800s. This reputation made him the obvious choice for the Quebec Bridge Company, who needed an engineer in 1898 to review a series of designs for a rail bridge across the Saint Lawrence River at Quebec. Cooper favored a cantilever plan recommended by the Phoenix Bridge Company of Phoenixville, Pennsylvania. When this proposal was adopted, Cooper was retained as the consulting engineer on the project. In 1900, the 61-year-old engineer recommended widening the main span of the bridge to allow pier construction in shallower water toward the shores. Work on the bridge began in 1903.

However, Cooper claimed his health was becoming increasingly frail and he was less inclined to visit the sites of his bridges. Although some say that Cooper never visited the Quebec Bridge site, he did so at least three times, though not during the two years before the 1907 collapse of the span's south cantilever.

The massive steel cantilever design was altered to create a longer central span at Cooper's suggestion, and when those plans were forwarded to Cooper, it was up to him to review them. Unfortunately for both the bridge and the hapless workers who would die on it, Cooper had bristled at having his computations checked by an engineer in the employ of the Quebec Bridge Company. Cooper created a situation in which he alone would have to confirm his own calculations. Lacking a staff and preoccupied by other bridge projects, Cooper relied upon the computations and design work of the Phoenix engineer Peter L. Szlapka, who actually made the changes. Szlapka, who added steel to the south cantilever span bridge to lengthen it, inaccurately calculated the additional stresses. As the south cantilever grew across the Saint Lawrence, its lower steel support chords began to buckle.

Cooper, apprised of this in the days before the cantilever collapsed, was alarmed enough to insist on explanations for the cause, but not so concerned that he demanded an immediate halt to the work. Cooper's confusion was based upon his erroneous assumption that the cantilever spans were properly designed, whereas they were not. Remaining in New York, Cooper corresponded by telegram with Phoenix engineers in Pennsylvania and onsite engineers in Quebec. Cooper was virtually the only engineer involved in the matter who seemed concerned about the bending of the steel components.

The problem became so pronounced that on August 28, 1907, Cooper wired a telegram to the Phoenix Bridge Company warning, "Add no more load to the bridge till after due consideration of the facts." The telegram's fate remains a mystery since it was either not received at Phoenix or not forwarded to Quebec as a result of a telegrapher's strike. In either case, the 600-foot south cantilever span collapsed at 5:30 P.M., August 29, 1907, killing 75 men.

Cooper, whose career had been one solid success after another and who had provided wise engineering techniques to other civil engineers in the area of bridge safety, found himself at the center of one of the worst bridge disasters in history.

A Canadian government commission eventually found Szlapka and Cooper responsible for failing to determine the stresses placed on the cantilever adequately. In his own defense Cooper told the commission: "I had and have implicit confidence in the honesty and ability of Mr. Szlapka . . . and when I was unable to give matters the careful study that it was my duty to give them, I accepted the work to some extent upon my faith in Mr. Szlapka's ability and probity."

Although Cooper maintained that ill health kept him away from the work site in the two years before the collapse, the lifelong bachelor lived until 1919, although he never again designed a major bridge. Some have maintained he was a man "broken" by the collapse, but Cooper had actually been contemplating retirement for some time. Nonetheless, it was said that he harbored bitter feelings toward the Phoenix Bridge Company until the end of his life.

See also EADS BRIDGE; QUEBEC BRIDGE.

covered bridge   See timber bridges.

Crowe, Francis Trenholm   See Hoover Dam Diversion Tunnels.

cut-and-cover tunneling   Although the danger and the excitement of burrowing underground tunnels make it a black art replete with human tragedy, massive expense, and exotic machinery such as tunnel-boring machines, many soft earth tunnels are still dug by the cut-and-cover method.

In essence, cut-and-cover tunneling involves the excavation of an open trench to the required depth and length. Once this excavation has been reinforced, a roof encloses the excavation, converting it into a tunnel. The advantages of cut-and-cover tunneling are obvious since this technique can be far

less dangerous and expensive than boring through rock or unstable soft earth.

In fact, most tunnels have been built in this way, including many of the subway lines in New York and Paris. During the 19th century and much of the 20th century, compressed air shield tunneling made soft earth and subaqueous tunneling possible, but the hazards of caisson disease made it risky to workers and the equipment and number of laborers required made it exorbitantly expensive. When relatively shallow tunnels had to be built on land the cut-and-cover method was a less expensive and safer alternative.

Although cut-and-cover tunneling for subway and utility lines often requires that a street be excavated in a metropolitan area, the work can be done relatively quickly, although not without following some rules. The walls of the excavation must be reinforced since collapses in these types of excavations are nearly a workplace epidemic in the United States. If the ground conditions are soft, pilings may need to be driven into the floor or invert of the tunnel for stabilization. Then the walls must be permanently reinforced, usually with vertical concrete walls.

After the reinforcement of the "cut" is completed, the "cover," in the form of a reinforced concrete roof, can be put into place. By following this procedure the street is repaved and life goes on for drivers and pedestrians.

Although many of the important tunnels of New York City's subway system were dug in soft earth by sandhogs in a compressed air environment, and others were blasted through solid rock with explosives, a large number were constructed by the cut-and-cover method. Roughly 52 percent of New York's subway tunnels started life as open trenches. The cut-and-cover method would have been used more extensively, but ground conditions and the need to maintain a level grade required tunneling in some sections of the city.

The cut-and-cover method was also employed extensively in the construction of the Paris Métro, which began in 1891. The project's chief engineer, Fulgence Bienvenue, realized he could not complete the system in time for the Universal Exposition of 1900 so he turned to cut-and-cover tunneling as well as other methods to expedite the process.

Cut-and-cover tunneling is still used on major tunnel projects and is an economical way of providing subterranean approaches to subaqueous tunnels.

A recent cut-and-cover project provided approaches to the immersed tube tunnel portion of the Ted Williams Tunnel connecting South Boston with Logan International Airport. The 1.6-mile combination cut-and-cover and immersed tube tunnel was opened for traffic December 15, 1995, and is part of Boston's Central Artery/Tunnel Project.

See also CENTRAL ARTERY/TUNNEL PROJECT; IMMERSED TUBE TUNNEL.

cutwater The piers of bridges, built as massively and as strongly as possible, need all the help they can get as they stand against strong currents and possible collision with floating ice. The installation of a V-shaped cutwater in front of the pier's foundation can split the flow of water and reduce the hydraulic force against the structure. The cutwater can also provide a knife edge to help divert floating ice away from the foundations of a bridge.

The term originally stood for the bow of a boat, its wedge shape designed to part the water in a cleaner fashion with less resistance. Cutwaters for bridges do the same job, with the exception that the water is moving while the bridge remains stationary. Sometimes the V shape is obtained by simply designing the foundations in the shape of a boat's prow. In rivers where the flow moves only one way (unlike in estuaries, where tides create flow in both directions) cutwaters need only face upstream.

Cutwaters tapered front and back can smooth the turbulence generated by water flowing past pier foundations and help reduce erosion of rock and sand at their bases. A square foundation's flat face generates massive turbulence, creating eddies and vortices that swirl away sand and rock away from the base.

Cutwaters differ from bridge fenders in that fenders give the piers and foundations protection from collision by ships and are built as separate structures. However, fenders can also serve to protect a pier's foundation from the erosive action of swiftly moving water.

See also FENDER.

D

Da Ponte, Antonio   *See* Rialto Bridge.

dead air   *See* ventilation.

Delaware Aqueduct (built 1848) *America's oldest suspension bridge*   Before there was much street traffic in America there was river and canal traffic, and one of the busiest waterway intersections was the confluence of the Lackawaxen and Delaware Rivers between Pennsylvania and New York. At this location during the 1840s was a massive traffic jam caused by huge rafts of felled logs navigated down the Delaware, where they invariably collided with the coal-hauling canal boats of the Delaware & Hudson Canal Company.

Wrecks, fistfights, and lawsuits made navigating that section of the Delaware River unpleasant and unprofitable. Around 1847, the Delaware & Hudson Canal Company decided there had to be a better way to move its canal boats across the Delaware to its New York–bound canals on the other side. The engineer John Augustus Roebling, who would later design the Brooklyn Bridge, recommended that Delaware & Hudson's canal boats be carried *over* the river by an elevated suspension aqueduct.

This aqueduct would carry a flume, or water channel, through which the canal boats would travel. Towpaths would be erected on the sides of the bridge so horses could pull the boats along. Delaware & Hudson engineers blinked at the novel plan, but after studying it, as well as Roebling's

completed suspension aqueduct over the Allegheny River at Pittsburgh, they granted their approval.

Armed with a $41,750 budget, Roebling went to work. His plan included a trio of stone piers seated in the Delaware's riverbed that would provide intermediate support for the four flume spans constructed of local pine. Mimicking his design of the Allegheny Aqueduct, Roebling made each span a beam rigid enough to support its own weight. The wire suspension cables running down each side of the bridge would have to support only the weight of the water inside the flume. Because of the nature of hydraulics, the aqueduct would never have to carry more than the weight of the water since the coal barges would displace an amount of water equal to their weight while in the channel.

Roebling tackled the task mindful of the hazards of ice and timber rafts, which could smash poorly designed piers. Roebling built his piers on solid stone islands in midstream and then built substantial timber cutwaters around the foundations of the piers with the sharp prows of the cutwaters facing upstream. Started in 1848 and completed the same year, the aqueduct solved the Delaware & Hudson's traffic problem. So pleased with Roebling's design were Delaware & Hudson engineers that Roebling built a total of four such aqueducts for the canal, which was one of America's most important routes of commerce.

Even as Roebling built the Delaware Aqueduct, railroads were being constructed, threatening the viability of American canal systems. With repairs to

The Delaware Aqueduct is significant because it is the oldest surviving suspension bridge in the United States and one of the earliest bridges built by the Brooklyn Bridge designer John Augustus Roebling. Although it once carried traffic over Pennsylvania's Allegheny River, it now belongs to the National Park Service and serves as a pedestrian bridge. *(Historic American Engineering Record, National Park Service)*

its vulnerable wooden flumes due to rot, the aqueduct served until 1898, when the Delaware & Hudson Canal closed down its operations.

The aqueduct itself had become an important crossing between Lackawaxen, Pennsylvania, and Minisink Ford, New York, and it passed to a series of private owners, the first of whom was Charles Spruks, who constructed a tollhouse on the New York end of the span. Drained of its water, the aqueduct then served as a toll bridge for more than 80 years. While in private ownership in the 1930s the last of its original pine timbers were removed. Because of its historical significance, the National Park Service acquired the aqueduct in 1980.

A Park Service engineering survey of the aqueduct in 1983 revealed that Roebling's wire suspension cables were in good shape and still capable of supporting the spans. Each of the 8 1/2-inch suspen-

sion cables is composed of seven strands, each containing roughly 307 individual iron wires. However, the exterior coating of tightly wrapped wire that provided a protective cover for the cables was replaced in 1985. The following year, Roebling's original plans were used in the complete reconstruction of the span's wooden superstructure. The timber cutwaters, long ago smashed by ice and log collisions, were rebuilt in 1995, as were the sidewalls of the flumes and the towpaths. Completely restored, the bridge is open to visitors, who can see exhibits related to its history in the tollhouse built by Spruks in 1900. It now serves only as a pedestrian crossing.

Another aqueduct was built across the Lackawaxen River a half-mile away by Roebling for $18,650 at the same time as the one across the Delaware. Unfortunately, the Lackawaxen Aqueduct fell into disuse after the closure of the canal system and is no longer in existence.

See also ALLEGHENY AQUEDUCT; CUTWATER; ROEBLING, JOHN AUGUSTUS; SUSPENSION BRIDGE.

depth to span ratio Bridge builders are obsessed with measurements and ratios, which tell the engineering story of a bridge. The thinner and lighter a bridge, the less expensive it is to build and the more attractive it is to those who view it. One of the most important ratios for bridges is the depth to span or girder to span ratio.

Since a bridge must not only be long enough to span a body of water but also rigid enough to resist bending, or deflection, its deck must be made substantial. Even suspension bridge decks must possess enough depth and weight to resist wind-induced movement.

As a result, road decks are strengthened by plate girders or the open reinforcement of stiffening trusses that appear as a latticework above or below the deck. When the truss or girder depth is compared to the length or the span of the bridge, the result is the depth to span ratio.

The Williamsburg Bridge, an all-steel structure that combines substantial truss reinforcement with suspension cables, is a bridge that possesses a low ratio since its massive reinforcing truss is so wide. Its depth to span ratio is only 1/40, which is unimpressive, considering the Golden Gate Bridge's depth to span ratio of 1/164. The Golden Gate Bridge therefore presents a far more slender image.

What may be the most striking depth to span ratio for a modern suspension bridge is that of the original Tacoma Narrows Bridge, which was 1/350, meaning that for every foot of deck depth, the span measured 350 feet. It should be pointed out that in the effort to build thinner and longer bridges during the 1930s, the Golden Gate Bridge was built excessively thin, requiring subsequent reinforcement. As for the Tacoma Narrows Bridge, it was so thin it was destroyed by wind in 1940, the same year it was completed.

See also GOLDEN GATE BRIDGE; TACOMA NARROWS BRIDGE; WIDTH TO SPAN RATIO.

Desmarest, Nicolas (1725–1815) *Geologic genius who first proposed an English Channel tunnel* Only 25 when he did it, the Frenchman Nicolas Desmarest made what some consider to be the first recorded plan to tunnel beneath the English Channel between his native France and the island nation of England 24 miles distant.

Desmarest was not the run-of-the-mill Channel tunnel nut who has so often cluttered the pages of history with outlandish, unrealistic schemes, but one of the founding fathers of modern geology. Desmarest's tunnel plan may not have been born of his own personal interest in designing a tunnel but in response to a competition for ideas of better ways of crossing the English Channel. The Academy of Amiens, which sponsored the competition in 1750, awarded Desmarest the prize for the most innovative idea.

Unconfirmed reports of ancient underwater tunnels aside, no one had attempted a major subaqueous tunnel in the history of humankind at the time of Desmarest's proposal. The dangers of flooding and cave-ins were issues that technology would not effectively defeat until the invention of Marc Brunel's tunneling shield in the 1820s. Nonetheless, Desmarest determined that a subaqueous tunnel was a sensible answer to crossing the often-treacherous waters of the Channel.

Desmarest, who correctly concluded that the basaltic rock on the Earth's surface was spewed out by volcanic action, correctly disputed a widely held 18th-century theory that all surface rock was deposited by ocean sedimentation. Desmarest's keen insights helped establish volcanology as a science, but he was not infallible. In attempting to explain the origins of lava, he incorrectly assumed that massive coal fires in the Earth's interior generated the fires that melted the rock.

Nonetheless, Desmarest correctly pointed out that the Dover Straits were once an isthmus connecting France and England that had been eroded by a great flood. Modern geologists have confirmed that a ridge of land existed between Dover and Calais a half-million years ago and that it was washed away during a spillover of the North Sea.

See also EUROTUNNEL.

Detroit-Windsor Tunnel See immersed tube tunnel.

drill-blasting Although modern tunnel-boring machines garner publicity for their roles in grinding out massive tunnels such as the Eurotunnel between Britain and France, tunnel excavation by filling drill holes with explosives is alive and well throughout the world, especially in tunnels with hard, quartzite rock. By carefully spacing shot holes of specific depths and using precise amounts of explosive, miners can excavate tunnels with superb precision. However, the explosives and methods used in this technique were far from safe when it was first introduced more than 300 years ago.

Although black powder blasting was used as early as 1643 by German miners, the earliest attempts to fill crude drill holes with the explosive did not take place until the late 17th century in Germany's Frieberg coal mines. By using the plug shooting method, a crude 2 1/4-inch-diameter hole roughly 40 inches deep would be drilled and then packed with black powder. A wooden plug notched to accept a match fuse would then be tamped into place. If this final exercise did not detonate the testy powder, the match would be lit. It was not uncommon for miners to be killed by the plug as it was blasted from the hole.

Various improvements followed, including the 1687 innovation of clay tamping instead of a wooden plug. In 1688, wooden tubes filled with blasting powder were introduced, making the process even safer. By 1700, the size of the drill holes was reduced by nearly an inch in diameter, reducing the amount of powder needed by half. The results were less expensive blasting and better results.

Hand drills and later water- and air-powered rotary drills introduced during the 19th century provided a faster and more accurate way to drill shot holes. In 1829, an American, Moses Shaw, developed a method of sending an electrical spark to a highly explosive silver fulminate detonator. This innovation allowed miners to set off numerous charges simultaneously.

The 1831 invention of the safety fuse in England by William Bickford was the first innovation that made a major dent in the number of blasting fatalities. Liquid nitroglycerin began to be widely used in tunneling after its discovery in 1846, but its terrific explosive energy was fatally unstable. Alfred Nobel's 1866 development of dynamite, which moderated nitroglycerin with a clay mixture, was a boon to the tunneling and mining industries. When dynamite was combined with his extremely safe blasting cap invented the year before, Nobel ushered in a new era of safety for those using the drill-blast method.

See also DYNAMITE; HOOSAC TUNNEL; NOBEL, ALFRED; ROCK DRILL.

dynamite (invented 1866) The first of the reliable and stable high explosives, dynamite was a godsend to miners and their bosses. It was safer to use underground than the less powerful black powder and the powerful but temperamental liquid explosive nitroglycerin. In addition, the blasting force of dynamite meant hard rock tunneling and mining could proceed at a faster rate than previously thought possible.

Dynamite's inventor, Sweden's Alfred Nobel, worked to turn nitroglycerin into a stable explosive in 1863 after his brother and four others were killed while producing the dangerously unstable liquid. Used in its pure form as an explosive, nitroglycerin developed a justifiable reputation for instability in one accidental explosion after another.

One of the most outrageous examples of its instability was coupled with the carelessness with which it was often handled. In 1866, a careless fool attempted to ship a container of nitroglycerin in a parcel that blew up a San Francisco Wells Fargo office, killing eight and causing an astounding $2.5 million in damage.

Nitroglycerin is a truly difficult compound to work with. When warmer than 70 degrees Fahrenheit it can be temperamentally explosive, and a mild bump is enough to cause its detonation. If it cools to 55 degrees Fahrenheit, it "freezes" into a solid and becomes about as explosive as mud, making it impossible to detonate. Even the production of nitroglycerin is inherently dangerous since the combination of glycerin, nitric acid, and sulfuric acid generates explosive vapors. Even Nobel's business of manufacturing nitroglycerin was made illegal in populated areas of Sweden, forcing the chemist and inventor to move his work to a barge in a harbor.

Methods that made the manufacture and transport of the explosive relatively safe were found. Ventilated production rooms were designed, and the nitroglycerin was then kept at specified temperatures before its use. Despite these measures, nitroglycerin remained a killer of "powder monkeys," as the explosives handlers who used it were called. The first large-scale use of nitroglycerin in the United States was in the long-suffering Hoosac Tunnel project in Massachusetts, in which working with it proved difficult and dangerous. Nonetheless, miners and powder monkeys continued to take their chances with the substance until something better came along.

Nobel finally hit upon an answer: By combining nitroglycerin with a form of diatomaceous clay known as kieselguhr, Nobel had a way of making the liquid insensitive to temperature fluctuations and physical shocks. The clay that absorbed the explosive liquid was composed of the silica-rich shells of ancient diatoms, tiny organisms that make up plankton and algae colonies. In this form, nitroglycerin could be handled, even roughly, and detonated by blasting caps, another Nobel innovation. Nobel named his explosive *dynamit,* the Swedish word for "power," derived from the Greek root of *dunamis.* Nobel was on his way to being one of the world's richest men because dynamite was an instant success. Bridge builders, miners, military forces, and even farmers—anyone who needed explosive energy to smash a rock or dig a hole—purchased it.

Black powder, which was composed of potassium nitrate, charcoal, and sulfur, had been around since the Chinese developed it in A.D. 1000. Whereas jarring the stuff did not cause an explosion, a stray spark could. Another problem of black powder was that inside an enclosed tunnel its detonation generated clouds of smoke and toxic gases. Most dangerous of all, the detonation of black powder created a fire that burned up to 2,000 degrees Fahrenheit that was capable of igniting flammable subterranean gases such as methane. On top of all this, the burn rate of black powder—the speed at which it detonated—was extremely slow and failed to generate the explosive force of nitroglycerin dynamite. Dynamite's detonation actually created a cooler flame, which was incapable of detonating mine gases. Although black powder had been employed in mining excavation since 1627 in Hungary, dynamite's attributes and competitive price made it the explosive of choice for underground work by the late 1800s.

A primary reason why dynamite was so valuable a tool for hard rock tunneling and coal mining is the nature of its detonation. An explosion is merely an extremely rapid combustion that disperses a large amount of energy. Dynamite can produce a pressure of up to 1 million pounds per square inch if the nitroglycerin content is high enough. Dynamite's release of energy is almost instantaneous, and the supersonic shock wave it creates travels at thousands of feet per second to shatter the hardest rock.

Although the image of workers using drills on the rock face of a tunnel or mine creates the impression that this is how the rock is removed, the drilling is only the first stage of the blasting process. The holes are carefully arrayed and drilled to a specific depth, depending on the condition of the rock and the amount of spoil the engineers want to remove. Dynamite is tamped into the holes, where black powder or pure, liquid "nitro" was once poured. The dynamite is then fitted with a blasting cap and detonated electrically by remote control or a lighted fuse. Experts can literally carve their way through mountains with a great deal of precision by using dynamite.

As the years progressed, dynamite was the subject of continuing development. The dynamite used today now contains ammonium nitrate and sodium nitrate to boost the power of the nitroglycerin. Wood pulp has replaced clay as the absorbing agent to moderate nitroglycerin's instability.

Other classes of explosives have come into use, including ammonium and fuel oil mixtures that can be produced in great quantities and poured by tank truck into hundreds of bore holes. Dynamite production is only a fraction of what it once was; however, dynamite still is often used to initiate the detonation of other explosives.

Although newer explosives have superseded dynamite, some tunneling projects still rely on blasting. Japan's Seikan Tunnel (completed in 1988) was almost entirely excavated by using high explosives to shatter 32.3 miles of tough rock, of which 14 miles lay 780 feet beneath the Tsugaru Strait.

See also HENRY, JOHN; HOOSAC TUNNEL; NOBEL, ALFRED; SEIKAN TUNNEL.

Eads, James Buchanan (1820–1887) *River salvor and bridge designer* Like many of the notable bridge designers of the 19th century, James Buchanan Eads did not limit himself to a single field of endeavor. And when he did design the only bridge of his career, it was considered one of the most impressive feats of bridge engineering of its day. He received his first and middle name from his relative James Buchanan, who was president of the United States between 1857 and 1861.

Born in Indiana to a family of dry goods merchants who later moved to Saint Louis, Eads would watch helplessly as the riverboat that carried them to their new home burned with all their possessions, including goods with which the family had hoped to stock their store. At 13, with his family suddenly impoverished, Eads began working as a store clerk and educated himself with books lent to him by his employer.

Hard-working and courageous, Eads soon followed the dream of many 19th-century youths by securing work aboard the steamboats that plied the Mississippi. The river was America's inland highway, stretching from Minnesota to the Gulf of Mexico, and riverboats were the primary means of transport.

At 22, he decided he could make his fortune salvaging goods from the wrecks of dozens if not hundreds of sunken boats that had fallen victim to the mighty river. He designed a purpose-built salvage boat and his own version of a diving bell. He then began descending into one of the world's most treacherous rivers to recover valuable cargo. During his salvage career, Eads made nearly 500 such dives. Marriage brought Eads up from the murky depths of the Mississippi to life on shore, where he purchased a glass factory in an effort to be a more conventional businessman.

The failure of the glass business sent him back to the river, where he regained his fortune by salvaging cargo and raising wrecks, which he repaired and sold. The years of diving and enduring the effects of decompression sickness, or "the bends," had taken their toll, and Eads's health began to fail. By 1860, diagnosed with tuberculosis, the 40-year-old Eads retired once again from salvage work.

Eads's retirement did not last long. At the outbreak of the Civil War in 1861, he proposed construction of ironclad gunboats to help the North gain control of the Mississippi by bombarding Confederate fortifications along its banks. A lethargic government half-heartedly approved the effort that Eads personally financed since government payments were slow in coming. He designed and built eight gunboats encased in iron armor—the first of their type in the world. The success of his gunboats was displayed February 6, 1862, when a bombardment from his boats drove Confederate defenders from Fort Henry on the Tennessee River.

At war's end, the business of America was once again business, and Saint Louis civic leaders sought to build a bridge connecting their city to the Mississippi's eastern shore. In 1867, the inventive Eads, who had never designed a bridge, decided to build one across the Mississippi. One good aspect of his ignorance of bridge design was that Eads was unconstrained by traditional construction techniques. As a result, he broke with the tradition of using iron in bridge construction and chose steel for the arches. Only recently available in quantity, steel was twice as strong as iron, convincing Eads it should serve as the material for his arches.

Eads also proposed using the latest pneumatic caisson technology to sink the piers of his proposed bridge to the bedrock beneath the Mississippi. Eads was eventually selected to be chief engineer for the project, because of his immense reputation as a salvage operator and builder of armored gunboats and his familiarity with the river he intended to bridge. Although Eads's health was problematical and he had long absences from the bridge site, he nonetheless took an active role in designing the bridge and overcoming the technical problems associated with the task.

Eads, the practical engineer, found that he also had to play the role of fund-raiser for the bridge by meeting investors in America and Europe. He also enforced stringent and precise quality control rules for the steel going into the bridge. The suppliers of his materials howled in protest, saying this was unprecedented, but Eads, for the most part, stood his ground. One of the engineers working for Eads even developed machinery that would test the strength and tolerances of the steel and iron components before they were accepted from the manufacturers.

The work was exhausting, and Eads attempted to quit the project in the summer of 1871, the fourth year of the effort. However, he was convinced by supporters to stay and resumed his role in building the bridge. On May 23, 1874, the bridge was completed and pedestrians were allowed to cross the privately owned span for a nickel apiece. The $6.5 million bridge was not a resounding success and it failed to repay its investors.

Disappointed at the fate of his bridge and fighting tuberculosis, which wracked him with bloody fits of coughing, Eads nonetheless had a final grand accomplishment involving the Mississippi River left in him. Alluvial silt had so clogged the mouth of the Mississippi at New Orleans that large boats were running aground. Eads contracted with the federal government to dredge the waterway through the ingenious placement of jetties that constricted the river and resulted in an increased rate of flow. Forcing the water to move faster removed the silt naturally to deepen the Mississippi at New Orleans from 8 to 30 feet.

Eads's last magnificent idea was to design a gigantic railway that would cross the Isthmus of Tehuantepec in Mexico separating the Pacific and Atlantic Oceans. Eads proposed that ships be hoisted from the water onto tracks and then towed from one ocean to the other across the isthmus. Eads, plagued by ill health, died March 8, 1887 in Nassau, Bahamas.

See also EADS BRIDGE.

Eads Bridge (built 1867–1874) *World's first steel arch bridge* The historic bridge spanning the Mississippi River at Saint Louis, Missouri, was conceived by a river wreck salvor, James Buchanan Eads. Though he had never before designed a bridge, Eads managed to create one of the world's most unique spans for its time and introduced steel to bridge construction. Not only was it the first arch bridge constructed mostly of mass-produced steel, which had just entered use in America, it also involved the deepest manned caisson work in history, and its three arches were the longest built up to that time.

Unfortunately, like many 19th-century projects constructed by the private sector, it was expected to be self-supporting. It never did turn a profit and fell into disuse and a shocking state of disrepair during the 20th century. Despite its failure as a financial enterprise, the bridge established a series of construction milestones in America and marked the first large-scale American use of European pneumatic caisson technology.

Eads's use of mostly steel structural components, the incredible depths of its deadly foundation work, and an attention to design precision and material quality set the standard for future projects. The Eads Bridge is often incorrectly viewed as the first bridge to incorporate steel components since steel eyebar chains were used in an Austrian bridge nearly 40 years previously. Nonetheless, the bridge was the first arch bridge to contain components of steel rather than iron, and its success alerted bridge engineers worldwide to the capabilities of steel.

Of the rivers in America, the Mississippi stood as the greatest obstacle of them all by virtue of not only its width but also treacherous hydrologic forces that could dig dozens of feet into the river bed and scour away the support around shallow bridge foundations. Although Americans in the latter half of the

The Eads Bridge soars across the Mississippi River connecting Saint Louis, Missouri, with Illinois. Designed by the self-taught engineer and riverboat salvor James Buchanan Eads, the three-arch bridge was the first to make large-scale structural use of steel. The caisson work on its foundations was the deepest and most dangerous conducted by that time in America. *(Historic American Engineering Record, National Park Service)*

20th century would take sturdy bridges for granted, bridging expanses of water like the Mississippi in the latter half of the 19th century was a fairly new effort. Although narrower portions of the Mississippi River had been bridged years before Eads began work on his span, the bridge he would build would dwarf the others in beauty, magnitude, and innovation.

Saint Louis had debated building a bridge across the Mississippi River since the onset of railroads in the United States during the early 1800s, but no real progress had been made. The city grew uneasy as Chicago—conveniently east of the Mississippi—prospered, served as it was at the end of the Civil War by 11 railroads. As rail bridges went across the Mississippi at such places as Dubuque, Iowa, and Quincy, Illinois, Saint Louis was still using ferries to move goods east and west. Most galling was the necessity to offload rail shipments at what is now East Saint Louis across the river in Illinois and then to ferry the goods the final 1,500 feet across the Mississippi at half what it cost to move them the previous 1,000 miles by rail.

Saint Louis, which had long been an economic rival of Chicago's, was being eclipsed as it sat behind its commerce-strangling moat. The river that once allowed it to dominate trade in the region through riverboat traffic was now an impediment to economic survival. River trade could only follow the water, and America was expanding beyond the reach of river traffic.

After years of quibbling about whether or not to build a bridge, by 1866 Saint Louis's leaders knew it was time to act. The Illinois and St. Louis Bridge Company, chartered by Illinois to bridge the Mississippi, had been made moribund by a lack of funding and by Saint Louis's inability to make a decision. Just as the effort to fund and plan a bridge got under way, the Chicago bridge builder Lucius Bolles Boomer arrived and finagled a charter for another bridge-building effort. Boomer, whose Gasconade River rail bridge collapsed in 1855, killing a number of prominent Saint Louis citizens, was distrusted not only because his engineering acumen was suspect but because no one could figure out whether he really wanted to bridge the Mississippi or to stall the effort. After protracted political infighting it was decided to merge the two bridge companies during the early part of 1868, a move that prompted Boomer to fade away.

James Buchanan Eads, who had unsuccessfully fought Boomer's attempt to form a second bridge company, had entered the fray and was made director of the Illinois and St. Louis Bridge Company. Although he had never built a bridge, the salvor had accumulated intimate knowledge of the Mississippi's hydrologic behavior while diving to salvage wrecks on its riverbed. Using this knowledge, he had created his own bridge design. A self-taught engineer with an inventive bent, Eads decided he wanted the bridge made out of the newly available Bessemer steel, which was twice as strong as iron. More importantly, he cautioned that any bridge built across the Mississippi must have its foundations emplaced far enough below the riverbed muck that they could rest on solid rock. Eads knew the Mississippi could scour tons of sand away from shallow foundations in minutes.

Though his general design of the bridge changed considerably as the project progressed, mostly because of calculations to make the bridge stronger and more streamlined, Eads provided an extremely modern concept with clean architectural lines. It would eventually take the form of a series of three arched spans built in a cantilever construction fashion. This meant cables would temporarily support both halves of the partially built arches until they met. Given the period as well as the experience of its designer, it was a daring plan.

The state of Illinois required that the bridge not be a suspension span or a drawbridge: the former was considered an unreliable type and the latter an impediment to navigation since it would allow the passage of boats only when its leaves were raised.

What Eads and his engineering assistants designed followed a concept developed 80 years previously by the American Revolution pamphleteer Thomas Paine—a long-span, metal arch bridge. Paine's own iron arch concept had been preceded by the 1779 construction of a 100-foot, single-arch bridge over England's Severn River at Coalbrookdale. Eads's bridge would consist of four masonry piers sunk as far as necessary to bedrock. These piers would support a series of three arched spans, with each made up of 16 nine-inch tubes of steel. Diagonal bracing would fill the spaces between the tubes forming the three arches. A lower deck would carry rail traffic, and an upper deck would serve pedestrians and horse-drawn conveyances.

Long-span bridges in America were still relatively rare in 1868. Though knowledgeable about the river, Eads knew his limitations when it came to bridge design. To assist him, Eads employed a team of European-trained engineers to hone the design and compute the complex mathematical calculations necessary to comprehend the stresses the bridge's components would endure. Among these were two German engineers, Charles Pfeiffer and Henry Flad. To ensure accuracy, the Washington University mathematician William Chauvenet would review the calculations of Flad and Pfeiffer. Eads, who survived years of submerged salvage work in the deadly currents of the Mississippi, knew a thing or two about precision and risk reduction and was taking no chances.

The bridge's three spans would measure a total of 1,524 feet with the center span running 520 feet in length. The arched spans on each side would measure 502 feet with approaches on masonry arches leading to the arched spans from both sides of the river. Eads, who began life as a store clerk and then worked aboard riverboats, was a man's man who carried a pistol and a knife on the job but who also moved comfortably among financiers and could speak the language of Europe's most progressive engineers. He would have to do all these things to turn his bridge idea into reality.

Although consumed with designing the bridge, Eads would find himself traveling to Europe not only to raise funds for the project but to pluck ideas from engineers familiar with the properties of steel and those experimenting with pneumatic caisson technology.

The east bank foundation of the Saint Louis bridge would go deeper than any built previously, striking bedrock 127 feet beneath the Mississippi's surface during high-water conditions. The caisson work remains the deepest excavations conducted by hand anywhere—and the deadliest. To do this, Eads would use a version of the caisson he had seen in France that was originally developed by William Cubitt and John Wright. The caisson was equipped with an ingenious way of hauling excavated muck to the surface by pumping the material through a pipe known as an air lift. Eads had such a caisson constructed, and by October 1869 it was positioned over the location of the east pier. To push the caisson down to the riverbed, masons began laying limestone blocks on its roof to sink it to the sandy bottom. Once it was in place, workers entered a nine-foot-tall excavation chamber dimly lit by smoky candles, where they began removing the muck. The deeper they went, the greater the air pressure was increased to forestall the entry of water that was also acquiring more pressure with greater depth.

As the pressure inside the chamber rose to 30 pounds per square inch, the men began to be affected by "caisson disease," or "the bends," in which nitrogen pushed into the bloodstream by pressure later expands with painful if not deadly consequences. Although some European physicians were beginning to understand the problem, engineers were baffled by its cause. In their ignorance, workers on the Eads Bridge took to wearing bracelets of a zinc-silver alloy, believing the bands would ward off the illness.

The caisson reached bedrock 93 1/2 feet beneath the sand of the Mississippi on February 28, 1870. By some miracle there had not been a death due to caisson disease up to then. The workers departing their shifts usually decompressed for only five minutes in the airlock. To allow the nitrogen to dissolve safely, those working in the caisson should have decompressed for as long as 133 minutes.

The deaths would soon begin. Less than two weeks after reaching bedrock, the air pressure inside stood at 44 pounds. One worker, James Riley, emerged from the caisson and fell dead minutes after announcing he felt fine. He was the first of 14 workers to die of caisson disease during this period. Of the 600 who worked below the surface of the water, at least 119 were made seriously ill with decompression illness, though many more probably suffered lesser symptoms from which they recovered. In what would today be a major workplace safety scandal, five more men died within days of Riley's collapse.

The workers—known as "submarines" before underground and submerged laborers were nicknamed "sandhogs"—were justifiably upset. Eads, who was stunned by the death of Riley, ordered his personal physician to undertake a study of the problem. The doctor himself fell victim to caisson disease after returning to his office from the pressurized chamber of the west pier's caisson. Although wracked by horrible pain, he recovered and recommended longer decompression periods, though nothing close to what was actually needed. Eads reduced the length of the working shifts and ruled that only men who were extremely physically fit would be allowed inside the caissons. Eads reflected the puritanical wisdom of the day by deciding that alcohol consumption and a dissolute lifestyle were also possible culprits, warning his workers to abstain from drink. Of the 14 who died, all but one perished as a result of working inside the deeper east caisson. Only one worker died on the west caisson, which was sunk to only 78 feet—nearly 50 feet shallower than its sister.

As the caissons sank deeper into the muck, so did their roofs, which required workers continually to raise the level of masonry to prevent the Mississippi from spilling into the caisson's innards. A near-tragedy struck the east caisson on April 13, 1870, when the Mississippi's rising waters gushed over the masonry wall and sent workers clambering out of the caisson. The airlock doors were left open in the rush, and with the loss of protective air pressure, the caisson's lower chambers were flooded. Repairs followed, and six weeks after the flooding the pier was finished, eliminating one of the project's toughest challenges.

Financing for the bridge was masterminded by no less of an icon of American business than Andrew Carnegie, the Scottish American whose love of money was a matter of religious intensity. Carnegie, who was already amassing firms that would later become Carnegie Steel Company, was also founder of Keystone Bridge Company in Pittsburgh, a firm that had already turned down the opportunity to serve as general contractor on Eads's bridge. Keystone's president, J. H. Linville, whose firm built some of the most impressive iron rail bridges in the East, believed Eads's design was impractical. Carnegie, however, was impressed with Eads and convinced Linville, Carnegie's partner, that they should build the bridge.

Carnegie-Kloman Company, another of Carnegie's businesses, would produce the iron fittings that would be emplaced on the faces of the masonry piers to accept the steel tubes of the arches. Since steel production was not yet Carnegie's forte, he relinquished the contract for the steel arch tubes to Butcher Steel Works in Philadelphia. Eads was to aggravate and enrage Carnegie and the subcontractors by his unswerving devotion to ensuring that only components of the required quality went into the bridge. Carnegie and the engineers at Keystone and Butcher (as well as all other fabricators of the day) were used to buyers' accepting components without question. This was not good enough for Eads, who installed his own inspectors at the factories to test the components before acceptance.

Eads's position dumbfounded the industrialists, who felt the salvor-turned-engineer was violating the "customs" of business. Eads, who was responsible for making sure a bridge being built on the doorstep of his own community did not collapse, cared nothing for the imprecision of gentlemanly tradition. Special machines in Saint Louis and Pittsburgh would test the steel components to ensure they met Eads's tough specifications, causing an exasperated Carnegie to complain, "Nothing that would and does please engineers is good enough for this work." Because of this corporate culture shock, not a single steel component from Butcher was found adequate during inspections at its factory.

A frustrated Eads finally stated that he did not care what sort of steel was used or what its chemical content was, as long as it was strong enough to pass inspection. The result is that tests done 100 years after its construction revealed that steel in the Eads Bridge varied wildly in its carbon, chrome, and manganese content.

As components finally began to arrive, work on the steel arches began, but many steel components still failed to meet Eads's specifications. As a result, iron components such as connecting bolts were used in the arches. Although it was heralded as an all-steel bridge, only about half the components in the bridge arches are steel.

By 1872, the piers were completed and arches began to sprout from their stone sides. The traditional method of building an arch bridge for thousands of years was to support the arch work by falsework, wooden framing that held the arch in place until it was structurally sound. This was impossible on the wide and wildly fluctuating depths of the Mississippi. The solution was to use steel bars as cabling to suspend the tips of the opposing arch segments. When they met, the arches would form a self-supporting structural unit.

As both segments of the first arch approached closure, the engineers made a horrifying discovery. During the summer of 1873, the ends of steel tubes extended beyond one another on the west arch, though by very little. This promised to be both an engineering and a fiscal catastrophe since an infusion of investment money needed to continue the project was dependent upon the closure of at least one arch. Eads, overworked and suffering from what was probably tuberculosis, sketched out a possible engineering solution before taking an extended recuperative leave in Europe. Eads left Flad and another engineer, Theodore Cooper, in charge, and both men ignored Eads's recommendation.

Flad and Cooper struggled to position the steel tubes so they would abut and allow collars to connect them. At wit's end, Flad chilled the steel tubes in 30,000 pounds of ice in an unsuccessful attempt to shrink them. Eads was consulted and recommended that the chords be cut to reduce the overlap and that interior diameters of their ends be tapped with opposing-twist threads. This would allow an insert threaded at both ends to be inserted and the chords drawn together as the insert was turned with a huge tool. Eads's elegant suggestion worked.

Riverboat interests, still enraged by an 1862 Supreme Court decision that upheld the right of anyone to bridge the Mississippi River, now took exception to Eads's bridge. Eads, who amassed his fortune clearing the river of obstacles, including sunken steamboats, was now the builder of a bridge that steamboat interests said was so low it provided the worst sort of hazard to riverboat traffic. The steamboat interests obtained the support of the corrupt secretary of war, William W. Belknap. Belknap was (suspiciously) on the side of the steamship interests and recommended a canal be dug around the abutments of the bridge at the bridge company's expense. This incredibly stupid suggestion would mean that another bridge would then have to be built over the canal!

Eads, whose ingenious development of armored river gunboats had gained the North several victories during the Civil War, had an ally in a former Union general who then happened to be president, Ulysses S. Grant. Meeting in the White House in 1873, Grant heard Belknap's flimsy argument on behalf of the steamboat industry and then listened to an appeal from Dr. William Taussig, the general manager of the Illinois and St. Louis Bridge Company. Ironically, Taussig, who was formerly county judge in Saint Louis, had declined to give the unemployed Grant a job as county engineer before the Civil War. Grant, apparently without malice toward Taussig, informed Belknap that neither would a canal be built nor would the bridge be dismantled. Opposition to the bridge had been dealt a mortal blow.

Shortly after he swatted down the riverboat owners' attempt to kill Eads's bridge, Grant visited the span in Saint Louis while it was still under construction. Soon after the arches were completed, work began on the lower deck for the rail line and the upper deck for pedestrian and horse-powered traffic.

On completion of the bridge in 1873, the Keystone Bridge Company announced it would allow pedestrians to cross the span on May 18 of that year. However, the avaricious Carnegie decided the night before to jettison the offer, believing it more prudent to hold the bridge as collateral against what he was owed for its construction. He ordered his workers to block the approaches to the bridge since allowing the public across it might be interpreted as a statement that the bridge had been turned over to the project's shareholders. Fortunately for the stingy Carnegie, a thunderous rain on the morning of May 18 reduced the turnout to a trickle of visitors, who were easily turned away. Carnegie's financial concerns were soon satisfied and the bridge was turned over to the bridge company, who had no qualms about allowing pedestrians to stroll across the bridge on May 24. On June 3, carriages and wagons were allowed across. Within days of this event, General William Tecumseh Sherman, then chief of staff of the U.S. Army, drove the final rail spike into a tie, indicating that the bridge was ready for rail service.

Confident of his bridge's structural soundness, Eads wanted a public demonstration that would prove its strength. On July 1, Eads invited local residents to watch as 14 steam locomotives and their tender cars, jammed with coal and filled with water, huffed their way across the bridge. The engines and cars crossed the bridge side by side on the rail deck's double tracks. What was most appreciated by the crowd that day was the appearance of a circus elephant striding slowly across the bridge as an additional gimmick.

Eads's demonstration of the bridge's strength has been substantiated by the structure's durability. The bridge has stood for more than 126 years when others—built by experienced bridge engineers—collapsed. However, investors in the project fared badly when the span failed to generate the tolls necessary to repay their bond investment. Though the bridge was a wonderful engineering accomplishment, no terminals or switching yards were built nearby to make the span attractive to rail companies. The railroads themselves snubbed the bridge, and nine months after it opened the unprofitable span was auctioned off to satisfy its creditors. The bridge was sold for $2 million, although it had cost more than $6 million to build. Like virtually every other major bridge project during the 19th century, the Eads Bridge exceeded its original cost estimate. What Eads originally thought he could build for $2 million eventually cost three times as much.

The bridge had become little more than an extraordinary example of engineering and design that was ahead of its time. Nonetheless, the bridge acquired a degree of fame that would be eclipsed only by the opening of the Brooklyn Bridge in New York on May 24, 1883. The Eads Bridge, originally bought at the auction by the St. Louis Bridge Company, was purchased two years later by the Terminal Railroad Association and kept in use as a rail line.

By 1974, modern freight trains had become far too heavy for the bridge's load limits, although it was still used for automobile traffic on its upper deck. However, the railway deck reentered service in 1993 for light rail passenger trains as part of Saint Louis's MetroLink system connecting Saint Louis with East Saint Louis. Use of the elevated roadway deck had continued through the decades until a 1991 inspection revealed that the upper roadway needed structural repairs. Although officials intended to repair the bridge quickly, additional study indicated the cost of renovation would be excessive. The roadway remained closed until 1999, when a $25 million renovation project was completed. By 2000, drivers were again able to use Eads's famous bridge across the Mississippi.

See also CAISSON DISEASE; EADS, JAMES BUCHANAN.

earth pressure balance machine The classic problems associated with soft earth tunneling are cave-ins and flooding. During the 19th century the use of a pressurized environment in subaqueous tunnels prevented water from rushing in through porous ground. Unfortunately for workers, a compressed air atmosphere posed numerous health risks, including caisson disease.

Even with the appearance of advanced soft earth, tunnel-boring machines by the 20th century, collapse and flooding still caused problems. Rather than force workers to endure a compressed air environment, Japanese engineers in the 1970s decided to pressurize a slurry of excavated material between the face of the tunnel and that of the machine for support. This prevented the face of the tunnel from collapsing and allowed the tunnel-boring machine to work more efficiently.

To maintain the pressure at the working face, water is mixed with the spoil to create the slurry. The spoil is removed by using an Archimedean screw, which allows the pressure to be maintained. This type of machine is most effective in clay soils, although less effective in soils with large sand grains or gravel since the slurry cannot suspend this type of material and prevent progressive failure of the tunnel face and ceiling.

See also PROGRESSIVE FAILURE; TUNNEL-BORING MACHINE.

East River Gas Company Tunnel (built 1892–1894) *New York's first completed underwater tunnel* It was not the world's biggest tunnel, but the driving of a 10-foot-diameter subaqueous tunnel beneath the East River separating Long Island and Manhattan was a major subsurface soap opera. The tragedies that befell the project were numerous: balking contractors, troublesome mud and rock conditions, sandhogs stricken with fatal bouts of caisson disease, and financial headaches.

Despite all, the brilliant English engineer Charles M. Jacobs took control of the project and led a reorganized crew of sandhogs to complete the tunnel while providing valuable lessons for subsequent projects in New York. The lessons learned from the East River Gas Company's effort were of particular importance to a pair of tunneling projects known as the Hudson Tubes and the Pennsylvania Railroad Tunnels, both of which were completed within the first decade of the 20th century.

Although most early subaqueous tunnels were built to facilitate travel, the primary purpose of the East River Gas Company Tunnel was to carry a gas pipeline from its plant on Long Island to Manhattan. Work on the gas pipeline tunnel began in 1892 with vertical shafts sunk 125 feet below the average low-tide level on both sides of the river. Sandhogs began digging horizontally after geologists assured them that borings showed they would follow a stratum of impermeable rock all the way across the East River. Unfortunately, all they struck was a muck that was more akin to mousse than to granite.

This disappointing situation meant tunneling through wet, unstable conditions in which the face of the tunnel often turned into slimy ooze as the sandhogs excavated. The conditions proved so worrisome to the original contractors that they sought additional funds from the East River Gas Company. A lawsuit followed, and the contractor departed the scene after threatening to shut down the pumps and air compressors that were preventing the tunnel from flooding.

Although the project began in 1892, Jacobs was authorized in early 1893 to organize a new contracting team and continue as both chief engineer and contractor. Jacobs, a coolheaded, innovative professional, proved equal to the task.

Although the use of tunneling shields was originally considered unnecessary, Jacobs commissioned the construction of two shields that would support the soft ground as sandhogs excavated beneath a protective steel hood. When the ground ahead was either blasted or dug out, depending on its hardness, hydraulic jacks would move the shields forward. Headings driven from Ravenswood on Long Island and the New York side would eventually complete the 2,516-foot tunnel.

As the tunnel progressed from both directions, the problem of water inflow became more pronounced, requiring an increase in air pressure that precipitated an increase in caisson disease. When the pressure was increased to 37 pounds per square inch, the project suffered the first of four deaths, when the foreman, Edward Ferris, collapsed and died of caisson disease on March 4, 1893. On March 10, Louis Doran died, and on March 16, the bends claimed Theodore Morris. The fourth and last death, that of Thomas Crimmins, occurred on January 2, 1894.

Eventually, a section of rock was struck by the sandhogs, who had been fighting mushy, leaking ground. Jacobs ordered the use of dynamite to blast through the hard rock while workers struggled with the jellylike walls that made up much of the tunnel.

At one point the sandhogs had to contend with rock on the lower half of the tunnel and mud on the upper portion. As a result, sandhogs both shoveled and blasted their way through.

Aside from the ordinary challenges of tunneling through unseen subterranean obstacles, the project was beset by dangerous mishaps that bordered on the bizarre. Early in the project a pair of workers got blind drunk on the job and allowed the steam engines running the water pumps and air compressors to stop in one heading. This resulted in a drop of air pressure that allowed water to rush into the tunnel and created a hole that sucked in the river from above.

In another incident, on May 16, 1883, a fire engulfed the project's Manhattan steam engine plant, which provided power to the pumps and air compressors. With a heading again threatened by flooding, temporary pumps were necessary to save the tunnel from inundation and structural damage.

Precarious financial markets also created a feast or famine environment for funding that forced work to be halted on the project for more than three months during the summer of 1893. During this period, the headings of the leak-prone tunnel deteriorated as a result of excessive water, but repairs were made and the work continued when additional investment funding was obtained.

Plagued as it was with blowouts, deaths, difficult contractors, and money problems, the tunnel was holed through just before midnight on July 11, 1894. Jacobs, who had proved his worth as a tunnel engineer, would subsequently design and direct the successful construction of the four tunnels of what became known as the Hudson Tubes across the East River between 1904 and 1909.

See also CAISSON DISEASE; HUDSON TUBES.

Effie Afton **case** (1857) *Abraham Lincoln's fight for bridges* Only three years before the Illinois lawyer Abraham Lincoln became president and preserved the Union with his leadership during the Civil War, he fought and won a landmark legal decision that cleared the way for railroads and anyone else to bridge America's rivers.

The late-1850s lawsuit was almost inevitable as America's powerful steamboat interests competed against railroads for the business of moving goods and people across an expanding nation. The main navigable waterway in America was the Mississippi River, and the riverboat interests considered the waterway theirs alone. When bridges started appearing over the river, riverboat captains became apoplectic over the appearance of new "navigation hazards" such as low bridge spans and bow-crushing support piers. The shipping interests also knew the bridges represented competition from the railroads.

The Chicago and Rock Island Railroad (later known as the Rock Island Line) was established in 1851 to build a railroad line from Chicago to Rock Island, which sat across the Mississippi from Davenport, Iowa. Railroads in the 19th century had to expand or die, and on reaching Rock Island in 1854 the company knew it would have to cross the river to Davenport. The expansion was cheered by Iowans, who knew that being linked by railroad meant the difference between economic life and death.

The Rock Island Bridge Company had been incorporated to build a Howe truss–type bridge over the river and began sinking piers into the riverbed. Once completed, the bridge would allow the Rock Island Line to link with the Mississippi and Missouri Railroad Company, which had been formed for that purpose.

The bridge was completed on April 22, 1856, and trains began carrying goods and passengers across the Mississippi almost immediately. However, a bizarre disaster involving the bridge soon erupted into a legal war between the railroad industry and the riverboat interests. On May 6, 1856—a mere two weeks after the bridge opened—the steamboat *Effie Afton* puffed its way down the river past the bridge. It seemed well on its way until it suddenly lost its ability to steer and drifted into one of the piers. The *Effie Afton,* like all riverboats of the period, was a firetrap and as a result of the collision immediately burned, incinerating part of the bridge. Some newspaper accounts of the time questioned whether the incident had been staged by riverboat interests eager to make their point about the hazards of building bridges across the Mississippi.

A lawsuit, *Hurd v. Rock Island Bridge Co.,* was filed with the owner of the *Effie Afton,* seeking $500,000 damages and the removal of the bridge. The real issue recognized by all concerned was whether the steamboat industry could halt future attempts to bridge navigable waterways.

Realizing the high stakes involved, the railroad brought in the attorney Abraham Lincoln to defend the railroad's case, which went to trial in 1857. The crafty and well-prepared Lincoln reduced the case to its essentials, making his points in a folksy and humorous manner. Along the way, Lincoln, who once had worked on riverboats, caught witnesses for the riverboat interests in numerous factual errors. Lincoln then zeroed in on the real issue, arguing that east-west railroad commerce had the same rights as north and south river traffic on the Mississippi. The jury in the first trial was deadlocked and could not reach a decision—a victory for Lincoln and the railroads.

The case was retried and a second jury ruled the bridge should be removed as a hazard to river traffic. The case was appealed and meandered through the justice system until finally reaching the Supreme Court of the United States. By this time Lincoln was president and no longer involved in the case. In 1862, the justices reversed the previous verdict, which had ordered the dismantling of the bridge, and agreed with Lincoln's arguments that bridges could be built over navigable rivers.

Eiffel, (Alexandre) Gustave (1832–1923) *Structural engineering genius*   His design of the Parisian tower bearing his name made Gustave Eiffel the world's most famous structural engineer but obscured his equally magnificent accomplishments as a builder of innovative and beautiful bridges. If the prolific and daring engineer had not excelled at the construction of such unique bridges as the Garabit Viaduct in France, he would not have had the practical experience necessary to design the Eiffel Tower or the supporting skeleton inside the Statue of Liberty.

Even compared to the extraordinary accomplishments and lives of other 19th-century bridge engineers such as America's John Augustus Roebling and Britain's Isambard Kingdom Brunel, Eiffel's career is marked by an extraordinary range of accomplishment and creativity. Aside from the impressive bridges that spanned deep gorges, Eiffel created small, prefabricated bridges that could be shipped anywhere for use in underdeveloped regions. Tragically, toward the end of his career, he was wrongly convicted of fraud in his role in the French effort to construct a canal across the Isthmus of Panama. He concluded his extraordinary life by accomplishing important studies related to aerodynamics and radio communications.

Born in Dijon, France, to a middle-class family, Eiffel was rejected in his attempts to gain admission to France's premier technical school, Ecole Polytechnique. He then entered Ecole Centrale des Arts et Manufactures, where he studied chemistry. Eiffel's original intent to work in a relative's vinegar factory was dashed by internal family squabbling, and he found employment as an engineer with a railroad company. It was there that Eiffel demonstrated an aptitude for designing rail bridges.

By the age of 28, Eiffel had built a 1,600-foot rail bridge across Garonne River near Bordeaux. Emboldened by subsequent successes, the 35-year-old founded his own engineering firm in 1867. Eiffel was a genius who would design anything and everything while adding his own special touch to the projects. Although Eiffel specialized in bridge design, he also created train stations, industrial plants, and retail store buildings. For the 1867 International Exhibition, Eiffel designed the Machinery Hall, the largest building of its day, enclosing an area 1,608 feet by 1,266 feet. It was supported by beautiful iron arches.

By this time, iron had become Eiffel's building material of choice. He understood its strengths and weaknesses, and, like his contemporary, John Augustus Roebling, he was wise enough to have a profound respect for the forces of wind on large structures.

Eiffel's bridges were artistic masterpieces that actually looked French. His mastery of iron construction led Eiffel to carefully design ornate iron latticework that supported his gravity-defying arches. From a distance, the sunlight through the artistry of the ironwork gave viewers of his bridges the impression that they could neither support a train nor stand up to a high wind. But the festive ornateness of Eiffel's structures was misleading. A closer inspection revealed iron components of massive size fabricated with technical innovations that enabled them not only to support the compressive weight of a railroad deck, but to endure horizontal wind forces. In Eiffel's artistry was engineering and in the engineering there was artistry.

His Rouzat Viaduct used a pair of 196-foot iron towers to support a trio of 196-foot truss girders across the Sioule River gorge. Built in 1869 near Gannat, France, two decades before his construction of the Eiffel Tower, the bridge exhibits characteristics Eiffel later incorporated into his famous tower. The legs of the Rouzat Viaduct's two towers are not simply formed by four square iron columns, instead, four iron braces flare from the lower third of each of

the latticework columns out to the ground to provide additional lateral reinforcement. The Eiffel Tower's legs would demonstrate the same type of design.

Eiffel reached the pinnacle of his bridge-building career in the design of what are referred to as his "crescent bridges," the Maria-Pia Viaduct in Portugal and the Garabit Viaduct in France's rugged Massif Central region. The Maria-Pia Viaduct over the Douro River has a 524-foot main span supported by a crescent-shaped arch that rises just above the horizontal line of the railway deck. Built in 1876, the bridge still stands.

His second crescent bridge—the Garabit Viaduct—was completed in 1884 over the Truyère River in France. Its central span is 540 feet, and its design differs slightly from that of the Maria-Pia Viaduct. The arch of the Garabit Viaduct touches only the bottom of the rail deck, giving the structure a cleaner, sparer appearance. In the same year that Eiffel saw the completion of his Garabit Viaduct, he also enjoyed the completion of the Statue of Liberty, for which he designed the interior iron structure.

Between 1887 and 1889, Eiffel oversaw his masterwork, the Eiffel Tower, which was nothing less than a compilation of his bridge designs with its four supporting base arches and interconnecting truss work. The 990-foot tower was the tallest artificial structure until New York's Empire State Building was completed 41 years later. A close inspection of the supporting arches at the base of the Eiffel Tower will reveal them to be miniature copies of his crescent arch from the Garabit Viaduct, as each arch just touches the horizontal beam above so that neither line is broken by the other.

Famous in his own time and massively respected for his ability to overcome engineering problems, Eiffel found himself handed the largest assignment of his career in 1889—design and construction of the locks for the Panama Canal. A French consortium had been attempting to build the canal since 1880, but the massively expensive project was mired in mismanagement, corruption, and tropical disease. When the project bogged down and the money ran out, the French government and investors wanted revenge. Eiffel, who had done absolutely nothing wrong, was subsequently found guilty of fraud, a conviction that was later overturned.

Nearing 60 with hundreds of iron and steel structures to his credit, Eiffel threw himself into scientific research. He founded the first institution for the study of aerodynamics in France and experimented with radio transmissions. In both fields of experimentation he used the Eiffel Tower as a high-rise laboratory by equipping it with wind sensors and antennae. He built a wind tunnel at the base of the tower for aerodynamic research and a larger one on the Rue Boileau in Paris in 1909. The second wind tunnel remains in use. Eiffel died in Paris December 27, 1923, just 12 days after his 91st birthday.

See also ARCH BRIDGE; GARABIT VIADUCT; IRON; TRUSS; VIADUCT.

electrochemical chloride extraction *Desalinating America's concrete bridges* The corrosive effects of salts, classed as chlorides, are so pervasive that an estimated $5 billion in bridge corrosion damage is caused each year in the United States. Although not readily obvious, even the steel reinforcement embedded in the concrete is attacked by chloride ions that seep through the porous cement that is concrete's binding component. When exposed to chlorides, reinforcing steel corrodes and expands. Not only does steel become weaker, but the corrosive swelling causes cracking and spalling of the concrete beams, decking, and piers. If steps are not taken to halt the corrosion and repair the damage to the bridge, the structure eventually requires replacement.

For decades the problem of the rapid corrosion of reinforcing steel in concrete bridges was an issue seen only in regions adjacent to salt water. By the 1960s, state, county, and municipal road agencies began the wintertime use of salt-based deicing compounds on bridges and roadways as part of a "bare pavement policy." The use of deicing agents was intended to increase roadway safety during icing conditions as the number of motorists on the road increased. The unfortunate side effect was accelerated corrosion of the steel components of inland bridges.

Perhaps most worrisome to bridge engineers about chloride-induced corrosion is the effect of salts on post-tensioned and prestressed concrete bridges whose concrete components are given tensile strength with steel cabling that is heavily tightened and locked into place. If these cables fail, the concrete component, which is often a beam, also fails.

In previous years, bridges whose steel reinforcement was compromised by chloride penetration would receive an overlay of new concrete. A modern solution to this expensive and potentially hazardous corrosion problem is the process of electrochemical chloride extraction, or ECE.

Although an electrochemical process like cathodic protection—a system that requires the installation of equipment for the life of a bridge—electrochemical chloride extraction is a short-term process. ECE might take between 6 to 10 weeks and causes the migration of the damaging chloride ions away from the reinforcing steel by electrically extracting salt that seeps through pores and cracks of the concrete.

Anodes, or positively charged electrodes, are attached to a concrete bridge, whose concrete surfaces are kept moistened so they will serve as an electrolyte. Electricity is sent coursing through the bridge, and the steel reinforcement serves as a cathode, or negatively charged electrode. The negatively charged chloride ions then move away from the negatively charged steel reinforcement into the positive anodes mounted on the exterior of the concrete. The process is similar to an attempt to push the two ends of a magnet of like polarity together, only to have them repel one another. However, one magnet's negative pole is attracted to the positive pole of another.

The ECE treatment has been found to remove anywhere from 50 to 90 percent of the chloride from concrete bridges, although the process can cause the concrete to lose up to 10 percent of its own hardness. After the application of ECE, future chloride penetration is prevented by the application of a special overlay on the bridge deck with low-porosity cement or a waterproof epoxy.

The Federal Highway Administration was continuing to sponsor tests of ECE systems around the country at the beginning of 2000 to determine how effective the concept is and under what circumstances it is most appropriate.

See also HEATED BRIDGE TECHNOLOGY PROGRAM.

Ellet, Charles, Jr. (1810–1862) *Bridge designer, war hero, and rival of John Augustus Roebling* Born on a farm in Penn's Manor, Pennsylvania, Charles Ellet, Jr., would turn away from the drudgery of farm life and cut a controversial and fiery swath across the field of American engineering. Brilliant, opinionated, and imbued with an iron will, Ellet had accomplishments that were many and varied.

Ellet managed to build one of America's first wire cable suspension bridges over Pennsylvania's

Schuylkill River in 1842 and spanned the Ohio River at Wheeling, West Virginia, with a breathtaking but flawed 1,010-foot suspension bridge. The latter project was one he fought hard to obtain in competition with his rival John Augustus Roebling, who would go on to design the Brooklyn Bridge.

In few ways was Ellet ordinary. His birthday of January 1, 1810, was indicative of his impatience to enter the New Year, and his background was unusual in that he sprang from both Quaker and Jewish roots.

At 17, Ellet obtained employment as assistant engineer on a survey crew laying out routes for canals, then considered the ultimate means of transporting goods and people. During this period, Ellet surveyed 170 miles of the Susquehanna's River's North Branch at a then-princely sum of two dollars a day.

Ellet then lobbied his family for permission to travel to France to attend the prestigious bridge engineering school Ecole des Ponts et Chaussées. By 1830 he had obtained his father's grudging permission and left for France, where letters of introduction allowed him to meet French soldier and politician Lafayette.

After observing the disorder of the French Revolution of 1830, Ellet suspended his studies and traveled throughout Europe to observe bridge and canal construction. He was amazed at the suspension bridges he observed and took copious notes of each. On his return to America in 1832, Ellet proposed a suspension bridge across the Potomac River in Washington, D.C., in response to a Treasury Department request for designs to replace an existing wooden span.

Ellet proposed a span measuring 572 feet between its towers with its road deck rising 46 feet above the Potomac. No matter the merits of Ellet's bridge proposal, his plans had arrived too late to be considered, and the government instead repaired the existing structure.

The year before the loss of his job, an important span considered the longest wooden arch bridge in the world and commonly known as the "Colossus" had been destroyed by fire. This prompted Ellet to lobby Philadelphia County to adopt his own plans for a suspension bridge as a replacement. At about this time, Ellet unleashed a torrent of effort to become a bridge builder.

He also published "A Popular Notice of Wire Suspension Bridges," one of a series of pamphlets he wrote to promote such bridges as well as himself. The propaganda was necessary because the suspension bridge spooked most public officials and investors since early versions invariably collapsed.

In his writings Ellet espoused the use of cables composed of individual iron wires, as opposed to links of iron bars, which yielded far less strength for the same amount of weight. Also during 1839, Ellet attempted to interest Saint Louis in a proposed span across the Mississippi River. Authorized by the city council to prepare a report on the project, Ellet did so, proposing a suspension bridge with a central span measuring 1,200 feet between its towers. Saint Louis's city fathers blanched at the daring plan and showed Ellet the door.

Despite his brilliance, Ellet exposed his own engineering shortcomings in his Saint Louis bridge plan since he wanted to place the two midstream piers on timber piling foundations rather than excavating them to bedrock. James Eads, who designed the first bridge spanning the Mississippi at Saint Louis, knew better and sank the piers of the Eads Bridge to bedrock. It is doubtful that Ellet's could have long survived the scouring effects of the Mississippi's powerful current.

Despite his writings and authoritative stance, Ellet had yet to build a bridge by 1839. However, his chance arrived that year with Philadelphia's acceptance of his design to replace the famed Colossus, a 340-foot wooden span over the Schuylkill River. When news spread of Ellet's successful bid, the German immigrant and engineer John Augustus Roebling wrote to Ellet on January 28, 1840, expressing his own enthusiasm regarding suspension bridges and inquiring about assisting Ellet in the project.

Ellet's reply had a condescending tone but did not specifically spurn Roebling's offer to help:

> It has given me much pleasure to learn that you have not neglected the subject of suspension bridges, in pursuing your professional studies abroad; and that you consequently appreciate the merits of that system of construction.

Although Roebling subsequently sent drawings and suggestions related to the suspension bridge's design, he heard no more from Ellet about employment, although the two continued to discuss bridge design. Although Ellet fully expected to be the one selected to execute the design of the bridge, he was angry to learn that a local contractor, Andrew

Young, had been selected to build the suspension bridge. Young, in turn, hired Roebling to assist him with the design and construction of the bridge.

Ellet, enraged by this turn of events, wasted no time in circumventing Young by offering to accept part of his payment in toll bridge stock and to finance part of the bridge with his savings and a considerable inheritance left to his wife. County officials were swayed by this offer and in June 1841 nullified their contract with Young, whose turn it was to be incensed. Roebling, for his part, had just lost his first bridge-building job.

The contract for the bridge at Fairmount called for construction of a suspension structure with a span of 357 feet. The bridge would contain a 26-foot-wide road deck that would accommodate an 18-foot lane for carriages and a pair of four-foot walkways. It was to be built at a cost of $50,000 but came in $3,000 over budget at the time of its completion in the spring of 1842. This cost overrun was not upsetting to Ellet, who savored the success of not only completing his first bridge but doing so on his own terms.

Ellet was now officially a bridge builder, and he had latched onto a surefire way to secure bridge contracts—by accepting stock in lieu of payment in the toll bridges he designed and built. The next major bridge project Ellet set his sights on was that of a suspension bridge across the Niagara River below the falls. By this time, Roebling had also established himself as a suspension bridge designer and builder.

The two rivals presented their proposals, but the savvy Ellet lobbied municipal leaders heavily and offered to accept $30,000 in bridge stock as his payment for building the bridge. Roebling tried the same tactic but lacked Ellet's financial resources, and Ellet was given the contract.

Ellet succeeded in spanning the Niagara Gorge with an iron cable that he crossed in a large bucket suspended from a pulley. Subsequently, Ellet constructed an iron wire footbridge, which he crossed while riding in a carriage. After these initial successes the project soured, as Ellet complained that the Canadian and American bridge companies involved were not making their payments. His response was to begin charging and keeping toll payments for the footbridge.

The companies dismissed Ellet as the chief engineer and obtained an injunction barring him from the bridge. As soon as the injunction was removed in

October 1848, the cantankerous Ellet posted on the bridge guards armed with a cannon. The resulting litigation was settled that December, when Ellet agreed to abandon the project. In 1851, the bridge companies offered the contract to Roebling, who completed the bridge.

While struggling with the Niagara span, Ellet had already embarked on a bridge-building job that would be not only his greatest but also his last—the suspension bridge over the Ohio River at Wheeling, West Virginia. Ellet had lobbied for the job since 1836 with the Wheeling & Belmont Bridge Company and for a while had worked on both the Niagara span and the Wheeling Bridge simultaneously. Unfortunately, the Wheeling Bridge that Ellet completed in 1849 suffered extensive wind-induced road deck damage five years later. Ellet designed the repairs to the bridge that still stands today.

As America drifted toward the Civil War in 1860, Ellet was using his considerable skill as a self-promoter to offer his engineering expertise to the United States Army. The indefatigable Ellet proffered a scheme for secretly resupplying besieged Fort Sumter and volunteered as an unpaid aide to Major General George B. McClellan. He even offered to lead a guerrilla force behind Confederate lines. Frustrated that his offers of help were unanswered, Ellet volunteered to serve as a balloonist who would ascend to observe Confederate troop movements.

Ellet eventually persuaded the Union Army to allow him to built a fleet of steam-powered rams, boats with hardened prows that could ram and sink Confederate boats blockading the Mississippi River. News that the Confederates had their own fleet of rams gained Ellet the permission he sought on March 27, 1862. By June 5, what became known as his Ram Fleet was engaged in a wildly successful river battle to take Memphis, Tennessee. Ellet's boats provided a decisive edge in the famous battle, but Ellet received a bullet wound to his knee while leading his fleet. Ellet died on June 21, 1862, as his boat came within sight of Cairo, Illinois.

It is possible that Ellet would have been little remembered except for the Wheeling Bridge, which still stands, had it not been for his daughter, Mary, who lived to be 91. She spent the remainder of her life attempting to ensure that her father's role in the Civil War was not diminished. In 1937, seven years after Mary Ellet's death, a U.S. Navy destroyer was Christened the U.S.S. *Ellet* in honor of Charles Ellet

Jr.; his brother, Alfred, and Ellet's son, Charles Rivers Ellet, who also served in the Ram Fleet.

See also EADS BRIDGE; ROEBLING, JOHN AUGUSTUS; ROEBLING, WASHINGTON AUGUSTUS; SUSPENSION BRIDGE; TIMBER BRIDGE; WALSH, HOMAN; WHEELING BRIDGE.

Ellis, Charles Alton (1876–1949) *Forgotten designer of the Golden Gate Bridge* Charles Alton Ellis should be famous as the designer of the Golden Gate Bridge, but he is little known. Born in Parkman, Maine, Ellis seemed bound for a career in mathematics or classical studies rather than bridge engineering.

Tall, polite, and appreciative of the irony of life, Ellis obtained a degree in mathematics and Greek language at the turn of the 20th century. Ellis took his mathematics skill to the American Bridge Company, one of the premier builders of bridges and buildings in the United States. He began to build a career as a structural engineer by calculating the stress factors presented by ground conditions to subway tubes under the Hudson River. Ellis obtained additional formal education in the spanking new field of structural engineering and wrote a standard textbook on the subject, *Essentials in Theory of Framed Structures.*

It seemed only natural that Ellis should find his way into academics, and he did so in 1908 when he became a civil engineering instructor at the University of Michigan. From there he joined the University of Illinois engineering school. Ellis, with no shortage of engineering ideas, published widely in professional journals and was a member of the American Society of Civil Engineers. In 1922, Ellis's life took another abrupt turn when he went to work in Chicago for Joseph Strauss, the inventor of an improved bascule bridge, the founder of the Strauss Bridge Company, and a man completely obsessed with building the Golden Gate Bridge.

Ellis's accomplishments and credentials were needed by Strauss—who had built only smallish drawbridges—in his quest to be the man who bridged the Golden Gate Strait north of San Francisco. While Strauss attended meetings and promoted the idea of the bridge in northern California, Ellis assisted Strauss in building bigger bridges elsewhere while tinkering with the design of the Golden Gate span. Strauss, an inventive and dynamic promoter, was no deep-thinking bridge guru. When Strauss admitted that his own bridge concept was inappropriate—it was an ugly and dubious combination of the cantilever and suspension bridge types—it was Ellis who labored to produce the sleek suspension bridge design that was ultimately built.

Working for Strauss was never easy. Mercurial and complex, Strauss was the type of boss who would fire and then rehire someone on the same day and who considered any idea of an employee to be his own.

Ellis's downfall occurred when he committed two major sins: Ellis, the respected academic, made a political error when he publicly divulged how much work he had done designing the bridge during a 1930 talk to the National Academy of Sciences. His second sin was to tinker too long with calculations related to the Golden Gate Bridge's crucial towers. Ellis was ordered to take a vacation in December 1931 by Strauss, who later informed him by letter that he was fired.

Ellis, unemployed in the Great Depression, sought work for three years as he continued to tinker with calculations to tweak the design of the bridge while maintaining a correspondence with the project's consulting engineers, Othmar Ammann and Leon Moisseiff. In 1934, Ellis finally managed to obtain a teaching position on the faculty of Purdue University. He retired in 1947 at the age of 72 and died in Evanston, Illinois, in 1949.

Friendly and respectful, Ellis would openly tell others that it was he who accomplished the detailed design of the Golden Gate Bridge, which became massively famous even before it was built. Strauss, maniacal about retaining credit as the mastermind of the bridge, never made an effort to credit Ellis's role, which was recognized by both Ammann and Moisseiff.

Although Ellis could have caused Strauss considerable embarrassment by taking his firing to the media and setting the record straight on his role in designing the bridge, Ellis preferred inspiring other engineering students for the remainder of his teaching career. The bronze plaque commemorating those with a role in building the Golden Gate Bridge contains the names of Robert Baermann Strauss, politicians, and other engineers but no mention of Charles Alton Ellis.

See also GOLDEN GATE BRIDGE; STRAUSS, JOSEPH BAERMANN.

English,   Thomas *See* Beaumont tunneling machine.

Eurotunnel, "Channel Tunnel," "Chunnel" (built 1987–1991) *How not to build the longest undersea tunnel* Tunneling across the English Channel was a project that obsessed levelheaded engineers, dreamers, and outright crackpots on both sides of the waterway for at least 250 years. And although those who sought to tackle this Holy Grail of tunnel engineering recognized it as a monumental challenge, none could have foreseen that politics, government regulation, and bureaucratic infighting would threaten this undersea passageway as much as the hazards of digging a 24-mile tunnel under water.

When this great engineering exercise was finally attempted, it was plagued by financial difficulties, ill will between the contractors and the tunnel owners, expensive government regulation, and heart-stopping cost overruns. The sum total of the experience in building the tunnel was to provide engineers and businessmen lessons on how not to tackle a huge infrastructure project.

When the tunnel was finally built—at twice the cost originally estimated—many were left wondering whether the project had been worthwhile. Was it, some critics wondered, a challenging dream that had been converted into an all-too-real sinkhole for nearly $18 billion?

Added to the technical challenges of burrowing beneath the English Channel, which were formidable, was the political swamp of connecting England with the continent that it had viewed with suspicion for hundreds of years. Talk of a Channel tunnel brought to the fore traditional English distrust of what some perceived as the untrustworthy minions of the Continent. Because of the history of warfare between England and France, general European conflict before 1900, and the small matter of two world wars, the politics of building a link between England and the Continent had already proved fatal to Channel tunnel schemes for more than 100 years.

Although opponents still made noises against a Channel Tunnel, resistance was slowly fading by the 20th century. In 1936, Member of Parliament Winston Churchill publicly supported construction of a Channel Tunnel, and engineers continued to present exciting drawings of tunnel schemes. But a pair of world wars squelched serious efforts until a group of Americans decided to revive the idea in 1957.

This effort, known by the simple name Technical Studies, was organized by American attorneys, Frank Davidson and Cyril Means, Jr. Davidson's brainstorm occurred while he, his wife, and their infant made a storm-tossed crossing of the English

One of the tunnel-boring machines used during the construction of the Eurotunnel is seen here after emerging into one of two crossover tunnels connecting the two main rail tunnels. The 24-mile subaqueous tunnel system was a dream of engineers for more than 200 years before it opened in 1991. *(QA Photos Ltd.)*

Channel. As the 1 1/2-hour trip stretched into seven seasick hours, Davidson recalled reading of previous plans to tunnel beneath the Channel. Davidson contacted Means, and both began to establish ties with the world's largest investment banks and engineering firms. They also formed an alliance with the Channel Tunnel Company, an entity granted a 99-year concession in 1881 by the French government to build a Channel tunnel. This consortium then became known as the Study Group.

The Study Group eventually spent more than a half-million dollars on geologic studies and produced a three-tunnel plan by the early 1960s. Bechtel Corporation, Brown & Root Construction Company, and Morrison Knudsen, three of the world's biggest construction concerns, assisted in the planning. Like Davidson and Means, all three companies were American. The final author of the three-tunnel plan was the septuagenarian Charles Putnam Dunn of Morrison Knudsen.

Dunn's design consisted of a pilot tunnel that would serve as a geologic scout in advance of, and parallel with, two larger rail tunnels. The pilot tunnel would eventually become a service tunnel and ventilation shaft. Dunn's 1959 proposal, with refinements, would be very similar to the actual tunnel system put in place nearly 35 years later.

A competing proposal to span the Channel with a bridge soon appeared in response and was diligently pitched to the French and British governments. Dunn's plan was nonetheless selected as the preferred concept. This physical linkage of England and France seemed close to reality until the French president, Charles de Gaulle, vetoed British membership in the Common Market in 1963. This diplomatic slight soured the Anglo-French tunnel project, although the French, while opposing British membership in the Common Market, really did want a tunnel.

Perhaps the most important thing the Study Group did during the late 1950s and early 1960s was to generate interest in the profit potential of an undersea link between Britain and Europe. There was money to be made by replacing long and sometimes arduous ferry rides with high-speed, nonstop rail travel. Of even greater consequence was the potential to eliminate the inefficiencies of cross-channel freight transport. Goods had to be loaded aboard ships for a trip across the water and again transferred to trains or trucks for inland transport.

With interest rekindled in a tunnel by the Study Group, a new era of tunneling plans emerged. By 1970, a trio of groups was attempting to advance competing Channel Tunnel concepts. Of these proposals, the government of the Conservative British prime minister Edward Heath backed one in 1973. The Anglo-French project was beset by inflationary cost increases during its planning in a period of dire economic conditions and turbulent British politics. When the Conservatives lost power in 1974, the incoming Labor Party under Prime Minister Harold Wilson dropped its support of the Channel Tunnel project only 16 hours before the excavation was to begin in 1975.

Despite the false starts during the 1970s, the reasons favoring a Channel Tunnel multiplied as trade and travel grew between France and Great Britain. Analysts accurately predicted a massive increase in cross-Channel traffic by 2000 which has occurred.

The return of a Conservative government in Britain ushered in a new era of receptiveness to the Channel Tunnel. The Conservative prime minister, Margaret Thatcher, and the Socialist French president, François Mitterand, despite their considerable ideological differences, were both wholeheartedly in favor of a Channel Tunnel. The tunnel game was on again when the two leaders signed a 1981 agreement to study its feasibility, though Thatcher steadfastly opposed spending a farthing of government funds on the project. The private sector would have to foot the bill.

Leading the effort to build a tunnel was the Anglo-French consortium known as Channel Tunnel Group/France Manche (CTG/FM). It was composed of 10 contracting firms split between French and British companies, and five banks, of which three were French. CTG/FM's design would transport freight and passengers as on an ordinary train while also carrying passenger cars and cargo trucks on specially designed rail cars. No one would drive a Renault or Jaguar through this tunnel system.

Although a 1984 report on the feasibility of a Channel tunnel supported the three-tube design (strikingly similar to Dunn's), several competing designs arose. One complex proposal, known as EuroRoute, called for bridges running offshore to artificial islands off both coasts where drivers would use spiraling ramps to enter an immersed tube tunnel for a drive along the floor of the Channel. The EuroRoute's backers claimed it would cost $9 billion. The vehicular portion of this plan was similar to the Chesapeake Bay Bridge–Tunnel in Virginia. A French proposal named the *Europont* ("Euro-bridge") suggested a bridge running the full 24 miles from Calais to Dover using support piers and synthetic fiber suspension cables.

An opponent of the Channel Tunnel who would later become a competitor with a fourth proposal was the Texan James B. Sherwood, who oversaw extensive shipping and ferry operations between Britain and the Continent as president of Sea Containers. Sherwood, whose own operations would face competition from a tunnel, mounted an aggressive advertising and public relations campaign against a Channel tunnel. When he could not beat the tunnel, he decided to join the fray with the Channel Expressway, one undersea tunnel for rail traffic and another for automobiles. People, Sherwood maintained, wanted to drive their cars, and a rail system that shuttled automobiles was no more than a ferry system.

The British and French governments made an official call for proposals on April 2, 1985. What they wanted was a link between the two countries with a lifespan of 120 years. The proposals suggesting vehicular tunnels were rejected because of the hypnotic effect that driving through a long tube has on drivers, and the bridges proposed by the Euro-Route and EuroPont plans were rejected because they would pose navigational hazards in waters plagued by bad weather and poor visibility.

Since the investment banks were unwilling to loan money to an organization that would both own and build the tunnel, the consortium split itself into two separate entities in 1986. Eurotunnel would own and operate the tunnel; the engineering and construction firms that formed into TransManche Link (TML) would do the digging. Eurotunnel signed contracting agreements with TransManche Link, but these agreements pleased neither side and were renegotiated over the years as the relationship between the two groups deteriorated.

Since the engineers had done the planning while being their own bosses, the invention of the Eurotunnel group was seen as an unwelcome layer of authority. They viewed Eurotunnel as merely the source of money for the tunnel the contractors would design and build. Another problem involved the builders' taking on the role of procuring trains and rolling stock, a responsibility that was far outside their ordinary realm of experience.

Although the project received approval in March 1986, complicated financial and planning problems delayed its start until 1987. A month before public shares in the venture were sold, the "Black Monday" stock crash of October 19, 1987, threatened to swamp the fund-raising. Capital was raised nonetheless and the project lurched forward. As it did so, the contractors began to chafe under what they thought were unfair demands from Eurotunnel that sought changes in the design. Eventually, Eurotunnel and the contractors developed a reciprocal hate so intense it unnerved banks and investors, making the project even more difficult.

Added to these issues were unrestrained and expensive changes in governmental safety requirements heaped upon the project as it progressed. One notable dispute was over the door width of passenger cars. When it was determined that 700-millimeter-wide doors were to be required, a large number of cars with narrower doors had already been built, wasting more than $80 million and causing a nine-month delay. Added to these woes were claims by contractors that Eurotunnel owed them hundreds of millions of dollars in expenses generated by unnecessary design changes.

The work, however, continued. At Folkestone in England and at Sangatte in France, massive shafts were dug and lined with reinforced concrete so workers and equipment could descend to begin the tunneling. Gigantic tunnel-boring machines were lowered in segments and assembled at the working face of the tunnel. These massive machines with huge circular cutting heads would grind toward the middle of the Channel from opposite directions until they rendezvoused beneath the Earth. Other boring machines would dig in the opposite direction to provide land tunnels leading to inland terminals.

Geological surveys indicated that a layer of chalk marl, a type of limestone formed by the deposition of the calcareous shells of tiny protozoa, ran the distance between England and France. Chalk marl was strong enough to support its own weight and soft enough to be easily excavated by the boring machines. In fact, these ideal characteristics are what made the Channel tunneling effort feasible from an economic standpoint.

Not all chalk is created equal, and upper layers of chalk near the Channel's seabed were porous and weak. Engineers began their tunneling deep enough to allow them to remain below a layer of chalk marl that was both impermeable and strong. Digging too high would invite flooding or collapse. Going too deep would penetrate the chalk floor and strike an oozing layer of Gault clay, a poor material for supporting the invert, or floor of the tunnel. This meant that tunneling required precise measurements to ensure the huge boring machines not only traveled in the proper direction but were at the right depth.

The French, anticipating wet, porous ground, constructed waterproof boring machines while the British, armed with incorrect geological data, employed machines that were designed for dry tunneling. The British were forced to modify their machines as water poured in on their tunneling effort in an early layer of unanticipated wet rock that persisted for nearly two miles. The French, who initially operated their boring machines improperly, made little headway early in the project. This predicament was made worse when national pride caused them to resist help from engineers on the far side of the Channel.

French and British tunnelers, while boring through the chalk, would immediately line the freshly excavated walls behind the boring machine with precast, reinforced concrete segments. Each of the curved segments was cast with holes so cement grout could be injected between the completed ring and the chalk marl to ensure a solid fit. The lining segments under land had to be far thicker to accommodate the weight of anything built above, so these were made 21 inches thick. Those supporting the tunnel beneath the Channel needed to be only 11.4 inches thick.

The reinforcing steel within the concrete segments needed to be protected from the corrosive effects of salt water, which would undoubtedly seep into any undersea tunnel. Should salt come into contact with the embedded steel, the resultant corrosion and swelling would crack and spall the concrete and weaken the lining. To preclude this damaging effect, the extremely fine and impermeable cement mixture that was developed not only resisted penetration by corrosive chloride ions in salt water but also made the segments monumentally strong.

A potentially disastrous event occurred when the boring machines struck segments of rotten chalk, allowing water to roar into the tunnel. Since this stratum was too weak to support itself, tunnelers had to abandon the use of concrete lining for far stronger, but exorbitantly expensive, cast iron segments at various points.

The boring machines ground their way forward 24 hours a day. During the peak of the tunneling, as many as 2,000 worked on each side of the Channel around the clock. At one point, more than 13,000 workers and engineers were involved in the project.

Engineers, anticipating the removal of a staggering amount of spoil from the excavation, designed massive fills for holding it on both sides of the Channel. On the English side the 5.3 million cubic yards of chalk marl removed from the tunnels by conveyor was used to create a park rising seven feet above the level of the Channel at Shakespeare Cliffs. A massive seawall measuring 2,221 feet in length and up to 38 feet thick and containing 235,000 yards of concrete was built to contain the spoil.

The French decided to store their spoil on land and built a dam using 2.4 million cubic yards of crushed chalk removed during preliminary excavation work. Within this dam they pumped 4 million cubic yards of chalk slurry by pipeline.

Inside the tunnel, workers labored in temperatures approaching 100 degrees Fahrenheit as boring machines and other equipment gave off large amounts of heat. When the tunnel became operational, heat would be generated by the frequent passage of trains traveling near 100 mph. Their sizable electric motors, the friction of the steel wheels on the tracks, and heat from other sources, including lighting, had the potential to raise the tunnel's temperature to above 130 degrees Fahrenheit.

Although the engineers were fully aware the tunnel needed a cooling system, such a system was intentionally stricken from the original proposal. This subterranean subterfuge was accomplished to make the tunnel less expensive and more enticing to investors. The cooling system, it was reasoned, could be added before the tunnel was opened for rail service. What was eventually developed was a $200 million water-chilling operation with facilities on both shores. The cooled water was pumped halfway through the tunnel from each side and the warmed water returned to the cooling plants.

Another atmospheric problem in the rail tunnels was the buildup of air pressure ahead of high-speed trains. The trains would act as pistons compressing the air ahead of them inside the snug tunnels that served as cylinders. This compressed air could generate enough resistance to retard the speed of the trains. Unless this pressure was alleviated, any attempt to drive the trains faster would generate even more air pressure. Engineers designed piston relief ducts every 820 feet that allowed air pushed ahead of a train to be forced into the other rail tunnel while sucking it out of the ducts behind the train.

Other amazing features of the tunnel include a cavernous pair of undersea crossovers. These cathedrallike spaces allow trains to turn around so they can travel in the opposite direction and were the largest subaqueous spaces ever constructed. The British crossover was 5.4 miles from the English coast and the French crossover was nearly eight miles offshore from France. The spaces were excavated and their roofs and walls reinforced with rock bolts and sprayed concrete.

A disaster seemed imminent when the foot-thick concrete lining in the ceiling of the British crossover began to crack. The problem was diagnosed not as a collapse but as the result of water under pressure trapped between the rock and the concrete lining. Holes were drilled to relieve the pressure and the problem was remedied.

In the fall of 1989, it appeared Eurotunnel might lapse into bankruptcy, but additional funding was eventually raised to keep the project alive. All the while, growing acrimony was the rule between Eurotunnel and the TransManche Link companies. Officials of Eurotunnel viewed the builders as fleecing it of money through chicanery; harried engineers and contractors groused they were being micromanaged by Eurotunnel's own engineers.

The builders also complained that engineering changes demanded by government officials and Eurotunnel were adding hundreds of millions of dollars in unnecessary costs. Many observers believed that the rancor between the two groups had a far greater chance of destroying the project than the physical collapse of the tunnels.

By the fall of 1990, the British and French tunnel-boring machines digging the 15.7-foot-diameter service tunnel neared one another underground. On December 1, 1990, a carefully staged, hand-dug breakthrough involving French and British workers took place on live television. Massive celebrations were thrown on both sides of the Channel with tunnel workers and engineers given a moment in the spotlight after spending precisely three years underground.

The 25-foot-diameter main rail tunnels would soon experience the same type of celebrations. On May 22, 1991, the north tunnel was linked, and on June 28, 1991, the south tunnel was completed. A total of 12 tunnel-boring machines were used in the project, six driving toward their counterparts under water and six more digging approach tunnels on land.

What followed were more arguments, fiscal disputes, and recriminations between Eurotunnel and the contractors' group, TransManche Link. By fall of 1992, these problems so unnerved the investment bankers that they considered withholding additional funding of the project. Last-minute maneuvering averted this fiscal catastrophe, and the investment banks provided the needed capital. On December 10, 1993, passenger trains from both countries crossed beneath the Channel, and a ceremony in which TransManche Link officially turned over the completed project to Eurotunnel was held.

The Channel Tunnel, which the abbreviation-loving British shortened to "Chunnel," is now known by the sleek official name Eurotunnel. Because of additional delays, Eurotunnel would not officially open until May 6, 1994. Despite even this "official opening," not all services were operational until the following December. Passengers can now be whisked beneath the Channel from portal to portal in 21 minutes at speeds approaching 100 mph.

On November 18, 1996, a massive fire raged through the tunnel when cargo trucks being carried aboard rail cars burned as the result of arson. This event, referred to by the media as the Channel Tunnel fire, caused no fatalities but inflicted $90 million in damage to the interior of the tunnel and sent seven people to the hospital.

Since its opening, the tunnel has experienced rough financial sledding and has had to restructure its massive debt several times with the 220 banks that funded the project. It was not until 2000 that Eurotunnel was able to announce that it had earned a slender profit.

See also CHANNEL TUNNEL FIRE; CONCRETE; DESMAREST, NICOLAS; GROUTING; MATHIEU, JACQUES-JOSEPH; SUBAQUEOUS TUNNEL; THOMÉ DE GAMOND, AIMÉ; TUNNEL-BORING MACHINE; TUNNEL LINING; WATKIN, EDWARD.

Fairbairn, William (1789–1874) *Self-taught Scottish engineer who helped invent the hollow-box girder* William Fairbairn, one of several great self-made Scottish engineers of the Industrial Revolution, became a collaborator with the bridge designer Robert Stephenson in the development of the tubular bridge concept that introduced the hollow-box girder or beam to structural design.

Stephenson, the son of the steam locomotive inventor George Stephenson, knew Fairbairn through his father since the two had worked together previously. It was the elder Stephenson who recommended in 1845 that the innovative Fairbairn help his son with the problem of creating a hollow-box beam that would allow the bridging of Wales's Menai Strait without the use of arches.

Fairbairn, who had mastered the art of iron shipbuilding, applied his skills to the problem and helped perfect what is today known as a hollow-box beam. This beam was to be so large that it not only served as a supporting structure for the bridge deck but was actually a square iron tunnel or "tube" through which a train could run.

The hollow-box beam provided a magnificent amount of strength for the weight involved since a solid beam of any material was so heavy as to be an inefficient structural concept. The amount of tensile strength provided by a hollow beam was supplemented by the property that it was far lighter than a solid iron or timber beam that must still carry its own weight or dead load. The self-supporting nature of a box beam was a boon to structural engineers and bridge builders, who could now support bridges with load-bearing beams weighing a fraction of the weight of solid beams.

Fairbairn would later feel slighted by Stephenson after the 1850 completion of the famed Britannia Bridge, which employed four sets of the innovative tubes. Soon after the bridge was built, Stephenson cast himself as the practical engineer who solved the problem with less-than-essential help from Fairbairn and a mathematician named Eaton Hodgkinson.

Although historical camps have divided over the issue, it is generally believed that Fairbairn's study of the problem and the practical solutions he provided for assembling and strengthening the Britannia's hollow beams made the innovation workable. Fairbairn was far from a fuzzyheaded theorist: He was an accomplished toolmaker who had even designed hydraulic hammers, allowing workers to seat the nearly 2 million rivets required in the assembly of the Britannia's hollow-box beams.

See also BRITANNIA BRIDGE; HOLLOW-BOX GIRDER; LIVE AND DEAD LOADS; STEPHENSON, ROBERT.

falsework The erection of falsework, or temporary supports, to hold up structures during construction is an ancient technique. This method began thousands of years ago to support unfinished structures, including stone arch bridges whose arches would not support them until all their stones were in place. The introduction of concrete arch bridges during the 19th century also called for the use of falsework since poured concrete had to be allowed to harden.

Falseworks are traditionally constructed of timber and are erected to follow the shape of whatever is being built. The falsework holds up the stone or concrete until it is completed and the cement bonding it has cured. The falsework is then removed.

See also ALCÁNTARA BRIDGE; MAILLART, ROBERT; SALGINATOBEL BRIDGE.

Federal Highway Administration *See* National Bridge Inspection Standards.

fiber reinforced polymer composites *Bridge materials for the 21st century* Most people know fiber reinforced polymer (FRP) composites as the lightweight but strong epoxy-impregnated fiberglass used in boat hulls and surfboards, but modern FRP composites have been refined to compete with concrete and steel as materials for the construction of bridges.

Some FRP composites contain a polymer resin mixed with glass fibers, but they may also contain aramid fibers (known by the trade name Kevlar) or carbon fibers, the strongest but most expensive of the three types.

The concept of FRP composites is nothing new. For thousands of years people built buildings of mud brick made stronger by the addition of straw. The straw provided tensile strength to the dried mud, which already possessed significant compressive strength. Alone, the materials have far less strength, but when combined they produce a good structural material. But it was not until the 1950s that resin-coated glass fibers came into wide use.

Modern technology has refined this concept to make lightweight FRP materials stronger than steel. Thermosetting polymers (plastic resins that can be permanently cured or hardened with heat) are the modern mud. The highly refined reinforcing fibers are today's "straw." Thermosetting resins are used because they are cured at very high temperatures, causing an irreversible cross-linking of their molecules. These resins not only are extremely hard but are also resistant to heat. Another type—thermoplastic resins—melt at an elevated temperature and harden when cool but can soften or liquefy if again exposed to heat. A thermoset resin retains its hardness forever.

The resulting material is highly resistant to heat, chemicals, abrasion, and corrosion. Steel, a widely used structural material, is susceptible to corrosion. FRP composites hold out the promise of a structural material that could be lighter and stronger than steel but capable of enduring rain and road chemicals without corroding.

Despite the advantages of polymer resins, researchers had to improve the capabilities of these polymers since many were prone to degradation by ultraviolet radiation. However, modern FRP composites are highly efficient materials that have been widely used in Europe and Japan since the 1990s.

Engineers are hopeful that FRP composites will be a useful tool in the war against an American epidemic of substandard bridges weakened by corrosion. American engineers, with the encouragement of the Federal Highway Administration, are experimenting with composite materials by using FRP beams to replace steel ones damaged by corrosion. Road decks of FRP composites are also being installed to remedy the corrosion that causes the destructive swelling of steel reinforcement inside concrete.

Damaged concrete road decks are also receiving layers of fiber reinforced polymer sheeting to repair damage and seal their surfaces to prevent further corrosion of their steel reinforcement. In some cases, weakened concrete or steel beams and columns have been wrapped in FRP composites to augment their strength and protect them from moisture and corrosive deicing chemicals.

Because of the high strength to weight ratio of FRP composites, bridges that receive structural beams or decks of the material can often be made capable of withstanding heavier loading while the dead load of the structure is significantly reduced.

The reinforcement of bridge components is especially important to spans in seismically active regions such as California. Making bridges safer through the process known as seismic retrofit is an endeavor in which FRP composites have made a great contribution. Wrapping cylindrical or rectangular bridge piers with FRP composites greatly enhances their ability to withstand shear forces that can shatter the columns and collapse a bridge.

Although bridges made completely of FRP composites are still fairly rare, their use is becoming more common. The Aberfeldy Footbridge over the Tay River in Scotland is one. Completed in 1990, the cable-stayed bridge has a deck of FRP composite panels and a supporting pylon made of the same material. The cable stays transmitting the load of the deck to the top of the pylon are made of woven Kevlar. The bridge spans a respectable distance of 371 feet.

See also ELECTROCHEMICAL CHLORIDE EXTRACTION; NATIONAL BRIDGE INSPECTION STANDARDS; SEISMIC RETROFIT.

Finley, James *See* Schuylkill Suspension Bridge.

firedamp For centuries coal miners have learned to fear and respect the methane gas created from the decomposition of coal known as firedamp. When mixed with air, it can ignite with massive explosive force and has been the cause of thousands of coal miner deaths in the world's worst mining disasters.

The principal component of firedamp is methane, an odorless, colorless, and tasteless gas. Less toxic than carbon monoxide, which is known as "whitedamp," methane can cause asphyxiation if it is present in high concentrations. The main dangers presented by firedamp are fire and explosion. When a 5 to 15 percent mixture of methane is combined with air, it produces a mixture that is highly flammable.

Canaries and specially designed lamps that would extinguish in the presence of firedamp were used during the 19th and the early part of the 20th centuries to detect this dangerous gas. The word *firedamp* is derived from the German word *dampf*, meaning "vapor."

Before the 20th century, miners lacking spark-free electrical lighting had to depend on open flame oil lamps or acetylene lamps to illuminate their work. The presence of open flames only added to the danger of working in an environment threatened by a flammable gas.

Ventilation has always been the best defense against firedamp buildup. Modern detection devices reveal the presence and concentration of firedamp, allowing mine operators to increase ventilation when necessary. If the levels are too high, an evacuation of the mine might be ordered until ventilation reduces the risk.

Mine operations must be extraordinarily vigilant against firedamp. Even modern techniques and equipment are of no help if they are not employed. The May 9, 1992, explosion at the Westray Coal Mine in Nova Scotia that killed 26 men tragically demonstrated this. A failure to pay heed to a methane monitor coupled with a buildup of explosive coal dust led to a methane fire that triggered a coal dust explosion. A digging machine that struck sparks off a chunk of pyrite originally ignited the methane.

The resulting scandal rocked Canada and prompted the firing of Canadian mine safety officials. Canadian prosecutors also launched an unsuccessful criminal prosecution of the operators of the mine.

See also COAL DUST; VENTILATION.

fire-setting *Hard rock mining using fire and water* Mining is an ancient enterprise that preceded the invention of modern explosives and mechanized digging tools by thousands of years. In soft soil, ancient miners could use antlers or wooden tools for digging, but hard rock was another matter. Although penetrating solid rock seems impossible without modern technology, techniques were developed 5,000 years ago in the Bronze Age to allow ancient miners to penetrate the hardest rock.

It was done with two of the most fundamental entities available: fire and water. When ancient miners encountered rock too solid for removal by hand, they built huge fires against the working face of a tunnel and heated the rock for long periods. After the fires died down, water would be dashed against the rock face. The rapid cooling of the rock would cause layers to spall off or create deep cracking that allowed removal with hand tools. The spoil would be mucked out and the process repeated.

This fire-setting method was so useful an innovation in hard rock mining that its use persisted in some parts of Europe until the 19th century, nearly 300 years after the introduction of black powder as a rock-removing explosive. Fire-setting was a simple procedure but was generally limited to regions where timber was readily available as fuel.

Firth of Forth Bridge (built 1882–1890) *Scotland's landmark cantilever railroad bridge* Cantilever bridges, although having extreme strength and rigidity, are not as pretty as slender suspension bridges or graceful arch designs. The Firth of Forth Bridge is a cantilever design much criticized for its hulking appearance, and yet, so large, so strong, and so mechanically ingenious is the Victorian span that to some it possesses a unique beauty.

It was financed by a consortium of Scottish and English railways that wanted to connect northern and southern Scotland across the Firth of Forth, or the "mouth" of the tidal estuary known as the Forth.

Although the bridge's appearance is often criticized, its structural soundness cannot be faulted,

This vintage postcard shows Scotland's Firth of Forth Bridge in all its overbuilt glory. Designed to be as strong as possible in response to the 1879 collapse of the nearby Tay Bridge, the cantilever truss span still carries rail traffic. The superstructure of the bridge was built entirely of steel. *(Author's Collection)*

since appearance was the last thing on the minds of its designing engineers, Benjamin Baker and John Fowler. Like many great engineering projects, the Forth Bridge, as it is commonly known, was shaped by a series of fatal rail bridge collapses in America and Great Britain. The bridge had to be structurally sound, unlike these scandalous past failures, in response to a growing culture of public safety in Great Britain.

As fate would have it, the most shocking of these rail bridge failures was the 1879 collapse of Scotland's Tay Bridge in hurricane-force winds that killed 75 passengers and rail crew. This disaster, a mere 35 miles north of the proposed Forth Bridge, also smashed the reputation of its designer, Sir Thomas Bouch, who had intended to design the Firth of Forth Bridge.

The Forth is more than a mile across and as deep as 218 feet. The width alone would have made the numerous piers required for a conventional pier and beam bridge outlandishly expensive. Also, the water at midstream was far too deep for the caisson technology of the day. Bouch's proposal of a suspension bridge across the Firth of Forth would have required only three main piers, making it one of the world's few suspension rail bridges. John Augustus Roebling's Niagara Gorge Bridge was the only other one of note during the 1800s.

With Bouch out and the Baker-Fowler team in, it should be pointed out that Fowler, who was 23 years older than Baker, had suggested a cantilever bridge across the Forth during the late 1840s. Can-

tilever sections would permit long, unsupported spans across the middle section of the Forth and also require only a trio of main pier systems. The trade-off was that a suspension bridge would have required far less structural material and would have been a far more svelte design.

Baker and Fowler's design was accepted and both men went to work spanning the Forth. Although the central spans would total only 3,420 feet, these would be connected to shore by truss-reinforced rail decks supported by masonry piers. The total length of the bridge would be 8,239 feet. The primary structural component of the cantilevers would be 12-foot-diameter steel tubes that would form diamond-shaped cantilever structures reinforced with a web of steel trusses. Each cantilever would rest upon three sets of four cylindrical piers seated on bedrock.

A trio of diamond-shaped cantilevers would rest upon their piers with each sporting two 680-foot spans being built simultaneously. With the spans of the outside cantilevers connected by elevated pier and beam decks, two pairs of spans would be built within 350 feet of each other. Rigid segments measuring 350 feet apiece would then bridge the two gaps. When the two opposing cantilever truss spans totaling 1,360 feet were completed by their 350-foot center spans, the two clear spans would measure 1,710 feet each. The 1,710-foot main spans would be longer than the 1,595 feet of the main span of the Brooklyn Bridge.

The foundation work for the piers, the toughest task, commenced in 1892. Although piers on the northern half of the bridge could be easily seated upon bedrock that lay near the surface, those on the south would have to be placed by using pneumatic caissons to allow men to work deep underwater. Each cantilever required a pier system consisting of four individual pillars driven to bedrock. The deep pier work was subcontracted to a French firm specializing in pneumatic caisson work. Placing piers in the deeper channels of the river would have been impossible, but outcroppings of bedrock at each side of the waterway and at the midstream island of Inch Garvie allowed placement of the three required foundations.

However, the French firm Coiseau et Cie hired to perform the deep caisson work endured one monumental glitch. One of the 70-foot-diameter cylindrical caissons being floated into position snagged an edge in the mud, tipping it upward. At

This view, looking down the rails of the massive Firth of Forth Bridge, reveals the massive strength of the design, which far exceeds any normal loading the bridge might experience. Built in response to the Tay Bridge disaster, it has stood solidly for more than 100 years. *(Author's Collection)*

high tide, water poured into the tilted caisson, requiring 10 months to refloat it. This incident aside, men working within the compressed air atmosphere of the caissons were quite successful, sinking foundations to depths ranging from 63 to 89 feet below the average high tide. Because of a growing understanding of "the bends," or caisson disease, no fatalities were recorded during the compressed air work.

However, the magnitude of the job and the techniques of the day meant that deaths were inevitable, and 57 died in incidents during the seven years required to complete the bridge. The amount of material required to build the structure was extraordinary: 58,000 tons of steel used in the tubular sections and truss girders alone. During the most intensive periods of construction on the bridge, 4,500 workers were employed.

When completed, the bridge was the essence of strength. Although other cantilever bridges were built, the political, cultural, and engineering environment of the day was such that the Forth Bridge was many times stronger and heavier than necessary.

The Forth Bridge remains in use and is considered a colossal feat of Victorian age engineering, and its powerful appearance has made it an enduring symbol of the region. A suspension span known as the Forth Road Bridge was built adjacent to the original Forth Bridge in 1964.

An enduring criticism of the Baker-Fowler design of the Forth Bridge is that it required far more material to accomplish the task than was necessary. To the untold thousands who have traveled safely across the bridge in the years since its construction, this complaint may seem like intellectual pettiness.

See also BAKER, BENJAMIN; CANTILEVER BRIDGE; COOPER, THEODORE; QUEBEC BRIDGE; TAY BRIDGE DISASTER.

Flagler, Henry Morrison   *See* Seven Mile Bridge.

Fréjus Tunnel   *See* Mont Cenis Tunnel.

Freyssinet, Eugène (1879–1962) *Developer of stressed concrete*   If you have driven across a highway overpass or any other type of concrete beam bridge, then you have unconsciously enjoyed the genius of the French bridge engineer Eugène Freyssinet, the man who figured out how to add stress to concrete to alleviate some from our own lives.

His development of prestressed concrete technology in 1928 is considered one of the most important structural engineering developments of the 20th century. His technique for making concrete capable of accepting heavy loads in tension—loading against the top edge of a horizontal concrete component rather than against its end, as in a column—allowed concrete to become a more useful structural material.

Although prestressed concrete technology was initially applied to bridges—including beautiful spans engineered by Freyssinet—the technology made concrete capable of supporting great buildings, storage tanks, and stadiums. It would be difficult to find a large modern structure of any type that does not contain prestressed concrete components.

Freyssinet's early years did not suggest that he would reshape the world of bridge and building construction. Born in France's relatively remote Corrèze Plateau, Freyssinet went to Paris for his education; there he was initially rejected in 1898 by France's prestigious engineering school, Ecole Polytechnique. (One must wonder about that university since it also rejected the bridge-building genius and Eiffel Tower designer Gustave Eiffel in 1855.) Freyssinet eventually gained admission and graduated 19th in his class before completing his education at Ecole des Ponts et Chaussées, a school for bridge engineers.

Freyssinet was a rookie bridge engineer for the government when, in 1907, he designed a bridge to span the Allier River at Le Veurdre. The young engineer was given the chore when he promised he could build three needed bridges for the price of one—as he eventually did.

The first bridge he would construct would be the Veurdre Bridge across the Allier River in a southwestern corner of France. Using a trio of reinforced concrete arches, Freyssinet also decided to place horizontal steel bars through the arches of the Veurdre Bridge. These bars would be anchored under tension in the abutments, supporting the ends of each arch and placing them under compression.

The bridge was completed in 1910, but by 1911 its arches were demonstrating a phenomenon later referred to as concrete creep and had dropped five inches. This was not a complete surprise to Freyssinet, who observed the same behavior in the concrete of a test arch built before the bridge was constructed. At the time, Freyssinet's findings contradicted the prevailing belief that concrete creep did not exist. Because of his earlier findings, Freyssinet was prepared.

Grabbing his bicycle, Freyssinet collected three workers and raced to the bridge in predawn darkness. Having left a gap at the center of each arch in case of this event, Freyssinet and his crew jacked the arches farther apart to increase the compression, which had slackened as a result of the creep of the concrete. The slumping was straightened and the Veurdre Bridge stood until its wartime destruction in 1940. The lesson was not lost on Freyssinet. Concrete could be compressed to alter its behavior and its strength. By 1928, he had perfected methods of placing steel tendons in concrete beams.

What Freyssinet invented and continued to perfect until the time of his death in 1962 was actually quite simple. The simplicity of his discovery was something he freely admitted: "The idea of prestressing is neither complicated nor mysterious; it is even remarkably simple." What Freyssinet did was pull high-strength steel cable or tendons to a specified tautness inside a concrete beam mold by using hydraulic jacks. The concrete was poured, and when it cured to adequate hardness the tension was released, transferring compression to the beam. The cable ends were held in place by conical stays or wedges that clamped to the cable and secured themselves against the concrete under tension.

Concrete beams that once groaned and then cracked under tensile loads now withstood their burdens with a slight camber arching upward toward the load. A steel tendon running through concrete beams like a taut bowstring accepted the weight, which was transferred to the ends of the concrete beams where the tendons were anchored. This technique effectively solved the problem of cracking.

Freyssinet, the structural engineering inventor, was nonetheless a bridge designer who went on to provide the world with some of its most graceful and innovative bridges. A graceful and modern structure spanning the Elorn Estuary near Brest,

France, is the Plougastel Bridge, built between 1925 and 1930. Although prestressed concrete was not a feature of the bridge, its design was nonetheless a bold one, employing a trio of hollow, reinforced concrete arches. Each arch supported a span of 592 feet.

Freyssinet survived World War II, during which he continued to build bridges when possible. At war's end, he took part in France's postwar building boom by designing replacement bridges for those destroyed by war. He applied his prestressed technology to such bridges as the Luzancy Bridge over the Marne River. A strong, simple span of prestressed concrete, it crosses only 180 feet of water but demonstrates a sleek and visually unencumbered design. He also designed bridges for projects around the world, including a trio of 492-foot bridges in Venezuela. Freyssinet's last bridge was completed three months before his 1962 death, the Saint-Michel Bridge in Toulouse, France. A French firm specializing in prestressed concrete technology still bears his name.

See also CONCRETE; MAILLART, ROBERT; PRESTRESSED AND POST-TENSIONED CONCRETE.

G

Ganter Bridge   *See* Menn, Christian.

Garabit Viaduct (built 1881–1884)   Better known in the United States for his creation of the Eiffel Tower, if not for his work on the internal support structure for the Statue of Liberty, Gustave Eiffel was primarily an engineer of magnificent bridges. A bridge considered his most innovative and beautiful is the Garabit Viaduct over the Truyére River in the mountainous Cantal region of France.

The Garabit Viaduct is known by students of bridge design as his second "crescent" bridge, because of its prominent semicircular arch supporting a railway deck across the river. The first crescent arch bridge designed by Eiffel was the Maria-Pia Bridge in Portugal in 1876. Although the bridges are similar in design and appearance, the Garabit span contains some structural and aesthetic improvements.

This Maria-Pia Bridge was on the mind of the French railway engineer Léon Boyer, who was studying ways to build a rail bridge across the Truyère River in France's Massif Central. The river was preventing establishment of rail service between Neussargues and Marvejols 45 miles to the south. Boyer's fellow French national railway engineers were cowed by the prospect of building towers tall enough to support the span across the deep gorge. Other government engineers wanted the train to follow switchback tracks into the gorge, then cross a low bridge, and then zigzag up the side of the opposing cliff. Boyer sought Eiffel's help on the project.

Eiffel's standing within engineering circles was so high that it was decided he should be granted the job of designing the bridge without competition. The contract essentially stated that Eiffel was the best engineer for the job. By 1879, Eiffel was working on the Garabit span, which would largely resemble the Maria-Pia work.

However, the iron truss arch beneath the Garabit Viaduct's railway deck would have a more pronounced curve, resulting in a nearly semicircular shape. On the side of the gorge that sloped more gently toward the river, four foundations, upon which four iron piers would stand, were built. On the steep side of the gorge, only a single concrete foundation was emplaced to support a single iron tower. Romanesque masonry arches supported the side spans of the bridge on both sides of the gorge.

The taut-looking arch was built by using a cantilever method in which supporting cables temporarily suspended both ends of the growing but unfinished arch. When the arch met, it became self-supporting and had an overlying railway deck. This method eliminated the problem of erecting expensive wooden falsework 170 feet from the floor of the gorge to the crown of the arch. This temporary cantilever suspension method had been used during construction of the arches on the Eads Bridge in 1873 since falseworks could not be erected over the Mississippi River.

On close inspection, the complexity of the arch becomes apparent. Eiffel designed it to be wide at its ends and to narrow severely as it rises to meet the

The graceful Garabit Viaduct is a blend of iron, aerodynamics, and the artistry of the structural engineer Gustave Eiffel. Known as one of his crescent bridges because of its arch design, it was completed in 1884 and remains in service in France's Cantal region. *(Robert Cortright, Bridge Ink)*

railway deck. By the time the arch and the deck met, they were the same width. The arch is made to appear even larger than it is by the optical effect of the arch's narrowing as it rises. A pair of short iron lattice struts reach up from each side of the arch's crown to provide additional support to the rail deck before it touches the rise of the arch.

Although Eiffel had allowed the arch of the Maria-Pia Viaduct in Portugal to curve up level to the rail deck, he brought the Garabit arch up only enough to kiss lightly the underside of the deck. In this way, the curved line of the arch and the horizontal line of the deck were left visually unbroken without any loss of support. The central span, between the opposing towers, measures 540 feet, and the entire bridge including its side spans totals 1,850 feet. Construction was completed in 1884.

See also EADS BRIDGE; EIFFEL, GUSTAVE; FALSE-WORK.

George Washington Bridge (built 1927–1931) *Big, busy, and beautiful* Connecting New Jersey and the north end of Manhattan with a bridge was a dream that consumed the famed engineer Gustav Lindenthal until his death in 1935. A crossing was needed, but Lindenthal's designs were too complex and expensive. It would be his protégé, Othmar Hermann Ammann, who had worked with Lindenthal on the Hell Gate Bridge, who would propose a realistic design.

Although the acceptance of Ammann's design would forever fracture the relationship between the two engineers, Ammann approached the problem with the pragmatism his former mentor had once displayed. Using modern materials and incorporating new concepts, his suspension bridge across the Hudson would be not only twice as long as any other similar span but also more slender and economical in terms of materials.

Ammann, a former employee of both Lindenthal and George S. Silzer, broached his plan to Silzer, who by 1923 was New Jersey's governor. Silzer recognized Ammann's plan as viable both politically and economically. At $59 million it was also 10 times less expensive than Lindenthal's dream bridge. By 1927, construction was under way. The bridge would connect Fort Lee, New Jersey, and Fort Washington on Manhattan Island, and good bedrock was located at or near the surface in both locations. This geological boon eliminated the need for pneumatic caissons for the tower foundation work, which would be costly in terms of worker health and money. Nonetheless, the bridge claimed the lives of 12 workers during its construction.

The approaching roadway would be carried over bluffs on each side of the Hudson River, providing a 213-foot clearance between the center of the suspended deck and the water level at mean high tide. Concrete anchorages seated on bedrock would rise to meet the side spans of the bridge.

The dimensions of the bridge were to be extraordinary: a central suspended span of 3,500 feet, a pair of side spans of 630 feet each, and two lattice truss towers rising 604 feet above the water. Bridges are often works in progress: For the George Washington Bridge, Ammann continued to oversee changes to the span as planning and construction progressed. The result would be an accidental masterpiece of design.

Ammann decided that an epidermal layer of concrete could be eliminated from the towers as structurally unnecessary. After 1929, when Depression era exigencies pressured all involved to reduce costs, Ammann also eliminated a granite façade that was to cover the towers, dashing his hopes of competing with the look of the stone towers of the Brooklyn Bridge. This accidental bit of "creativity" resulted in a bridge that unapologetically appeared to be a product of the 20th century, instead of having its functionalism camouflaged by stone.

The bridge would also have a magnificently slender deck. This innovation was based on the deflection

With its open steel latticework towers the George Washington Bridge is one of the world's most beautiful suspension spans. It was the first great bridge across the Hudson River and was designed by Othmar Hermann Ammann, whose cost-efficient design was selected over a far more expensive one submitted by his former boss, Gustav Lindenthal. *(Historic American Engineering Record, National Park Service)*

theory touted for use on suspension bridges by the consulting engineer Leon Moisseiff. The concept theorized that long-span suspension bridges needed less stiffening reinforcement than originally believed.

In lieu of a deep stiffening truss like that of the Williamsburg Bridge, the road deck would be stiffened by a girder-truss system that utilized a continuous plate girder on its sides along with transverse girders underneath. Stiffening this arrangement was a system of trusses. Despite its strength, from a distance the bridge deck appeared wafer thin. The completion of a second deck below the first in 1962 involved the use of a deep stiffening truss whose triangular members connected to the bottom of the original deck and the top of the new one.

This roadway would be suspended by a potent twin-cable arrangement swooping up from the anchorages to the towers, then down again to the center span. Each of these cables would weigh 28,450 tons and contain a total of 107,000 miles of 0.196-inch-diameter wire. The cables would perform a disappearing act by passing through the roadway to the anchorages below.

Ammann's bridge, often referred to as "The George," stands as a monument to efficient engineering and construction. It was completed October 25, 1931, a half-year ahead of schedule. The bridge is still the subject of awe, respect, and admiration. In March 2001, the *New York Times* praised the bridge as "an astonishingly spry septuagenarian."

The bridge now contains eight traffic lanes on its upper deck and six on the lower deck and generates $249 million in annual toll revenues for the Port Authority of New York and New Jersey, the owner of the bridge. The authority claims the bridge is the busiest in the world, and the yearly toll revenues accumulated by the end of the 20th century seem to back up this assertion. The initial investment in the span was a wise one since the bridge requires only $60 million yearly for upkeep, allowing the remainder of the funds to support other transportation and infrastructure projects.

See also AMMANN, OTHMAR HERMANN; LINDENTHAL, GUSTAV; MOISSEIFF, LEON SALOMON.

geothermal gradient   Those who tunneled into the Earth soon learned that the deeper one descends, the hotter the surrounding ground becomes, providing support for those claiming that Hell is a subterranean place. For sandhogs and miners who might descend hundreds or thousands of feet into the Earth, this natural occurrence is more than a curiosity: It is a hazard.

Because the Earth is composed of a layer of cooled crust wrapped around a ball of molten rock, geothermal heat migrates outward. In some parts of the world this heat ascends close to the surface in the form of hot springs or the lava flow from volcanoes. The geothermal gradient is the rate at which heat increases as one descends or drills into the Earth. On average, the temperature rises 48 degrees Fahrenheit for every mile of depth achieved. This figure is highly variable, less in some areas or far greater in others according to local geologic conditions.

Miners who penetrate subterranean hot springs can be scalded and severely injured, and workers have been forced to retreat from rock faces that were so hot that the working environment was unbearable. In regions where there are young volcanoes, excavating a few inches into the soil yields enough heat to cook canned food.

In the case of dry heat radiating from walls, ventilation or the application of a water mist can provide cooling, whereas drainage systems must be put in place to deal with the hot water from subterranean springs.

See also SIMPLON TUNNEL; VENTILATION.

Golden Gate Bridge (built 1933–1937) *Engineering milestone that spanned the Golden Gate Strait* The Golden Gate Bridge, a symbol of the nation's industrial and technological prowess during its darkest economic days, is shrouded in more mythology than almost any other engineering project except the construction of Egypt's pyramids. The man widely thought of as the designer of the bridge did not design it, and the man who did is virtually unknown. And the graceful and proportional lines of the bridge as built bear no resemblance to the hideous original concept for the span.

Far from being a straightforward engineering project, the bridge was the dream of a controversial San Francisco public official that was expropriated by a self-promoting bridge builder from Chicago who sought to make the project a monument to himself.

The combination of cutthroat Northern California politics, well-financed ferryboat interests, and Cassandra-like claims that the bridge would collapse made the evolution of the bridge from dream to

The Marin County anchorage of the Golden Gate Bridge is visible in this photograph, which looks south toward San Francisco. Perhaps the most controversial bridge project since the construction of the Brooklyn Bridge 50 years before its own completion, it has become an American icon. *(Historic American Engineering Record, National Park Service)*

hard-won reality a panoramic drama. Before it could be built, California's courts would play a role, and the man who designed the bridge would be crudely dismissed into obscurity. Finally, after dozens of political skirmishes and the deaths of 11 men, it would be discovered that the bridge, once built, had a potentially fatal flaw.

San Francisco's Board of Supervisors passed a resolution in 1918 calling for a bridge to link San Francisco to the south with Marin County on the north side of the Golden Gate strait, but it was an Irish immigrant, Michael Maurice O'Shaughnessy, who would get the ball rolling. Brought in by reformers to rebuild the city's infrastructure after the April 18, 1906, San Francisco earthquake, he was the undisputed lord of San Francisco's public works, much like New York's public works czar Robert Moses in the 1930s.

Eager to have a bridge across the Golden Gate, O'Shaughnessy had been asking engineers to suggest how it could be done and for how much. Some were doubtful it could, and those willing to make a recommendation estimated a cost as high as $250 million. When O'Shaughnessy posed the question to Joseph Baermann Strauss, who had built one of his Strauss bascule bridges in San Francisco in 1916, Strauss shot back that it could be done for $25 million. This figure would prove to be only $8 million below the actual cost of the bridge when completed 21 years later.

O'Shaughnessy liked the answer, and Strauss, the builder of small bridges who was hungry to build a mighty bridge, in 1920 began a 12-year campaign to promote the bridge and himself as its chief engineer.

Strauss conjured up a design by June 1921 that was as frightening in appearance as it was unworkable in practice. Instead of designing a long-span bridge of the cantilever truss type like the Quebec Bridge or the suspension type like Roebling's Brooklyn Bridge, Strauss presented a design that was an amalgamation of the two bridge concepts. The spans closest to shore would be of cantilever design with suspension cables tied at the apex of the cantilevers to support the central span of the bridge. It resembled the cartoonist's Rube Goldberg's devices—overly complex contraptions that serve no useful purpose except that of being a source of mirth—and it was one ugly bridge.

Strauss tuned his estimates once again, submitting an even more attractive figure of $17,250,000 for spanning the strait. By contrast, the famed New York bridge designer Gustave Lindenthal had estimated that the bridge could cost as much as $77 million.

After receiving Strauss's unattractive but affordable bridge proposal in June 21, 1921, O'Shaughnessy kept it secret for 18 months. In the meantime, Strauss popped up at public meetings in the counties north of San Francisco drumming up support for the bridge he considered his own.

The result of this stumping by Strauss was the Coombs Bill, passed by the California Legislature on May 25, 1923. The measure established a district that would raise money for and supervise the construction of a toll bridge across the Golden Gate.

If Strauss's official credentials as an engineer were a bit thin (he had a bachelor of arts degree from the University of Cincinnati), an engineer hired by Strauss to help with the Golden Gate Bridge project more than compensated for Strauss's academic shortcomings. Charles Alton Ellis was a New Englander from Maine, an expert in structural engineering and a prodigious mathematician. Ellis's practical and academic credentials were superb. He could look into the dark heart of physical forces that could destroy a bridge and then calculate precise engineering remedies.

Ellis would play a central role in designing the bridge and ensuring its stability with his calculations. Strauss would shield Ellis, the academic, from gritty municipal politics as Ellis tinkered with the delicious engineering challenges of bridging the Golden Gate. Strauss would be the front man, pressing the flesh at chamber of commerce luncheons, wallowing in the back alleys of power politics, and fighting the political brushfires that ceaselessly threatened the project. But Strauss always saw himself and the bridge as one, and employees like Ellis were underlings serving his cause. In reality, modern bridge projects are the product of the acumen of several, if not dozens, of engineers. If one were to look for one of the central designers of the Golden Gate Bridge, one should probably look no further than Charles Alton Ellis.

For all the engineering problems facing the bridge project, including the width of the strait, bedrock conditions, and tides, the political struggle would prove just as difficult to overcome. O'Shaughnessy, whose broad powers had made him unpopular with a new generation of San Francisco politicians, had become a liability to the bridge effort. When it came time for San Francisco's city council to join the bridge district, dislike of O'Shaughnessy prompted the city supervisor, Ralph McLeran, to call for new tests and surveys in addition to those ordered by O'Shaughnessy. This measure would cost $150,000 and delay the effort to bridge the strait. McLeran's dislike of O'Shaughnessy failed to take into account the popularity of the bridge project with the public.

In the face of a public uproar, McLeran retreated, saying a new survey was less important than ensuring that San Francisco dominated the Bridge District's board. McLeran's new blunder ignited a political firestorm among the less heavily populated counties to the north, which joined the district with the understanding that the City of San Francisco would not have undue control over the project. It prompted a threat from State Representative Frank Coombs (who introduced the legislation establishing a bridge district) to instead build the bridge to Oakland.

In addition to the political fallout, residents in six of 21 nearby counties that had joined the district were spooked by McLeran's questions about the analysis of the bridge site's geologic characteristics and regretted their decision to join. Mendocino County voted to withdraw from the Bridge District, and citizens there signed a petition to counter the earlier one for inclusion in the district. When this petition was presented in early 1926, this second petition prevented California from certifying the formation of the Bridge District, throwing the matter into court. The curious legal issue was whether a signature on a petition could be withdrawn, and the ruling was that it could not, affirming the legality of the Bridge District.

As efforts to recruit more counties into the Bridge District continued, more clouds were gathering. The Joint Council of Engineering Societies of San Francisco held a meeting, during which the council proclaimed the bridge project had not been "adequately investigated" and expressed disapproval of the way the Bridge District was organized.

The Joint Council also recommended that an engineering commission investigate the geologic characteristics of the rock on which the bridge would sit and review the costs associated with constructing the bridge. These studies would probably cost no less than $500,000 and halt engineering work on the bridge. The proclamations reflected a growing distrust of the way the bridge was peddled to the public. This was when Strauss came close to overstaying his welcome. The public and its elected officials realized the woods were full of bridge builders and were asking why they should settle for Strauss and his bridge. The cart seemed to be before the horse. Should not the decision to build a bridge be *followed* by the selection of its designer? By the spring of 1926, the public suspiciously perceived Strauss and the bridge as an unnecessary package deal.

The public wanted more answers about the project, and meetings were held throughout the six counties making up the Bridge District. Thanks to inflation as well as a clearer idea of what it would take to build the bridge, its estimated cost was creeping upward. Strauss revealed the cost of the bridge and right-of-way purchases totaled $27 million ($21 million was the cost of the bridge itself). Soon after, a damning report prepared by a trio of consulting engineers hired by Sonoma County citizens, claiming the bridge would cost $112 million, was released.

As 1927 approached, the San Francisco board of supervisors decided to sneer at the pledge to depoliticize the Bridge District when it appointed three of its sitting members to the district's board of directors. It also appointed the former supervisor Richard J. Welch, then a congressman. Welch, however, was a natural choice since he had sponsored the 1918 resolution calling for a bridge and had supported it ever since.

The problem was that the Bridge District had been sold to the rural counties as a means of constructing a toll-supported bridge devoid of local political shenanigans. This was a lofty and possibly unattainable goal. By lodging four local politicians on the board, San Francisco had done something the other municipalities had refrained from doing and had thrown more gasoline on the fires of public opinion. More hearings were held, and Strauss fought, with Ellis and the consulting engineer Leon Salomon Moisseiff at his side, to deflect the endless criticism of the bridge. In the first month of 1929, California Assemblyman R. R. Ingels unsuccessfully sought legislation that would eliminate the Bridge District. Soon after, a bill recognizing the legality of the Bridge District was passed.

By the summer of 1929, Strauss found he was just another bridge builder hat in hand seeking the job of spanning the Golden Gate Bridge. The Bridge District, under fire for being too close to Strauss, had listened to its critics. The job of designing the bridge was now offered to every notable bridge builder in America. Strauss, the man who almost single-handedly sold the bridge to the public, was no shoo-in. Among the short list of those queried about taking on the bridge were the famed bridge engineers Leon Moisseiff and Othmar H. Ammann. It was Ammann who designed the George Washington Bridge (along with Moisseiff) and would later design the mighty Verrazano Narrows Bridge in New York. Strauss knew that the accomplishments of these engineers in large bridge construction exceeded his own. Strauss, the successful builder of small utilitarian bridges, was now competing with the most brilliant and innovative bridge-engineering minds in the world.

Strauss decided that if you could not beat the competition, then you had to hire them. He fired off letters to Moisseiff and Ammann, then chief engineer for the New York Port Authority. Moisseiff was ready to climb on board, but Ammann was put off by Strauss's offer since he understood he would be working for Strauss and not the District. When Moisseiff figured this out, his enthusiasm also dimmed. Both men were not fully on board until their contracts stated that the Bridge District was their employer. Both believed their recommendations should not be tainted by filtering them through the builder of the bridge. As a nod to the local engineer-ing community, Strauss had also recruited Charles A. Derleth, a respected engineering professor at the University of California at Berkeley.

The Bridge District board of directors announced on August 15, 1929, that four engineers had been hired: Ammann, Derleth, Moisseiff, and Strauss. Strauss was named "chief engineer" by the board, giving him what he had wanted all along. Although the consultants would be well paid (Ammann alone would receive $50,500), Strauss's financial reward would be relatively slender. Instead of gaining a standard fee equal to 7 percent of the bridge's construction cost, Strauss agreed to 4 percent. Out of this amount, Strauss would have to pay the salaries of all engineering and design personnel.

O'Shaughnessy, whose own banishment to a political wilderness had been in the making for some time, took Strauss's appointment badly. The city engineer would have no role in the bridge's construction and, incredibly, became its most vocal critic, falsely predicting it would cost four times more than Strauss had claimed.

Moisseiff, Ammann, Ellis, Derleth, and Strauss began to study the fine details of what would be needed to bridge the Golden Gate strait. Early on it was decided it would be necessary to construct a support pier on the Marin side of the strait, though the U.S. Army Corps of Engineers had been promised that only one pier on the San Francisco side would be necessary. The army would have to be persuaded to provide permission for a second pier.

Moisseiff, the engineering wizard who led a movement to make bridges thinner and lighter, determined that if designed as a suspension bridge, the Golden Gate span could be flexible and therefore capable of absorbing the horizontal and vertical forces of the winds in the notoriously blustery strait. Moisseiff's conclusions won over his colleagues, who recognized that a lighter, thinner structure was superior to Strauss's heavy design, which depended on the brute strength of rigid steel beams and thousands of tons of unyielding structure. It would be Ellis and his staff who would do the detail work on the design of the bridge, which, it began to appear, could be built cheaper, lighter, and more quickly than originally thought.

Then, in late 1929, as the bridge went from a dream to a project in its embryonic stages, another brushfire threatened to burn all of Ellis's blueprints to ashes. The arsonist holding the match was none other than Major General Edgar Jadwin, chief of the

U.S. Army Corps of Engineers, the bureaucratic "owner" of all navigable waterways in the United States. On his last day in the army, Jadwin notified his superiors that he opposed construction of the bridge despite his previous approvals of the project in 1924 and 1927. A meeting held soon after with the secretary of war—the precursor of today's secretary of defense—reassured the Bridge District board member William P. Filmer that the government stood by its original approval.

And critics still sniped at the bridge. Although the distance between the supporting piers of the bridge was to be 4,200 feet and the space between the road deck and the water would be 220 feet, some decried the bridge as a hazard to navigation. Steamship companies complained about the clearance beneath the bridge, but the days of tall-masted commercial clippers had long since ended, and ships were actually becoming longer rather than taller. This issue persisted until the bridge was completed seven years later, when a sizable flotilla of navy ships glided effortlessly beneath the span on its opening day.

The elements of the bridge began to come together. Moisseiff, the grand master of stress loading and wind effects on suspension bridges, had an able partner in Ellis, who used Moisseiff's findings to create entirely new calculations for the Golden Gate Bridge. The bridge would be built strong enough to survive a 50-mph wind, the strongest ever recorded in the strait. A wind of this magnitude would place pressure equal to 6.5 pounds per square foot on the bridge's structure. The road deck would be designed to endure 30 pounds of pressure per square foot, and the tall (746-foot) towers were designed to accept 50 pounds per square foot of

This view of the Golden Gate Bridge looking north from San Francisco toward Marin County in 1953 provides an eastside view of the famous bridge. The relatively slender reinforcing truss can be seen as well as a steel arch supporting a final segment of approach span. The bridge and its locale combine to create one of the most recognizable sights in the world. *(National Archives and Records Administration)*

pressure. The towers, Ellis believed, should be similar to those of the Benjamin Franklin Bridge. Completed in 1926, the Philadelphia–Camden Bridge's towers were incredibly strong, made of individual steel cells riveted together until the interior of the towers were honeycombed with mutually supporting boxes.

In the days when slide rules and mechanical computation machines were state-of-the-art tools for engineers, Ellis's work load was crushing. The bridge was a unique, almost experimental project, and his calculations required checking and double-checking. Strauss, who spent his time promoting the bridge and basking in the glow of the effort, was growing impatient with the time necessary to overcome engineering issues that were beyond his experience. Ellis, ordered by Strauss to turn over the computations for the towers to a subordinate, did so but later discovered they were filled with errors.

Strauss was a man of action, accustomed to erecting small bascule bridges quickly and moving on. Ellis, the academic, was striving for perfection. As if Ellis were not busy enough solving complex engineering issues, he was also attempting to provide cost estimates on every aspect of the bridge's construction to allay public fears. As requested by Strauss, all the cost estimation work was completed before the November 4, 1930, bond election. When the ballots were cast, approval to finance the bridge passed in a landslide.

As 1931 dawned, the Bridge District discovered that the Great Depression had caused construction costs to plummet: When a call for bids was announced, 27 companies eagerly sought to build the bridge. When the contracts were let, they totaled $23,843,905—well below the $27 million estimate. Unfortunately, accomplishing the sale of the construction bonds was going to prove tougher than anyone expected. Many questions had orbited around the legality of the Bridge District and its ability to levy taxes should the toll revenue be inadequate to pay off the bond debt. The banks and investment firms that would have been the market for the bonds announced a boycott until the district could gain a clear-cut ruling that it had the power to tax.

Only one bid for the bonds was initially received, and that was deemed unacceptable, putting the district officials and the whole project on tenterhooks. When an acceptable offer was finally received, it was qualified with the request that the

Bridge District obtain a California Supreme Court ruling that it was a bona fide taxing entity. This bit of legal chicanery was intentionally choreographed between the bond house and the district. As a result, the district's secretary, W. W. Felt, was to make the bonds nonnegotiable by refusing to sign them. Felt was then to be sued by the district to force a court ruling that the bonds were legal and negotiable.

This maneuver opened the door to perhaps the most powerful opponent the bridge would face, in the person of the attorney Warren Olney, Jr., who marched into court on August 15, 1931, on behalf of a list of unnamed clients seeking to dismiss the district's suit against itself. Olney hoped to block judicial validation of the Bridge District. Olney actually represented Southern Pacific–Golden Gate Ferries Company along with 91 other "taxpayers." The politically powerful Southern Pacific had opposed the bridge from the start because its massively profitable ferry service was threatened.

To the dismay of Southern Pacific, the court ruled against Olney's motion in November 1931 and then ruled that the district had the right to levy taxes and that its bonds were perfectly good with or without Felt's signature. In July 1932, a U.S. District Court judge upheld the district's right to operate and issue bonds, clearing the final legal hurdles to financing and building the bridge.

In December 1931, Ellis was fired by Strauss, who claimed Ellis's work was substandard. This occurred after Ellis revealed the extent of his role in designing the bridge during a public speaking engagement. For all practical purposes, Ellis was now a nonperson in the history of the bridge as presented by Strauss.

On January 5, 1933, work began on the bridge as crews began digging and blasting their way into the rock on each side of the strait. They were preparing two excavations to house the 64,000-ton concrete anchorages that weighed 6,000 tons more than a 900-foot-long *Iowa* class battleship. These anchorages hold up the bridge deck through the bridge's two suspension cables.

On the north side of the strait, just offshore from Marin County, workers drove steel plates into the seabed to form a cofferdam for the construction of one of the two pier foundations. Once completed, the water-filled enclosure was pumped dry to allow excavation down to solid bedrock. Because of concerns about earthquakes, the foundation penetrated 35 feet into the bedrock. Concrete then filled the excavation until the foundation rose 44 feet above the surface of the water.

On the San Francisco side, where the foundation for the pier was to be 1,200 feet offshore, the work would be far more difficult. A steel trestle was erected to the site of the south pier by driving pilings into rock blasted loose by explosives. Once completed, the trestle allowed materials to be carried to the site of the pier. Conditions were tough: The floor of the strait lay 100 feet beneath the water and tidal action was so severe that helmet divers working on the bottom had to dodge huge boulders moved by the currents. Explosive charges were sent through a pipe to the seabed, where they were detonated to crumble the seabed's rock, which was lifted by cranes with clamshell buckets.

The south pier was nothing but trouble to the Pacific Bridge Company, the job's general contractor. In the foggy predawn darkness of August 14, 1933, the cargo ship *Sidney M. Hauptman* strayed off course and knifed into the trestle. The ship backed away, taking a 120-foot section of the trestle with it. The collision had also damaged forms being assembled underwater to accept concrete for a 300-foot-long protective fender being constructed around the pier.

The fender was a massive project in itself. Shaped like a ship's hull to protect the pier from collision as well as from scour and wave action, it had a wide section left open on the bay side to allow a caisson to be floated in for the excavation to bedrock. Concrete, which can cure and harden underwater, was poured into the forms of the fender, and it slowly began to rise from the floor of the strait. But more trouble followed. On December 13, 1933, a powerful storm smashed all but 600 feet of the 1,200-foot trestle that provided access to the site, forcing workers to build it a second time.

Although the bridge was a hard-nosed engineering project, aesthetic considerations were far from forgotten. Strauss had the foresight to hire the local architect Irving Morrow, who added important features to the bridge. A San Francisco resident, Morrow had long studied the colors of Marin's hills overlooking the strait. As a painter, Morrow believed the color of the bridge should blend rather than clash with the reddish hues of the hillsides. That color he preferred—in the face of strong objections—would be International Airways Orange. Many of the engineers wanted aluminum paint, a silvery coating often used by industry to protect metal

components from corrosion. Although the red color was repellant to some, including the influential Ammann, Morrow persisted until he gained its approval. As a result of Morrow's efforts, the Golden Gate Bridge is not gold, as some who have never seen it might think. During World War II, some marines and soldiers departing on troopships beneath the bridge for combat in the Pacific assumed the span was only wearing a coat of red primer paint while awaiting its final coat of "golden" paint.

Strauss's growing confidence in Morrow resulted in the architect's being given the task of designing dimensional effects on the towers. Morrow, a life-long student of light and color, employed vertical fluting on the towers' cross braces. The fluting would create shadows as the sun rose and set to make the appearance of the towers seem changeable. Morrow's final contribution to the bridge's beauty was the lighting design. It was muted lighting employing the then-new sodium vapor lights that could penetrate the region's famous fog. Morrow wanted the bridge lighted not in a garish fashion by dots of white incandescent bulbs like a party boat, but softly, with lights positioned to showcase aspects of the bridge's design. The roadway light fixtures themselves—also Morrow creations—were sleek and modernistic.

About this time, Strauss disappeared without a word after an April 1933 meeting with Bridge District directors. After several months, the Bridge District board began to investigate his whereabouts and was told by a spokesman that Strauss had suffered a nervous collapse and was under a doctor's orders to sail leisurely from California to New York. The supposedly mentally shattered Strauss managed to pull himself together enough to grant newspaper interviews while in New York and, during this period of recuperation, to marry. The situation became similar to that involving Washington Roebling, the chief engineer of the Brooklyn Bridge, who spent the last 10 years of the project as a recluse who was made a near-invalid by caisson disease. Strauss, like Roebling, assured the directors that everything was well in hand despite his absence.

The Marin tower's pier was completed; the prefabricated steel components for the tower that would stand upon it began arriving, accompanied by the arrival of ironworkers from across depression-stricken America to rivet the huge components together. The completed Marin tower was "topped off" by ironworkers on May 4, 1934, with the raising of an American flag.

With the south pier's elliptical fender completed in October 1934, it was time to tow the largest caisson ever built through the fender's opening. Once inside the fender, the expensive caisson was to be sunk to allow excavation to the bedrock. But neither the weather nor the caisson would cooperate. Swells and high winds tossed the 185-foot-by-90-foot box into the fender, threatening to damage both. The scheme of moving the four-story box inside the fender was abandoned. The fender's back door would be closed with concrete so it could do double duty as a massive cofferdam. With its interior pumped dry, the huge enclosure would hold back the surrounding bay, allowing workers to excavate down to the bedrock, below which the concrete base for the pier would be poured as if it were being done ashore. The uncooperative and now useless 10,000-ton caisson was eventually towed out to sea and sunk with explosives. Once the excavation work was done, inspection shafts were bored deep into the bedrock to reassure critics that it was solid. The resident engineer Russell G. Cone entered the shaft and struck the cohesive rock with a loud ring. The hole was filled with concrete, and the concrete foundation was poured to allow construction of the south tower.

After seven months' work, the San Francisco tower was completed in June 1935. Soaring 746 feet above the strait, it finally faced its sister tower in Marin County, which had looked down disapprovingly at its tardy twin for nearly 14 months. With the anchorages in place on both sides of the strait, it was now time for the venerable John A. Roebling's Sons firm to spin the massive cables that would support the roadway.

The cables, which were each composed of 61 strands that were in turn composed of 452 individual wires for a total of 27,572 wires apiece, had phenomenal strength. The size of the job dictated that Roebling's Sons Company would have to spin cables at a far greater rate then it had previously. By adding a third wheel to each of two carriages unreeling the wire across the strait, Roebling's engineers managed to spin 271 tons of wire a day. The George Washington Bridge, completed in 1930, received a maximum of 61 tons of cable wire a day. The resulting cables measured 36 3/8 inches apiece, the largest yet spun. Each cable was capable of supporting 61,500 tons.

The spinning effort began in August 1935 and ended on May 20, 1936, two months ahead of schedule. A celebratory three-day beer bust was sponsored by Roebling for the workers, but the party was darkened by a dispute over quality control on the bridge's components.

When cast steel collars that would connect the vertical suspender cables to the roadbed were found to have cracks and other imperfections, Strauss's team cried foul and demanded new components. Roebling's own engineers, who considered the clamps capable of surviving any foreseeable loading, deemed the clamps acceptable. Nonetheless, the Roebling Company provided new collars, albeit under protest.

With the newly arrived clamps bolted every 50 feet on both cables to hold the vertical suspender cables, the deck began to take shape as the truss-reinforced sections were assembled and secured to the suspenders. Once the road deck was in place, concrete was poured to form the road surface and add weight to the stunningly thin deck with an extremely thin depth to span ratio of 1/164.

Miraculously, no one had been killed during the first 40 months of the job. There had been injuries and falls but no fatalities. Even a sizable earthquake in June 1935 that shook but did not damage the south tower failed to create a fatality, as the workers clung to their perches on the superstructure of the bridge.

When construction of the road deck commenced, Strauss ordered that a massive safety net stretching the length of the bridge's underbelly be erected. The net (which was actually four nets) cost $125,000 and was constructed of rope woven into six-inch squares. Hung beneath the road deck, it would catch not only men but tools larger than its mesh. It served the bridge workers well, catching 19 before they fell to their deaths. One was Al Zampa, whose fall was only partially broken by the netting but who lived to tell of it anyway. Unfortunately for Zampa, the netting was dangerously loose and he connected with the rocks below a side span, smashing four of his vertebrae. Survivors of a fall like Zampa's composed what was known as the "Halfway-to-Hell Club."

Death first touched the Golden Gate Bridge in an ironically grisly way. As work progressed on the road deck during 1936, a crane fell, forcing workers to scramble to avoid its boom. One who did not make it was Kermit Moore, 23, who was decapitated in full view of his brother-in-law, Jack Turnipseed. The net was of no value in this freak accident, and its effectiveness would be negated in a subsequent and deadly incident.

After the truss-reinforced deck was installed, the skeletal structure was given a temporary wood floor of form lumber to allow the pouring and finishing of the concrete road surface. As this progressed across the bridge, workers with Pacific Bridge Company clambered into a pair of platforms connected to steel beams beneath the bridge to remove the forms. The 60-foot platforms each held roughly a dozen workers and had been in use since the road deck's installation. These platforms were crafted at the site with materials at hand. Connected to a steel deck beam by wheels, the carriage was moved by using a winch and a cable. When halted, workers clamped the platform to the beam to prevent it from sliding either way down the cambered center span.

On February 17, 1937, the overhead wheel system of one platform carrying 13 men lost its purchase on the steel beam above. The five-ton platform, known as a stripper, tore loose and plunged into the safety net, where two other men were working 220 feet above the water. The 3/8-inch Manila rope net could not take the weight of the platform and its strands popped explosively. The platform held fast only for a moment, and then the net pulled loose from its moorings and began to cascade downward. One bridge worker, Tom Casey, leaped from the tumbling platform and clung to a beam for all he was worth. Below, 12 others were falling. In an amazing coincidence, a newspaper photographer snapped a photograph of the tragedy in which six of the falling men appear as specks.

Of the 12 who plummeted into the strait, only two survived. One was Slim Lambert, the crew foreman, whose ability as a swimmer allowed him to remain afloat. His fellow workers saw him below and looked in vain for a life preserver. As Lambert swam in the outgoing tide, he saw the feet of drowned coworkers floating past and then the terrible sight of a friend's face bob to the surface briefly before sinking. Lambert then located his fellow worker Fred Dummatzen, whom Lambert pulled along. Though he assumed his coworker was dead, Lambert thought the man might be revived if help arrived soon enough. Eventually a U.S. Coast Guard boat passed within 300 feet of Lambert, but its crew did not see him among the swells. Lambert and Dummatzen, who was in fact dead, were pulled

from the water by a passing fishing boat. The boat's skipper continued to look for survivors and pulled one bridge worker, Oscar Osberg, from the water. The 51-year-old's hip and leg were broken and he had suffered massive internal injuries. Both Lambert and Osberg recovered.

Like most accidents, the one involving the carriage was preventable. Safety officials revealed that they had warned Pacific Bridge officials about the contraption on the very morning of the accident. The movable scaffolding was so rickety that State Industrial Accident Inspectors were treading across the bridge to inspect it once again when one of the two rigs crashed into the bay. It was a state inspector who helped to pull Casey to safety. Although a sizable blame game followed, a grand jury investigation found no criminal wrongdoing in the deaths. A plaque bearing the names of those who died in its construction remains at the bridge. Casey, who held on to the bridge for dear life with his smoking pipe still clamped in his teeth, left the bridge that day and never returned.

Much has been made over the years about the expected death toll on construction jobs and on the Golden Gate Bridge project in particular. A rule of thumb often quoted is that a worker was expected to die for every one million dollars of a project's value. By this standard, 33 men should have died on the Golden Gate Bridge, which eventually cost $33 million. Had it not been for the one accident that killed so many, the death toll of the bridge might have stood at the incredibly low number of one.

The bridge was ready for traffic by May 1937, and it was decided to imitate the opening day of the Brooklyn Bridge, its historical and engineering predecessor. When the Brooklyn Bridge opened in 1883, it was given over to a day of pedestrian traffic. The entire celebration of the bridge's opening—the Golden Gate Bridge Fiesta—would last from May 27 to June 2. At 6 A.M. on May 27, 1937, nearly 20,000 pedestrians were waiting for the bridge to open. Before the day ended, an estimated 200,000 people—roughly 30 percent of San Francisco's population—had crossed the bridge on foot.

Stunts, so popular during the Great Depression, were the craze that day as people arrived to walk across the bridge as the first to do so in numerous zany categories. Some wanted to walk backward across the bridge, and others wanted to sprint across. Sisters Carmen and Minnie Perez put on their roller skates to be the first to roll across the bridge. One woman's stunt so aroused concern among officials that they stopped her during her walk. As it turned out, she was only attempting to be the first person to cross the bridge with her tongue protruding. Most fittingly, one who crossed the bridge was the 74-year-old Henry Broder, who as a 20-year-old had rubbed shoulders with the thousands of New York and Brooklyn citizens who thronged to walk across the Brooklyn Bridge on its opening day 54 years earlier.

During the festivities, a sizable portion of the U.S. Navy's Pacific Fleet steamed beneath the bridge after navy carrier aircraft conducted a flyby overhead. The bridge, which achieved celebrity status even before its completion, was welcomed with open arms by the citizens who voted it into being. The following day, May 28, 1937, automobile traffic was allowed on the bridge, and thousands crossed the span for no other reason than to say they had.

To this day, Strauss is still officially considered the designer of the bridge by the Golden Gate Bridge and Highway District. He basked in the glow of his affiliation with the bridge and died on May 16, 1939, a mere 12 days short of the bridge's one-year anniversary. The name of Charles Alton Ellis is not on the tower plaque commemorating the politicians and engineers who had a hand in bridge's construction. After three years of unemployment, Ellis eventually returned to academic life in 1934 as an engineering professor at Purdue University.

The nasty and petty politics that tarnished so much of the effort to build the bridge did not end with its completion. The Bridge District board of directors fought and scrapped with each other so virulently after the bridge was completed as to alarm the counties belonging to the district. The one-way 50-cent toll was tough for many to pay during the hard times of the late 1930s, and the district never seemed capable of reducing unnecessary expenditures, adding to criticism of its operation.

Russell Cone, the gutsy and astute engineer who had overseen the actual construction work on the bridge as resident engineer, was about to find himself engulfed in the district's political cesspool.

After the bridge was built, Cone became the district's engineer, overseeing the span's maintenance. On February 9, 1938, a powerful gale blew through the strait, striking the bridge at a 90-degree angle, and Cone was the first on the bridge to observe the effects. What he saw disturbed him deeply. The bridge, which is designed to sway 21 feet side to side, was undulating like a blanket being snapped at one end. Sheltered behind the legs of a tower, Cone saw physical evidence that the bridge had a structural flaw—a weakness in its road deck reinforcement—a mistake that had somehow eluded the engineers. He dutifully reported his findings to the district's rancorous and thin-skinned board, who—predictably—decided to keep mum about the problem. Some on the board construed Cone's honest concern as rebellion and a threat.

On November 7, 1940, while the board was still keeping the potentially dangerous behavior of the Golden Gate Bridge a secret, a relatively mild 40-mph wind struck the Tacoma Narrows Bridge in Washington. Since the suspension bridge collapsed as it was being filmed, newsreel accounts of the event struck fear in the hearts of the board members, who were withholding from the public information about similar effects seen in their own bridge.

Rather than confront the problem, the board grew more suspicious of Cone, who might—because of his penchant for honesty and bluntness—reveal their dirty little secret. Eventually, the district became the target of a grand jury investigation into their inability to reduce unnecessary expenditures. Cone, much to his surprise, was asked to testify by the grand jury, to whom he revealed that the district failed to save money by not canceling expensive and worthless liability insurance for the bridge. Cone revealed that the Tacoma Narrows Bridge collapse had so spooked insurers that they charged massive premiums for minimal coverage in the event of collapse. This enraged a majority on the embarrassed board, who fired Cone. The Bridge District was now without the services of an engineer familiar with the construction of the bridge. Cone's dismissal provoked one concerned board member to release Cone's suppressed report about the bridge's flaws.

As a result, instruments for measuring the wind's effects on the bridge's movements were installed, but little else was done. In addition, the district sought the opinion of Clifford Paine, the engineer who replaced Ellis after his dismissal, who assured them the bridge was in fine condition. However, a December 1951 gale closed the bridge to traffic and once again generated wild gyrations on the road deck. Paine immediately altered his previous diagnosis and recommended that another 5,000 tons of steel reinforcement be added to its deck. Since then, the bridge has behaved quite nicely.

The Golden Gate Bridge served the public faithfully and repaid the bonds that financed it in 1967—30 years after its completion. It was provided little additional maintenance until 1975, when a one-year project replaced the 254 suspenders connecting the deck to the suspension cables. However, even the mighty Golden Gate Bridge could not continue indefinitely without refurbishment. Between 1982 and 1986 the steel decking of the bridge was also replaced, and a lighter concrete roadbed with an epoxy surface was poured, thus reducing the dead load of the bridge by 12,300 tons. In 1986, the 50th birthday of the Golden Gate Bridge was celebrated by allowing the citizens once again to stroll down its 6,450-foot-long, 90-foot-wide roadway. For the previous half-century, pedestrians were restricted to the bridge's 10-foot-wide sidewalks.

Although the bridge was not damaged by the destructive 1989 Loma Prieta earthquake centered 60 miles to the south along the San Andreas Fault, plans are under way to retrofit the Golden Gate Bridge with quake-resistant improvements. The work will include strengthening of the towers, the cable saddles, the submerged piers, and the anchorages on shore. Although the work will cost nearly $180 million—six times the original dollar figure for its construction—it is far cheaper than the estimated $1.4 billion required to build an earthquake-resistant bridge from scratch.

See also AMMANN, OTHMAR HERMANN; DEPTH TO SPAN RATIO; ELLIS, CHARLES ALTON; MOISSEIFF, LEON SALOMON; STRAUSS, JOSEPH BAERMANN; SUSPENSION BRIDGE.

Greathead, James Henry (1844–1896) *Perfected the subaqueous tunneling shield* Born in South Africa and sent by his family to be educated in England at the age of 15, James Henry Greathead eventually worked under the tutelage of the famed engineer Peter W. Barlow. Barlow had designed improvements to Marc Brunel's early 18th-century tunneling shield by making it circular instead of boxlike and greatly reducing its weight.

When Greathead went to work for Barlow he began improving the shield with his own innovations. By 1869, the improved shield, known as the Barlow-Greathead shield, was being used during a 14-week project to drive the 900-foot Tower Subway Tunnel to completion, a stunning achievement for its time. The improved shield was pushed forward by hydraulic jacks using cutting edges to slice through the soft ground, which was then removed.

Barlow's original design also specified that flooding at the face of the excavation could be controlled by closing the doors leading to the face of the tunnel and that compressed air could be injected between the front of the shield and the excavation. This process would buy time for the shield to be moved forward, squelching the leak.

One innovation of the Barlow-Greathead shield was the use of cast iron lining plates that were bolted together inside the shield and then put in place once the shield advanced. Once assembled, these plates provided immediate support, although it was necessary to inject grout between the curved plates and the irregular walls of the tunnel to ensure a solid fit and to prevent leakage. Previous tunneling efforts depended on masons to lay a brick lining as quickly as possible. The relatively small size of the Tower Subway Tunnel meant the ring segments were relatively easy to handle, but later versions of the shield cut larger tunnels. The larger rings required a hydraulically operated arm at the rear for lifting the heavier liner segments, a feature that remains in use on some modern tunnel-boring machines.

Greathead was in charge of the tunneling shield drives for the Tower Subway Tunnel and for the 1884–1890 work to construct tunnels for the City and South London Railway. For this work, Greathead made further improvements to the shield, including the development of a compressed air system for injecting cement grout between the cast iron lining and the tunnel.

By 1886, Greathead, while driving tunnels for the City and South London Railway, had pioneered the use of compressed air to reduce water leakage in wet ground. Although a great innovation, it exposed workers to the still-misunderstood effects of caisson disease, which could be cripple and kill.

By the time of his death Greathead had perfected his tunneling shield. It employed high-pressure water jets for dissolving the soft ground ahead of the shield, when practical, and allowed for the introduction of compressed air at the face to control flooding. It also allowed miners to enter the excavation chamber at the front of the shield to dig a two-foot advance by hand. Hydraulic jacks were used against the iron lining at the rear to move the shield forward two feet at a time to make room for the next lining ring of the same width. When pressurized water was used to turn the soil into slurry, it was pumped away from the face. Harder clay dug by the miners was removed in rail-mounted carts.

Since Greathead died in 1896 at a relatively young age, he did not live to see his tunneling shield used on many of the world's great subaqueous tunneling projects, particularly subway lines. His tunneling shield design remained virtually unchanged until the appearance of soft ground tunnel-boring machines in the late 1960s. These massive devices could cut, remove muck, and advance themselves while allowing the workers to remain under normal air pressure.

See also EARTH PRESSURE BALANCE MACHINE; SUBAQUEOUS TUNNEL; TUNNEL-BORING MACHINE; TUNNELING SHIELD.

Greathead shield   *See* Greathead, James Henry.

ground-freezing *Cold, hard way of stabilizing tunnels*   Tunneling in soft ground usually means tunneling in wet ground, and although easier in some respects than blasting or grinding through solid rock, soft ground tunneling presents stability issues both above and below the excavation.

Soft ground tunneling is almost never attempted without the use of a tunneling shield or a tunnel-boring machine since soft ground must be supported either during excavation or immediately after by rigid linings. Soft ground tunneling is also complicated in urbanized areas because buildings, bridges, and other structures might suffer damage if tunneling underfoot caused movement in their foundations.

One solution to this danger is to inject various types of consolidation or stabilization grouting into the surrounding ground to preclude this from happening. Although the wetness of the ground can contribute to weakness and instability, freezing this sogginess not only provides engineers a way of stabilizing the walls and crown of a tunnel but also prevents flooding. In addition, this moisture provides the means of stabilizing the ground above an excavation to prevent structural damage.

The answer for these types of situations is ground-freezing, which converts the ambient water in the ground into a form of ice cement. Cooling elements inserted into the walls of a tunnel or into the ground from the surface chill it to below-freezing temperatures. When the groundwater that surrounds the minute particles making up the soil freezes, it

locks them together into a mild form of self-supporting "concrete." Anyone who has tried to dig in permafrost can attest to the rigidity of frozen soil.

Although the process is seemingly a modern method of stabilizing ground for tunnel construction, F. H. Poetsch first patented ground-freezing in Germany in 1883. Some claim the method was used in tunnel construction in Wales as far back as 1862. All this was made possible by refrigeration technology patented in 1834 by an American engineer, Jacob Perkins, who was working in England. Ground-freezing caught on quickly and was used in Michigan in 1884 to stabilize the soft ground surrounding a tunnel shaft. Its most famous, large-scale use occurred during construction of Washington state's Grand Coulee Dam, which was completed in 1942. The threat of landslide from adjacent soft ground to the work site was eliminated when the contractor froze the earth in place.

Ground-freezing has matured into a complex art in which ground moisture levels and the type of soil involved must be taken into consideration. Engineers determine the location and depth of refrigeration piping so full coverage of the work site is accomplished. Sometimes injection grouting and ground-freezing are used in conjunction to ensure that surface structures experience no settlement or movement.

See also CENTRAL ARTERY/TUNNEL PROJECT.

grouting *Injecting strength and dryness into tunnels*   The history of hard rock tunneling is the unending story of the fight against flooding and collapse. Incompetent rock too unstable to sustain its own weight during and after excavation poses the nightmarish risk of collapse. Even a stratum of solid rock might contain a fracture through which a subterranean river could enter a tunnel.

Tunneling, in many ways, is a form of warfare waged by miners and engineers against the Earth. When wounded by the excavation of rock and soil, the Earth reacts as a living thing trying to heal itself. The ground may try to eliminate the artificial void by imperceptibly squeezing itself back together or by simply collapsing without warning. When the Earth bleeds, it does so in the form of flooding with water that can be ice cold or scalding. In a sense the Earth has its own immune system for foiling the human infection digging away at its insides. As a result, miners must be ready to outwit their foe at every turn.

To preserve their own lives and to drive the tunnel, the miners defended themselves against collapse before the 20th century by erecting curved iron ribs and timber lagging. Drainage systems and pumps dealt with groundwater infiltration until masonry linings could be installed. Should a fissure gush water from a subterranean spring into the tunnel, miners would have to construct a bulkhead or wall into the tunnel's crown, sides, or floor.

A new weapon against both collapse and flooding emerged during the early 20th century in the form of an old construction tool: cement. Able to harden under water, cement grout is a dream for tunnelers since it can flow like pudding into crevices and become as hard as rock within a handful of days.

Miners simply drill a hole into the unstable or leaking rock and pump in the grout under pressure. By watching gauges they can place a predetermined amount into the surrounding rock, and the amount of grouting holes drilled is dependent on the conditions in the tunnel. However, some tunnels possess such poor rock conditions that grouting is done in a systematic pattern throughout their length. Grout is also used to fill the voids behind concrete tunnel linings and the tunnel for a cohesive fit.

A massive grouting operation was conducted during construction of the Seikan Railway Tunnel beneath Tsugaru Strait, the deepest undersea tunnel ever constructed. Because the tunnel was 780 feet below the surface at its deepest point, the pressure of the water above was immense. Had the surrounding rock been solid and cohesive, this would have been of little concern, but the rock was riddled with fractures.

Although grouting stabilizes the surrounding rock while making it more impermeable to leakage, it is often not the final step since the tunnel must also be reinforced with iron or concrete linings.

See also GROUND-FREEZING; ROCK BOLT; SEIKAN TUNNEL.

Hawk's Nest Tunnel (built 1930–1932) *America's deadliest tunnel project*   There was to be no light at the end of the Hawk's Nest Tunnel for an estimated 700 to 1,500 miners who perished from the ghastly lung disease silicosis. The tragedy was all the more deplorable because it was completely preventable with safety techniques available at the time.

After reviewing evidence of what had taken place during the tunneling project, the U.S. senator Rush Holt of West Virginia said in 1936, "This is the most barbaric example of industrial construction that ever happened in this world."

It is still considered America's worst workplace disaster and often held out as an example of environmental racism since most of the victims were black. Workers died in such large numbers that they were often interred in mass graves near their company-owned shacks. Many of the deaths were hushed by officials, who pretended the workers had moved elsewhere when family members made inquiries.

The Hawk's Nest Tunnel was bored through Gauley Mountain in southern West Virginia to carry water from a dam at New River to the turbines of the New-Kanawha Power Company's hydroelectric plant. The private project was designed to provide cheap electricity for the Electro Metallurgical Company, which produced metal alloys. When completed, the hydroelectric project was purchased by Electro Metallurgical, then owned by a subsidiary of what became the Union Carbide Corporation. West Virginia was eager for the development, and state

mine safety officials looked the other way when the problem of silica dust—the culprit in silicosis—became apparent.

Despite the horrible deaths suffered by the miners, the massive tunnel project seemed to be a resounding engineering success—at least on paper. The 3.2-mile tunnel's diameter ranged from 32 to 46 feet, with 10,230 feet of its interior lined with reinforced concrete. Work began March 30, 1930, and ended almost exactly two years later. At a cost of $4.2 million, the tunnel's rapid progress was an extraordinary feat for the time.

But the price of speed was fatally high. Miners drilling shot holes for dynamite were generating clouds of dust from the pure quartz silica rock they were penetrating. Quartz had been pinpointed decades previously as one of the sources of dangerous mineral dust that causes the nefarious lung disease silicosis. The disease and its relationship to rock dust had actually been documented for centuries among stonecutters, but modern drilling equipment and explosives that arrived during the 19th century generated far more dust and made the illness an epidemic among hard rock miners.

The health threat of silica dust can be controlled through the use of filtration masks (readily available at the time) and the well-known remedy of applying water to the rock face through the use of "wet drills" to keep silica dust to a minimum. However, wet drilling slowed the penetration of the rock, said George Robison, a dying African-American miner who coughed out his testimony on conditions in the mine to the U.S. House of Representatives in 1936. Barely able to breathe and predicting his own imminent death, Robison said dry-drilling with a pneumatic drill could produce three holes for dynamite charges in the time it took to drill a single wet hole.

"When drilling, the hole would go straight down and the air [from the pneumatic drills] would then force the [dust-filled] air back into one's face," Robison told congressmen investigating health conditions of workers employed in public utility construction. Robison's testimony provided a chilling description of conditions in the mine, where the dust hung in the air so heavily that black men appeared white when they left the tunnel. Said Robison, "When they would bring in water to drink the dust would settle on top of it and one would have to drink that dust too."

The Great Depression, which dealt a body blow to the nation's economy hit hardest at African Americans, struggling as they were on the economic periphery of a nation still far from extending them equality. Robison revealed that few white workers could be induced to enter the tunnel choked with the powdery, white dust. Hoping to forestall starvation, African Americans eagerly took jobs excavating the Hawk's Nest Tunnel. The rapid pace of the construction led to inevitable accidents, and Robison also saw men crushed "beyond recognition" in rock falls just inches from where he stood. The ailing worker had helped to bury 35 coworkers both white and black during the construction of the tunnel.

The horror of the Hawk's Nest Tunnel disaster is that the silica dust was so pure and so thick that workers began falling ill within months of arriving for work. This was grim proof of the deadly conditions in the tunnel since silicosis ordinarily took 15 to 20 years to develop. The disease is triggered when silica is absorbed by macrophages, scavenging cells that absorb infection and foreign matter. The silica causes them to die and clump within the lymph nodes. Collagen, a connective tissue, then forms as a scar tissue around these clumps, causing pulmonary fibrosis, which eventually constricts breathing. Those afflicted with silicosis are often prone to deadly lung infections including tuberculosis.

The former coal mining town of Vanetta became a company town for the miners, and within months of the tunnel project's initiation, it was filled with coughing and wheezing men. When they arrived home, their clothing carried the deadly dust to their wives and children to breathe. Robison told Congress that many workers were buried within three hours of dying. The Rinehart & Dennis Company, the contractors for the mine, had eagerly paid all funeral expenses.

Even sickness produced no respite from their deadly duties of working in shifts around the clock to complete the tunnel. An obese company rouster, "Cap" McCloud, patrolled Vanetta armed with a pair of pistols and a blackjack. This tunnel truancy officer would track down coughing workers who said they were too ill to work and, despite their pleas, force them back into the mines. Some were roughed up or even shot, said Robison, who once watched a youth beaten with an ax handle by a foreman for refusing to work. When silicosis had devoured the lungs of workers so fully that they could no longer stagger into the tunnel, they were fired and evicted along with their families from company housing.

With the completion of the tunnel in 1932, company officials arrived to evict all the unemployed miners from their shacks. A confidential 1934 report prepared by a West Virginia statistician, Leon Brewer, revealed a frightful aftermath. By 1934, Vanetta—which once housed around 800 workers—became a near ghost town of only 101 residents. Of those, 14 were children, 44 were adult females, and 43 were adult males. According to Brewer's report, 33 of the men were afflicted with silicosis, and of that number only 15 could find work. Only one of the men able to find work had no symptoms of silicosis. To fend off starvation, these weakened men had to walk 36 miles to and from a road construction job they barely had the strength to perform.

As if dying from silicosis were not bad enough, the men soon discovered that West Virginia, whose major industry was the mining of coal, did not even provide worker's compensation for the mining-related disease of silicosis. When workers attempted to sue Rinehart & Dennis and Union Carbide for their silicosis illness, they learned the statute of limitations on civil suits had expired, a circumstance upheld by West Virginia's Supreme Court. Later, workers would find themselves being told by their attorneys to sign a settlement agreement in which they were paid one dollar with the promise of more money at a later date, money Robison told Congress he had not seen by 1936. State legislation later made it possible for workers injured by silicosis to receive some compensation.

Brewer's secret 1934 report pointed out that the workers were in such dire straits that West Virginia was obligated to help them financially and to assist them in returning to their original communities. "Relief should be adequate despite the protests of white people," wrote Brewer, recognizing the bias of the time. However, Robison was one of those who remained in Vanetta in 1936 and credited the donation of food and clothing by white residents living nearby with helping the remaining African Americans survive.

Many of the African American miners desperate for work flocked to the Hawk's Nest project from jobless southern states only to be sickened in the atrocious working conditions. Said Senator Holt during testimony before Congress in 1936, "They [Rinehart & Dennis] brought in these transients, especially from the South, and treated them worse than dumb animals should be treated. The company openly said that if they killed off those men there were plenty of other men to be had."

See also DRILL-BLASTING; VENTILATION; WET-DRILLING.

Hazard, Erskine   *See* Schuylkill Bridge.

heading   The heading of a tunnel is the limit of where the tunnel has been excavated and has a dynamic connotation since miners are always working to advance this heading, which is also referred to as the "working face."

Although tunnels can be excavated with a single heading, this method is usually extremely uneconomical for a number of reasons. Because of the astronomical cost of subterranean construction and the limited working space at a tunnel heading, the general rule is to excavate a tunnel with as many headings as possible.

When tunnels are dug below relatively shallow overburden, or overlying rock and soil, two end portals as well as intermediate shafts may be built to allow miners to excavate a tunnel with numerous headings or faces. In this way more crews or tunnel-boring machines can drive toward each other, expediting the project.

Some major tunnels deep underwater or beneath mountains have only allowed excavation from both portals, providing two headings for each bore. Eurotunnel beneath the English Channel is one, and the 12.3-mile Simplon Tunnel completed in 1905 in Switzerland is another. Deep water precluded the sinking of intermediate shafts on the Eurotunnel project, and hundreds of feet of overlying rock made it impossible to do the same at Simplon.

The term *heading* is also significant in a directional sense because miners and engineers can alter the direction of the tunnel excavation to avoid bad ground conditions or to ensure they rendezvous with an approaching heading.

Heated-Bridge Technology Program   *Breaking the ice*   Bridges and overpasses are structures that not only allow traffic to flow over them, but by the very nature of their design also allow cooling air to flow beneath them, presenting engineers with the hazardous problem of snow and ice accumulation.

Roads and bridges absorb radiant heat energy during the day, but roads are in contact with the ground that provides residual warmth and insulation. The result is that snow and ice have a more difficult time accumulating on the ordinary road surfaces. The effect of winds at freezing or subfreezing temperatures on a bridge is to chill the exposed structure rapidly, robbing it of whatever radiant heat it may have accumulated from the sun. As a result, snow or ice may remain on a bridge's deck far longer than on the approaching roadway. However, if a bridge does manage to warm enough from sunlight to melt snow and ice, the resultant water quickly refreezes after sunset.

The result is that bridges, particularly those in areas with cold winters and substantial amounts of snow or other types of precipitation, may become dangerously slick. Accumulations of snow on cold bridges can block travel or produce dangerous conditions if not removed.

Highway departments spend huge sums every year patrolling their bridges in winter to determine whether they remain passable. If snow and ice accumulations have reached dangerous levels, maintenance supervisors must dispatch snow plows or order the application of deicing agents containing salt, which is harmful not only to the steel elements of the bridge and passing cars but also to the environment.

Engineers in at least five states have initiated the installation of innovative systems designed to warm bridges by a variety of methods to eliminate icing, enhance driver safety, prolong the lifespan of bridges, and reduce maintenance and environmental damage. Though little known to the public, the program could have a major impact on their ability to drive safely across bridges in winter.

Between 1992 and 1997, the Federal Highway Administration (FHWA), under legislation known as the Intermodal Surface Transportation Efficiency Act (ISTEA) of 1991, funded tests of various bridge-heating systems. The act was passed to fund the testing of new highway technologies and encourage the use of those that are successful. States participating in the federal heated-bridge project were Nebraska, Texas, Virginia, West Virginia, and Oregon. The three main methods tested for heating the road decks of bridges included the use of heated fluids, heated vapor, and electrical resistance cables roughly similar to heating elements on electric stoves.

Also tested with the heating methods were mechanisms that sensed the presence of ice and activated the bridge-heating system. Without the sensors, it would be necessary to leave the heating systems on full-time or to dispatch personnel to the bridge, both unacceptably expensive options. These sensors include pneumatic models that determine when ice has blocked the roadway and other sensors that electrically detect moisture on the bridge. The moisture sensors are used in conjunction with temperature sensors to trigger a heating system when moisture and freezing temperatures are present.

When using the heated fluid (hydronic) system, fluid is pumped through a bridge deck to keep its surface too warm for the formation of ice. Piping to carry the warmed liquid must be installed within the concrete of the bridge deck, and the liquid must be warmed either by a boiler, which adds to the cost of the system, or by geothermal heat obtained from the Earth.

In Amarillo, a pair of two-lane bridges carrying traffic on U.S. 287 are heated by a mixture of polyethylene glycol and deionized water that is warmed by traveling through 50 geothermal wells. Each bridge deck contains piping circulating 3,000 gallons of heated fluid. Operation of this system costs $7,500 yearly. This system is considered relatively economical because no energy is consumed to warm the liquid, although electrical pumps must be activated to move the liquid from geothermal wells to the bridge.

All the bridge-heating concepts were relatively successful, with some of them combining heated fluid and electrical resistance heating systems. Although the installed systems are still being studied, many needed modifications to ensure proper opera-

tion. It was noted during the pilot programs that it was far less expensive to install a heating system during bridge construction than to rebuild the road deck to accommodate a system.

See also ELECTROCHEMICAL CHLORIDE EXTRACTION.

Heathrow Tunnel collapse   A spectacular engineering failure stunned Great Britain with the collapse of a trio of subterranean tunnels on October 22, 1994. The bores had been under construction near a terminal at Heathrow Airport. Although no one was injured, the collapse prompted an intensive investigation and resulted in some of the stiffest government fines ever levied against a British contractor for a workplace mishap. The tunnels under construction were part of a $340 million project to extend rail service from London's Paddington Station to Heathrow.

The tunnel was being built using the technically complex New Austrian Tunneling Method, which initially supports a tunnel excavation with a sprayed concrete lining in soft or unstable ground. Performing the work was the prestigious construction firm of Balfour Beatty with the engineering consulting firm Geoconsult.

This rail line tunnel had fallen behind schedule, threatening the profit margin of the contractor, and as managers prodded workers to accelerate work on the tunnel 60 feet below the ground, corners were cut and the invert, or floor of the sprayed concrete lining, was improperly constructed. When a joint of this invert began to fail around 12:40 A.M. on October 22, workers evacuated the tunnel and watched as the tunnel collapsed. The failure of the tunnel then spread to two adjacent tunnels under construction, sucking a pair of nearby parking lots into a cavernous sinkhole. Plastic foam was pumped into the ground to stabilize the subsidence, and concrete bulkheads were constructed to protect the remaining sections of the tunnels from collapsing.

The mishap snarled traffic and parking operations at the airport and prompted an investigation by Britain's Health and Safety Executive, an agency akin to the Occupational Safety and Health Administration in the United States. The matter eventually made its way to court, where Balfour Beatty was assessed a $1.7 million fine and Geoconsult received a fine amounting to $700,000. In 1999, a criminal court determined that the contractor had failed to recognize the impending hazard of a tunnel collapse because it was rushing to make up for being behind schedule. Balfour Beatty and Geoconsult both pleaded guilty to violating Britain's Health and Safety at Work Act.

See also NEW AUSTRIAN TUNNELING METHOD; SHOTCRETE.

Hell Gate Bridge (built 1912–1916) *Gustav Lindenthal's deceptive rail bridge*

Gustav Lindenthal, recognized for never building the same type of bridge twice, broke new ground with his design of his rail bridge over New York's Hell Gate strait. The steel arch bridge was designed to carry four rail lines, making it the strongest such structure of its kind. In addition, the 977.5-foot bridge was the longest steel arch at the time of its construction.

Lindenthal's bridge is actually an optical illusion of sorts. Strong and secure, the bridge remains in service after 85 years, but to casual observers the form of its superstructure is misleading, and intentionally so. The upper chord of the arch is essentially ornamental, and the two massive stone towers at each end of the bridge do nothing more than hide the fact that the ends of the upper chord are attached to nothing at all. Take away the towers and the upper chord, and one sees the skeletal functionalism of Lindenthal's design in the load-bearing lower chord of Lindenthal's steel arch.

Lindenthal's use of stone towers as a nonstructural, visual effect framed the graceful steel arch of the railroad bridge as it curved above the East River channel named Hell Gate. The bridge, capable of bearing the weight of 60 locomotives on its deck, owes its final appearance more to artistic whim and psychologic effect than to engineering. The final arch design was picked because it looked stronger and more dependable than other designs.

Lindenthal was not trying to cover up any inherent weaknesses or flaws in his concept but was bowing to the historical aesthetics of bridge design while attempting to create emotional trust in the bridge. During the 19th century, rail bridges had demonstrated a propensity for collapsing as designers struggled to understand the relationships between materials, form, and loading. Although Lindenthal knew how to make the Hell Gate Bridge work by using a minimal amount of structural material, he also had to give the public a sense of its strength.

The Dutch navigator Adriaen Block named the Hell Gate channel of the East River between Manhattan and Long Island in 1614. Hell Gate is an anglicization of the original Dutch term *Helle Gadt,* which can be construed to mean either "bright stream" or "hell's gate." Since treacherous currents and hull-ripping rocks are hallmarks of the channel, the latter meaning may be the more appropriate.

The bridge began making its slow journey to becoming a reality in 1896, when Lindenthal began serious discussions with the Pennsylvania Railroad engineer Oliver W. Barnes. A rail bridge over Hell Gate would allow the movement of passenger and freight trains between New York and New England. Within a year the New York Connecting Railroad was established to construct the necessary 3.2-mile bridges and viaduct system of which the Hell Gate Bridge would be the crown jewel. Not until 1904 would Lindenthal be appointed chief engineer on the project. Lindenthal reviewed potential designs and considered a crescent arch bridge akin to the Garabit Viaduct created by the French designer Gustave Eiffel.

Eventually, Lindenthal settled on a bowstring arch design with twin chords. Among the nearly 100 engineers assisting Lindenthal, then considered one of America's greatest bridge designers, were two who would find fame in their own right within a few years: Othmar Hermann Ammann and David Barnard Steinman. Ammann was Lindenthal's chief assistant; Steinman was Lindenthal's special assistant, charged with making the complex calculations related to loading on the bridge. Ammann, with Lindenthal's permission, would write an award-winning professional paper on the design and construction history of the bridge.

The reinforced concrete piers at each end of the bridge were sunk 90 feet beneath the surface to provide a springing for the lower chord of the arch and a foundation for the majestic though nonfunctional towers that would rise 250 feet above Hell Gate. The steel arch was constructed in separate halves from the opposing piers in a cantilever fashion with steel cables holding them aloft as the halves moved toward each other. When the halves were completed, the tiebacks were relaxed to allow them to meet.

The deck of the rail bridge is 135 feet above the channel. The upper chord of the arch rose 305 feet above the water. A quartet of rail lines were installed on the deck of the bridge to carry the rail traffic of New Jersey's Pennsylvania Railroad; the New York, New Haven and Hartford Railroad in the Bronx; and the Long Island Railroad in Queens. By the time work began on the bridge, Pennsylvania Railroad had become the owner of the New York Connecting Railroad, responsible for building the bridge system. To accommodate the four rail lines, the width of the bridge was set at 100 feet.

Lindenthal's Hell Gate Bridge provided a 977.5-foot clear span that was the world's longest steel arch. The Hell Gate Bridge served as a model for Australia's Sydney Harbour Bridge, which was completed in 1932. The clear span of the Sydney Harbour Bridge measured 1,650 feet.

By the beginning of 2000, Hell Gate Bridge had served more than 80 years with only minimal maintenance. During the intervening years one of its four sets of railway tracks had been taken out of service, but the bridge continues to carry the trains of Amtrak, CSX, and Norfolk Southern on the remaining three. Its carbon steel arches had not been repainted in decades and the appearance of the bridge had suffered. The United States senator Daniel Patrick Moynihan of New York began to

raise a stir about the appearance of the historic bridge, and Congress eventually appropriated $55 million to give the bridge a facelift. This work was completed in 1996 and included a fetching dark red coat of polyurethane paint.

See also AMMANN, OTHMAR HERMANN; BAYONNE BRIDGE; LINDENTHAL, GUSTAV; STEINMAN, DAVID BARNARD.

helmet men *America's early mine disaster rescuers* The frightful death toll in American coal mining accidents led to the creation in 1910 of the Bureau of Mines, which immediately established rescue teams equipped with early self-contained breathing apparatus. Because of the large diving-type helmets that were part of this equipment, the teams became known to miners as "helmet crews."

Their breathing devices were developed by the Draeger Company in Germany and featured a compressed air tank and a valve system that gave res-

This team of tough-looking helmet men is equipped with the self-contained breathing apparatus in a photograph taken during the early 1900s. Special trains rushed these skilled men to mine disasters—usually fires or explosions—for the grisly task of bringing out the living and the dead. *(National Archives)*

cuers the ability to work for two hours in poisonous air after a mine explosion.

The teams would not only search for casualties but also make an early reconnaissance of a mine to determine whether any remaining fire or explosion hazards existed. The robotic appearance of these crews changed when new breathing systems were developed with gas mask–type face pieces.

Early crews organized by the U.S. Bureau of Mines in 1910 were stationed around the United States and rushed to coal mine disasters aboard trains donated by the Pullman Car Company. When not dashing into collapsed or burning mines, they trained miners in the art of rescue work and the use of modern self-contained breathing equipment to create additional groups of rescuers. This elite group of "helmet men" were famous for their courage and their technical knowledge of dangerous mine conditions. Although many died performing their duty, they also rescued a large number of miners who would not have otherwise survived.

A Mine Safety Health Administration web site provides a detailed history of the helmet men and the work they performed at http://www.msha.gov/century/rescue/rstart.htm.

See also AFTERDAMP AND BLACKDAMP; CARBON MONOXIDE ASPHYXIATION.

Hennebique, François (1842–1921) *Reinforced concrete pioneer* A stonemason by trade, François Hennebique helped usher in the modern age of reinforced concrete construction, a building system that is today part of virtually every substantial structure in the world. Although the technique of embedding metal reinforcement into concrete was predated by the work of a fellow Frenchman, Joseph Monier, Hennebique's mass-marketing of his system made its use common practice. Hennebique's system was later applied to a number of notable bridges that he claims to have designed. However, since Hennebique granted franchise rights to use his reinforced concrete system, some are still pondering who designed what.

Hennebique was a 19th-century blue-collar worker who started his professional life apprenticed to a stonemason. When he became a mason in his own right in 1867, he moved to Brussels, where he pursued the reconstruction of churches, an endeavor that became his specialty. Ignorant of Monier's 1867 patent, which established a system for reinforcing concrete with enmeshed iron, Hennebique began constructing buildings with hefty iron beams encased in concrete, believing he had developed a new system.

Hennebique's first reinforced concrete structure was for a customer who wanted a fireproof house. Hennebique's experimentation led him to decide that concrete's brittleness made it a poor structural material unless combined with iron—and later steel. The metal embedded in the concrete would endure the tensile forces, and the concrete would endure the compressive forces.

In 1892, Hennebique founded a wildly successful business based on his theory of reinforced concrete construction and licensed firms and engineers around the world to use his method.

Although he solidly established the practice of using reinforced concrete in everyday construction, Hennebique's critics have long contended that his concepts were not far advanced and that the concrete mostly served to protect the iron from corrosion while the iron or steel reinforcement did all the real work. Even Hennebique's 1879 Belgian patent on his reinforced concrete system was overruled in favor of Monier's own concept. Nonetheless, bridges attributed to Hennebique, particularly the triple-arch Châtellerault Bridge over the Vienne River in France, demonstrate a lightness of form not previously seen. Completed in 1900, it was the first true reinforced concrete bridge. Although its structure was bulky in comparison to structures designed just a few years later by the Swiss reinforced concrete bridge designer Robert Maillart, its construction forged the path for future concrete bridge construction.

Despite his detractors, Hennebique knew how to make a buck and franchised his system to builders just about everywhere. He even published a periodical discussing the applications of his product, *Béton armé,* or appropriately enough, "Reinforced concrete." Hennebique's system was so well thought of that anyone possessing a franchise was almost assured plenty of business. Hennebique, lacking the formal training in mathematics and structural theory afforded other bridge designers, relied on experimentation and his experience as a stonemason—a trade replete with its own technical complexities. Hennebique also led the way in the use of steel as a reinforcing material to replace iron.

That Hennebique did not base his bridge designs solely on computation is actually fortunate, since 19th-century theory had not progressed enough to fully comprehend the abilities of reinforced concrete arch bridges. Hennebique, like Maillart, found that his bridge concepts baffled engineers, who were not sure that the bridge designs would succeed and were left scratching their heads when they did.

Advances in reinforced concrete bridge design and construction meant that Hennebique's system would become outdated. Hennebique's firm would not last long after his death, making Hennebique another footnote, though an important one, in the history of bridge building.

See also CHÂTELLERAULT BRIDGE; CONCRETE; MAILLART, ROBERT; MONIER, JOSEPH; REINFORCED CONCRETE.

Henry, John *"Steel-drivin' man"* Although most who labored under harsh and dangerous conditions to build America's railroad tunnels during the 19th century have been forgotten, the legend of the steel driver John Henry, who is said to have worked in West Virginia's Big Bend Tunnel, has been immortalized in stories and song.

Was this former slave capable of simultaneously swinging a pair of 10-pound steel hammers based on a real person? Was he so powerful that he defeated a compressed air drilling machine in a contest to prove which was better, man or machine?

Although it is doubtful that any man could accurately wield two hammers while driving steel drill bits into rock, the legend of John Henry is full of details indicating that a flesh-and-blood person may have been at the center of the tale. The story emerged from the 1870–1873 effort to build the Big Bend Tunnel, which ran a little more than a mile through the Allegheny Mountains.

If nothing else, John Henry represents the working men of America, especially those who entered the collapse-prone, gas-filled, explosion-ridden death-traps that were America's rail tunnels and coal mines during the 1800s. A look at the legend of John Henry provides a snapshot of the job of the average steel driver, his place in history, and the eventual elimination of his heroic job through technological advancement.

The premier means of advancing tunnels through solid rock during the first seven decades of the 19th century was the use of explosives placed inside hand-drilled holes. Although some imagine miners as using a pick, shovel, and sledgehammer to advance a tunnel, this method would have been slow beyond comprehension. Placing the explosives (black powder during the first half of the 1800s and later nitroglycerin and dynamite) meant a two-man team drove holes into the rock where the blasting agent could be inserted.

The punching of these shot holes required a high level of skill, mutual trust, fearlessness, and physical stamina. A drill bit would be held against the rock by a shaker, a man who would hold the bit steady for the blow delivered by the steel driver. After a blow, the shaker would then shake the bit and turn it (accounting for his also being known as a turner) to ensure the bit did not become pinched inside the hole. Rotating the bit also allowed a different aspect of the drill bit's head to meet the rock.

The work was arduous even by the tough standards of the day, when muscle and not machines did most of the work. The drill bits dulled quickly and a steady stream of the tools were carried by youths back to a blacksmith, whose sole purpose was to sharpen them. Piles of fresh bits would be available so the shaker and the steel driver could continue without pause. One slip in this ballet of ringing steel and flying chips could maim or kill the shaker, who sometimes held the drill bit between his legs as well as with his hands. These holes would penetrate the face of a tunnel for up to 14 feet before they were tamped full of explosives and detonated.

Those working at the face of the tunnel were most vulnerable to injury since this section of the excavation was usually not shored or reinforced, making it vulnerable to cave-in or flooding. This rough-and-tumble work would finally disappear during the late 1800s with the invention of the steam-powered rock drill, which could work far faster than a human steel driver. It is this collision of labor-saving technology with the livelihoods of working men that has made the legend of John Henry endure for more than 130 years.

As the legend goes, John Henry was employed building the Big Bend Tunnel for C&O Railroad near Talcott, West Virginia, and his prowess as a steel driver was unparalleled. John Henry's hand-drilling partner was reputedly named Phil Henderson. With the arrival of the steam-powered rock drill, some believed that the redoubtable John Henry had met his mechanical match.

Charles Burleigh of Fitchburg, Massachusetts, had perfected a compressed air–powered rock drill that the management of the Big Bend Tunnel project wanted to buy. According to the legend, John Henry boasted he could beat the steam drill in a competition. This was arranged and the seven-foot-tall John Henry, supposedly swinging two sledgehammers, began pounding a drill bit into the rock. The former slave managed to beat the steam-spewing mechanical drill by three inches but at a terrific cost: John Henry fell dead from his superhuman exertions.

The legend of John Henry is poignant in that it gives America a hero who is an underdog's underdog. Not only is he doing one of the toughest jobs in a time when all the jobs were tough, but he was a former slave working as a free man five years after the Civil War. In addition, this former slave suddenly finds his own value as a workingman endangered by a soulless machine just as his human value was challenged by slavery. John Henry stands up to the machine, proving his own worth, but this mythic struggle ends with his death because the machine feels no pain, has no thirst, and is devoid of a heart that can fail. The legend is a powerful one that still resonates.

The Appalachian folk song "John Henry" celebrates the legend of the steel driver, a tale that is no doubt heavily embellished. However, interest in the legend of John Henry was on the rise more than 100 years after the Big Bend Tunnel was completed. A statue depicting the African-American hero today stands near the Big Bend Tunnel in Summers County, West Virginia. Another indicator that interest has not waned in the mythic workingman is the 1997 issuance of a United States Postal Service stamp commemorating John Henry's story.

The ballad that immortalized John Henry speaks eloquently of his defiant, workingman's pride:

> John Henry said to his captain
> A man ain't nothin' but a man
> Before I let that steam drill drive me down
> I'll die with this hammer in my hand, Lord, Lord
> I'll die with this hammer in my hand.

See also ROCK DRILL.

hinge A bridge, although made of steel and concrete, is nonetheless a living thing in its own way. The steel and concrete components of bridges expand and contract with heat and cold, respectively, and even the earth on which bridges are anchored is often far from motionless. Subtle (and sometimes not so subtle) movements within the bridge and the ground on which it stands can cause the components of a bridge to shift position with potentially harmful results.

For concrete arch bridges in particular, these movements pose special problems to the arch or the multiple arches supporting their roadways. Concrete, which lacks the tensile abilities of steel, tends to crack as a result of expansion rather than bend. Since arches curve, any compression or expansion can create stress or even cracking near the base of the arch (its springings) where it attaches to an abutment. Arches are also prone to cracking along the crown or apex.

Because of these factors, bridge arches of iron, reinforced concrete, or any other relatively brittle material must be able to rotate with physical stresses to avoid damage. The engineering answer was to connect the arches to their foundations or abutments with what is known as a hinge. Some early hinges were exactly those, steel hinges that allowed a bridge to rotate slightly according to expansion or contraction. Another type of hinge used early in the 20th century connected an abutment with the arch, using concrete rods with an expansion joint separating the abutment from the arch.

The use of hinges became particularly important as reinforced concrete arch bridges grew popular at the beginning of the 20th century. Strong, relatively inexpensive, and easy to erect, concrete arch bridges were nonetheless stiff and incapable of flexing with movements of any magnitude.

The Austrian engineering professor Fritz von Emperger wrote in the early 1900s that the proliferation of concrete arch bridges made it imperative that not just two but a trio of hinges be built into concrete arch bridges. Tests of concrete arch shapes subjected to pressure reveal they will crack at the crown as well as at the springings of the arch. Hinges, Emperger said, should be placed at both bases of the arch and a third at its center.

Although it might make someone nervous to drive over a bridge "wobbling" up and down on hinges, the structure is no less stable because of this ability to flex slightly. A lack of hinges would mean unrelieved stressing and the resultant cracking of bridge arches, a consequence that is far less desirable.

As a result, concrete bridges of all types, including beam and segmental bridges, have been fitted with hinges to absorb daily stresses as well as thermally induced contraction and expansion at many points within their structures.

See also ARCH BRIDGE; BEAM BRIDGE; BEARINGS; PRESTRESSED CONCRETE; SEISMIC RETROFIT.

Historic American Building Survey/Historic American Engineering Record *Documenting and digitizing America's engineering heritage* Known as HABS/HAER, the Historic American Building Survey/Historic American Engineering Record is a unique government and private sector program that has been compiling a photographic and written historical record of America's architectural and structural engineering accomplishments since 1933.

Although little known to the public, this program is administered by the Library of Congress, and is the repository for photographs, drawings, blueprints, and histories of notable structures, including bridges and tunnels.

The Library of Congress has already digitized thousands of images and histories from the HABS/HAER collection and has placed them online. Paper documents that have yet to be converted into electronic form can be viewed at the Library of Congress. It is a unique program that by 2001 was in its 68th year.

Started in 1933 by the National Park Service landscape architect Charles E. Peterson, the HABS project was designed to provide employment to architects displaced by the Great Depression. They would employ their skills in documenting America's architectural heritage in a program also sponsored by the American Institute of Architects.

In 1969, the National Park Service and the Library of Congress along with the American Society of Civil Engineers launched the Historic American Engineering Record to document construction and structural engineering projects aside from buildings. Shortly afterward, the American Society of Mechanical Engineers, the Institute of Electrical and Electronic Engineers, the American Institute of Chemical Engineers, and the American Institute of Mining, Metallurgical and Petroleum Engineers agreed to sponsor this history program.

Those records already digitized have been placed online and are available to the public for viewing or downloading. The images are available in low-resolution or in extremely high-resolution, print-quality files. Other records not yet digitized can be ordered for a fee from the HABS/HAER collections of the Library of Congress. Photographs can also be ordered. Although most images are in the public domain, some have been donated from private collections and may be governed by use restrictions.

The program no longer provides temporary employment to displaced architects but instead gathers much of its materials through a summer internship program involving engineering and architectural students.

The HABS/HAER web site is http://www.cr.nps.gov/habshaer/. The site contains a link to a search engine for the HABS/HAER records, information on the program, and procedures for ordering copies of images and documents. It is a magnificent resource not only for professional engineers and architects but also for students and other Americans interested in the story behind some of America's most fascinating structures.

holed through Because tunnels are usually excavated from each end, the excavation phase of tunnel construction is considered complete when both teams meet one another underground. When the final wall of rock has been blasted or ground away, this means the miners or sandhogs have "holed through." Another term commonly used is *breakthrough*.

Besides being an actual engineering accomplishment, holing through is as symbolic as the "topping off" event celebrated by ironworkers who complete the superstructure of a tall building.

Although the events are usually greeted with a great deal of fanfare, particularly by management, holing through is actually bad news for the ordinary workers since it usually means their employment will soon be ending.

Nevertheless, holing through ceremonies are great public relations gimmicks, something not lost on those promoting and building tunnels. When the English Channel Tunnel's service tunnel was completed on December 1, 1990, television camera crews were invited beneath the Earth to film the event. The carefully staged image of the tunnel workers Graham Fagg of England and Philippe Cozette of France shaking hands through a thin wall of chalk marl was broadcast around the world as it happened.

Even before the age of electronic media, the hype surrounding holing through events in great tunnel projects was in evidence. An artist sketched the February 29, 1880, breakthrough in Switzerland's Saint Gotthard Tunnel after eight dangerous years of blasting and digging in dank and diseased conditions.

The accuracy of the drawing is questionable. It shows two groups of burly and bearded miners encountering one another at an opening dug between the two tunnels. A shirtless, torch-holding, clean-shaven miner—who looks suspiciously like a statue of the Greek god Dionysus—lights the scene. The rendering of the event is made more comical since two of the rough-and-tumble miners are toasting their accomplishment with long-stemmed wineglasses. The central activity in this idealized drawing of the event is a kiss on the lips between two heavily bearded men.

During the January 30, 2001, breakthrough of a tunnel carrying traffic to Boston's Logan Airport as part of Boston's Central Artery/Tunnel Project, management wanted to record this event, one of several breakthroughs during this long and difficult project. The 21st-century American sandhogs posed a bit more stiffly while facing the camera and there were no wineglasses or toasts.

Despite the 121 years separating the breakthrough in the Saint Gotthard Tunnel and the completion of a freeway tunnel, the sandhogs looked as tough and dirty and wet as their 19th-century Swiss counterparts. And although none of the Boston sandhogs was kissing another, one hardhat-wearing worker blew a kiss to the camera, proving that American tunnelers could be just as affectionate as their Swiss colleagues.

See also CENTRAL ARTERY/TUNNEL PROJECT; SAINT GOTTHARD TUNNEL.

Holland, Clifford Milburn (1883–1924) *Chief engineer of the first long vehicular tunnel* Plagued with a weak heart, a condition inherited from his father, Holland should have probably sought a different line of work, since, during the 1800s and early 1900s, many of the great tunnel engineers had died in the line of duty. More often than not, the implement of their demise was not a cave-in or flooding but 24-hour workdays, stress, unending problems, and the life-and-death decisions that had to be made on an hourly basis.

Raised in Somerset, Massachusetts, Holland obtained his engineering degree from Harvard and eventually was supervising tunnel construction beneath the bustling streets of New York City. Holland developed a reputation as a capable, hands-on engineer who inspired the confidence of his superi-

ors. He eventually became one of the world's authorities on subaqueous tunneling.

It was Holland who convinced many officials and members of the public that a vehicular tunnel beneath the Hudson River was a feasible project. In 1919, Holland was made chief engineer of the effort and committed himself to the task.

By 1924, Holland was a broken man emotionally and physically. He departed work that year on a leave of absence. Although the record is vague on this point, Holland was described as having what was tantamount to a nervous breakdown. Soon after this he suffered a fatal heart attack at the age of 41. The tunnel he worked so hard to complete was finished in 1927 and was named in his honor. In 1975, his hometown of Somerset dedicated a plaque to Holland during its bicentennial celebration.

See also HOLLAND TUNNEL.

Holland Tunnel (built 1920–1927) *First long vehicular tunnel* Although digging and reinforcing a tunnel are usually the hardest parts, when it came time to build the first long vehicular tunnel, the real challenge was ensuring that this otherwise helpful crossing did not become a lethal gas chamber. Although soft earth tunneling had been perfected by the early 1900s, engineers had yet to design a system that would exchange millions of cubic feet of foul air for fresh in a tunnel filled with lethal carbon monoxide and other automobile exhaust gases.

A crossing between New Jersey and Manhattan had been a subject of discussion for years, but it was not until 1906 that the New York State Bridge and Tunnel Commission and the New Jersey Interstate Bridge and Tunnel Commission began serious discussions on the issue. A bridge was originally considered, but such a long span might require midstream piers, something deplored by shipping interests. In addition, aboveground approaches for an elevated span would cost a prohibitive amount of money. By 1913, a tunnel was considered the best possible option for allowing traffic to cross the Hudson River.

The need for a New Jersey–Manhattan rail tunnel was apparent, but as the debate continued, the automobile began to compete with rail, horse trolleys, and subways as a means of personal travel. When serious planning on a tunnel began after World War I, it was obvious that the unhealthy exhaust gases of autos would have to be dealt with in what would be the world's longest vehicular tunnel.

Public officials like to be reassured that an expensive project, particularly one employing untried technology, is not only feasible but also unlikely to cause them political embarrassment. An inspiring cheerleader for the project arrived in the person of Clifford Milburn Holland, a dynamic engineer who had dedicated his career to building rail tunnels beneath the streets of New York. Holland enthusiastically and carefully explained how the tunnel would be excavated, reinforced, and then ventilated.

The officials and the public were reassured, and in 1919 Holland was appointed chief engineer of the Hudson River Vehicular Tunnel project. Helping Holland in the project were two chief assistants, Milton H. Freeman and Ole Singstad.

Work began October 12, 1920, on a pair of tunnels commencing from the Manhattan side of the Hudson. Soon, a second pair of tunnels was being dug from the New Jersey side to expedite the work. Cofferdams had been sunk and the pneumatic tunneling shields were aimed on their way through a pair of gigantic portals. The shields contained doors through which workers would attack the face of the tunnel with hand tools. At times, the doors could be opened to allow the soft clay to ooze through for removal.

As the shield progressed, pushed as they were by massive hydraulic jacks, workers would bolt together iron rings to line and reinforce the tunnel. Cement grout was then pumped behind the rings to fill any voids.

Although Holland is frequently credited with designing the innovative ventilation system for the tunnel, Singstad tackled this obstacle with input from Holland and others. Since little if any health research had been conducted into the effects of gasoline combustion engine exhaust in enclosed spaces, Holland's engineers turned to the U.S. Bureau of Mines and Yale University to study the safe limits of automobile exhaust. The Bureau of Mines was a heroic organization in that it pushed for safety in America's most dangerous workplaces, its coal mines, and had long experience in reducing the hazards of working in conditions where poison gases were present.

With this research in hand, Singstad and his colleagues designed a system that would allow the air in the tunnel to be exchanged every 90 seconds with-

out turning the tubes into wind tunnels filled with hurricane-force winds.

Singstad solved this problem by installing a false floor containing a ventilation shaft beneath the twin tubes through which fresh air was forced through curbside vents. In the crown of the tunnels were gratings leading to overhead exhaust conduits that vacuumed up the rising air to carry away the deadly carbon monoxide and other gases. In this way, Singstad was able to provide a massive volume of fresh air without creating a windstorm. A quartet of aboveground ventilation buildings provided a two-way flow of air to make this system work. The air was injected into the tunnel with 42 fans and drawn out by an equal number of exhaust fans. Officials liked to boast that the air was fresher in the tunnel that it was on the streets above.

The sandhogs and engineers encountered conditions that had plagued other Hudson River tunneling efforts, including a mixture of fractured rock and the easy-to-dig clay. This required a combination of blasting and digging that was time-consuming and dangerous: The digging had to halt during blasting, and the blasters had to be extremely careful to prevent creating a blowout in which the compressed air would escape the tunnel. Although the tunnel was as deep as 93 feet beneath the riverbed, other portions of the tunnel were far closer to the water above. At one point on the New York side, where rock needed to be blasted out of the way, only 14 feet of clay and muck stood between the tunnel and the river.

In 1924, after four years of virtually living in his office or the tunnel supervising the project, Holland suffered an emotional collapse and left to recuperate. Soon after, he died of heart failure. Ironically, his death occurred the day before sandhogs met beneath the Hudson to connect the first of the tunnels. Freeman stepped in to command the project but lasted only five months before he too died. Singstad then supervised the tunnel's completion and managed to survive the task.

During the early part of 1927, the sandhogs holed through to complete the second tunnel. This was the usual contrived event in which the opposing sandhogs were carefully selected for the moment. In this case, the two shaking hands through the completed tunnel belonged to brothers. The second tunnel was soon holed through, and work to complete the tiling and the exhaust system proceeded rapidly. President Calvin Coolidge was present on November

12, 1927, to officiate at ceremonies opening the tunnel to the public.

The tunnel then saw 52,000 vehicles pass through its tubes on its first day of operation and proved to be a self-supporting project, generating toll revenues that quickly repaid its $54 million cost and even providing money for other infrastructure projects.

In 1931, the Port of New York Authority (today's Port Authority of New York and New Jersey) took possession of the tunnel, which had operated almost flawlessly since its completion, carrying an increasingly heavy amount of traffic.

However, operations were marred by a 1949 fire that roared out of control in the tunnel, collapsing tiles and concrete from the roof's interior. The incident occurred at 8:48 A.M., May 13, 1949, when a drum of carbon disulfide fell from the bed of a tractor–trailer rig carrying 80 of the 55-gallon drums of the dangerous amber-colored liquid. A police officer in the tunnel telephoned an alert and helped motorists leave their cars.

At the time the drum fell from the truck, the vehicle was already 2,900 feet inside the tunnel. The touchy carbon disulfide—a toxic and highly flammable liquid—burst into flame, producing clouds of toxic fumes. The mishap caused the driver of the rig to come to a stop as the flames then ignited the remainder of his cargo. Another 10 trucks were completely destroyed in the blaze and 13 more were damaged. Panicked drivers unable to drive forward or backward ran back through the tunnel toward New Jersey.

Several police officers were overcome by smoke at the scene of the fire and firefighters rushed to spray water on the blaze. They wanted not only to extinguish the fire but also to cool the interior of the tunnel since it was feared the heat might weaken its lining. Because of the fear of collapse, fireboats were called out to look for a telltale whirlpool or a rush of bubbles indicating the tunnel was collapsing. Additional fire fighters arrived to battle the stubborn blaze.

At 9:45 A.M., Singstad's ventilation system was put to the ultimate test when its 84 fans were accelerated to their highest possible speed to clear smoke from the tunnel and give the fire fighters as much breathable air as possible. Some fire fighters, overcome by the ghastly fumes from the chemical fire, were able to push their faces against Singstad's curbside air ducts to be revived by fresh air.

Although 66 people were injured, and 27 of those required hospitalization, the tunnel was not damaged structurally, but wiring at the site of the fire as well as several ventilation fans were knocked out. As fire fighters worked to clean up the mess, the vicious remnants of carbon disulfide reignited, forcing them to inundate the scorched section of tunnel with fire-suppressing foam. The superb and farsighted ventilation system in the tunnel was credited with removing the deadly smoke and fumes of the fire and saving dozens of lives.

Despite this event—the only serious fire to date in the tunnel—Port Authority officials were able to reopen the Holland Tunnel a mere 56 hours after the fire was extinguished. The tunnel was completely repaired at a cost of $1 million in 1949 dollars. With inflation, repair of the damage would today cost more than $7.5 million.

See also COFFERDAM; HOLLAND, CLIFFORD MILBURN; SINGSTAD, OLE; SUBAQUEOUS TUNNEL; TUNNELING SHIELD.

Holland Tunnel chemical fire   *See* Holland Tunnel.

hollow-box arch   Among the many innovations of the Swiss bridge designer Robert Maillart is the development of the reinforced hollow-box arch. Instead of merely building a massively heavy concrete arch to support the Züoz Bridge in the Swiss city by the same name, he made the arch hollow, filling it only with a dividing wall.

The result was an arch with a concrete floor, arched reinforced concrete sidewalls, and the concrete road deck as its roof. The internal, centerline wall running through the hollow arch divided it and provided additional reinforcement. The design is lighter and therefore easier and cheaper to build while providing superior strength for the materials used.

Maillart, who was often stirring controversy with his innovative bridges, was assailed because his designs outstripped the mathematical skills of engineers to calculate how the bridges worked.

The hollow-box arch was no less a puzzle in that computations available at the time of the construction of the 1901 Züoz Bridge were inadequate to explain its workings fully. The hollow-box arch was later used in other of Maillart's designs, including the breathtaking Salginatobel Bridge in western Switzerland.

See also MAILLART, ROBERT; PRESTRESSED CONCRETE; SALGINATOBEL BRIDGE.

hollow-box girder or beam   The innovation of the box girder is a child with many fathers, although it took a shipbuilder, a bookish mathematician, and one of Britain's most famous engineers to understand the concept fully and turn it into a reality.

With the introduction of iron to shipbuilding by the 19th century, stronger, lighter ships were now possible. When an iron hull was combined with an iron deck the result was an incredibly strong structure resembling a box, though one tapered at the keel, or bottom edge of the ship.

Carpenters had been building what would later be called box girders for centuries in the form of rectangular, wooden coffins, although the structural abilities of this type of design had never been fully realized.

When the 19th-century bridge designer and engineer George Stephenson learned that an iron ship was accidentally suspended above the water during an abortive launching, he realized that it might represent a structure for a bridge that could sustain its own weight as a bridge span without crumpling or deforming. At the suggestion of his inventor father, Robert Stephenson, George obtained the assistance of the iron shipbuilder William Fairbairn, who helped further develop the concept. Both Fairbairn and Stephenson realized that a hollow girder in tubular or square shape could provide terrific rigidity and strength.

With the help of a London University professor of mathematics named Eaton Hodgkinson, who calculated the mathematical basis of the concept, the first structural hollow-box girder was developed. It was in fact a rectangular tunnel of iron reinforced with parallel iron girders on its roof and floor. This arrangement allowed the iron tube to serve as one of four prefabricated sections of a railway tunnel that would be raised atop their piers. Since the construction of an arch bridge did not meet the approval of the British government, the box girder had solved the problem of how to throw a bridge across the Menai Strait in Wales, a project completed in 1850.

Using an iron plate floor, walls, and roof, the men designed a structure that was remarkably rigid. Known as a tubular bridge by some, this design was robust enough to carry trains but was far more expensive and heavy than the iron and steel arch bridges that superseded it as a more economic type of design.

However, with the capabilities of the box girder identified, the concept was soon put to use in virtually every type of large structure, including bridges. Rather than serving as the actual conduit for rail or vehicle traffic, box girders served as supporting members within bridge decks. The concept of a box girder is that each side of the rectangular tube stiffens the structure and complements the strength of the other sides.

Box girders provide a massive amount of strength for their weight when compared to a solid girder such as a steel beam or solid concrete beam. Solid horizontal structural beams are extremely heavy and must support not only external loads but their own weight as well. The introduction of post-tensioning and prestressing of concrete saw this strengthening technique applied to concrete hollow-box girders, making them better able to endure loading from above.

The Swiss bridge designer Robert Maillart tweaked the use of the hollow-box girder at the turn of the 20th century when he created a curved hollow-box arch of reinforced concrete for the design of his Züoz Bridge. This hollow-box arch provided more resistance to twist and proved the versatility of the hollow-box girder concept.

An ugly legacy of this superb innovation was the fact that Stephenson hogged the limelight for himself. Fairbairn and Hodgkinson felt slighted and justifiably so. Fairbairn would later resent Stephenson for not sharing credit for the hollow-box girder concept more equitably. Wrote Fairbairn, "While I freely award to Mr. Stephenson the honour of this conception, I claim for myself the credit for having rendered it practical."

See also BRITANNIA BRIDGE; GANTER BRIDGE; MAILLART, ROBERT; STEPHENSON, GEORGE.

Hoosac Tunnel (built 1851–1875) *Massachusetts's tunneling nightmare*   A tunnel through Hoosac Mountain in western Massachusetts was originally conceived during the canal craze of the early 1820s as a means of carrying canal traffic. The mania for canals gave way to the mania for railroads by the time a decision was made to build the longest tunnel in North America. It would be railroad tracks and not a canal that would run through what would eventually be a 4.75-mile tunnel.

Working next to a jumbo bearing an array of compressed air drills, miners drill shot holes for nitroglycerin charges during construction of Massachusetts's Hoosac Tunnel. A financially ruinous endeavor, it took nearly 25 years to bore 4.75 miles through Hoosac Mountain. *(Culver Pictures)*

It would be an epic project. And because it was constructed in the 19th century, it would be an agonizingly slow, dangerous, and controversial undertaking. It was a scheme initiated perhaps two decades in advance of technologies that could have made the project far less costly in terms of time, lives, and money. It would take 24 years, 195 lives, and would nearly bankrupt the state of Massachusetts, which eventually took over the tunnel and financed its completion.

The reason for the Hoosac Tunnel was the fever to invest in railroad enterprises that began to grow in the 1830s, a frenzy rivaled only by the unbridled investment in high-tech enterprises during the late 20th century. Massachusetts business interests felt isolated from the states to the west and south and began building their own railroads. Promoters dreamed of a railway reaching its tendrils west from Boston and slithering toward the Berkshire Mountains to cross into upper New York state 120 miles to the west.

The beginning of what was arguably America's most agonizing construction project was in 1825, when the civil engineer Loammi Baldwin was asked to select a canal route. This route logically followed the contours of the terrain and was considered ideal except for one major obstacle, the five-mile-wide lump of rock called Hoosac Mountain. Even this was not an obstacle, insisted Baldwin, as long as a tunnel could be punched through it.

Baldwin, considered the father of American civil engineering because of his professional approach to a discipline then in its infancy, was as highly regarded as any engineer America possessed. He was a designer of forts and canals and even the monument celebrating the Battle of Bunker Hill. Baldwin predicted this tunnel would cost no more than $920,000. When it was finally completed a half-century later, the tunnel's cost would be 18 times greater than Baldwin's estimate, adding to the great tradition of tunnel cost estimates that were more optimistic than accurate.

The daring plan by Baldwin eventually was metamorphosed into a rail tunnel and found a ready promoter in the self-made millionaire Alvah Crocker, a Massachusetts paper manufacturer. By 1845, Crocker had built a series of railroads heading westward from Boston to Fitchburg, north to Brattleboro, Vermont; and then south to Greenfield, Massachusetts. The western path of his railway drive into New York State was blocked by the Berkshire Mountains, or more specifically by Hoosac Mountain, which hulked within the range.

Crocker obtained a charter for the Troy and Greenfield Railroad, which would tunnel through Hoosac Mountain, creating a route into New York State. Work began on the tunnel January 8, 1851, but was slowed by two problems that would continually plague the project: a lack of money and inadequate tunneling technology. Despite enthusiasm for the project by businessmen such as Crocker, the conservative Yankees of western Massachusetts were not about to dive willy-nilly into railroad investing. When only $3,000 was raised through the sale of stock, Troy and Greenfield Railroad pleaded for a $2 million loan from the Commonwealth of Massachusetts, which was rejected.

Work continued with the funds available and the railroad purchased an early tunnel-boring machine designed by Charles Wilson. It was believed that the 90-ton machine would grind away the stone with its cutting head at such a rate that the tunnel could be completed in five years. Unfortunately, the iron cutters of the steam-driven machine were unequal to the task.

The machine began its work in March 1853 and its early progress was encouraging, as the machine ground 4 1/4 inches into the east face of Hoosac in a quarter of an hour. Working day and night at this rate, the tunnel could be bored in less than two years! This was not to be. The machine bored only 12 feet into Hoosac before it became stuck inside the mountain. Hoosac had eaten its first drilling machine.

The Hoosac Tunnel's chief engineer at the time was A. F. Edwards, who soon realized that traditional drill and blast methods would have to be employed. Men swinging sledgehammers weighing 10 or 20 pounds would drive star-shaped drill bits (also known as drill steels) into the rock to make room for black powder charges. These would be detonated, the crumbled spoil mucked out, and the process repeated. It was monumentally tough work,

painfully slow, and ridiculously dangerous for both drillers and the powder men. It was, however, state-of-the-art tunneling technique.

In the meantime the politically adroit and well-connected Crocker continually sought state funds, arguing that the state had previously assisted Massachusetts's Western Railroad with a $2 million loan. When the Amherst College president and state geologist Edward Hitchcock told legislators that a tunnel through Hoosac would be free of flooding and require only minimal reinforcement, his assessment encouraged passage of the 1854 Tunnel Aid Bill, which would provide $2 million for Crocker's tunnel. However, state money would become available only after considerable progress in penetrating the mountain had been made and only if $650,000 in private funds were raised through the sale of stock.

The New York City contractor Edward W. Serrell signed on to build the Hoosac Tunnel and would be the first of a series of builders singed in tangling with both the mountain and the Commonwealth of Massachusetts. Before Serrell could begin work, difficulties in raising additional money through stock sales caused his contract to be voided in 1855. In July 1856, H. Haupt & Company, of which Serrell was a partner, signed a $3,880,000 contract to build the Troy and Greenfield Railroad line as well as the tunnel. Haupt, who was the senior partner, had been chief engineer of the Pennsylvania Railroad and was regarded as an expert on rail bridges. Headings on the east and west sides of the mountain were begun in 1856. In the east, the slate that was initially encountered fell easy prey to drill bits and black powder. Soon, however, the miners found themselves straining to drive drill bits into unyielding granite, which was far less susceptible to black powder charges. On the west side the miners blasted through a layer of water-filled limestone, unleashing a flow of clay that pushed into the tunnel and then solidified. Known as running ground, this soft rock and clay mixture can flow like plastic and then solidify, or it can absorb ambient moisture and flow once again. It is one of the most intractable and expensive conditions for miners to deal with, requiring extensive drainage work and reinforcement. State Geologist Hitchcock, it appeared, had been wrong.

Desperate to increase his progress, Haupt also decided to experiment with another tunnel-boring machine, again designed by Charles Wilson. The

$25,000 machine was unleashed on the face of the tunnel August 25, 1857, but was soon defeated by the mountain and removed. The miners were soon back at the face of the tunnels with drill bits and black powder. By May 1, 1857, Haupt's workers had clawed their way through a mere 80 feet of tunnel. To increase the number of faces being worked by the miners, Haupt began work on a seven-foot by 14-foot shaft that would descend 318 feet, allowing the workers to blast their way east and west toward the two original portals.

Money woes plagued the project since Haupt had been unable to sell enough shares to trigger the release of state funds and claimed he had spent $220,000 of his own money to finance the work. Crocker's plea for additional state funds was vetoed by Governor Henry J. Gardner in 1857. The defeat of Gardner by a tunnel supporter, Nathaniel P. Banks, gave Crocker hope of state assistance.

Haupt's work on the rail line and the tunnel segment during his tenure was generally quite sloppy, although this may have been due to the severe rush and lack of funds. Throughout his efforts Haupt, as well as Crocker, were hectored by the paper manufacturer Frank W. Bird, a political ally of the competing Western Railroad. Bird railed against the tunnel through vicious pamphlets decrying the project as a boondoggle and castigating its supporters.

By 1860, Haupt had made enough progress to convince legislators to free the Troy and Greenfield Railroad from the necessity of raising $650,000 before state funding could be released. Ezra Lincoln was appointed state engineer to gauge the progress of the work and to meter the release of state funds. Lincoln had proved himself a friend of the tunnel project, but he fell ill, and Haupt, hounded by the pamphleteer Bird, quickly moved to obtain Lincoln's resignation and the appointment of Lincoln's assistant (who was also amenable to the project) as state engineer.

In what was almost a tragicomedy, this plan unraveled with the departure of Governor Banks for Union army service in 1861. Banks had become so preoccupied with writing his departure speech that he failed to sign a $100,000 draft desperately needed by Haupt. After Banks's departure from office, his successor, Governor John A. Andrews, was reticent to approve the funds.

In the meantime, Bird had succeeded in having Haupt's hand-picked choice as the new state engineer fired and replaced with a tunnel opponent, William S. Whitwell. When Andrews turned to Whitwell for his opinion on providing funds to the tunnel, Whitwell truthfully pointed out that what progress had been made on the tunnel and the rail line was far from finished work. Andrews went along with Whitwell's recommendation and refused to release the $100,000.

Denied the badly needed state funds, Haupt ceased worked on the project on July 12, 1861. After five years of effort, only 3,060 feet of tunnel had been opened inside the mountain, or roughly one eighth of the distance needed to complete the 25,000-foot tunnel. Broken financially, Haupt fought his own unsuccessful public relations campaign to recoup the $1.2 million of his money and that of private investors. Initially unsuccessful in this endeavor, Haupt entered the Union army as an engineering officer to build bridges.

Buried under crushing debt, the Troy and Greenfield Railroad became insolvent, and the rail project and the uncompleted Hoosac Tunnel became the property of the state on September 4, 1862. Work on the tunnel was halted for a year, and a trio of engineers, Charles Doane, James Laurie, and Charles Storrow, was commissioned to study the feasibility of completing the project.

One of the most important assignments was given to Storrow, who was dispatched to Italy and France to observe work on the Mont Cenis Tunnel. On this project its chief engineer and contractor, Germain Sommeiller, had converted an unsuccessful steam-powered rock drill into an efficient compressed air drill for which he designed a water-powered compressed air system.

Mostly on the basis of Storrow's observations of the technical innovations used on the Mont Cenis Tunnel, a state commission decided the Hoosac project was possible. An outrageous error was discovered when it was learned that Haupt's opposing tunnel headings were out of alignment, meaning the west heading would have to be rebuilt. This outrageous error was the fault of the Hoosac Tunnel's original engineer, A. F. Edwards, whose incorrect survey was utilized by Haupt.

The newly formed Massachusetts Tunnel Commission then selected Doane as the chief engineer on the project. The commission was to oversee the effort through the office of the state engineer in an effort to complete the tunnel and recoup the $955,000 in state funds already spent on the project. Work slowly began during the summer of 1863.

Although a capable engineer, Doane did not have a free hand in the project and was ordered by Whitwell to sink a 1,000-foot shaft to the level of the tunnel's route to provide additional working faces. Doane was opposed to this approach since it was a monumental and expensive task that might reap few rewards.

As Sommeiller had across the Atlantic, Doane dammed up a river to provide a reservoir of water that would power the turbines of air compressors as a source of power for the pneumatic rock drills he intended to use. Coincidentally, these were Sommeiller-type drills redesigned, lightened, and improved by Charles Burleigh, a resident of nearby Fitchburg, Massachusetts.

Miners digging through the west portal of the tunnel were soon living in a dank subterranean world that was soaked with water from underground springs ruptured by drill bits and explosions. Pumps waged a 24-hour struggle to keep the heading clear of water. So concerned about the flooding was Doane that he sank four shafts along the centerline of the tunnel's route to determine how much more water awaited the miners. An adit was dug from the base of one of the vertical scout shafts to drain water from the main tunnel. Pumping water from the tunnel faces soon became as important as drilling and blasting.

Doane continued in his methodical and competent way to extend the tunnel beneath Hoosac Mountain, but the 1865 cessation of the Civil War unleashed two enemies of the project—an inflationary spiral that hiked the cost of construction and the vituperation of the pamphleteer Frank Bird. As costs to build the tunnel rose, so did the ire of miners demanding more money for dealing with postwar inflation. A strike ensued and some buildings were burned in protest, but the work stoppage ended and the tunneling continued.

Doane was still fighting all the problems of the tunnel, including the running ground that flowed into the tunnel like a living thing. To tame this phenomenon, Doane turned to a rock-solid contractor, Bernard N. Farren, to build a brick lining in the tunnel to defeat what was then called "demoralized" rock. The job was so massive that Farren chose to build a massive kiln near the west portal of the tunnel, where he manufactured bricks for the work.

To accomplish his task of reinforcing the tunnel, Farren used heavy timber bracing to hold the ground in place as the brickwork was laid. Eventually Far-

This Stereoview image of the Hoosac Tunnel's west portal was taken at the point when miners had penetrated 4,500 feet into Hoosac Mountain in this difficult project. The brick lining of the tunnel has been completed at least at the portal, although much finish work remains. A group of young girls stands dangerously close to the edge of the tunnel entrance, probably at the insistence of the photographer, to show the scale of what was one of the most famous and controversial American civil engineering projects of the 19th century. *(Author's Collection)*

ren's men would line 7,537 feet of the tunnel with 2 million bricks bonded with cement to create a masonry tube.

Despite small victories in advancing the tunnel, Doane needed to accelerate the excavation. In 1866, Doane saw an ad in *Scientific American* magazine advertising a safer method of using nitroglycerin touted by the English chemist George M. Mowbray, who had immigrated to America to take part in Pennsylvania's oil boom. Nitroglycerin was a more powerful explosive, but its touchiness and toxicity scared away potential users.

Mowbray recognized that nitroglycerin released twice as much energy as black powder when properly manufactured and used. He also knew that reservations about its hazards meant its employment depended upon a safe system of producing and handling the chemical. By the mid-1860s Mowbray had created methods to accomplish both aims.

Doane was convinced by Mowbray's claims, and by the fall of 1867 the chemist was at Hoosac

Mountain constructing a two-story building near the tunnel works. There, Mowbray employed substantial ventilation to reduce the toxic and explosive vapors generated during the mixing of the nitric and sulfuric acids. He also employed copious amounts of chilled water to keep the nitroglycerin stable. Once a batch of nitroglycerin was produced, it was poured into rubber tubes and frozen until ready for use. Thawed to a liquid state with warm water, it was placed in the shot holes and detonated.

Although the "nitro" was at first considered only as effective as black powder, Mowbray pointed out that the shot holes were being drilled 12 inches shorter than he had specified. Workers punched their shot holes to 42 inches and found the newer explosive was indeed more powerful.

By 1867, a routine had set in at Hoosac Tunnel with improved water pumps keeping the works relatively dry, efficient Burleigh drills punching shot holes at a rapid clip, and nitroglycerin removing more rock than black powder.

However, Doane's careful management of the difficult project and the political backbiting that swirled around it had taken their toll. In July 1867, Doane quit in disgust over the halting way in which Massachusetts conducted the work and the growing death toll among the workers. In addition, the vertical shaft ordered by Whitwell had reached only 500 feet, or halfway to the tunnel's grade. This vertical hole was by then known as the Central Shaft.

The Central Shaft was a monumental construction effort in itself. Its opening was covered by a building known as the hoist house for the winch inside, that lowered and raised men and equipment. Instead of drilling forward, the men drilled and blasted downward. An added hazard of working in the shaft was that carelessly dropped objects from above became high-velocity missiles that killed and maimed those at the bottom. A Burleigh drill once fell from the top of the shaft, drove through a worker's head, and continued through his torso until it clanged against the rock floor.

As if it were not dangerous enough, the shaft had been equipped with a lighting system that used the flammable liquid naphtha, but this had proved too dangerous and had been abandoned. On October 17, 1867, a worker who did not realize the reason why the lighting system was in disuse, began to tinker with the defunct system, releasing volatile naphtha fumes, which were somehow touched off. The resulting explosion set the hoist house afire,

driving away the hoist operator and stranding 13 men mucking out the shaft more than 580 feet below. The real tragedy occurred when the hoist house became so weakened by fire that it began to collapse. Initially, 300 steel drill bits showered down the shaft like a rain of deadly spikes. Then the hoist and tons of flaming timbers poured into the shaft.

At three o'clock on the following morning, a worker named Mallory was lowered by rope into the pitch-black hole carrying oil lanterns for illumination. When the lanterns sputtered out, indicating a lack of oxygen, an unconscious Mallory was hauled up. When Mallory regained consciousness, he reported seeing nothing but burnt timbers floating on deep water that filled the shaft as a result of the failure of the pumping system. Some of the bodies floating on this rising column of water were recovered, but it was not until October 1868 that the remaining bodies were brought out.

In 1868, the pamphleteer Bird ran for governor, vowing to shut down the deadly and wasteful Hoosac Tunnel project. His criticisms had become too shrill and the Hoosac Tunnel's harpy was roundly defeated. However, the Massachusetts legislature had had enough of the controversial and costly experiment and voted to stop funding the project. Crocker, knowing his beloved tunnel faced a death knell, lobbied the legislature to seek bids on the project from outside contractors. Opponents of the tunnel included the provision that the contractor would have to post a $500,000 completion bond, a draconian requirement they believed was adequate to kill the project by scaring away contractors.

Hoosac Tunnel opponents, however, had not counted on a firm so eager to get the job it would meet the requirement by offering to do $500,000 worth of work before it received a penny of the state's money. That firm was W & F Shanly, a Canadian construction firm owned by brothers, Walter and Francis Shanly. Having outfoxed the tunnel's opponents, they were granted a contract in the amount of $4.6 million to complete the tunnel. Opponents of the tunnel sat back and waited for the Shanlys to fail. For their part, the Shanlys hoped to eke out a substantial profit through careful management of the tunneling effort.

In March 1869, the Shanlys resumed work on the tunnel, but calamities persisted. A heavy rain flooded a stream that spilled into the tunnel, causing an adit that was draining the west heading to become clogged with debris. The 100 workers inside the tunnel soon found themselves facing rising water. They ran ahead of the flood to a shaft, where a hoist operator methodically lifted groups of the men to safety. The flooding stopped only 18 inches from the roof of the tunnel.

The Shanlys' dream of making a tidy profit began to unravel after the flooding since they had to repair unexpected damage to the tunnel's brick lining and sections of the aboveground rail line washed out by the flood. The contract they signed with the state had not exempted them from such costs, making the Shanly brothers an insurance company for the state's project.

On August 13, the Central Shaft finally reached the point where miners could begin blasting two horizontal headings 1,028 feet below the mountain. This was a welcome victory after the flood damage, which had taken nearly a year to repair. Unfortunately, miners struck even more underground springs, causing flooding that threatened to drown the workers.

Walter Shanly had unsuccessfully sought permission to abandon the western heading and concentrate on connecting the east heading from the Central Shaft with the heading driven from the east portal. Once the east section of the tunnel was holed through, Shanly would be able to allow the tunnel to drain through the completed section. The state's engineer had refused to allow this, forcing Shanly to purchase additional pumps costing $200,000. Fed up with this intransigence, Shanly finally appealed directly to the governor's office, pointing out he would eventually be removing far more water than rock from the west heading. Shanly's request was eventually granted, although he was by then in possession of expensive pumps he did not need.

Deadly events that would have stymied other projects simply became the norm on the long-suffering Hoosac Tunnel project. When a work train carrying a quarter-ton of nitroglycerin exploded in 1870, vaporizing the locomotive as well as a group of miners, Walter Shanly immediately ordered a new engine. Although the Shanlys had a reputation for fair dealing and compassion, they were now in a race against financial ruin. Time was money, and a new train was needed sooner than later.

As 900 men labored in and around the tunnel in 1873, workers finally holed through the rock separating the east portal and the Central Shaft. This allowed the water to drain through the east heading and the resumption of the excavation in the troublesome

west heading. By November 1873, the miners on both ends of the western half of the tunnel were able to hear the efforts of their counterparts. Walter Shanly selected Thanksgiving Day, November 27, 1873, as the day the miners would complete the tunnel.

Dignitaries received handwritten letters inviting them to witness the detonation of a final nitroglycerin charge. On that day Mowbray himself entered the tunnel to connect the final charges to their electrical detonator. When detonated, a five-foot by five-foot hole roared open between the two headings, completing the 25,081-foot bore through the mountain. A newspaper reporter, Manley M. Gillam, walked from the east portal to the west portal on the day of the ceremonial breakthrough to be the first person to travel the length of the tunnel.

The Shanlys exited the project in early 1874 by subcontracting completion work on the tunnel. In addition, the reliable Bernard Farren accepted a contract directly from Massachusetts to widen the tunnel's narrower sections to their final width in November 1874 for $300,000. This work was completed by early 1875, and on February 9 the first train to pass completely through the tunnel did so. The tunnel eventually accommodated a pair of tracks, and traffic through the tunnel grew steadily.

Massachusetts, whose public coffers had been bled by the interest on the Hoosac Tunnel bonds, decided to rid itself of the largest item in its budget by selling the Troy and Greenfield Railroad and its tunnel. Despite the decades of carping over the tunnel's exorbitant cost—which may have been somewhere around $17 million ($363 million in 21st-century dollars)—the deal struck by the Commonwealth of Massachusetts stands as a textbook example of how to get ripped off. In 1887, Fitchburg Railroad purchased the tunnel and the railway for $5 million in railroad bonds and what the state believed was $5 million worth of Fitchburg's common stock. The common stock was valued at $100 a share for the purpose of the sale, but the actual market value was closer to $20 a share. The result was that Fitchburg Railroad bought itself a tunnel for a third of its actual value. It seemed that the Hoosac Tunnel would always have the last laugh on the Commonwealth of Massachusetts.

Despite all that had happened, the Hoosac Tunnel was the longest tunnel in North America and would remain so until 1916 with the construction of the 28,986-foot Connaught Tunnel in the Canadian Rockies. The Hoosac Tunnel would remain the longest tunnel in the United States for 11 more years, until the 1927 completion of Colorado's 32,795-foot Moffat Tunnel.

A comparison of the Moffat Tunnel and the Hoosac Tunnel is interesting since the hard rock Moffat Tunnel, despite being 7,000 feet longer, was completed in one fifth the time. However, the Moffat Tunnel utilized the techniques and tools that were pioneered in the Mont Cenis Tunnel and perfected in the Hoosac Tunnel. The workhorse explosive of the Hoosac Tunnel—nitroglycerin—was also used, although in the form of dynamite. In 1866 Alfred Nobel rendered safe the tricky liquid sledgehammer by mixing it with clay.

Acrimony among those who built the tunnel lived on long after it was completed, as the Shanlys and Haupt demanded payments they claimed were owed them by Massachusetts. Haupt eventually received an additional $500,000. Records show the Shanlys collected only $288,000 of the $500,000 the state agreed to pay them for work not covered by their contract, according to Carol R. Byron in his succinct history of the tunnel, *A Pinprick of Light.* Walter Shanly, states Byron, later referred to Massachusetts's legislators as "the most contemptible scalawags in the Commonwealth."

Oddly, one factor seemingly ignored by all the engineers building the tunnel was the problem of ventilation. The soot-filled exhaust gases belching from the coal-fired steam engines were so heavy that crews did not even know whether the trains were moving and had to lie down to gasp enough air. A steam-driven ventilator fan with wooden blades was installed at the Central Shaft but did little good. Worst of all, the dark and smoky tunnel led to collisions when parked trains were struck from behind.

In 1910, the tunnel was electrified so electric engines could haul steam locomotives (with their fires banked) through it. In addition, a more efficient electrically driven ventilator fan was installed atop the 1,028-foot Central Shaft. At this time as many as 48,000 rail cars were passing through the tunnel each day. Electrification of the section was removed in 1946 as diesel-electric locomotives entered widespread use, followed by a drastic decline in the number of trains using the tunnel. In 1957, the twin track gave way to a single track centered in the tunnel to provide extra clearance for piggyback rail cars carrying semitrailers.

However, the Hoosac Tunnel's days are not over. Work completed in 1999 excavated the roadbed of the Hoosac Tunnel to create an additional 15 inches of clearance to allow the passage of double-decker containers known as stacks. The tunnel is now owned by Guilford Transportation Industries.

See also ADIT; DYNAMITE; MONT CENIS TUNNEL; NOBEL, ALFRED; ROCK DRILL; SOMMEILLER, GERMAIN; SQUEEZING GROUND.

Hoover Dam diversion tunnels (built 1931–1932) *Black Canyon's man-killing hellholes* The monumental task of digging four gargantuan tunnels to divert the Colorado River and allow the construction of the Hoover Dam rivaled the building of the dam itself as an engineering achievement. A ruthlessly low bid by the contractors coupled with financial incentives encouraging early completion of the tunnels invited management to disregard worker

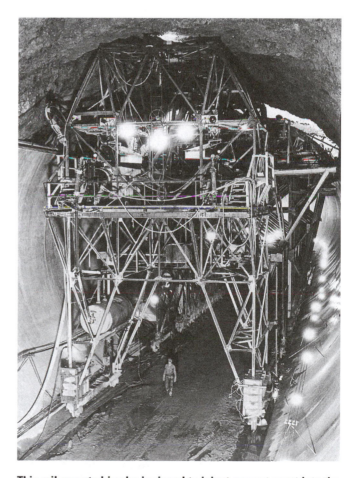

This rail-mounted jumbo is rigged to inject cement grout into the roof of the Hoover Dam's diversion tunnels. The tunnel project was essential to reroute the Colorado River before dam construction could begin, and jumbos like this accelerated the effort. *(National Archives)*

safety during the effort. As a result, this engineering victory would be tainted with the deaths of at least 35 men in and around the tunnels.

Westerners had begged for the project for years to control Colorado River floods and to capture water for agricultural irrigation. By 1902, the federal government began conducting studies and accumulating data on where and how to dam the Colorado. A site was selected in Black Canyon, a desolate locale with annual rainfall of only four inches and summer temperatures that could reach 125 degrees Fahrenheit. In 1928, Congress appropriated funds for the construction of what was then referred to as Boulder Dam.

The dam would rise 727 feet from the riverbed to the roof of Black Canyon. It would be 660 feet wide at its base, tapering to 45 feet at its summit. The arch-shaped bridge would measure 1,244 feet across the canyon and create 110-mile-long Lake Mead, the largest reservoir in the United States. Before pouring a spoonful of the 4.4 million cubic yards of concrete needed to build the dam, the Colorado River would have to be dewatered upriver from the site. The solution was to bore four massive diversion tunnels through the shoulders of Black Canyon, two on the Nevada side and two on the Arizona side.

These tunnels would be excavated with a diameter of 56 feet and lined with three-foot-thick concrete walls. Each would be roughly 4,000 feet in length. And although each was not long by 20th-century tunneling standards, these bores would be built with terrific speed. Once completed, a gigantic cofferdam would stand as a bulwark against the river and direct the water into the diversion tunnels. The chore was nothing short of herculean and was made financially dicey by the requirements: Tunnel construction would start in May 1931 and be completed no later than October 1, 1933, or the contractor would incur a $3,000-per-day penalty.

The contractor was a corporate amalgamation known as Six Companies Inc., formed because the job was too big for any single company then in existence, or more specifically, because no single contractor could post the required $5 million completion bond. The consortium would be composed of the biggest names in American construction, including Morrison-Knudsen, W. A. Bechtel Co., and Henry J. Kaiser.

The former United States Bureau of Reclamation dam builder Francis Trenholm Crowe (1882–1946) was a six-foot, three-inch, ramrod who had already built six bridges and would build Hoover Dam. He had left government service during the 1920s to work for Morrison-Knudsen when the federal government decided it would let private contractors do its dam building.

Crowe was the construction superintendent over the entire project. It was also Crowe who had studied the plans for the dam and recommended that Six Companies bid a remarkably low $48.9 million for the project. The government leaped at giving the contract to the consortium since the Six Companies bid was $24,000 below the government's own estimate. A Canadian who had graduated from the University of Maine in 1905 with a degree in civil engineering, Crowe developed a love affair with dam-building and an obsession for innovative construction techniques. He was also determined to bring the project in ahead of the deadline.

In normal times finding workers in the desolate and undeveloped area would have been difficult, but it was 1931 and the Great Depression sent thousands of workers scrambling to the wasteland along the Nevada-Arizona border looking for work. Early arrivals lived in a series of makeshift camps where the heat was so oppressive it killed even the wives of the workers.

Although touted as a great engineer and a superb motivator of workers, Crowe was also considered somewhat ruthless in his drive to complete the tunnels. As it would turn out, working as fast as possible enabled Crowe to earn a $250,000 bonus for completing the tunnel ahead of schedule. The men began drilling and blasting their way through the rock, expending a ton of dynamite for every 15 feet of penetration. The heat inside the tunnels, filled as they were with the residual gases of explosives, rock dust, and vehicular exhaust, soared to near 140 degrees Fahrenheit.

Crowe took the concept of the jumbo, a rig allowing multiple drillers to prepare shot holes, to a new extreme. Mounting a multistory steel pipe frame on the bed of a heavy truck, he made it possible for 22 drillers assisted by 21 drill bit handlers, known as chucktenders, to punch the shot holes in a tunnel face. Crowe employed eight of these drilling jumbos during the project. The drillers would then reposition themselves until they had drilled the 126 holes necessary for a "shot." The broken rock was then hauled away by the 100 trucks racing at breakneck speed along the winding roads to the tunnels.

Although the truck-mounted jumbos were efficient, the method was also morbidly dangerous since the gasoline engines of the trucks belched carbon monoxide into the sweltering and poorly ventilated tunnel. Although the use of internal combustion vehicles inside tunnels under construction was outlawed by Nevada regulations, enforcement was argued away with the claim that the job was a federal project unfettered by state codes.

During the first summer of work in the tunnels, 14 men died of heat illness and five others died in rock falls, one of which killed three men on October 17, 1931. Another man was run over outside a diversion tunnel by a dump truck whose driver, like everyone else, was told to race to get the job done. Workers later recalled the frantic pace of the job and

that even stopping to check on the condition of a man who fell from a fatal height earned a worker a rebuke for wasting time.

The ways in which a worker can die on a tunneling job are innumerable. Some men in the diversion tunnels died in mishaps with explosives, and others were electrocuted by dangerous wiring. Many were killed in falls, and others died when struck by falling equipment. Perhaps as many as 35 workers died inside of or just outside the diversion tunnels. Official records show that 96 workers died of "industrial accidents" during the entirety of the dam project, but this figure excludes those who died as a result of heat- or atmosphere-related illnesses in the tunnels.

What was obvious is that Crowe's primary interest was in completing the tunnels as early as possible, which he did by beating the deadline by more than a year, earning Crowe his handsome bonus.

A worker at the right is silhouetted near the entrance of one of the four 50-foot-diameter water diversion tunnels bored through solid rock during the Hoover Dam project. Summertime temperatures inside the tunnels were so intense that at least 14 men died of heat illness during the first year of construction. *(Bureau of Land Management/National Archives and Records Administration)*

As the tunnels were blasted and 1.5 million cubic yards of rock and earth was mucked out, an invert, or floor, was built of concrete. Then, curved steel forms for the concrete tunnel lining were erected along the sides of the tunnels and held in place by jacks. Workers poured concrete progressively higher behind the forms through openings. The arch or roof of the tunnel was then built by pumping concrete behind overhead forms by using a cement gun.

After the concrete lining of the roof was in place, it would be coated with an asphalt material to retain moisture and help it cure as strong as possible. To complete the lining job, grout holes were drilled into the concrete walls and a cement mixture was pumped in to provide a solid fit between the rock and the lining. Once the three-foot-thick lining was in place, the 56-foot tunnel had a completed diameter of 50 feet.

The Boulder Dam would be renamed after President Herbert Hoover in 1930, but his successor, Franklin Roosevelt, would rescind this and reinstate the dam's original name three years later. Congress reversed this bit of pettiness in 1947 when it renamed the dam after Hoover.

Once the tunnels had completed their job of diverting the Colorado River, two of them were closed with gigantic concrete plugs and a third was also fitted with a concrete bulkhead containing four six-foot-diameter tubes equipped with valves to allow water to flow through. On February 1, 1935, the last open tunnel was closed off with a 1,000-ton steel door.

A U.S. Bureau of Reclamation web site provides extensive information about the Hoover Dam project and the construction of the diversion tunnels. It can be accessed at http://www.hooverdam.usbr.gov/History/index.htm.

See also CARBON MONOXIDE ASPHYXIATION; DYNAMITE; JUMBO; MUCKING.

Howe, William (1803–1852) *Inventor of the Howe truss, which revolutionized rail bridges* William Howe began his career as a builder of houses and mills in Massachusetts but stumbled into fame and fortune in 1838 when asked to build a railway bridge in Warren. As a carpenter, Howe understood well the concept of trusses, a repetitive, triangular arrangement of wooden members that provide stiffness and strength to buildings. Although other truss designs, stiffening bridge decks and increasing the loads they could sustain were in existence, Howe applied an important innovation to a truss he was building into his bridge at Warren.

Howe added substantial strength to the truss by employing vertical iron rods that connected the opposing vertices of the triangles formed by the wood diagonals. Threaded at one end, the iron rods connected at the vertices of the wooden diagonals and could be tightened by bolts to maintain the rigidity of the truss. In this way the rods accepted loading through tension, and the wooden diagonal members assumed the load in compression. He patented this design in 1840 and by 1842 had obtained a patent for incorporating a timber arch in conjunction with his truss.

Instead of pursuing a career as an active bridge builder, Howe lived off the royalties of his design, which became one of the most commonly used truss types in 19th-century bridge construction. A reason for its popularity was that Howe's brother-in-law, Amasa Stone, another New England carpenter, paid $40,000 for exclusive rights to Howe's truss in New England. Stone became a prolific railway contractor and employed Howe's truss in dozens of railroad bridges. The Howe family was an enterprising clan: Howe was an uncle to the sewing machine inventor Elias Howe.

Despite Stone's use and promotion of his relative's truss, Howe's design became a favorite among other bridge builders, who often added their own refinements. It was an excellent innovation for the railroad age since bridges were carrying heavier loads. Commonly used in railroad spans, it was also employed in roadway and pedestrian bridges.

See also STONE, AMASA.

Hudson Tubes (built 1874–1908) Wealthy and confident, the California railroad engineer DeWitt Clinton Haskin wanted to build one of the 19th-century's early subaqueous tunnels, among the most difficult technical challenges to an engineer. And although his own attempt would result in death and failure, it would kick off a 32-year effort to link New Jersey with Manhattan with a mile-long system of tunnels beneath the Hudson.

Fresh from his successes as a railroad promoter and engineer who built tunnels and bridges in California, Haskin appeared in New York during the early 1870s, when he managed to attract investors for the construction of a railroad tunnel beneath the Hudson River between New Jersey and Manhattan.

Having formed the Hudson Tunnel Railroad Company in 1874, Haskin began work on a pair of tunnels that would traverse a mile-wide section of the river, 65 feet beneath the water. In a common occurrence during the 19th century, the work had to be halted to turn aside legal challenges from ferry operators, who invariably opposed competition from bridges and tunnels with specious legal arguments.

Whatever his previous engineering successes, Haskin displayed unbelievably bad judgment when he began digging beneath the velvety soft muck of the Hudson River's bed. Instead of employing a tunneling shield to prevent the collapse of the tunnel—such as one used nearly 50 years before on London's Thames Tunnel—Haskin used only compressed air to prevent water from infiltrating the works. He should have used both.

By 1879, full-scale work was able to begin as Haskin's crews fought to burrow beneath the Hudson River. Without a shield the men had to dig as rapidly as possible while others hurriedly installed quarter-inch linings of iron plate, which were then reinforced by a 36-inch lining of brick. Compressed air kept the tunnel relatively dry, but even that posed a danger since it pushed against the weak walls of the soft ground tunnel. Should the internal pressure become too great, a blowout could occur.

At 4:30 A.M. on July 21, 1880, a blowout did occur, causing flooding that killed 20 sandhogs. The man in charge of the crew at the time was the assistant superintendent, Peter Woodland. As water rushed into the tunnel, Woodland yelled to his men, "Make for the lock!" Although nine sandhogs were able to obey Woodland's command, a timber killed one before he could reach safety.

Woodland helped close the airlock door but remained outside. The others watched through a glass porthole as he struggled through waist-high water in a futile attempt to plug the blowout. Soon the tunnel filled with water. Only seven bodies were recovered from the incident, including Woodland's. By 1882, this disaster, among other problems with the project, had effectively ended Haskin's attempt to link New Jersey with New York by 1882. The workers had completed 600 feet of one tube and a second tunnel had been pushed 1,600 feet beneath the river.

In 1889, the Hudson Tunnel Railroad Company under the attorney John Dos Passos awarded Haskin's ill-fated project to the British firm of Pear-

sons & Sons. Using a tunneling shield that operated in conjunction with compressed air, workers made progress but soon struck sections of bedrock interleaved with silt. Sir Benjamin Baker, the renowned English engineer, and James Henry Greathead, the inventor of the modern tunneling shield who pioneered compressed air tunneling, were both consultants to the project.

Despite this expertise and the good tactics they employed, difficult conditions underground stalled the project, which was halted once again by funding problems. With only 1,000 feet left to excavate before completing the tunnel, the project was abandoned for the second time and the tunnels intentionally flooded in 1891.

In 1892, an attorney and public transportation enthusiast, William G. McAdoo, moved to New York from Chattanooga, where he had pushed through the electrification of Knoxville's horse trolleys. He also believed that tunneling beneath the Hudson to operate passenger trains was a profitable idea. From Dos Passos and later Jacobs, McAdoo learned of the uncompleted tunnels that ran under the river for nearly a mile. McAdoo then set out to organize backers for a new attempt to finish the tunnels.

Not until 1902 would McAdoo acquire funds with which to complete the Hudson tunnels. It was well known that an English engineer, Charles M. Jacobs, had in 1894 masterminded the East River Gas Company Tunnel—New York's first successful subaqueous tunnel—and he was the obvious choice to complete the Hudson Tubes. The newly formed Hudson & Manhattan Railroad Company, consisting of the various business entities supporting the project, made Jacobs the chief engineer on the project.

Jacobs ordered the water pumped from the abandoned tunnels and after an inspection estimated the project would cost $4 million to complete. As Baker and Greathead did, Jacobs would combine compressed air with tunneling shields in the effort.

As the tunneling shields were pushed ahead, it was realized that in many sections of the tunnels digging was unnecessary. When the excavation doors at the face of a shield were left open, ribbons of clay would squeeze into the rear of the shield, where it could be chopped off and mucked out. With 11 hydraulic jacks exerting 2,500 tons of force to push the shield, much of the progress was made in this way. Steering the shield became only a matter of leaving a door closed on one side or another so the shield would turn in the direction where it encountered the least resistance. With this method, a record of 72 feet was excavated in a 24-hour period.

Not all the tunneling went so smoothly. Most of the problems encountered were due to the shallow, 14-foot layer of silt overhead, which would often give way because of the compressed air in the tunnel. Blowouts became numerous as the work progressed. At various times the sandhogs had to strip off their clothing to plug holes through which air escaped, and once the bare back of a tunnel worker was used to block a hole. In March 1904, a door in the shield face was opened as it advanced in wet ground and a blowout occurred, shooting a powerful stream of silt shot into the heading, burying one sandhog alive while others scrambled for safety.

Sandhogs eventually cleared out the silt, recovered their coworker's body, and tried to block the hole ahead of the shield with tons of clay dumped from above. When this failed, the hole was covered from above by an expanse of weighted canvas sailcloth. This was discovered to be a failure when the bulging sailcloth began to enter the tunnel through the hole in the riverbed! Eventually the hole was plugged.

The silt that composed the walls and crown of the tunnel was anything but solid. Jacobs ordered hundreds of gallons of kerosene into the tunnel to fuel blowtorches, whose heat hardened the soft ground. Working ahead of the tunneling shield, sandhogs heated the clay, excavated, advanced the shield, and repeated the process. This method would probably raise the eyebrows of today's safety experts.

Mud was not the only obstacle that Jacobs and his men had to tunnel through. Hard rock ledges that required blasting were also encountered. Although the use of explosives in soft ground tunneling was considered dicey work, Jacobs employed methods he perfected in the East River Gas Company Tunnel, and the work progressed. On March 11, 1904, the first of the tubes was holed through, and the second bore was connected at midstream on September 24, 1905.

Eventually, Hudson & Manhattan Railroad would build 17 miles of railroad tubes from New Jersey, under the Hudson, and then across Manhattan at a cost of $70 million. The Port Authority of New York and New Jersey took control of the Hudson & Manhattan Railroad, operating it as the Port Authority Trans-Hudson System, or PATH.

See also EAST RIVER GAS COMPANY TUNNEL; GREATHEAD, JAMES HENRY; TUNNELING SHIELD.

I

immersed tube tunnel *Build first, sink later* Although the construction of subaqueous tunnels in the soft ground beneath waterways presented a monumental and dangerous challenge to engineers, conquering the soft, leaking ground was essential in placing tunnels beneath waterways. However, a way of building a tunnel without having to burrow through rock or unstable clay was discovered and tested by 1810 and led to a series of tunnels that were more akin to submarines.

Known as immersed tube tunnel technology, the concept is simple. Build a tunnel above the water, float it into position, and then ballast it so that it sinks into a trench dredged into the riverbed. Connect additional sections until the underwater crossing is complete and open the tunnel to traffic. Though not nearly as simple as it sounds, the concept was developed by Charles Wyatt during the early 1800s as a way of completing the problem-ridden Thames Tunnel.

Attempts by the British engineer Charles Trevithick to tunnel beneath the soft bed of the Thames had been plagued by flooding and cost overruns in 1808. Desperate to complete the tunnel and to prevent flooding problems, Trevithick proposed what amounted to a cut-and-cover tunnel in which a riverbed trench would be excavated in the dry confines of a series of cofferdams. Cast iron tunnel tubes would then be inserted into these trenches and covered. The cofferdams would then be flooded and moved to where another tunnel segment could be built. The Thames Archway Company, which owned the tunnel, rejected the idea but was interested in another by the engineer Charles Wyatt (1751–1819).

Wyatt proposed that 50-foot tubes be built of brick and made into watertight vessels with temporary closures at both ends. The tubes would be fitted with cock valves and pumps. The completed tubes would be buoyant until the valves were opened to allow the entry of water that would enable them to settle into predug trenches. Additional segments would be built, floated into position, sunk, and then connected to the others. The valves would be closed and the water pumped out. The bulkheads enclosing the segments would then be removed to create a tunnel.

Ordered by the Thames Tunnel Company to turn the concept into reality was none other than the famed English civil engineer and inventor John Isaac Hawkins, who invented the first upright piano, the polygraph device for writing multiple letters, and trifocal eyeglass lenses.

Born in 1772, Hawkins was highly respected on both sides of the Atlantic for his inventiveness and mastery of civil engineering techniques. Hawkins attacked the task of building the prototype immersed tube tunnel sections with gusto in 1810 and took just over a month to produce each of two tunnel tubes out of brick and cement. The experiment was a complete success, but money woes plagued the company and a tunnel would not cross the Thames until 1843, when after 17 years, Marc Brunel accomplished the feat by using his invention of the tunneling shield.

Nonetheless, the concept of the immersed tube tunnel was considered sound and was resurrected in October 1865, when it was chosen as a way of creating a rail crossing beneath the Thames. Engineers planned to sink 235-foot tubes constructed of riveted 3/4-inch-thick iron plates, but money woes halted the project in 1866 before any of the tubes could be built.

The first immersed tube tunnel in the world came about because of the difficulties inherent in tunneling beneath the Detroit River, where large pockets of sulfurous gas and troublesome underground rivers threatened any subterranean effort. Michigan Central Railroad needed a reliable means of moving its trains between Detroit and Windsor, Ontario, Canada, since rail traffic between the two cities depended on massive ferries that were often obstructed by heavy winter icing.

The Detroit River Tunnel Company was formed to build a tunnel beneath the river, and an immersed tube tunnel design was selected since it would avoid pockets of deadly gas and the problem of flooding from the high-pressure artesian rivers below the riverbed. The rail tunnel's total length was 6,000 feet, although the immersed tube segments consisting of pairs of 3/8-inch-thick steel tubes would constitute only 2,668 feet of the tunnel. Work began in 1906 and was completed in 1910.

The tubes were fabricated by the shipbuilding firm Great Lakes Engineering Works of Detroit and were fitted with watertight bulkheads at each end. Like submarines, the segments were ballasted and deballasted by using compressed air until they were precisely positioned. The process was essentially the same as that devised by Wyatt and Hawkins nearly 100 years earlier, although helmet divers were essential in guiding the work below the water. Instead of using a fixed scaffold for lowering the tube segments into trenches as did Hawkins, anchored barges were employed. Once nestled in their trenches, the segments were covered with concrete to seal their joints fully, then gravel and clay were used to backfill the remaining depth of the trench.

The success of the tunnel played no small role in the construction of the world's second immersed tube tunnel, the Detroit-Windsor Tunnel, completed in 1930 to carry vehicular traffic. Both of the world's first immersed tube tunnels remain in use today.

Although concrete poured beneath the water using a tremie pipe was the early means of sealing the joints between tube segments, the use of rubber-type gaskets for ensuring a completely waterproof seal between tube segments has become popular. This has been made possible with the elegant use of water pressure to push the segments together. When the space between two sections is pumped dry, eliminating the water pressure, the ambient hydrostatic pressure forces the segments together with extraordinary force to ensure an excellent seal. This system was first used during construction of a Hong Kong rail tunnel beneath Victoria Harbour between 1976 and 1979.

Although steel seemed to be an ideal material with which to build immersed tube tunnels, Dutch engineers developed the first concrete immersed tube tunnel segments and have employed them with considerable success. The Dutch built the first immersed tube tunnel outside the United States in the 1937–1942 construction of the Maas Road Tunnel. The relatively low cost of concrete compared to that of steel coupled with the phenomenal strength of reinforced concrete made it a good alternative for the thrifty Dutch engineers.

Since some immersed tube tunnels cannot be buried because of the hardness of the riverbed, the immersed tubes must be placed on the surface. Although this structure exposes the immersed tunnels to possible damage from passing ships, two such immersed tunnels, built in Yugoslavia in 1978 and Norway in 1982, were designed to transport coal and natural gas, respectively. Officials and engineers would approach exposing a vehicular tunnel to potential damage by passing ship traffic with greater caution.

One variant of the immersed tube tunnel being seriously discussed by the 1990s was a submerged floating tunnel, proposed by the Norwegian Submerged Floating Tunnel Company. The concept is nothing new, as an Englishman, Sir Edward James Reed, patented it in 1886. Reed proposed a buoyant but rigid rail tunnel anchored to the seabed by weighted chains.

The Norwegian proposal envisions a tunnel anchored 80 feet under water in the Hogsfjord near Stavanger to provide a crossing that measures nearly 4,600 feet. Variations on the plan recommend suspending the tunnel with anchors as in Reed's design; another plan suggests a pontoon arrangement suspending the tunnel underwater like a submerged pontoon bridge.

See also CENTRAL ARTERY/TUNNEL PROJECT; CONCRETE; DETROIT-WINDSOR TUNNEL; THAMES TUNNEL; TREMIE.

incompetent rock   *See* grouting, rock bolt.

Inspired Carpenters, the   *Palmer, Wernwag, and Burr*
A trio of bridge designers who were famous in 19th-century America for their strong, innovative timber bridges have acquired the label the "Inspired Carpenters." Timothy Palmer, Louis Wernwag, and Theodore Burr all were experts in truss arch timber bridge design.

Palmer (1751–1821), like his colleagues, was a self-taught carpenter and bridge builder. Born in Newburyport, Massachusetts, he established himself as a capable artisan in the tricky realm of bridge-building beginning in 1780. By 1797, Palmer had patented a trussed arch design that he employed in his bridges. This truss-reinforced arch was capable of spanning longer distances more efficiently than standard post and beam construction.

Palmer is best known for the Permanent Bridge over the Schuylkill River in Philadelphia, an imposing structure measuring 1,300 feet in length. Built in 1805, the bridge spanned 485 feet of the waterway using three truss arches. It was also to be the first covered bridge in America. Although the bridge designer and historian David Steinman points out that Palmer was also one of the first to promote the construction of covered bridges, it was a public official in Philadelphia who recommended that the Permanent Bridge be roofed to increase its lifespan, and Palmer concurred.

Wernwag (1770–1843) was a German immigrant to America, where he began a prolific career as a bridge builder. At the time of his death in 1843 he had designed no fewer than 29 bridges and patented three truss configurations. His bridges were truss arch designs in which the arch performed as a rigid but inverted suspension cable in the way it accepted loading.

His most famous bridge was the Upper Ferry Bridge, commonly known as the "Colossus," built near Philadelphia over the Schuylkill River in 1812. At 340 feet, 3 1/2 inches in length, the bridge was one of the longest timber arch bridges in America at the time. Wernwag's bridges were beautiful and sound although fire, a common foe of wooden bridges, destroyed his Colossus in 1838.

The most influential of the Inspired Carpenters was Theodore Burr (1771–1822). He also used the truss arch method on his bridges and in 1806 had also patented his own arched truss design. This truss was notable in that the arch reinforced the triangular truss construction rather than the other way around.

Although Burr's truss design and bridge work were not generally as stunning as the master works of his two contemporaries, Burr's mark on American bridge building was assured by the Burr truss, which was used in the majority of American covered timber bridges. Because of the reverse strengthening of the arch by the truss, the design is sometimes referred to as the Burr arch.

Burr was so popular a bridge builder that between 1812 and 1820 he was occupied with the construction of five bridges across the Susquehanna River. One of those spans built during this frenzy of activity, at McCalls Ferry, Pennsylvania, is perhaps his most impressive. Completed in 1815, the bridge had a span of 360 feet, making it the longest timber arch bridge ever built. The span was destroyed two years after it was built, though not by fire. River ice carried away its piers.

Oddly enough, this highly respected bridge builder died penniless under unexplained circumstances at the age of 51 and is buried in an unmarked grave.

See also COLOSSUS; PERMANENT BRIDGE.

iron   *Beginning of the new bridge materials*
Humans began using iron around 1,200 B.C., but it took them 3,000 years to figure out that it would be a superb structural material for the building of bridges and the reinforcement of tunnels. By the time it was applied to bridge and tunnel construction, mass-produced steel was on the horizon, and iron's heyday as a structural material would last little more than 100 years. However, its use resulted in stunning bridges, of which many remain.

Nonetheless, iron, both wrought and cast, was the material of choice for many bridge designers during the 18th and 19th centuries and formed some of the sturdiest and most creative bridge designs of the Industrial Age. Though outmoded as a material for bridges by 1900, cast iron continued to be used as a lining component for tunnels through the mid-20th century. The Brooklyn-Battery Tunnel built by New York City during the 1940s was lined with cast iron.

Iron, naturally available in its purest form in meteorites, which were once sought by ancient artisans, is among the most abundant elements. Once refined, the ore can be molded (cast iron) or hammered and forged into a shape (wrought iron). Because pig iron contains a higher percentage of carbon and other impurities than does steel, it is less ductile. Pig iron is the initial form of manufactured iron that absorbs carbon in a process called carburization when it is melted through the burning of coke, a substance derived from coal.

The first bridge of iron was designed and built by British ironmasters, Abraham Darby III and John Wilkinson, with the help of an architect, Thomas Pritchard. It measured 100 feet and was built between 1777 and 1779 over England's Severn River. The bridge was attractive and functional but initially had few imitators except the would-be bridge designer and American patriot Thomas Paine, who designed and built a model iron bridge in England in 1790. However, Paine's dream of spanning the Thames River with his bridge did not materialize.

Thomas Telford took note. He turned to iron with the 1796 construction of the Buildwas Bridge over the Severn River. The iron bridge-building boom then caught on for a number of reasons: Despite the seeming heaviness of iron, bridges made of it were far lighter and less bulky than stone structures. They were also easier to construct since the iron components made the bridges exercises in modular construction. This made them less expensive and more attractive to municipalities and entrepreneurs financing such projects.

One of the most notable iron bridges constructed is the Garabit Viaduct built in France in 1884. Gustave Eiffel, who became one of the world's foremost experts on the structural uses of iron, designed the bridge. The Garabit Viaduct, like a number of iron bridges, remains in use as a railway span. Although the first bridge to use steel components was the Eads Bridge in Saint Louis in 1874, Eiffel and others continued using iron as a bridge material throughout the 19th century.

However, the strength of iron—half that of the steel available during the 19th century—and its tendency to crack instead of bend meant it would soon be superseded by steel. Steel, which was little more than iron whose weakness-inducing impurities were oxidized and removed by the forced introduction of oxygen, was stronger and could flex without failing. The 1879 collapse of Scotland's Tay Bridge and the resulting deaths of 75 aboard a train sounded the death knell for iron bridge construction.

See also EIFFEL, GUSTAVE; GARABIT VIADUCT; STEEL; TAY BRIDGE; TELFORD, THOMAS.

Iron Bridge at Coalbrookdale *See* Telford, Thomas.

J

jumbo The installation of numerous drills, concrete pumping guns, or any other tunneling or mining equipment on a movable vehicle or frame has acquired the name *jumbo*.

Germain Sommeiller's 1863 installation of numerous pneumatic rock drills on a rail-riding carriage during his construction of the Mont Cenis Tunnel created what was probably the first jumbo. Rock drill jumbos would become extremely elaborate, allowing miners to drive vehicles to the face of a tunnel and simultaneously drill numerous shot holes for blasting.

During the 1931–1932 construction of the Hoover Dam's four 50-foot-diameter water diversion tunnels, what may stand as the world's biggest tunneling jumbos were used. Heavy trucks mounting multilevel steel-pipe frames held 50 workers manning 30 drills. These same jumbos held the explosives workers, who packed the drill holes with dynamite. These jumbos were so large, they looked like multistory buildings without walls.

When the tunnels were fully excavated, other jumbos rode on tracks along the invert, or floor, of the tunnel carrying grouting crews who pumped a stabilizing cement mixture behind the tunnel's lining.

Jumbos are still being refined; some new ones contain drills or jackhammers while their drivers sit inside glassed-in, air-conditioned cockpits. The operation of modern drilling jumbos can be automated with the use of computers, and some firms offer computerized drilling jumbo simulators so workers can hone their skills on the machinery before entering a tunnel or a mine.

See also HOOSAC TUNNEL; MONT CENIS TUNNEL; SOMMEILLER, GERMAIN.

K

Kilsby Tunnel   *See* Stephenson, Robert.

L

Lake Pontchartrain Causeway (1955–1956, 1967–1969) *World's longest bridge* At the beginning of the 21st century the longest bridge was 24 miles in length and crossed not a strait or a dramatic gorge but southern Louisiana's shallow but massive Lake Pontchartrain. This bridge (composed of two side-by-side causeways) is so long that someone standing at one end cannot see its far end because it curves below the horizon.

More of a tidal lagoon than a lake, with its brackish water serving as a rich commercial fishing ground, Lake Pontchartrain was an obstacle that for years that could be crossed only by boat. New Orleans sits on Ponthartrain's south shore, and the need for a causeway across the lake was obvious to everyone. The 624-square-mile lake is Louisiana's largest and sits just north and east of the Mississippi River delta. The smaller Lake Borgne, another tidal estuary, links it to the Gulf of Mexico. Ringed by swamp, thick forests, and bayous, Lake Pontchartrain averages only 15 feet of depth but is barrier enough to block road traffic between New Orleans and northern Louisiana.

Although it is romantic to think of a Cajun in southern Louisiana poling or paddling a pirogue beneath a canopy of Spanish moss, Louisianians were not interested in crossing a 24-mile body of water by canoe. They and the tourists crossing the state wanted to drive on a high-and-dry roadway like the rest of America's automobile culture. The pirogue, still a treasured part of Louisiana's heritage, did have its limitations. In 1952, state politicians acted to make a bridge across Lake Pontchartrain a reality. First, the state's constitution had to be amended to allow the parishes of Jefferson and Saint Tammany to create a joint toll bridge authority that could issue revenue bonds. The Greater New Orleans Expressway Commission, formed in 1954, was also empowered to use auto registration taxes from six nearby parishes, including the Parish of Orleans, to finance the bridge. The commission's goal was construction of a toll bridge and approaches that would be known as the Greater New Orleans Expressway.

Revenue bonds amounting to $46 million were sold and combined with $5 million in auto registration fees as seed money for the project. The commission contracted with the huge Texas-based construction firm Brown & Root, Inc., to build the bridge. The two-lane causeway would run 23.87 miles from New Orleans to Mandeville on the north shore of Lake Pontchartrain.

The design of the bridge would be that of a simple and reliable beam bridge, one of the oldest of all bridge types. Since the lake was too shallow to support large commercial ships, the majority of the bridge needed to be only 14 to 16 feet above the surface of the shallow lake to allow the passage of skiffs under its roadway. At five points along the original bridge, allowances were made to provide clearance for larger watercraft using three "humps," or gradually elevated sections of the roadway, and a pair of bascule bridges. These crossing points coincided with existing deepwater channels dredged in the lakebed.

The bridge would be marched across the lake with the use of 56-foot prestressed concrete beam panels laid across tees supported by hollow, prestressed concrete piles driven nearly 75 feet into the lakebed. The concrete components were cast and prestressed on shore and then carried to the bridge site by barge. A consultant in this prestressed concrete work was none other than a French firm that obtained its prestressed concrete techniques directly from their inventor, Eugène Freyssinet. It seems fitting that a French engineer's technique would make possible a record-setting bridge constructed in a region that was once owned by France. The bridge would be built across a lake named by French explorers after Louis XIV's minister of naval affairs, Count Jerome de Pontchartrain. In a curious bit of historic irony, the first portion of the count's last name is *pont,* the French word for "bridge."

Cranes lifted the horizontal caps, or tees, atop the piles to await the placement of prestressed concrete panels that were also placed atop them by crane. Segment by segment, the bridge marched across Lake Pontchartrain. The original two-lane span consisted of a 28-foot roadway bordered by 18-inch curbs. Construction on the causeway began January 20, 1955, and was completed 14 months later. From the air, the finished bridge appeared as a slender, gray, concrete line running to infinity on a watery canvas. It possessed no magnificent supporting towers, no graceful suspension cables, and no stupendous arches. It was a workingman's bridge, simply a way of getting from point to point, and hands down, it also happened to be the world's longest. Its construction required 2,243 prestressed concrete panels.

The bridge was an immediate success, and motorists found using it far more convenient than detouring around the perimeter of Lake Pontchartrain. The presence of the bridge helped promote growth north of the lake, one of the original goals of its construction. The public liked the bridge so much that by 1965, traffic on the two-lane causeway exceeded the 5,000 car-a-day design limit, prompting officials to construct a second span. This would double the traffic capacity crossing the lake, but more importantly, it would create far safer driving conditions since both bridges would carry one-way traffic instead of two-way traffic on an undivided roadway.

Brown & Root again entered the low bid and undertook the $30 million effort to construct a second bridge 80 feet east of the first. And, thanks to improvements in prestressed concrete technology

during the preceding decade, the second bridge would cost far less. The prestressed beams of the newer bridge allowed the use of 84-foot beams—50 percent longer than those used originally, slashing the requirement for supporting pilings and caps by half. The second bridge was built between March 1, 1967, and March 23, 1969, using 1,500 roadway panels, 743 fewer than the original span. Both bridges are connected by crossovers at seven locations to provide detour options in case of accidents or construction work.

As part of the 1967–1969 project, the South Channel bascule bridge of the causeway was converted into an elevated span. The bridge now contains four elevated spans for the passage of water traffic and a double-leaf bascule bridge on its northern end. The South Channel elevated span provides 150 feet of horizontal clearance and 50 feet of vertical clearance. The remaining bascule bridge at the North Channel provides 125 feet of width to marine traffic and a vertical clearance of 45 feet. A trio of humps, smaller elevated spans located at both ends of the bridges and near its middle, each provide an opening 56 feet wide and 25 feet tall.

The ordinary sections of the original bridge, which is now the southbound leg of the causeway, provides 51 feet of horizontal clearance and 16 feet of vertical clearance for smaller watercraft. The newer bridge containing the northbound lanes provides 79 feet of clearance side to side and 14 feet of vertical clearance.

The Greater New Orleans Expressway Commission boasts of the safety record of the longest bridge in the world, pointing out that its fatality rates are lower than those of similar bridges in the region. Recent safety improvements include 102 emergency call boxes that, when used, activate flashing messages on large signs to alert drivers that a vehicle may be stalled on the roadway. Causeway officials hope this device will reduce collisions with stalled cars, the cause of 13 percent of all accidents on the bridge. Another safety feature was the installation of an improved radar system at the North Channel bascule bridge in February 1989, to provide ample warning if a ship or boat is attempting to pass when the bridge's movable span is in the down position.

Although the bridge has been in operating since 1956, the commission did not employ its first rescue vehicle until 1987. That mission had been left to local municipal rescue units on each side of the bridge. A second rescue unit was added in 1989 so one could be stationed at each end of the bridge to reduce emergency response time. The rescue crews are trained to provide emergency medical services, cut people from mangled autos, and even pull out those who crash through the railings of the bridge into the water.

Crossing Lake Pontchartrain on the bridge requires payment of a toll that stood at three dollars per round-trip in 2000. This is a relative bargain for a timesaving shortcut across 24 miles of water when one considers the $48 fee for Japan's 2.4-mile-long Akashi Kaikyo Bridge between the islands of Honshu and Awaji.

Of course, the Pontchartrain Causeway's initial construction costs were minuscule when compared to those of the multibillion-dollar Akashi Kaikyo Bridge, which required massive foundations and gargantuan anchorages for its lofty suspension cables. Although it crossed a greater distance, the world's longest suspension bridge also required foundation construction in water 200 feet deep. Although the Pontchartrain Causeway is no more spectacular than an ordinary beam bridge crossing a 100-foot creek multiple times, the challenges of crossing Lake Pontchartrain called for no more nor less than the engineering solution that was used. A lofty arch bridge would have been a waste of money, and a suspension bridge would have been impossibly costly and pretentious. Louisianians are rather proud of their unassuming but capable bridge.

Like many of the world's most interesting bridges, the relatively plain Pontchartrain Causeway has its own "live cam" providing the online public a 24-hour view not of a statuesque bridge but of traffic conditions at its north shore tollbooths. It can be accessed at http://www.thecauseway.com/default.html.

See also BEAM BRIDGE; FREYSSINET, EUGÈNE; PRESTRESSED CONCRETE.

Leonard P. Zakim Bunker Hill Bridge   *See* Central Artery/Tunnel Project.

Leonhardt, Fritz (1909–1999) *German engineer considered one of the 20th century's greatest bridge designers* Fritz Leonhardt's decision to pursue a career in civil engineering was prompted by his desire to work on a wider range of building projects than he might have in following his father into architecture. Leonhardt's love of variety is revealed in his accomplishments, which range from one of the earliest and most successful cable-stayed bridges to the use of concrete in television towers. He also designed light but strong bridge decks and taught generations of engineering students the necessity of making style and beauty the equal of structural integrity.

Leonhardt made the progression from engineering student to government engineer before establishing himself as a consulting engineer. He played a vital role in rebuilding Germany's transportation infrastructure after World War II and eventually went on to become a professor of engineering and a believer that a beautiful design is worth the extra money it may cost.

Born in Stuttgart, Germany, Leonhardt initially obtained a degree in civil engineering at Stuttgart University. He then traveled to America to attend Indiana's Purdue University as an exchange student during 1932–1933. A political neophyte, Leonhardt was unable to enlighten his American classmates about Adolf Hitler's rise to power and would have remained in the United States had he been able to secure a job in depression-plagued America.

On his return to Germany he found work as a civil engineer, first in the Reichsautobahn department designing the Stuttgart section of Germany's famed highway system and later for the Reichstransportation department. At the age of only 25, when most engineers were struggling with the small but important details of bridge design under the tutelage of senior engineers, Leonhardt was handed the task of overseeing the design of a suspension bridge at Cologne. The inexperienced Leonhardt was surprised by this assignment, a selection based on his study of bridge engineering in America, where so many superb examples of that type of bridge had been built. He tackled the task by designing the Rodenkirchen Bridge, a suspension span that was completed in 1941.

This sleek suspension bridge employed a plate girder to stiffen its thin road deck and boasted a clear span of 1,240 feet. The bridge was destroyed during World War II. Leonhardt would later note the similarity of his bridge's appearance and Othmar Hermann Ammann's masterpiece of design, New York's Bronx-Whitestone Bridge. Even before the Rodenkirchen Bridge was built, Leonhardt had obtained a doctorate in engineering in 1939 and had opened his own engineering office in Munich.

During World War II, major bridge construction ground to a halt, but the smashing of 15,000 bridges in Germany during the conflict created an unparalleled need for replacement spans. One solution was the use of easy-to-erect cable-stayed bridges, and Leonhardt provided a stunning example of this type of span with his design of the Theodor Heuss Bridge, completed in 1957. The bridge utilizes two pairs of steel pylons carrying the weight of the steel road deck through steel cables. This stark design explored not only the beauty of the cable-stayed type but also its structural possibilities and helped to fuel interest and confidence in cable-stayed spans, which have become popular around the world.

Leonhardt was to prove himself a master of virtually every form of bridge design and was unafraid of innovation. Although not the inventor of the hollow-box beam, Leonhardt designed a sleek 606-foot span at Cologne over the Rhine built in 1948 that utilized an extremely thin example of this structural concept. A sizable number of his bridges were constructed of reinforced concrete, showing Leonhardt to be a master of one of the most humble and yet efficient structural materials.

Leonhardt established himself as a master of virtually every type of span. His contributions were many, including development of the orthotropic deck, an innovation that allowed the construction of lighter bridge decks reinforced by steel ribs. The orthotropic deck reduced weight and costs while providing adequate rigidity with the use of less structural steel.

Championing the installation of beautiful bridges even if creative visual designs added to the cost, he was critical of American municipalities and designers for their lack of boldness in pushing for designs that were as visually stunning as they were practical. Leonhardt authored an influential book aimed not only at the engineering profession but also at the public, *Bridges: Aesthetics and Design*. The book extolled what he perceived as the necessity of combining art with engineering.

Despite his interest in promoting beautiful architectural creations with bridges and his philosophy that architects and engineers must work together more closely, he became critical of the excesses of design for design's sake near the end of his life. He told an interviewer just before his death at 90, "Engineers have to stop architects from creating nonsense."

Leonhardt's accomplishments were not restricted to bridges. He designed massive and innovative tele-

vision transmission towers by using reinforced concrete and applied suspension bridge technology to the cable net structure that provided the roof for the 1972 Munich Olympic Stadium. Leonhardt also took part in research on the behavior of reinforced and prestressed concrete between 1958 and 1974. He served for two years as president of Stuttgart University in 1967 and 1968 and continued to make occasional appearances at his internationally renowned engineering firm Leonhardt, Andrä and Partners until the age of 89.

Lindenthal, Gustav (1850–1935) *Pragmatist and dreamer who designed the Hell Gate and Queensboro Bridges* If allegations that he bridged over some bare spots in his educational background are correct, Gustav Lindenthal may have been the last of the great self-taught American civil engineers. Despite questions about his academic credentials, Lindenthal's real-world achievements in bridge design propelled him into the ranks of America's most famous civil engineers.

He designed the visually stunning Hell Gate Bridge, the world's longest steel arch railroad bridge at the time, and his fertile mind also produced plans for two other New York spans: the Bayonne Bridge and the Queensboro Bridge. Even before these accomplishments Lindenthal gained fame for his design of Pittsburgh's unique Smithfield Street Bridge with its fascinating lenticular trusses. During his long career, Lindenthal also served as a mentor, teacher, and employer of two of America's most prolific bridge engineers: Othmar Hermann Ammann and David Barnard Steinman.

Born in Brünn, Austria (now Brno, Czech Republic), as Gustavus Lindenthal, he was originally employed as a mason and carpenter. Lindenthal's official biography claims he attended technical universities in Brünn and Vienna, but a journalist writing a story on Lindenthal decades after his death was unable to find documentation that Lindenthal attended either university.

Although Lindenthal did receive an honorary doctorate in engineering from Dresden's Polytechnic College in 1911, it is not certain that he ever attended a class there, although he has been credited with doing so in some accounts. Whether Lindenthal performed a bit of engineering on his resume may be unanswerable. What is known is that Lindenthal worked as an assistant in the engineering office of an Austrian railroad and later as an engineer overseeing the location and construction of rail lines for Switzerland's national railway.

Lindenthal, a burly six-foot-tall man accustomed to physical labor, emigrated to the United States in 1874 and immediately found work as a journeyman mason at Philadelphia's Centennial Exhibition. He was soon promoted to assistant engineer in the construction of buildings for the exhibition. In 1877, Lindenthal began designing railroad bridges for the Keystone Bridge Company, the firm started by Andrew Carnegie to construct Saint Louis's Eads Bridge. He then embarked on a successful career as a railway engineer building numerous bridges.

The project that allowed the Pittsburgh-based Lindenthal to break away from the growing pack of 19th-century railway engineers was his design of the Smithfield Street Bridge over the Monongahela River in Pittsburgh. Completed in 1883, the span utilized opposing top and bottom arched trusses whose profile appears to have the lenticular shape of an eye. Lindenthal's paper for the American Society of Civil Engineers on the bridge earned him the society's Thomas Fitch Rowland Prize. Lindenthal was famous for seldom repeating the designs of his bridges as some engineers are prone to doing, believing that each bridge was a unique engineering challenge requiring specific solutions.

After spanning the Allegheny River near Pittsburgh with an iron chain suspension bridge, Lindenthal set his sights on New York City, where the biggest and most famous bridges were being built. Relocating his engineering business to that city by the late 1880s, Lindenthal also took up the cause of building a bridge across the Hudson River, a waterway then commonly known as the North River. Almost until his death, Lindenthal would propose a series of plans to bridge the river between New Jersey and Manhattan.

Lindenthal's great dream would never be realized for a series of reasons, including his infamous disdain for paying homage to the labyrinth of New York municipal politics. Although his apolitical nature and commitment to solid engineering principles gained him an appointment by the reform mayor Seth Low as New York's commissioner of bridges, his 1902–1903 tenure was short. Lindenthal, whose achievements in designing sensible and capable bridge systems were widely recognized, became obsessed with spanning the Hudson River with a series of increasingly ornate and massive proposals.

By the 1920s, he was touting gargantuan suspension bridge designs with a pair of hideously massive towers that were composed of nothing less than skyscrapers. The deck of a 1923 bridge proposal measured an unbelievable 225 feet in width, its upper level carrying 20 lanes of subway and auto traffic and its enclosed lower deck carrying 12 lanes of rail traffic. Lindenthal fought for his bridge as he continually made it more expensive and ungainly. He vociferously but unsuccessfully opposed construction of New York's Holland Tunnel beneath the Hudson, estimating (correctly) that the tunnel would cost far more than was claimed.

Ammann, his loyal lieutenant, would eventually break with the old master over the Hudson River crossing and eventually design the George Washington Bridge, built in 1931. Nonetheless, the old master was still pushing his gargantuan bridge until his death in 1935.

Despite his setback over his Hudson River bridge, Lindenthal had a singular accomplishment with his design of the stunning Hell Gate Bridge. During the construction of that bridge, Ammann was his assistant chief engineer and Steinman ranked just below Ammann. The 977-foot span would stand as a record for an arch bridge until the completion of Ammann's arch-supported Bayonne Bridge in 1931 with its 1,653-foot main span.

Another miscalculation by Lindenthal was his devotion to steel eyebar suspension chains, which afforded less strength for the weight involved. Even John Roebling had eschewed iron eyebar cables in favor of bundled wire cables before 1850. New York's famed Manhattan Bridge as designed by Lindenthal was to have an eyebar chain suspension scheme, a proposal shouted down on technical and aesthetic grounds. The slender trademark towers designed by Lindenthal would remain, but they would be draped with wire suspension cables.

By the early 1900s, Lindenthal had become America's reigning authority on bridge design, his opinions supported by superbly written papers in professional publications. Although he may have been primarily a self-taught engineer, he had proved himself with dozens of bridge designs and suffered only a handful of political setbacks and engineering defeats. Lindenthal represented the new breed of bridge designer, who fully grasped the tragic lessons of 18th-century bridge design to ensure his own structures did not flounce in the wind or collapse outright.

Standing behind the uncompleted Hell Gate Bridge in 1916 are three of America's most important bridge engineers. Front row, center, is the Hell Gate Bridge's burly chief engineer, Gustav Lindenthal, with his white hair and beard. To Lindenthal's right is his chief assistant, Othmar Hermann Ammann, and to Ammann's right is David Barnard Steinman in a light-colored suit. *(Library of Congress)*

Of note are two other successful bridge designs by Lindenthal that include a pair of cantilever spans, the Queensboro Bridge across New York's East River, completed in 1909, and the Sciotoville Bridge, a cantilever railroad span built in 1917 at Fullerton, Kentucky. Lindenthal died two months after his 85th birthday.

See also AMMANN, OTHMAR HERMANN; BAY-ONNE BRIDGE; HELL GATE BRIDGE; QUEENSBORO BRIDGE; SMITHFIELD STREET BRIDGE; STEINMAN, DAVID BARNARD.

live and dead loads When engineers design bridges, they realize the structure must carry two kinds of weight, the mass of its own structure and the weight of traffic in the form of automobiles, pedestrians, or trains. The weight of the bridge itself is known as the dead load while the weight of traffic is known as the live load.

Although it seems a bit strange to distinguish these weights, engineers must make myriad calculations to ensure a bridge is structurally sound under all conditions. Their complex task continually demands that stronger bridges be made with lighter materials. Although a bridge might be massively strong, the loading it can sustain might be mostly consumed by its own weight. If a bridge's dead load is 1,000 tons and the entire structure can hold only 2,000 tons total, then the live load of traffic on the bridge is restricted to 1,000 tons.

It is relatively easy for designers to know how much their bridge will weigh, but the live load, the weight of traffic to be carried on a bridge, is tougher to figure out since the volume of traffic ebbs and flows. There are times when the Golden Gate Bridge may bear the weight of only a handful of cars and other times when traffic it supports is quite heavy. As a result, the live load of a bridge can fluctuate wildly. Although engineers know the dead load of a bridge cannot increase and is therefore a constant, they must wrestle with what sort of live load capability to build into a bridge. Because of the fluctuations in the traffic crossing a bridge at any given time, engineers are forced to build in a massive safety margin.

When Charles Alton Ellis designed the Golden Gate Bridge, he engineered the suspension span to carry a live load of 4,000 pounds per linear foot, or nearly four times more than what might realistically be expected. He based his calculations on the weight of the heaviest passenger cars of the 1930s jammed bumper to bumper on all six lanes of the bridge. Since most cars on the bridge would be far lighter than the massive Cadillacs incorporated into Ellis's computations and the vehicles would have at least a car length between them, Ellis ensured the bridge could sustain the live load required.

All modern bridges are built with live load limits, which explains why there are often signs at smaller bridges warning of the load they can sustain. If exceeding the load limit of a bridge does not make it collapse instantly, doing so generates stresses that accelerate its deterioration.

London Bridge (A.D. 43?) *Bridge that was "falling down"* Although no one alive has ever seen so much as a drawing of the original "London Bridge," built by Britain's Roman conquerors around A.D. 43, there is probably no English-speaking child who has not heard of it thanks to the famous nursery rhyme. But the children's song has history a bit wrong. London Bridge did not fall down. Viking warships pulled down the bridge, or, more precisely, a descendant of the original bridge built in the same place, in a bloody and convoluted episode of 11th-century British politics.

The story of the original London Bridge began after the Romans arrived in force in A.D. 43 to make their way up the Thames River and establish a fortress city originally known as Londinium. With protective walls encircling three sides of the city, the Romans used the Thames as a barrier against attack on the unwalled side. Because the Romans needed to cross the Thames to extend their influence and allow for trade, these master bridge builders constructed a wooden bridge across this ribbon of water. The first London Bridge may have been a pontoon bridge whose roadway was supported not by pilings or piers but by boats. Subsequently, the Romans possibly constructed a more substantial bridge built upon wooden posts driven into the riverbed. It was not one of the impressive stone arch bridges for which the Romans are so famous, but a crude wooden span was functional on the frontier.

With the passage of time, the timbers of the bridge burned, rotted, or were damaged. As a result, these timber bridges were continually repaired if not replaced altogether during the nearly 1,200 years that a wooden bridge spanned the Thames at London. By A.D. 407, the Roman legions had departed Britain to defend a crumbling Roman Empire, and the task of rebuilding the essential bridge fell to the Britons, who were to experience centuries of civil wars, invasions, and bloody political intrigue. One of these subsequent wooden bridges was to play a critical role in the history of Britain that would spawn the nursery rhyme that made it famous.

In A.D. 793, Danish Vikings began a series of invasions of Britain that would continue for more than 400 years. Peace was often purchased with Danegeld, or protection money, paid to the Danes to prevent them from raiding the country. King Aethelred, a notoriously inept British ruler, had eagerly paid more Danegeld than just about any other English king. Aethelred was known as "The Unready," a double-edged moniker meaning that he was the recipient of poor counsel and also never quite ready to deal strongly with the Danes.

Aethelred, who was rumored to be a bit unstable emotionally, ordered the slaughter of thousands of Danes who had settled in England in A.D. 1002. This not-so-shrewd move prompted yet another destructive invasion by Danish Vikings and forced Aethelred from Britain. By A.D. 1012, Aethelred had regained his nerve enough to consider retaking his throne and driving the Danes from London. To do this, he turned to a group of Viking mercenaries led by Olaf Haraldson, the 17-year-old stepson of a Norwegian Viking king.

Aethelred sought the services of Olaf, who had already established himself as a ruthless and capable commander since he had begun taking part in Viking raids since the ripe old age of 12. Retaking London with a naval force using the Thames had always been a problem because of the presence of the wooden "London Bridge" that had long stood across the river. Maritime forces sailing up the Thames River estuary from the North Sea would inevitably find themselves pelted with rocks from above as they neared the bridge. If they did not want to endure this type of damage, the attackers would have to ground their boats and continue the attack on foot.

Olaf, however, had a plan. To protect his men, the young Viking rigged protective roofing for his boats. According to the detailed *Chronicle of the Kings of Norway*, a collection of Norse historical sagas, Olaf ordered the wooden roofs pulled from houses along the Thames and assembled over his ships as protective barriers. Using this method, Olaf protected his Viking crews from the barrages of rocks and other weapons as they approached the bridge. However, this was only half of Olaf's plan. Once his Vikings fought their way to the bridge defended by the Danes, Olaf's men tied ropes to its wooden piers. The Norwegian Vikings then grunted mightily against the oars of their boats, known as knorrs and drakkars, until they had pulled down the bridge and drowned its defenders.

The nursery rhyme so familiar today was derived from a Norse ballad in the *Chronicles* to celebrate Olaf's shrewd victory. The similarity of the ballad and the nursery rhyme is apparent:

> London Bridge Is Broken Down
> Gold is won, and bright renown
> Shields resounding
> War-horns sounding
> Hildur shouting in the din
> Arrows singing,
> Mailcoats ringing
> Odin makes our Olaf win!

The less bellicose nursery rhyme is different but generally follows the same meter:

> London Bridge is falling down,
> Falling down, Falling down.
> London Bridge is falling down,
> My fair lady.

The wooden London Bridge pulled down by Olaf was rebuilt once again and would remain a crossing over the Thames (with the necessary repairs and replacements) until the last wooden bridge was constructed in 1163. In 1209, a stone arch bridge was completed at the same location where so many wooden ones had stood. Known as the "Old London Bridge," it remained in service for just over 600 years, until construction of the "New London Bridge" in 1831. Ironically, the 1831 bridge itself lacked proper foundations and began to sink into the Thames riverbed. This prompted the construction of the present-day London Bridge, which was completed in 1972.

See also NEW LONDON BRIDGE; OLD LONDON BRIDGE.

Lowe's Motor Speedway walkway collapse (May 20, 2000) Footbridges seldom make news, but when they collapse beneath the feet of pedestrians, as one did at a North Carolina speedway May 20, 2000, the injurious results are a reminder of how dependent people are on the structural integrity of such spans.

The incident occurred as thousands of spectators were flowing out of the Lowe's Motor Speedway at Concord, North Carolina, after the finish of the NASCAR Winston stock car race. Spectators were streaming across a pair of 320-foot walkways over U.S. Highway 29 to parking areas when a loud snap was heard. At least 107 people tumbled 17 feet to the highway when an 80-foot section of one of the prestressed concrete walkways snapped at its center. The broken deck fell, forming a V that blocked the four-lane highway.

Paramedics at the speedway, whose job was to care for injured stock car racers, rushed to the scene of the walkway collapse to treat scores of moaning fans. Witnesses said they heard a pair of sharp cracks as people began to tread across the bridge around 11:15 P.M. and saw the bridge and its occupants slammed to the roadway below.

The walkway failure took only a second to occur but hurt 40 people so seriously they required hospitalization. The bridge, whose walkway deck was supported by prestressed concrete, was only five years old at the time of the collapse. The failure was surprising since the prestressed concrete was supplied by one of the nation's leading producers of such components. Investigators on the scene quickly noted warning signs that something was not right with the bridge. North Carolina transportation officials spotted cracks in portions of the intact concrete sections and also saw signs of corrosion in the steel tendons that were used to prestress the concrete.

Early analysis of the walkway's concrete components initially indicated that the steel tendons inside the concrete suffered corrosion from exposure to external moisture or corrosive elements that may have entered the concrete beam. Some investigators said that the corrosive additive calcium carbonate sometimes mixed with concrete to speed its hardening had accelerated corrosion of the prestressing tendon. North Carolina's Department of Transportation prohibits the use of calcium carbonate on prestressed structures.

Prestressed concrete beams depend upon the tightly stretched high-strength steel tendons embedded inside to provide tensile strength. An upright concrete beam can sustain massive compressive forces, but when used horizontally without prestressing, concrete has only a fraction of its compressive strength. Should the prestressed steel tendon corrode and snap, the concrete looses much of its ability to withstand tension.

Immediately after the accident the firm that constructed the bridge began a national investigation of other prestressed concrete spans to determine whether there were any other signs of potential failures.

See also CONCRETE; PRESTRESSED CONCRETE.

Ludendorff Bridge, "Bridge at Remagen" (built 1916–1918) *World War I bridge that helped end World War II* Far from being one of the world's largest spans, the Rhine River's Ludendorff Bridge in Germany, popularly known as the "Bridge at Remagen," is one of the world's most historically significant. Its capture by U.S. troops surprised both the attacking Americans and the defending Germans during the final weeks of World War II and allowed a massive and crucial penetration of Nazi defenses in western Germany.

Its historical significance stands in contrast to the uselessness that marked most of its existence. It was among a series of railway bridges ordered built across the Rhine during World War I by General Erich Ludendorff, chief of staff of the Imperial German Army. Ludendorff wanted the railway bridges to allow the movement of troops and war supplies to the Western Front. After his 1916 order, work began on the bridges, including the largest one connecting the village of Erpel on the Rhine's east bank to the west bank town of Remagen.

The Mannheim architect Karl Wiener designed the Ludendorff Bridge, which was built by the firm of Grun and Bilfinger. Because of wartime labor shortages and the urgency of Ludendorff's directive, the bridge was rapidly built with the forced assistance of Russian prisoners of war. Despite the circumstances, the bridge was not a shoddy effort. The span was extremely strong and exhibited a good deal of architectural style. The 1,069-foot span was of a steel arch design with truss-reinforced side spans. The arch curved over the center span with its hinges resting on steel beam piers anchored in concrete-and-masonry foundations sunk into the riverbed. The center span measured 513 feet; while each of the two truss-reinforced side spans extending from the riverbanks was 278 feet in length. Its center span was 48.5 feet above the river.

Despite the urgency of wartime, local citizens in 1916 stirred up a bit of a ruckus over the bridge's being built near their homes. It seems they believed the structure would be an eyesore that would spoil the countryside. This early example of visual environmentalism was shrugged off by the German high command, and the bridge was completed in 1918. What resulted was a bridge—though a touch Gothic in appearance—that was considered one of the most attractive spanning the Rhine River.

The construction of the railroad bridge was also complicated by the necessity to tunnel 1,200 feet through the Erpeler Ley—a cliff of basalt rising 600 feet above the east bank of the Rhine opposite Remagen. Engineers also built a pair of medieval-looking guard towers at each end of the bridge. As it turned out, the bridge at Remagen was the most expensive and impressive of the wartime Rhine bridges ordered by Ludendorff. The bridge, not surprisingly, would be named after the general who ordered its construction.

Because of the vicissitudes of war, the Ludendorff Bridge was built with repositories for explosives inside the foundations. In case capture was eminent, explosives packed inside these hollow spaces were to atomize the piers and drop the central span into the river to halt the enemy.

With the end of World War I on November 11, 1918, German forces still held the territory west of the bridge. The intact bridge was subsequently inspected by victorious French troops, who discovered the demolition receptacles in the piers and filled them with cement, and life settled back to normal in the rural areas surrounding Remagen. The bridge, which contained two sets of rails and a pedestrian

As the swollen Rhine River rushes below, U.S. Army combat engineers struggle in vain to restore structural stability to the crucial Ludendorff Bridge captured during the waning days of European combat in World War II. Tragically, the bridge collapsed just days after this photograph was taken, killing 28 U.S. soldiers. *(National Archives)*

walkway, saw little rail activity and was of little use other than as a footbridge.

The lonely Ludendorff Bridge did have some excitement in 1928, when a sizable fire consumed wooden portions of the bridge. When the smoke cleared, the German government decided it was worth preserving and funded its restoration.

The onset of World War II and the movement of Nazi military forces from Germany into France revitalized the bridge with new purpose since it was now needed to help troops and munitions cross the Rhine. Realizing as they did in World War I that friend and foe alike can march in either direction on a bridge, German army engineers made it ready once again for demolition in case of capture. An electrical

system for detonating explosives was installed and the explosives themselves were stored nearby. As the war dragged on, the circuits that would set off the explosives were tested on a regular basis.

In the aftermath of the Allied invasion of Europe on D-Day and the subsequent Battle of the Bulge in Belgium, the Allies pushed steadily eastward across Europe in their final drive toward Germany. Bridges like the Ludendorff were bombed frequently by Allied planes hoping to stall German troop movements, but, given the abysmal accuracy of World War II bombing techniques, the Ludendorff Bridge suffered inconsequential damage.

The bridge, so purposeless in the years after its construction, had become an important conduit for

German troops, weapons, and supplies, especially as the Allies moved across Belgium's border toward the Rhine. No Allied commander expected to capture a bridge over the Rhine since German troops conscientiously blasted spans at the slightest hint of Allied attempts to do so.

As the Allies moved eastward, German soldiers knew that failure to destroy a bridge in danger of capture would be punished by death. By March 1945, the only bridge across the Rhine not destroyed by Allied bombing or the retreating Germans was the Ludendorff Bridge. It had initially funneled troops into battle and subsequently provided an escape route for German forces fleeing Allied encirclement. Both the approaching Americans and the defending Germans knew the span would eventually be "dropped" into the Rhine by explosives—or all assumed it would. This military reality was so well accepted that the Americans had made no plans to take the bridge.

On March 7, 1945, an armored reconnaissance company charged into Remagen as part of the Allied effort to encircle retreating German forces. Oddly enough, the commander of the company was Karl H. Timmermann, a German American born in the region, the son of an American World War I veteran and his German bride. Timmermann and his men—members of the combat-weary 9th Armored Division—exchanged machine gun and small arms fire with Germans at the far end of the bridge. They were amazed to see the span still standing but assumed it was going to disappear momentarily in a huge explosion.

A German army captain across the river made the same assumption. In fact, he ordered the bridge demolished. When the detonation mechanism was activated, his effort was greeted by silence. For some mysterious reason the tested and retested firing mechanism had failed. Although historians are still guessing about the cause, it is believed a lucky explosion or bullet miraculously severed the buried electrical connection during the firefight. A German sergeant courageously sprinted out to the bridge to ignite the fuses of the explosives manually. An explosion rocked the bridge, lifted the center span off its piers, and then dropped it back down, damaging one chord of the span's steel arch. Amazingly, amid the smoke and the gunfire, the bridge was still intact. The German soldiers were dumbfounded. Timmermann was ordered to rush across the bridge and seize the far bank, which he and his

troops did. The hapless German captain and his men were captured, saving them from execution for their failure.

As American tanks and infantry flooded across the bridge, U.S. Army combat engineers arrived to patch the battered span. The bridge, damaged by air raids and the German attempt at demolition, shuddered under the tanks and trucks that raced across it. Horrified German commanders ordered artillery to blast the bridge into the river. The Germans even dispatched their amazing jet fighter planes to attack the rugged span. When all else failed, explosive-equipped frogmen clambered into the river in an unsuccessful commando raid to demolish the bridge. While American combat engineers continued to repair the fractured span, nearly five divisions crossed the Rhine and swarmed through the Erpeler Ley Tunnel to defend a growing bridgehead.

Adolph Hitler's usually unpleasant temperament knew no bounds when he heard American troops were pouring across the bridge. Within hours a hasty court-martial found five officers guilty of failing to blow up the span. Of these, four were executed immediately in the Westerwald Forest not far from Remagen.

On March 17, 1945, precisely 10 days after Timmermann's men charged across the weakened bridge, the injured span collapsed without warning as dozens of American engineers attempted to preserve it. Of those on the bridge, 28 were killed and another 63 injured. Some of the bodies were never found in the fast-moving waters of the swollen Rhine. By this time, however, combat engineers had assembled several temporary bridges alongside the original span. Possession of the bridge allowed Allied troops to enter a section of Germany lacking a concentration of forces that could have beaten them back, throwing German defensive efforts off balance. The resilience of the Ludendorff Bridge undoubtedly shortened the war.

The importance to the Allies of a bridge like the Ludendorff cannot be overstated. The Rhine, Europe's longest river, posed a massive obstacle to a final plunge into Germany. Without an intact bridge, troops in that sector would have had to cross the river by boats or by slow-moving amphibious vehicles. Worse yet, the troops would have had to construct a temporary span such as a pontoon or Bailey bridge under withering enemy fire. Crossing the Rhine for the final push into Germany by these methods would have been far bloodier.

During Germany's failing fortunes near the end of World War I, Ludendorff ordered his retreating forces to use a "scorched earth" policy, which required the destruction of anything usable to advancing forces. Ludendorff, of all people, would have been shocked to know his own philosophy had been betrayed by the capriciousness of war in the matter of a bridge that was his namesake.

When it was decided in 1976 that the remaining masonry piers of the bridge needed to be demol-ished, Remagen's mayor, Hans Peter Kuerten, was seized by an idea. He arranged to have the disassem-bled stonework sold in chunks as historical sou-venirs. The money raised went to a memorial museum to be housed in the bridge towers near Remagen. The public eagerly purchased the stones, and, in 1980, the museum was opened. By 1999, more than 440,000 had visited the site where the Ludendorff Bridge once stood.

M

Mabey, Marshall *Sandhog who survived the impossible* Nothing illustrates the dangers facing compressed air tunnel workers more completely than the ejection of Marshall Mabey and two other sandhogs from a subaqueous tunnel beneath the Hudson River in 1916.

Mabey was working beneath the Hudson River in a tunnel under construction by the Brooklyn-Manhattan Transit Corporation in 1916 when the scream of escaping compressed air signaled that a blowout had occurred. The compressed air, which had been equal to the water pressure outside the tunnel to prevent seepage, had pushed through an unlined segment of the tunnel.

Two men working with Mabey were killed outright as high-pressure air shot through the collapsed portion of the tunnel. Miraculously, Mabey was blasted through the hole on a jet of pressurized air and then pushed through a dozen feet of riverbed. The geyser that carried Mabey to the surface blasted him 40 feet above the Hudson River.

Mabey survived and was retrieved by a workboat. Undeterred by the terrifying incident, Mabey returned to tunnel work for another 25 years and saw two of his sons enter the dangerous profession.

See also BLOWOUT; SANDHOGS.

Mackinac Bridge (built 1954–1957) *Michigan's great bridge* David Barnard Steinman's design of what is known as the Mighty Mac became one of the greatest bridges of the 20th century. The Straits of Mackinac (pronounced "Mackinaw") is a body of water as wild as open ocean situated between Lake Michigan and Lake Huron. Anyone building a bridge there would face foul weather, delays, and danger.

Since the 19th century many had pontificated on the need for a rail and carriage bridge across the straits connecting Michigan's Upper and Lower Peninsulas. But it would take more than pronouncements to build this bridge, which would have to be

Reaching across the storm-tossed Straits of Mackinac is the Mackinac Bridge, one of America's most daring and impressive spans. The bridge with its approaches and suspension span crosses 3.4 miles of open water and connects Michigan's Upper and Lower Peninsulas. *(Michigan Department of Transportation, Mackinac Bridge Authority)*

five miles long and cross nearly 3.4 miles of open water.

And the work would have to be done in tumultuous wind and wave conditions vicious enough to smash huge Great Lakes ships. Talk of a bridge became even more serious with the arrival of the automobile culture during the 20th century. In 1923, ferry service was inaugurated across the straits and was soon overwhelmed by a growing number of impatient motorists waiting their turn to cross. Local leaders cried out for a bridge but lacked the estimated $30 million needed at the time to turn their desire into reality.

As the Great Depression gripped the nation, federal willingness to pay for job-producing infrastructure projects revived interest in a bridge. In hopes of lassoing funding, the Michigan Legislature in 1934 established the Mackinac Straits Bridge Authority. Despite two requests for federal financing of a bridge during the 1930s, the Public Works Administration refused to open its pocketbook to the Michiganders.

Between 1937 and 1940 the Bridge Authority conducted studies and built a 4,200-foot causeway southward into the straits from the Upper Peninsula. World War II put most bridge-building out of business for the duration, but the end of hostilities saw only the elimination of the Bridge Authority by the legislature in 1947. By 1950, the Bridge Authority had been resurrected, but the outbreak of the Korean War that year caused such severe steel shortages that plans for building the bridge were delayed once again.

By 1953, a private consortium agreed to finance a toll bridge across the straits, but a weak money market made it necessary for the Michigan Legislature to agree to pay the span's maintenance costs to sweeten the bonds for the buyers. Economic conditions soon improved, and by June 1953, just over $95 million in bonds had been sold. In May 1954 work began on the bridge.

Steinman, who had long lobbied for the project, was chosen as the chief designer of the bridge, giving him the opportunity to build one of the world's grandest suspension spans. At first glance Steinman's bridge did not seem to measure up as a record-breaker since its suspended center span would be only 3,800 feet, or 400 feet less than the then-unsurpassed center span of the Golden Gate Bridge. However, the 8,614-foot combined length of the Mackinac Bridge's three suspended spans and two

472-foot backstay spans between the anchorages was almost 1,900 feet longer than the Golden Gate's length between anchorages. The approaches to the Mackinac Bridge were extremely long, making the total length of the bridge 26,372 feet.

Weather and water conditions would add challenges to those building the bridge since a sizable fleet of barges, workboats, and floating cranes would be needed for the effort. Between September and April, strong gales can sweep the straits, creating waves up to 22 feet and generating winds up to 104 mph. In the dead of winter, thick ice covers the straits, halting all shipping and suspending efforts of any waterbone work force. Because the crews preparing the piers and erecting the superstructure would often be working from a fleet of boats and barges, they were dangerously susceptible to rough water and high winds.

Under these conditions, workers had to sink a total of 34 piers, including the two gargantuan main piers for the twin suspension towers. They would also have to construct a pair of massive offshore cable anchorages. At one point during the project, a wind of 76 mph whipped the straits, forcing workers to hang on to whatever they could to avoid being blown into the chill water. Massive floating caissons waiting to be submerged were pushed free of their moorings, and bad weather often meant that workers seldom got a full day on the job during blustery fall weather.

Nonetheless, work progressed. The concrete approach piers were sunk to bedrock up to 105 feet below the overburden beneath the straits. Barges then carried preassembled steel truss deck sections from distant steel plants to be placed gently upon the finished piers. Massive open caissons necessary to build the two cable anchorages were sunk to bedrock by giant excavation cranes. When the excavations for the anchorages were completed, workers began pouring the first of the 91,600 cubic yards of concrete that each required. Each of the two main piers consumed 80,600 cubic yards of concrete.

The finished anchorages were shaped like monolithic goal posts to accept the spans that rested between their concrete uprights. By July 1955, the main piers were completed, and the erection of the 552-foot suspension towers began. The modular tower sections were delivered by barge and lifted into place by creeper derricks that jacked themselves up between the legs of the rising towers. That

November the towers were completed, and soon after, the storm packing the 76-mph winds struck.

After another wintertime work interruption, the towers were fitted with catwalks so workmen could begin suspension cable spinning July 18, 1956. Each of the two 24 1/2-inch suspension cables contained 12,580 wires that, if combined, would stretch 42,000 miles. By the time the cable spinning ceased on October 19, 1956, the completed cables weighed a total of 11,480 tons. By July 1957, the prefabricated, suspended truss deck segments were connected to the cable suspenders, and paving of the 54-foot wide roadway began. Although the stiffening truss of the bridge originally was to have been a massive 46 feet deep, an improved aerodynamic design of the superstructure reduced its susceptibility to wind motion and the truss depth was trimmed to 38 feet. Steinman maintained that such design improvements shaved $35 million from the cost of the bridge, so that it ultimately cost only $100 million.

The labor requirements for the effort were impressive. There were 2,500 men employed at the site, and another 7,500 labored in steel mills, fabricating shops, and concrete plants. The calculations and design work for the bridge required 350 engineers with Steinman at the helm supervising his concept. The bridge was to be the last major effort for Steinman, who died in 1960 at the age of 73.

The million-ton Mackinac Bridge has endured and remains equal to the challenge of the windy, wave-tossed straits it soars across. Despite its bulk, the span's roadway is designed to flex as much as 35 feet in each direction in strong winds. Each north- and southbound side of the bridge is individually designed to carry one ton per lineal foot. This gives the 19,243-foot bridge the capacity to carry 38,486 tons of vehicular traffic.

The span was the longest suspension bridge in the world at the time of its construction, an achievement that must have pleased Steinman immensely at the time. However, both Japan's Akashi Kaikyo Bridge (12,826 feet) and Denmark's Great Belt Bridge (8,614 feet) superseded its overall length in 1998. Still, the Mackinac Bridge remains the longest suspension bridge measured anchorage to anchorage in the Western Hemisphere.

The official web site of the Mackinac Bridge with a link to a live camera image of the span and a superb historical photo gallery is at http://www.mackinacbridge.org/. Steinman and his coauthor John T. Nevill wrote a book, long since out of print, about the bridge, *Miracle Bridge at Mackinac,* giving a full account of the project.

See also AKASHI KAIKYO BRIDGE; SUSPENSION BRIDGE; STEINMAN, DAVID BARNARD.

Maginot Line (built 1930–1940) *The subterranean defensive line of France* What was probably the most massive fortification effort undertaken since the construction of China's Great Wall was the building of what became known as the Maginot Line, a 175-mile defensive line between France and Germany. Still shrouded in myth and misunderstanding, the remnants of the Maginot Line dot the French frontier, the gun ports of abandoned bunkers now empty eye sockets staring into the past.

Although publicity generated during the building of the fortifications led many in France and elsewhere to believe the Maginot Line was a vast, uninterrupted system of deep tunnels and shoulder-to-shoulder forts, it was in fact a series of individual regional forts. These forts were usually composed of a large central underground bunker surrounded by smaller underground positions containing machine guns, artillery, or observation posts. These satellite positions were connected by tunnel systems to a larger blockhouse containing sleeping quarters and equipment for generating emergency electricity and purified air in case of attack.

These fortifications were actually spread out over large areas, taking advantage of the highest terrain so observation turrets could have the best view of approaching enemy forces.

Newsreels of the period showed French troops sleeping, dining, and relaxing deep underground in these supposedly vast subterranean cities. Although some of the fortifications were sizable, they were not as massive as stories of the period portrayed. The troops did live underground for long periods if necessary, but the accommodations were Spartan and the decor mostly concrete.

Nevertheless, the fortifications both large and small were impressive. Some were equipped with special electric-powered trains that carried munitions, troops, and supplies through the tunnels to the bunkers and gun positions. Another significant aspect of the project was the extension of France's electrical grid to remote regions to provide power for these strongpoints known as *ouvrages,* or "works." At five miles from the fortifications the power lines would be buried a half-dozen feet under ground to protect them from sabotage and shellfire.

The huge project was spurred into reality by French fears of another German invasion. This concern was exacerbated by an expected shortage of men eligible for military service beginning in 1935. The destruction of much of France and the decimation of the male population in World War I caused the deaths of 1,357,800 soldiers and stunted the country's birth rate. France had neither the stomach nor the manpower for another war and wanted the Ligne Maginot to stymie a German invasion through the Alsace-Lorraine region.

Some officers of the French army, including Charles de Gaulle, argued for what is today known as an "active defense," based on air power, armored fighting vehicles, and maneuver. Others wanted a fortification system similar to that of the fortified city of Verdun, a strongpoint that German troops were never able to conquer fully during a horrific costly battle in World War I. Eventually, a decision was made to implement both philosophies.

The idea of a fortified line caught on as early as 1919 when politicians and military officials began debating the matter. The project seemed to stall until a French war hero and politician, André Maginot, took up the cause during the 1920s. Highly decorated for bravery and afflicted by a stiff walk due to a knee shattered in combat, the former sergeant was a tireless champion of the rights of war veterans. Maginot's family hailed from the Lorraine region on France's eastern border, a place that had often felt Germany's bellicose breath upon its neck. Maginot was consumed by the idea of protecting his homeland from invasion.

Outmaneuvering opponents of a defensive line, Maginot appointed the Frontier Defense Commission to make recommendations on where the fortifications should be positioned. Maginot gained right-wing support by touting the security benefits of a defensive line while encouraging leftist support by pointing out how many jobs the project would generate.

Work began on the system of regional strongpoints in 1929, and most of the facilities were completed by 1936. By the mid-1930s, France had established 108 fort systems from Belgium to the Alps. The tunnel systems for these fortifications totaled 62 miles, an amount roughly equal to the entirety of the Paris Metro of the period. The forts were technological marvels employing mushroom-shaped steel turrets that were raised and turned electrically, communication systems, and underground kitchens. Specially designed racks containing artillery ammunition were carried from the underground magazines to the gun turrets by a monorail arrangement suspended from the ceiling.

The speed at which these fortifications were built meant they were less refined than they could have been and dampness and ventilation problems—common issues for tunnels—were unpleasant for the troops. As Germany militarized the Rhineland in violation of the Versailles Treaty in 1936, heightened tensions expedited additional work to alleviate these wet conditions and make the fortifications more habitable.

Although often referred to as the "Great Wall of France" and the "impregnable" Maginot Line, the system of fortifications was nothing of the kind. Each of the strongpoints was designed to delay a German advance until army units could be rushed in. The fortifications, burrowed into the ground and sheathed in concrete and armor plate, would serve as anchor points for a defensive battle. France, with its limited manpower, could not fortify its entire border with Germany without putting every ablebodied man in uniform. Soldiers and politicians alike recognized the military axiom that one who defends everything defends nothing.

Maginot himself would not live to see his magnificent defensive line completed. The six-foot, six-inch-tall Maginot, famed for his love of wine and fine food, fell victim to oysters possibly contaminated with typhoid. The politician and bureaucrat revered by so many of his countrymen died at age 55 on January 7, 1932.

The concept of the defensive line was sound in principle, but the failure to fortify the Belgian border was a fatal miscalculation. In 1940, Adolf Hitler's Wehrmacht crashed through the dense Ardennes Forest, sliced through Belgium, and outflanked the Maginot Line. Despite this, the fortifications placed under frontal attack by German aircraft, artillery, and infantry troops held up well against the assaults. So strong were these defenses that the fall of France did not end the resistance of the vast majority of the troops inside to German invaders. The Fortress Troops, as they were known, surrendered only after French officers arrived to tell them to do so.

During their occupation of France, German troops confiscated the armaments inside the Maginot Line and used sections as headquarters facilities and supply dumps. The Germans even used portions of the fortifications against advancing Allied troops during the liberation of France in 1944. After World War II, some of the fortifications were incorporated into the North Atlantic Treaty Organization (NATO) since their concrete and steel bunkers and air purification systems were considered ideal protection from nuclear attack.

By the late 1960s, the fortifications had fallen into disuse, their generators, rail lines, tunnels, and turrets left to the elements and the scrap dealers. Without the pumps necessary to control water seepage in the tunnels, the fortifications slowly filled with water and muck. Over the years, vandals stripped much of the wiring and machinery from the fortifications. Although most of the fortifications are off-limits to the public since they have been sold to private owners or remain on government property in remote areas, some communities have begun to adopt the strongpoints and convert them into historical attractions.

An added benefit for those who tour parts of the restored Maginot Line is an unusual but sometimes beautiful display of amateur art left behind by troops who manned the strongpoints during the 1930s. The artwork ranges from painted "windows" with picturesque and colorful views of the "outside" to intricate comic scenes.

On the French border with Luxembourg, four miles north of the town of Thionville, local citizens have restored a fortification known as Immerhof, which is open to visitors on a limited schedule. Among two of the largest fortresses were the Hackenberg and the Schoenenbourg, the former located nine miles east of Thionville and the latter near Hunspach. These have also been restored and retain their compact underground rail cars. Online tours of the Immerhof, Schoenenbourg, and Hackenberg fortifications can be viewed at http://www.geocities.com/Athens/Acropolis/9173/index.htm#anglais, http://www.ligne-maginot.com/, and http://www.lignemaginot.com/ligne/index-en.htm.

See also NATIONAL REDOUBT.

Maillart, Robert (1872–1940) *Artistic pioneer of reinforced concrete bridges* Robert Maillart had two talents that ranked him among the world's great bridge engineers: He was an innovative thinker regarding the structural properties of reinforced concrete, and he combined this engineering wizardry with an artistic appreciation of form. Combined, these abilities allowed him to build unbelievably sleek bridges with far less structural bulk than those of his contemporaries. Maillart, little appreciated during his career, became a powerful influence not only on modern bridge engineers in their quest to design daring, efficient, and fiscally realistic bridges, but on designers and artists.

Concrete, which had been used by ancient civilizations including the Roman, disappeared from use with the fall of the Roman Empire only to be rediscovered in the 19th century. The building material was soon used in making bridges, initially in the creation of artificial stone, and later as a homogenous structural material reinforced with iron and steel. These early concrete bridges were overly heavy since they were little more than replicas of stone arch bridges. It would be Maillart who would fully understand reinforced concrete's capabilities to allow bridges of the material to become far more graceful and exciting. A truly elegant bridge weighs less, costs less, and provides superior reliability even as it is a visual wonder. Maillart, with reinforced concrete as his medium, created bridges that met all of these conditions.

The bridges Maillart designed during the last half of his career defy those ignorant of their age to guess accurately when they were built. Slender and modernistic, they impart a sense of visual movement as if they were leaping across the rivers and gorges of his home nation of Switzerland. So fantastic is their design that they look like fictional creations crafted as props for the set of a science fiction movie, instead of useful bridges built during the first 40 years of the 20th century.

Born in Berne, Switzerland, in 1872, Maillart graduated from the Federal Polytechnical Institute in Zurich at the age of 22, whereupon he obtained employment working for other engineers. Maillart, always dissatisfied with commonplace designs as he began to refine techniques for using reinforced concrete, started his own engineering firm in 1902. In mountainous Switzerland, where bridges and tunnels made the nation whole, Maillart's daring designs frightened away some officials, who voted against his proposals. Others were impressed with the strength, elegance, and low cost of his bridges, and Maillart was soon spanning gorges and rivers with his creations. His boldness, however, meant he would never become as famous or financially successful as designers of less daring bridges, and he would struggle both to meet his financial obligations and to remain true to his concept of design.

Maillart's knowledge of reinforced concrete construction actually led him to pre-Revolutionary Russia in 1912, where he oversaw the construction of reinforced concrete buildings, an endeavor that imploded with the fall of the Romanov family in 1917. Maillart's experimentation led him to design integral mushroom-shaped supporting columns in concrete warehouses and to define more efficient ways to use reinforced concrete.

Pragmatic and artistic, Maillart returned to Switzerland and fought to build bridges based on empirical testing rather than calculations alone. Maillart's daring did not compromise safety. He wanted to know whether something would actually work and was uninterested in whether a computation indicated that it might. Actual experience counted with Maillart. When cracks formed in the concrete wall covering the spandrels of his 1901 Züoz Bridge—the spandrel is the triangular space formed by the outside curve of an arch—Maillart eliminated this unnecessary façade to expose the guts of the concrete arch in future designs. With this move he succeeded in reducing the dead load, or the weight, of subsequent bridges as well as their cost while adding to their stark beauty. He initially made the Züoz Bridge stronger and lighter by incorporating a hollow-box arch design into it, the first such hollow-box arch ever used. It took the place of a heavier, solid concrete arch.

An example of Maillart's combination of engineering and artistry is the Salginatobel Bridge in western Switzerland. Built in 1930, the supporting arch of the bridge is a seemingly wafer-thin, shallow arch ridden by a pair of twin arch walls to form a hollow-box arch, Maillart's innovation, introduced in his Züoz Bridge design. From the side, the hollow-box arch appears to be a homogenous element that melds with the equally thin, horizontal roadway.

Another of his superb designs is that of the Schwandbach Bridge near Hinterfultigen, Switzerland. A deck-stiffened roadway is tied to the sleek supporting arch strengthening the bridge, another innovation claimed by Maillart. The roadway is connected to the bridge by trapezoidal cross walls whose wide bases connect to the arch and whose narrow ends join the underside of the road deck. Seen from the side, the bridge is much like the Salgi-

natobel span, but from above one sees that the roadway is severely curved to follow the winding mountain road. The two views reveal the bridge as something of an optical illusion, but this visual trickery resulted in a deceptively strong structure praised by critics as a work of art.

Maillart was virtually unknown outside Switzerland until given praise in the writings of the Swiss architect and noted design technology critic Siegfried Giedion in 1941. With the horror of World War II fading, New York's Museum of Modern Art mounted an exhibition of Maillart's work in 1947. The showing was a precedent, the first devoted to the efforts of a single engineer.

Maillart's work inspired numerous bridge designers, most notably a fellow Swiss engineer, Christian Menn, to use reinforced concrete in the creation of stunning, modernistic bridges with simple beauty. Menn's Ganter Bridge (1980) is one of the world's most unique in terms of engineering with its stayed-box girder design. As for its visual impact, like Maillart's designs, it has a nearly hypnotic effect.

Maillart built 47 bridges, of which 45 remained in use by the end of the 20th century, an impressive testament to his engineering prowess. (The Tavanasa Bridge, an early example of his innovative thinking, was swept away by an avalanche in 1927.) That his bridge and building creations subsequently became symbols of modern design is an example of his equally important talent as an artist. In 1991, Maillart's Salginatobel Bridge was designated a "world monument" by the American Association of Architects and Engineers.

The Garstatt Bridge, the last of his designs built, was completed in 1940, the year of his death at age 68. The straight lines of the supporting arch beneath the Garstatt Bridge reveal Maillart's tendency to abandon pure curves near the middle and end of his career.

See also ARCH BRIDGE; CONCRETE; DECK-STIFFENED ARCH; GANTER BRIDGE; HINGE; HOLLOW-BOX GIRDER; MENN, CHRISTIAN; REINFORCED CONCRETE; SALGINATOBEL BRIDGE; TAVANASA BRIDGE.

Manhattan Bridge (built 1901–1909) *Controversial crossing* The Manhattan Bridge is not New York's longest bridge, but it may qualify as the city's most controversial and politically charged one, as its troubled tradition hotly pursues it into the 21st century.

Passenger rail interests had sought a railway suspension bridge running from Manhattan to the Bronx at least since 1892, but serious work on designing the span did not begin until 1900, when city fathers approved the issuance of $1 million in bonds for the bridge. An early design by R. S. Buck was not well received, but as the debate over the design proceeded, contracts were signed for the foundation work.

The famed bridge engineer Gustav Lindenthal, who had recently been appointed commissioner of New York's bridges, submitted a new design in 1902 changing Buck's bridge from a suspension bridge using wire cables to one using eyebar suspension chains.

A board of highly regarded engineers was appointed by New York's mayor, Seth Low, to review Lindenthal's eyebar suspension system. However, Wilhelm Hildenbrand, who had prepared plans for the Brooklyn Bridge and who was then employed by Roebling's Sons Company, which manufactured and erected wire suspension cables, warned that eyebar chains could add millions to the cost of the bridge. When the board of engineers made their final report it was far from a resounding endorsement of Lindenthal's design and left the debate unresolved.

The far bulkier appearance of an eyebar chain suspension span when compared to that of a wire cable bridge prompted the Canadian engineer George W. Colles to criticize roundly the ugliness of such a bridge in the highly regarded publication *Engineering News* (later *Engineering News-Record*). Like Hildenbrand, Colles opposed the use of eyebar chains and not only for aesthetic or fiscal reasons: Colles pointed out that eyebar chains were prone to cracking, corrosion, and failure.

Colles's argument was sound since a bundled wire suspension cable is well protected from the elements by oil, an external wrapping of wire, and paint. The thousands of wires used also provide redundancy so that the snapping of a single wire is virtually inconsequential. Usually, no more than a handful of eyebars run side by side, so the loss of one could be catastrophic. This was demonstrated with fatal finality in the collapse of the chain-supported Silver Bridge over the Ohio River in December 1967.

The debate raged until politics intervened. Low was defeated in his bid for reelection, and in 1904 George Brinton McClellan took office. An avowed enemy of the Tammany Hall machine and a sup-

porter of public infrastructure projects, McClellan was also the namesake son of the famed Civil War general. McClellan moved quickly to replace Lindenthal as bridge commissioner with George E. Best.

Best ordered the Department of Bridges chief engineer Othniel Foster Nichols to redesign the bridge, replacing the eyebar chains with a wire cable suspension scheme. Nichols called in the well-established bridge designer Ralph Modjeski to assist with the design and took the controversial step of applying the deflection theory to the design of the bridge.

This theory held that suspension bridges were more stable and rigid than originally believed, therefore allowing thinner and lighter decks with narrower stiffening trusses. Originally devised by the Austrian engineer Josef Melan, the deflection theory was applied to suspension bridges by the Department of Bridges engineer Leon Moisseiff, who urged that the theory be applied to the Manhattan Bridge.

Despite the changes, Lindenthal's impact on the appearance of the bridge would remain substantial. The slender towers of the bridge, with their lattice-braced twin legs, were a distinctive touch that was virtually left intact. Although the towers were originally designed to flex slightly where they connected to the piers, their mountings were stiffened.

The new design was adopted, and by 1908 the Roebling company was spinning the wire cables across the East River. Nonetheless, controversy still surrounded the bridge. Lindenthal sniped at the project, questioning Moisseiff's thinner and lighter approach, as opposed to his own structural overkill. Lindenthal also ridiculed Tammany Hall politicos for engaging in graft and interfering with the timely construction of the city's bridges, including the Manhattan.

Indeed, Lindenthal may have had the last laugh. Although the bridge would eventually carry eight rail lines, it would be discovered four decades later that the bridge's deck moved so much in reaction to passing trains that the span was being damaged. The lightness of the two-level bridge deck meant the deck deflected up to eight feet when a train passed. The bridge was modified to carry only seven lanes of rail traffic and repairs were made.

By the late 1970s, the superstructure of the bridge had sustained so much damage that a debate grew over whether a replacement should be built. In 1982, the New York City Department of Transportation made the decision to repair and renovate the bridge. Even this project ran into difficulties when an initial $40 million effort to repair the bridge was delayed when substandard steel used in the renovation had to be replaced. In 1999, another project that involved strengthening and stiffening the road deck began. This $350 million project was to be concluded in 2005. The work will also involve seismic retrofit efforts to make the bridge resistant to earthquake damage.

Adding to the traditional controversy that has surrounded the bridge, a city Department of Transportation engineer overseeing the project was arrested in 2001 and charged with soliciting a bribe from a contractor working on the project. The bribe was to be in exchange for allowing the contractor to employ more workers and reap more profit on a $127 million segment of the rebuilding project.

The Manhattan Bridge today carries six lanes of vehicular traffic and two passenger rail lines. A total of 78,000 vehicles and an estimated 350,000 people cross the bridge on average each day. As part of the refurbishment of the span, a hike-and-bike lane was added to the bridge on May 16, 2001.

See also BUCK, LEFFERT LEFFERTS; LINDENTHAL, GUSTAV; MODJESKI, RALPH; MOISSEIFF, LEON SALOMON; SEISMIC RETROFIT; WILLIAMSBURG BRIDGE.

Maracaibo Bridge   *See* cable-stayed bridge.

Mathieu, Jacques-Joseph (1700s–1800s) *Early designer of an English Channel tunnel* Sometimes identified as Albert Mathieu-Favier, this 19th-century mining engineer made what is considered one of the first serious proposals to tunnel beneath the English Channel. Unfortunately for the French engineer, he presented his plan at a time when France and Britain were eying one another as enemies. As for many plans since, nothing came of his idea except the encouragement of other engineers.

Mathieu knew that a tunnel could be excavated beneath the English Channel and in 1802 proposed a plan that included a main tunnel for carriages and a service tunnel for ventilation. His plan was something of a historical mystery for nearly two centuries since no records of the proposal could be found. A fanciful sketch of his tunnel design, which was misplaced for nearly 200 years, has survived and depicts an underground passage ventilated by chimneys rising above the waters of the English Channel.

Keeping Mathieu's contribution alive was Aimé Thomé de Gamond, who said he was inspired by Mathieu's concept. Thomé was another Frenchman who dedicated his life to developing schemes for crossing the Channel by a tunnel from the 1830s until the 1860s.

Thomé de Gamond, who was something of a promoter and given to embellishment, was not considered the best possible source of information about Mathieu. However, Mathieu's contribution to the history of Channel tunnel design was given an injection of credibility in 1993 with the chance discovery of an ink drawing of Mathieu's tunnel. The drawing was given to a university in Dunkerque.

Tinted by watercolors, the drawing bears an early 1800s date and an inscription in French that reads, "Plans of an underground communication between Calais and Dover by an Engineer of Mines, J. J. Mathieu."

See also DESMAREST, NICOLAS; EUROTUNNEL; SUBAQUEOUS TUNNEL; THOMÉ DE GAMOND, AIMÉ; WATKIN, EDWARD.

McCullough, Conde Balcom (1887–1946) *Oregon's master bridge builder* During the first half of the 20th century, when the nation's preeminent bridge builders were often immigrant engineers trained in European universities, America was beginning to produce its own homegrown bridge designers. One of the most notable of these was Conde Balcom McCullough.

Born in Redfield, South Dakota, 13 years before the arrival of the 20th century, McCullough was the son of a physician and minister, John McCullough, and his wife, the former Lena Balcom. Soon after his birth, the family moved to Iowa, where his father was crippled in an accident. John McCullough subsequently died, forcing a teenaged McCullough to support his mother while he attended school.

After high school, McCullough spent a year as a railroad surveyor's assistant before enrolling in Iowa State College. After obtaining a civil engineering degree in 1910 McCullough spent a year working for the Marsh Engineering Company of Des Moines, a well-known firm whose founder, James B. Marsh, promoted the use of reinforced concrete bridges.

During this period another engineer filed a lawsuit against Marsh, claiming Marsh had infringed upon the engineer's patents. McCullough was ordered to research concrete bridge techniques used around the world to demonstrate the hollowness of the engineer's claims. Marsh prevailed in the lawsuit, and some credit McCullough's research assignment with his enthusiasm for and willingness to use European concrete techniques often eschewed by American engineers.

By 1911, McCullough was working for the Iowa State Highway Department. After five years there he had been promoted to the position of assistant state highway engineer. McCullough's move to Oregon in 1916 occurred when he accepted a teaching position in the civil engineering department of Oregon State College. By 1918, the brilliant and industrious McCullough was chairman of the department. In the following year, McCullough took the job as chief of the Bridge Division for the Oregon State Highway Department, a department he built with a handful of engineering graduates of Oregon State University.

What McCullough gave to the job was a devout faith in the strength, durability, and economy of prestressed and reinforced concrete bridge construction. These techniques were pioneered by European engineers like the Frenchman Eugène Freyssinet, who developed means of strengthening concrete through prestressing, and the Swiss Robert Maillart, who elevated reinforced concrete bridge design into an art form. McCullough's dedication to concrete would prove to be politically incorrect in Oregon, where powerful logging interests wanted every bridge in the state, no matter its size, to be constructed of timber.

Although the idea of using timber in one of the nation's greatest timber-producing regions seems sensible, the expense of replacing wood rotted by moisture or destroyed by fire meant that the use of timber in large bridges was a false economy at best and backward at worst. However, in regions of the state where the climate was friendly to timber bridges, such spans were built. McCullough's rule of thumb in determining what type of bridge to build was farsighted in that it looked at not only the initial cost of a bridge but also the long-term maintenance expenses. For McCullough, imagination and engineering mixed with concrete would be the way for Oregon to bridge its myriad waterways and bays.

McCullough took the job as Oregon's top bridge designer at an auspicious time since the state was then struggling to complete a highway system along its Pacific coast. The spans McCullough would build for what eventually became the Oregon Coast Highway would make the state famous for its beautiful bridges. McCullough's design touch and his willingness to use European reinforced and prestressed concrete methods shunned by his colleagues gave the

bridges a distinctive appearance seasoned with a hefty dose of American practicality.

Between 1919 and 1935, McCullough designed or oversaw the design of hundreds of bridges. At least 162 bridge designs were completed during his first two years as chief of the bridge division. As if he were not busy enough, McCullough also continued his education by attending law school, and he passed the Oregon State Bar exam in 1928. As a result, the certified civil engineer also became a licensed attorney. McCullough obtained a legal education to help him understand fully how legal issues impacted on the engineering profession. He would later author a text, *Engineer at Law,* with his son and write numerous professional articles on bridge construction.

Although Oregon was relatively impecunious compared to more populous states, its politicians realized its economic development was being stymied by an inadequate road system that depended on inefficient ferries at numerous water crossings. The state's first halting attempts at constructing a highway along Oregon's coastline began in 1914; not until the end of World War I did defense concerns prompt federal support of what was at first known as the Roosevelt Coast Military Highway. The construction of this highway and other roadways was considered so important that by 1920 Oregon had established the nation's first gasoline tax to finance road and bridge work.

The Yaquina Bridge, which is part of the scenic Oregon Coast Highway, is indicative of Conde B. McCullough's style and expertise. Using a combination of concrete and steel arches for the viaduct portion of the bridge, McCullough's design culminates in a bowstring arch that complements its natural surroundings. (*Historic American Engineering Record, National Park Service*)

The highway would be essentially complete by 1932 and receive the friendlier title "Oregon Coast Highway." By then the roadway was so heavily traveled that irksome congestion was being created at the remaining snaillike ferry crossings. In the southwest corner of the state the highway was interrupted by five waterways: Coos Bay, Alsea Bay, Yaquina Bay, the Umpqua River, and the Siuslaw River. This situation would present McCullough with an opportunity to design five of his most impressive bridges, each bearing the name of the waterway it crossed. Federal money became available to bridge these waterways as the Depression era Public Works Administration helped fund projects through grants and loans.

This windfall of money threw McCullough and his assistant engineers into a maelstrom of work to conduct surveys of bridge sites and develop designs. The lumber interests shouted for timber bridges—impermanent spans the federal government would not finance—and the debate threatened the federal funding. The issue was finally settled in favor of the reinforced concrete and steel bridges that McCullough had smartly embraced as ultimately more economical.

What is perhaps McCullough's most impressive span is the Coos Bay Bridge (1936), a cantilever truss bridge with graceful reinforced concrete arch approaches. Although the cantilever truss form is far from the lightest and most graceful, McCullough's thoughtful design minimized the girth of the structure, making it one of the state's most impressive. The cantilever's clear span of 793 feet over the bay was an impressive achievement, but the stylish concrete arches that support the side spans of the bridge give it a gently distinctive look. In the year after McCullough's 1946 death of heart failure, Oregon officials renamed the Coos Bay Bridge after McCullough, making it one of the few bridges in the world named after its designer.

Another of McCullough's achievements was the concrete Alsea Bridge, also built in 1936. The beautiful span fell victim to corrosion of its reinforcing steel and had to be demolished in 1991. A trio of concrete arches that utilized vertical hangers to suspend the concrete roadway supported the bridge's three main spans. The side spans leading up to the center spans were supported by a series of concrete arches.

The bridge fell victim to the seaside environment of the Oregon coast when corrosion due to sea salt caused expansion of the embedded steel reinforcement. The bridge was eventually demolished in 1991 by both high explosives and machinery. Another concrete bridge with a single arch over the navigable channel was built alongside McCullough's, mildly mimicking the style of the first.

In 1935, after the designs of the five major spans for the Oregon Coast Highway were completed, McCullough and an assistant and former student, Raymond Archibald, departed for Costa Rica. There the two designed a series of badly needed bridges along the route of the Inter-American Highway, which was to run from Alaska to the southern tip of South America. While in Costa Rica, McCullough became an eager student of the Mayan culture and later incorporated Mayan elements into bridge ornamentation.

After two years in Central America, McCullough and Archibald returned to resume their duties in Oregon, although McCullough was by then the assistant state engineer and was an administrator far less involved in bridge design. A lifetime of heavy smoking and an even heavier work load had taken their toll on McCullough, who died after a stroke a month shy of his 59th birthday.

The Oregon Department of Transportation maintains a website providing a photo gallery and information about that state's collection of unique bridges with particular emphasis on those designed by McCullough. It is at http://www.odot.state.or.us/eshtm/br.htm.

See also FREYSSINET, EUGÈNE; MAILLART, ROBERT.

Menai Bridge (built 1818–1826) *Prototype for modern suspension bridges* When the Menai suspension bridge was built in 1826, its daring design was to be the forerunner of all modern suspension bridges by proving that such a long-span bridge could not only be gracefully thin but be (reasonably) solid as well. In many respects it is the father of modern suspension spans, including the Brooklyn and Golden Gate Bridges. That the Menai Bridge was even built is quite surprising since, at the time, suspension bridges had fallen into disfavor because of their propensity for collapsing.

But the Menai Bridge was built as a result of a desperate need and the massive reputation of its designer, the self-taught Scottish civil engineer Thomas Telford, a national icon in Great Britain for his construction of innovative bridges, harbors, and canals. Having recognized the engineering elegance

The Menai Bridge spanning the Menai Strait was the most important bridge of its time since it renewed faith in the concept of the suspension bridge when frequent collapses generated distrust of this type of span. Although the bridge by the famed Scottish engineer Thomas Telford had its flaws, subsequent repairs and strengthening have kept the bridge in use since 1826. *(Godden Collection, Earthquake Engineering Library, University of California, Berkeley)*

offered by a suspension span—relatively low cost and ability to support a long span—Telford successfully lobbied for a suspension bridge across the Menai Strait.

By the early 1800s, England was in the midst of the Industrial Revolution, which prompted an explosion of roadway and railroad construction that required newer and stronger bridges. As part of this, the government decided it was necessary to provide a direct route from London to the port city of Holyhead on the Welsh island of Anglesey. This would perform the twin tasks of assisting commerce and strengthening England's links with Ireland, whose capital city, Dublin, lay across the Irish Sea 60 miles to the west. The major obstacle to the project, obvious to anyone with rudimentary knowledge of local geographic conditions, was the Menai Strait.

Although one of the most developed nations, Great Britain still had many regions isolated by rivers and straits, with slow and unreliable ferry-boats the only means of crossing such obstacles. Such was the case of the sizable island of Anglesey, jutting into the Irish Sea off Wales' west coast. It is believed Anglesey was once a peninsula protruding from Wales but that the end of the last ice age 10,000 years ago raised the level of the seas. This flooded the valley between the island and the Welsh mainland, creating the Menai Strait, which runs roughly north from Conwy Bay, south to Caernarfon Bay. Because the strait is used by maritime traffic, it was essential that any bridge spanning it not require foundations in the middle of the waterway. One of the narrowest portions of the strait was at the city of Bangor, Wales, where the opposite shores were roughly 900 feet apart.

Thomas Telford understood the obstacles facing any bridging effort across the Menai Strait since he had taken part in the Holyhead Road Survey of 1810–1811 that laid out the best overland route from London to Anglesey. His engineer's mind had long been pondering the bridging problem. Another engineer eager to bridge the strait was Telford's contemporary, John Rennie, Sr., who proposed a massive iron arch bridge at Bangor. The cost for Rennie's bridge was estimated as high as 290,000 British pounds, an amount equal to tens of millions of dollars today. The British government winced at Rennie's price tag and waited for a less costly plan.

In the meantime, Telford became acquainted with Captain Samuel Brown of the Royal Navy, who had invented an improved flat iron chain link, known as an eyebar because its ends contained holes to allow the links to be interconnected. Telford believed Brown's eyebars were the means to an improved suspension bridge chain. Telford proposed a suspension bridge for the Mersey River using the bar links but was turned down. Undeterred, he subsequently recycled his plans and proposed a bold suspension bridge for the Menai Strait, the longest such bridge built up to that time. Desperate for a bridge across the strait to complete the Chester-Holyhead Turnpike, a government commission approved Telford's plan in 1817.

The bridge promised to be spectacular. The summits of its two main towers would rise 153 feet above the water at high tide. (The tidal action could cause the level of the strait to fluctuate by 20 feet.) The roadway of the bridge would be a minimum of 100 feet above the waterway to allow the passage of ships. The total length of the bridge would be a staggering 970 feet. The wrought iron suspension chains composed of Brown's eyebars would support an unprecedented 579-foot central span between the two towers. The Bangor approach would measure 170 feet and be supported by a trio of masonry arches, and four masonry arches would support the 220-foot Anglesey approach.

Not surprisingly, critics assailed the bridge's novel design and use of newly developed eyebar iron links. Many predicted it would crash into the strait, unable to support the weight of its insanely long span. This prediction would be partially correct, but for reasons that neither the prognosticators nor Telford would foresee.

Unlike that of the Brooklyn Bridge, which would be built nearly 50 years later, construction of the Menai span's supporting towers would be relatively easy thanks to the shallow depth of bedrock on each side of the strait. Workers needed to penetrate a mere six feet beneath the muck of the strait on the

west bank to reach bedrock. Incredibly, no excavation was needed on the Anglesey side of the strait, where rock protruded above ground. In 1819, work began in earnest, but despite the accessibility of the bedrock, the job of building the bridge would take nearly seven years.

In another break with precedent, work on both main towers commenced simultaneously and continued from 1820 to 1822. Also, foundations for the approach span towers were anchored, and masonry columns began to rise from them to form the arches required on both sides of the strait. The tasteful tone that the towers and arches would comprise was extremely durable gray limestone quarried on Anglesey Island itself.

The most exotic components in the bridge were the iron bars that would compose the supporting chains of the structure. To ensure the bridge was built to the highest possible standards, Telford decided to "proof," or test, each eyebar to ensure it could sustain the loads expected on the bridge. Each square inch of the wrought iron eyebars had to be able to withstand five tons per square inch of pressure or it was discarded. A total of 935 eyebars would be needed to form 16 chains bundled into four groups. A pair of chain bundles ran along the outside edges of the bridge, the second pair was draped over the towers to travel down the center of the bridge.

With the towers and masonry work completed by 1825, Telford was ready to drape the iron chains over four cast iron saddles atop the masonry support towers. The process of pulling the massive chains from the mainland over the tops of the towers began on April 1825. Workmen turning large capstans, or manual winches, would pull the 16 eyebar chains weighing a total of nearly 2,000 tons across the strait one at a time. Hawsers were initially tied to the chains that would be anchored on the Bangor side and slowly pulled across the strait. Once in place, the chains would be tensioned so they formed the required curvature.

During the pulling of the chains, a band of fifers played music to provide a rhythm for the workmen. The workmen turned the handles of the 50-ton-capacity capstans moving the chains slowly across the strait, a process that took 16 weeks. The chains were then anchored to solid rock on the Anglesey side of the bridge. When the cabling was completed, but before the installation of the road deck, which seemed a comparatively minor feat, a party was held for the workers. A pair of bridge workers, fueled by celebratory rum and the exhilaration of the accomplishment, shinnied across the chain links to be the first ever to cross the strait by bridge, albeit by an uncompleted one.

Although flexible, steel wire cables would prove far more efficient in future suspension bridges, the iron chains provided the unique feature of allowing the disconnection of any of the 16 redundant chains for needed repairs, a feature Telford tested after the chains were in place. After a final study of the curvature of the 16 chains, Telford installed four additional eyebar links in each as a minute adjustment to the curve of the suspension chains.

The final process of construction the bridge involved the roadbed that was fabricated by using three layers of strong fir timbers. A layer of felt would be applied above and below the middle layer of planking, and the road deck would be secured to the suspension chains with 444 vertical iron rods hinged to the connecting points of the eyebars above. The roadway provided two nine-foot-wide lanes that passed through double archways built into both suspension towers. A separate walkway was provided for pedestrians.

Safety railings and the requisite tollbooths were the last items necessary to make the bridge complete, and this work was finished during the winter of 1825–1826. On January 30, 1826, Telford and anyone else who could find a perch on a horse-drawn mail coach clopped across the bridge for a test of the span. The public was then allowed on the bridge, and people turned out in droves to make their way across the historic span. The bridge endured the weight despite its thin and light appearance compared to that of contemporary stone and iron bridges. London's more conventional iron arch Vauxhall Bridge of similar dimensions built over the Thames River in 1816 required three times as much iron for its arches as the Menai Bridge needed for its chains.

It was a precedent-setting design: far lighter than traditional bridges that would have required massive towers in the middle of the strait and a heavy stone roadway. Its gracefulness was unbelievable to observers of the day. Said one, identified only as Mr. Roscoe in Samuel Smiles's biography of Telford, "The Menai Bridge appeared more like the work of some great magician than the mere result of man's skill and industry."

Although considered a masterpiece of early 19th century engineering, Telford's bridge had its Achilles

heel. Within a week of its completion, the 579-foot center span experienced powerful undulations as a strong wind blew through the strait. The roadway fluttered and shook as a result of aerodynamic forces in much the same way the Golden Gate Bridge did in strong winds on February 9, 1938. The Menai Bridge survived the gale, but a January 1836 storm packed winds strong enough to cause the road deck to flail so wildly that it snapped six iron suspender rods connecting the chains to the roadway. Telford, who died in 1834 at the age of 77, missed the heartbreak of seeing his bridge fall victim to nature.

Disaster struck again in January 1839, when a potent storm's winds caused the bridge to undulate so violently that the movement broke one third of its suspender rods, sending the shattered road deck crashing into the water. The bridge was soon repaired, but nature had put the engineers on notice. The bridge was built at a time when Telford and others knew too little about the effect of aerodynamics on bridges, especially the slender and light suspension spans. The Menai span's long road deck served as a lifting surface, much as the wing of an aircraft does. In a high wind, suspension bridge roadways tend to be lifted and then slammed down with great force, like bed sheets snapping horizontally in the wind. If the motions are strong enough, the road deck can pull itself loose from its chains and suspender rods. Engineers would later understand that defeating these aerodynamic forces meant stiffening the bridge with a ventilated truss deck or lateral bracing or even affecting the wind flow pattern around the bridge by ventilating the road deck or altering its shape.

After the 1839 storm, this was precisely what the engineers did. They installed four truss railings on the roadbed to provide additional stiffness and weight to counteract aerodynamic forces—the forerunner of heavier and more complex truss systems that would become a standard design element of future long-span suspension bridges. The bridge held firm for the next 100 years. The only other changes to the bridge included the addition of a steel road deck in 1893 to replace the one of timbers. In 1939, as war loomed over Europe, the iron chains were replaced with chains of steel that had twice the strength of the original eyebars. Except for the elimination of the center pair of chain bundles, made unnecessary by the increased strength of the replacement steel chains running down each side of the bridge, the span looks essentially the same as it did when originally built.

The Menai Bridge still towers above the tidal waters of the Menai Strait, where it now carries automobile traffic in lieu of horse-drawn carriages.

The Menai Bridge is not Telford's only one in the area. Just 12 miles to the north is the Conwy Bridge, another suspension span designed by Telford, which was completed in the same year as that across the Menai Strait. Its center span measures 320 feet, and it crosses the Conwy River beneath a castle by the same name. Refurbished, the bridge retains its original iron chains but was closed to automobile traffic in 1958.

See also BRIDGE AERODYNAMICS; BROOKLYN BRIDGE; BROWN, SAMUEL; GOLDEN GATE BRIDGE; SUSPENSION BRIDGE; TELFORD, THOMAS.

Menn, Christian (1927–) *Designer of daring bridges in the tradition of Robert Maillart* If Switzerland is famous for making watches, it should be equally famous for producing some of the world's most imaginative and capable bridge designers. The Swiss engineer Christian Menn is one of the 20th century's most prolific designers of bridges that are not only innovative in structural terms but also imbued with a powerful simplicity.

Not surprisingly, Menn has been heavily influenced by Robert Maillart, a fellow Swiss bridge designer, who showed the world that concrete was not only a strong and economical building material but one that could be used to construct stark, graceful bridges.

Menn was born in Meiringen, Switzerland, 13 years before the 1940 death of Maillart. In 1950, Menn graduated from Zurich's Federal Technological Institute with a degree in civil engineering, although his career was interrupted briefly by a bout of tuberculosis from which he recovered in 1953. Menn continued his education under Professor Pierre Lardy, who helped introduce Menn to Maillart's design philosophy, although Menn was already aware of Maillart's work.

Menn's father, Simon Menn, who was also a civil engineer, had worked on at least two of Maillart's bridge projects and was friends with Maillart. The younger Menn obtained his Ph.D. in civil engineering in 1956 and worked as an engineer on the UNESCO buildings in Paris. In 1957, Menn opened his own engineering office in Chur, a Swiss city on the banks of the Rhine, where he specialized in bridge design.

By 1960, Menn's first bridge designs were being constructed across the Averserrhein River at Cröt and Letziwald, with both arch bridges revealing the influence of Maillart. In 1962, Menn's Reichenau Bridge across the Swiss Rhine was completed. The 328-foot bridge was his longest span at that time and was supported by a concrete arch, although a prestressed, hollow-box concrete girder stiffened its deck. The clean lines of the concrete bridge again show the influence of Maillart. However, the necessity for scaffolding in the construction of arch bridges added to the costs, which had risen with labor rates.

Driven as all engineers are to find a less expensive but structurally viable way of building bridges, Menn designed the Felsenau Bridge so it would not require falsework or scaffolding. The bridge would be made up of a trio of prestressed concrete deck sections supported by a hollow-box girder. After three pairs of 206-foot-tall supporting columns were erected, the segments of the bridge were poured and then stressed with steel cable tendons. Not only was the roadway of the bridge curved, adding to the difficulty of constructing it, but Menn opted for a single large box girder instead of two hollow-box girders used side by side. This design eliminated a pair of superfluous concrete walls.

Although the general concept of building the bridge outward from its supporting columns was developed in Germany, Menn refined it with his own touches. The bridge, built between 1972 and 1975, has a total length of 3,661 feet. Its longest single span measures 511 feet and was at the time the longest in Switzerland. From the side, the bridge is a graceful and slender sight rising above the Aare River north of Berne.

A year before the completion of the Felsenau Bridge, Menn was asked by local Swiss officials to recommend a bridge design for the Simplon Road. Menn proposed a design similar to that of the Felsenau Bridge but with even longer spans. The design was so economical and stylish that Menn was granted a commission to design the bridge without competition.

The result was a hybrid cable-stayed design. Like the Felsenau Bridge, the Ganter Bridge would utilize hollow concrete girders in the form of cantilevers. However, because of the extreme length of the prestressed concrete deck spans—nearly 571 feet—Menn decided also to use cable stays radiating to the road deck from the columns that would extend above deck. The stays, contained in triangular concrete walls, would transmit the loading from the long spans to allow thinner hollow-box girders.

The total length of the bridge is 2,224 feet, its main span measuring just over 570 feet and two side spans of 416 feet each. The bridge is one of the most visually striking in the world. Construction on the bridge began in 1976 and was completed in 1980. The three supporting columns of the bridge vary in height with two shorter columns flanking a massive, 492-foot pillar rising from the valley below.

Menn's success at blending form and function has made him one of the reigning wizards of bridge engineering. Menn expanded his use of cable-stayed bridge designs, becoming a leading authority on the design type. Although he gave up his engineering office in 1970 when he was appointed a professor of structural design and construction at the Federal Technological Institute, where he originally studied, he remained busy as a consulting engineer.

As debate raged over the type of structure that should replace Boston's old "High Bridge" over the Charles River during the 1990s as part of the Central Artery/Tunnel Project, Menn submitted a cable-stayed design in 1994. His two-tower cable-stayed bridge design was deemed far more attractive than the originally proposed bridge, a multilevel monstrosity. Completion of the breathtaking Leonard P. Zakim Bunker Hill Bridge is expected in 2003.

See also CABLE-STAYED BRIDGE; CENTRAL ARTERY/TUNNEL PROJECT; MAILLART, ROBERT; PRESTRESSED AND POST-TENSIONED CONCRETE.

microtunneling *Avoiding big problems with small tunnels* Microtunneling is a technique that grew out of the patented Mini Tunnel system developed in Great Britain in 1973. This system allowed the digging of small tunnels no taller than a standing man in heavily urbanized areas where trenching or cut-and-cover techniques would disrupt surface activities.

The Mini Tunnel system developed by Michael A. Anderson involved the use of a cylindrical tunneling shield up to 4.2 feet in diameter. Only three workers were required to operate the system, one worker excavating the tunnel by hand and also erecting the reinforcing lining. A second drove a tiny train with muck carts to and from the rear of the shield, and a third operated a crane lifting the filled carts from the access shaft.

By 1975, a miniaturized tunnel-boring machine was developed by Anderson that could excavate the soil with a rotating cutting head as a man inside controlled the operation. A conveyor belt moved the muck to a waiting cart. By 1979, Iseki Poly-Tech Inc. of Japan had developed a tunnel-boring machine that could be operated remotely from the surface; that utilized earth pressure balance technology, that is, application of water pressure against the face of the tunnel to prevent flooding and collapse in soft ground.

This device, known as the Tele-Mole, was a breakthrough that formed the basis for future microtunneling machines. The Tele-Mole contained a closed-circuit television camera to monitor the digging and to ensure the tunnel was following the correct alignment. Instead of erecting concrete or iron or concrete lining segments by hand, pipe was then jacked hydraulically into the completed excavation to line and reinforce the tunnel. Because the machine utilized the slurry shield method of tunneling, in which pressurized bentonite slurry is maintained between the working face of the tunnel and the cutting head, muck is removed by pumping. This technical breakthrough meant no humans would be exposed to subterranean gas, flooding, or cave-ins.

Although the advantages of microtunneling were soon obvious to engineers and municipal officials, the technology seemed too exotic and expensive to conservative engineers. Microtunneling was essentially designed for soft ground tunneling, and the variability of ground conditions in the United States meant American engineers were slow to accept the technology.

However, microtunneling entered the United States in 1984 when it became necessary to install a six-foot-diameter, 615-foot pipe beneath an interstate highway in Miami without digging a trench in the roadway. The project was a success, but it would take a massive infrastructure project in Houston, Texas, to initiate the wide-scale use of microtunneling in America. Houston's need to rebuild its sewer lines resulted in a massive project from 1987 to 1991 that bored nearly 12.8 miles of microtunnels beneath the city, which sits on nothing but soft ground. The city's table-flat topography and its rock-free, soft earth conditions were perfect for microtunneling.

Construction of the sewer lines that crisscrossed Houston's 581 square miles was scarcely noticed by the city's 1.6 million inhabitants and convinced other municipalities that microtunneling was an excellent method of digging tunnels without creating traffic tie-ups. Texas soon became an epicenter of microtunneling with dozens of other projects conducted in Dallas and Austin.

See also PIPE-JACKING.

Millennium Bridge (built 1998–2000) *Britain's wobbly symbol* It was to be an enduring symbol of Britain as the island nation began its next 1,000 years. But the Millennium Bridge—a gleaming steel and aluminum pedestrian bridge across the River Thames—proved itself a wobbly embarrassment to the nation and a political headache to the ruling Labour Party, which touted the project.

The bridge, opened for only three days, beginning on June 10, 2000, treated the 150,000 who crossed it to a swaying and sometimes frightening experience. The bridge moved so violently to and fro that pedestrians were forced to cling to its railing to prevent themselves from falling. By June 12, the bridge was ordered closed pending an engineering review of its unanticipated instability.

Strikingly beautiful, the unconventional suspension span measures 13 feet wide by 1,082 feet long. It crosses the River Thames between London's financial district and the far less affluent Southwark borough on the Thames's south bank. It is the first footbridge built over the Thames in nearly 100 years. Each end of the bridge is greeted by two of London's landmarks: the 17th-century St. Paul's Cathedral on the north and the new Bankside Tate Gallery on the south.

As it turned out, the Millennium Bridge was another carefully planned, painstakingly constructed, and expensive bridge that was both beautiful and unstable. The instability of the bridge managed to embarrass those who designed and erected it as well as the politicians who pushed to spend almost $28 million to have it built. Although the experts maintained that the bridge was never in danger of collapsing, it appears the Millennium Bridge, like an anorexic supermodel, may have lost a tad too much weight for the sake of beauty.

The Millennium Bridge found itself at the mercy of the most common problem created by pedestrians, which is movement of the bridge as a result of the tendency of people to walk in step. The result of this compensation by pedestrians is even more motion accentuated by people who land on the same foot at the same time. This well-known effect on suspension bridges led to centuries of concern over the effect of marching troops on bridges, prompting the placement of signs warning troops to "Break step" during the 19th century.

Although suspension bridges are supposed to move horizontally and vertically in response to wind, temperature changes, and the weight of the pedestrians or vehicles crossing them, smaller and lighter bridges exhibit exaggerated motion.

The design of the Millennium Bridge is unique, although it contains many of the standard features of a suspension bridge. The two piers built for the bridge were fitted with 70-ton Y-shaped steel arms fabricated in Finland to hold the suspension cables suspended roughly even with the deck, in contrast to larger suspension bridges with cable suspension towers that stand hundreds of feet tall.

The deck of the structure is supported not by heavy stiffening truss systems like that of New York's Verrazano Narrows Bridge, but by two lengths of structural tubing that run beneath either side of the bridge deck. Attached in a perpendicular fashion to the underside of this deck are arms that emerge from each side to point slightly upward to accept the suspension cables.

Critics of the bridge have assailed its lightness, pointing out that bridge builders of the 19th century would have added weight and stiffness to a bridge any way they possibly could. Even the handrail of the Millennium Bridge, it has been pointed out, is not continuous and does not provide additional stiffening.

Its slender design is revealed in the numbers that show its relative thinness to its length. The Golden Gate Bridge, considered a marvel of thin bridge design during the 1930s, possesses a depth to span ratio of 1/164, whereas the Millennium Bridge boasts a depth to span ratio of 1/443. Essentially, this means the bridge is 443 times longer than the thickness of the deck.

After 17 months of construction the public swarmed to walk the bridge on June 10, 2000. The span immediately began to bounce and sway, prompting immediate closure. It was later reopened to a limited number of pedestrians but was closed again three days later, pending an investigation.

One man crossing the bridge described the sensation of walking along the flouncing bridge as akin to "being on a boat at sea." Said another who experienced the unwanted motion of the bridge: "If it is possible to tighten the bridge, they would be better to do it sooner than later."

Soon everyone had a consulting engineer, including the Southwark Town government. Arup, the engineering firm that designed the bridge, finally announced that the excitation could be controlled by the installation of tuned mass dampers and shock-absorbing dampers on the structure. The dampers would absorb the force of the footfalls and reduce the motion of the bridge to an acceptable level. The dampers, Arup engineers pointed out, have the added benefit of fixing the problem without altering the appearance of the bridge, an issue that remained nearly as important as its stability.

If the work proceeds according to plans announced in November 2000, it would take as long as eight months to complete. The price tag of the work—an estimated $7.5 million—stunned the public since the amount is more than 25 percent of the total cost of the $28 million bridge. By winter of 2001 the work had not been completed.

However, the electrical engineer Eric Thompson, who noted that a tiny suspension bridge in northern Scotland failed to exhibit the slightest movement as hundreds of sheep crossed it, recommended what was dubbed the woolly remedy. Thompson suggested that people wander like sheep across the bridge.

See also BRIDGE AERODYNAMICS; DAMPERS; GOLDEN GATE BRIDGE; LIVE AND DEAD LOADS; SUSPENSION BRIDGE; TACOMA NARROWS BRIDGE.

Mine Safety Health Administration *See* rock burst.

Modjeski, Ralph (1861–1940) *Innovative Polish-American bridge designer* Born Rudolphe Modrzejewski in Krakow, Modjeski had a background and upbringing slightly different from those of the ordinary structural or civil engineer, or anyone else for that matter. The man who would be the chief engineer on what was once America's longest suspension bridge was an accomplished classical pianist and son of the 19th century's most celebrated Shakespearean actress, Madame Modjeska, and her husband, Gustav Modrzejewski.

Although their altered last names differed, mother and son were both extraordinarily talented. Madame Modjeska's fame and wealth allowed her

son to spend much of his early life training as a pianist. A fellow piano student of Modjeski's was Ignace Jan Paderewski, who would become Poland's beloved concert pianist as well as its prime minister. At four, Modjeski demonstrated an interest in engineering by disassembling the lock of a door. When Modjeski was 15, he and his mother toured the Americas. During this trip the pair traveled by train across the Isthmus of Panama, where Modjeski declared his intent to become a civil engineer and construct a canal linking the Pacific with the Atlantic.

Modjeski was true to his word. Though he would play piano religiously for the rest of his life, his professional pursuit would be civil engineering. To accomplish this goal he took an examination to gain admission to what was then the world's most prestigious bridge engineering school, France's Ecole des Ponts et Chaussées (School of Bridges and Roads). Modjeski, despite his brilliance, initially failed to score high enough to be admitted. He took the examination again and gained admission. In 1885, Modjeski repudiated his bad showing in gaining entry by graduating first in his class.

Modjeski then made straight for the United States. Although Europe was training most of the world's engineers during the 19th century, it seemed America was where all the exciting bridges were being built. Modjeski went to work for the self-taught engineering genius George Shattuck Morison. On his arrival in America, perhaps taking a hint from his mother, Modjeski changed his name to make it easier to pronounce.

Despite his adolescent ambition to work on a Panama canal, Modjeski concentrated instead on building bridges, and by 1893 he had started his own engineering practice in Chicago. Modjeski was a prolific bridge designer, engineering the designs for dozens of bridges across the East and the Midwest. He also gained a reputation for the artistry of his designs.

His first major job as a self-employed consulting engineer was to design a replacement superstructure for a railway bridge leading to the federal Rock Island Arsenal in Illinois. The bridge was composed of a series of through-truss spans as well as a swing span to allow the passage of Mississippi River traffic. Owned by the federal government, it remains the oldest rail bridge in use between Iowa and Illinois. The span was built in 1895 on six piers designed and built in 1872 for the original bridge.

Modjeski's reputation for solid bridges touched with a dash of verve was well known. He subsequently designed a series of important rail bridges in Washington state and Oregon and was a consulting engineer on the controversial Manhattan Bridge project. Gustav Lindenthal had offered two designs that were rejected between 1901 and 1904, of which the second was an ugly eyebar chain suspension span. A new design was sought in 1904 under the microscope of public, artistic, and political scrutiny. Modjeski worked with Othniel Foster Nichols, chief engineer of the New York Department of Bridges, to review the design of the bridge.

Modjeski, who had helped organize publicity for his mother during her theatrical engagements, was cognizant of public relations, and photographs abound of Modjeski at his bridge projects in poses that might be considered silly by today's standards. As work was under way on the ill-fated Quebec Bridge, which had previously collapsed during construction, Modjeski and other engineers sat astride a gigantic connecting pin during the 1917 reconstruction. He and his fellow engineers also posed inside the open cells of one of the huge chords of the span. Modjeski, a small, goateed man invariably attired in a coat and tie, looked like a serious engineer, but his engineering genius cohabited with a touch of theater.

In fact, Modjeski's appointment to Canada's Government Board of Engineers to oversee the construction of the second Quebec Bridge was indicative of how far Modjeski had come. The original cantilever truss bridge, which had collapsed in 1907 during construction, was to be rebuilt with a new design and under new engineers, of whom Modjeski was one. The success of the second bridge only added to his growing reputation as one of the great American engineers of the first half of the 20th century.

Modjeski's most famous accomplishment is perhaps the Benjamin Franklin Bridge. The suspension bridge was built between 1922 and 1926 to connect Philadelphia with Camden across the Delaware River. Originally named the Delaware River Bridge, it was renamed after Franklin in 1956.

The Benjamin Franklin Bridge measures 3,356 feet between its anchorages and has a main suspended span between its two 385-foot towers of 1,750 feet. Touted as the longest clear span suspension bridge in the world at the time it was built, the bridge was supplanted in this category in 1929 by the Ambassador Bridge in Detroit. The Franklin Bridge has a similar appearance to and shares certain design elements with the Manhattan Bridge.

Leon Salomon Moisseiff, the leading proponent in America of the deflection theory, which mathematically justified using thinner and lighter road decks on long suspension bridges, had worked on the Manhattan Bridge. He had also been asked by Modjeski to serve as a consulting engineer on the Benjamin Franklin Bridge. By 2001, the bridge had been up to its task of carrying seven vehicular lanes and a pair of commuter rail tracks for 85 years.

Another of Modjeski's famous bridges, the Mid-Hudson Bridge, is a suspension span that began construction in 1925 and was completed in 1930. Its central span between anchorages is slightly less than that of the Benjamin Franklin Bridge at 1,500 feet. The overall length of the Mid-Hudson Bridge from anchorage to anchorage is 3,000 feet. At the time of the opening, New York state's first lady was Eleanor Roosevelt, who stood next to Modjeski during the ribbon cutting ceremony.

The Mid-Hudson Bridge has a sleek appearance with a slender road deck and graceful 315-foot towers. The bridge is made rigid and solid by a stiffening truss that runs as a railing on the roadway level. The two-lane bridge has had a third contraflow lane added for use during peak travel hours. The span underwent extensive renovation work during the 1980s and 1990s and was renamed the Franklin D. Roosevelt Mid-Hudson Bridge in 1994.

Modjeski was also chairman of the board of engineers appointed to review and provide input to the design of the San Francisco–Oakland Bay Bridge, which was completed in 1936. Of the several engineering offices that bore Modjeski's name during his career, the firm of Modjeski and Masters, founded in 1924, remains active in bridge design with eight offices throughout the United States.

See also LINDENTHAL, GUSTAV; MANHATTAN BRIDGE; MOISSEIFF, LEON SALOMON; QUEBEC BRIDGE; SAN FRANCISCO–OAKLAND BAY BRIDGE.

Moffat Tunnel   *See* Hoosac Tunnel.

Mohawk ironworkers *Native American bridge builders* Since 1886, and perhaps as far back as the 1860s, members of the Mohawk Kahnawake tribe in Canada's Ontario province, as well as other Native people, have been performing a life and death balancing act working as bridge builders across Canada and the United States.

The Kahnawake tribe's own history maintains that their men were drawn into bridge construction when 19th-century Canadian bridge contractors sought workers with the nerve to clamber about the rising superstructures of the iron bridges coming into vogue. The Mohawks, eager for work and more than happy to prove their courage, began taking the jobs that terrified others.

By the late 19th century, Kahnawake Mohawks had developed a reputation as superb ironworkers, willing to travel any distance to risk their lives on the "high iron." One of the early 20th century's most impressive bridge projects was the Quebec Bridge across Canada's Saint Lawrence River, which also became the scene of one of the tribe's worst calamities.

The bridge was located only a few miles from their home, the community of Kahnawake. On August 29, 1907, poor engineering led to the collapse of the partially completed span. Of the 92 workers on the bridge, 75 were killed. Of those who died, 33 were Kahnawake Mohawks, whose tribe then numbered 1,400 members.

Among the dead Mohawks was 42-year-old Joseph Orite D'Aillaboust, the father of 11 children. In all, 53 children lost their fathers that day. As a result, tribal members made a group decision that its men would never again work on the same project so as to prevent a similar tragedy. Steel crosses and memorials dedicated to the memory of the 33 who died on the Quebec Bridge can still be seen in Kahnawake.

The need for work did not end with the tragedy, and the Kahnawake Mohawks still sought work on bridges on both sides of the border. As the bridge-building boom of the 20th century stalled and the skyscraper boom began, Mohawks took their skills as ironworkers south into the United States to work on the Empire State Building, which was under construction in 1930–1931. Although not the only ironworkers traveling to faraway jobs to build bridges and buildings, the Kahnawake tribe's members had a willingness to travel long distances to work on first iron and then steel bridges that made their migration something of a diaspora of a closely knit people.

When bridge jobs became available during the 1920s and 1930s, the Mohawks, who originally worked mostly in America's Northeast, were eventu-

ally drawn to California's Golden Gate Bridge, where three worked during the mid-1930s. They would later work on Michigan's Mackinac Bridge in the late 1950s. The imposing Throgs Neck Bridge built at roughly the same time as the Mackinac Bridge also attracted the Mohawks, who had not worked on a major New York span since the 1930s.

With most of the major bridges built, the Mohawks turned to other forms of ironwork, and many of the Canadians bought or rented homes in a Brooklyn neighborhood. Today, most of the Mohawks still working in New York City now reside in boarding homes or communal apartments during the week and make the six-hour weekend commute to their tribal land across the Canadian border.

Although the Mohawks readily admit they possess no special skills in "skywalking," as they call their work, most attribute their ability to pride. At least one sociologist who conducted research among the Mohawks living in New York during the 1950s believed that the dangerous job of structural ironwork was a way for the Mohawks to express their traditional warrior ethic. One Mohawk ironworker told the *New York Times* in a March 16, 2001, article that they were not naturally better than ironworkers of other ethnic groups: The Mohawks, he said, simply worked hard to develop their skills.

And although the Mohawks are the most famous of the Native ironworkers, others hailing from different tribes have also entered the profession. Most of these individuals have been members of the six Iroquoian-speaking tribes who constitute the Iroquois League, a confederation created between the 17th and 19th centuries that included the Mohawks.

See also QUEBEC BRIDGE.

Moisseiff, Leon Salomon (1872–1943) *Influential bridge engineer tarnished by the Tacoma Narrows Bridge collapse* Leon Salomon Moisseiff had a hand in designing many of the most impressive bridges built in America during the first half of the 20th century. His triumphs include work as a consulting engineer on the Golden Gate Bridge, the San Francisco Bay Bridge, and Detroit's Ambassador Bridge, to name just a few. Virtually all his bridge designs were marvels to behold as each often represented a bold leap in bridge design and technology. Although he was the chief engineer on only one bridge, Moisseiff built his reputation as a consultant who stepped in to apply his expertise to bridges that bore the names of others.

Although he was one of the most successful bridge engineers in America, he also suffered one of the most ignominious events that can happen to a bridge designer. The one bridge that Moisseiff could claim as his own personal design, that of the Tacoma Narrows Bridge, collapsed on November 7, 1940, after only four months of operation.

The brilliant Moisseiff's intensive work with the deflection theory, which mathematically determined the relative stresses on a suspension bridge's cables and road decks, paved the way for suspension bridges to become thinner, lighter, and more graceful. As were many of the great bridge engineers who influenced design in the United States during the first half of the 20th century, Moisseiff was an immigrant. Born in 1872 in Latvia, Moisseiff attended the Baltic Polytechnic Institute in Latvia's capital city, Riga, from 1889 until 1891, when he immigrated with his family to New York City. Some accounts indicate that his political activism prompted the family's move.

Moisseiff then entered Columbia University and emerged in 1895 with a degree in engineering. After graduation in 1898, he secured a job with the New York Department of Bridges, the organization that for decades built and maintained the most progressive and impressive spans of any American city. Moisseiff demonstrated so much talent that he rose quickly in the department, playing a major role in virtually all the bridge building in the city during his tenure. By 1910, he was chief engineer of the agency. So successful were Moisseiff's concepts that in 1915 he left to enter private practice as a consulting bridge engineer.

While working for the bridge department, Moisseiff played a key role in designing what is considered the first of the truly streamlined suspension bridge—the Manhattan Bridge over the East River—the first designed by using the deflection theory. This theory, in part, held that as the dead load of the bridge's total structure gained heft, its deck could be made thinner since the increased weight of the bridge could provide additional stability. The center span between its towers is 1,470 feet, making it the longest suspension bridge of its day. Its steel towers were slender for the period, and its relatively shallow reinforcing truss gives the bridge an appearance of lightness when compared to the powerful physique of the much-overbuilt Brooklyn Bridge.

Both spans are within 500 yards of each other on New York's East River.

Moisseiff was convinced through his calculations that bridges could be built thinner and lighter than previously believed. Although he considered the effects of horizontal wind pressure on structures, Moisseiff and other designers had yet to consider fully the subtle and sometimes disastrous effects that the aerodynamic shape of the bridge would have or to factor in such a phenomenon as wind vortices. His work on the Golden Gate Bridge with the engineer Charles Alton Ellis provided an opportunity to make the longest suspension span in the world one that was supported by the thinnest truss decks seen up to that time. Thinner, lighter bridges were less expensive and easier to build. They were also far more beautiful.

His star had risen so high by the 1930s that the bridge builder Joseph Strauss found it necessary to recruit Moisseiff as a consulting engineer in order to obtain the commission as chief engineer of the Golden Gate Bridge.

When asked to be the chief engineer on a bridge for Washington state's Tacoma Narrows—his first and only such assignment—Moisseiff was warned that a limited budget of only $6.4 million was available for construction of the bridge. He also knew that a heavier and more expensive design had already been rejected. Moisseiff went to work with his calculations and came up with the thinnest and lightest long-span bridge of his career, and, more importantly, one that fell within the extremely tight budget, before it ultimately fell into the Tacoma Narrows. Although consulting engineers on the project worried whether a narrow stiffening girder would be enough to control the bridge's up and down movements, Moisseiff assured one and all it could. Unfortunately for the project, Moisseiff's prestige as a bridge builder was a powerful defoliant to criticism. Completed on July 4, 1940, the span lasted only four months, until a moderately strong wind of 42 mph demolished it on November 7, 1940.

Although exonerated of blame by a commission consisting of the bridge designers Othmar H. Ammann and Glenn B. Woodruff and the aerodynamics expert Theodore von Kármán, the 68-year-old Moisseiff now had a cloud over him. The commission held that Moisseiff's design was in line with prevailing bridge engineering practices (that Moisseiff himself had helped establish). Engineers who have investigated the matter have concluded the bridge was far less robust than it should have been. The famed and outspoken American bridge designer David B. Steinman went a step further: actively criticizing Moisseiff and other designers for foolishly turning a blind eye to the long history of potentially destructive effects of wind on flexible, long-span bridges.

Ammann also recognized that Moisseiff's lack of a complete understanding of aerodynamics was a problem confronting the entire civil engineering community during the 1930s. Ammann and Moisseiff had worked together building the Bronx-Whitestone Bridge in New York using the deflection theory. That bridge had a relatively narrow roadway and narrow solid plate girders for reinforcing the span as had the Tacoma Narrows Bridge. Ammann was later required to retrofit and strengthen the Bronx-Whitestone Bridge with truss guardrails running its length.

As a final blow to Moisseiff's drive for thinner and lighter structures, a February 9, 1938, gale made the road deck of the Golden Gate Bridge develop waves like a rug being shaken. This action was similar to the wind effects on the Tacoma Narrows Bridge. This unexpected behavior by the structure prompted the Golden Gate Bridge's refitting with additional reinforcement after another oscillation-producing gale in December 1951.

The engineering pendulum swung back the other way after the loss of the Tacoma Narrows Bridge, as engineers opted not only to design bridges with features that would allow them to pose less resistance to the wind, but also to make them stronger and heavier. As a comparison, the weight of New York's 1927 George Washington Bridge was 31,590 pounds for each foot of its center span, whereas the Tacoma Narrows Bridge—built 14 years later—boasted a mere 5,700 pounds of weight for each foot of its center span.

Moisseiff died in New York in 1943, his brilliant and successful bridge-designing career at a standstill after building what some called Tacoma's "jinx bridge."

See also BRIDGE AERODYNAMICS; MANHATTAN BRIDGE; TACOMA NARROWS BRIDGE.

Monier, Joseph (1823–1906) *Inventor of reinforced concrete* Joseph Monier was a maker of flowerpots and planting tubs who, as an entrepreneur, was looking for a way to improve his product. He experimented with cement, a building material rediscovered in England in 1824, but was disappointed that items made with it cracked easily. Monier was not about to give up on the gray powder that cured as hard as stone, so he continued tinkering.

Embedding iron wire in his cement flowerpots made them virtually unbreakable, and Monier patented his creation in 1867. Monier, the flowerpot maker, then jumped into the most exciting engineering field of the day—bridge-building. He decided—correctly—that reinforced concrete was an ideal material for building bridges. Ironically, Monier, who was smart enough to figure out how to turn concrete into a structural material, found himself shut out of designing bridges in France. Monier did not have a diploma from a French engineering school and could not legally design bridges on which the safety of hundreds might depend.

Undeterred, Monier sold his patent for *béton armé* (reinforced concrete) and his bridge ideas to the German engineering firm Wayss, Freytag and Schuster. The firm immediately began constructing bridges and buildings of all types across Central Europe, including the first reinforced concrete bridges in Bremen, Germany, and Wiggen, Switzerland.

A fellow Frenchman, François Hennebique, who conjured up his own method of concrete reinforcement, began constructing buildings with his method before obtaining a Belgian patent on his version in 1892. However, a legal battle that demonstrated that Monier's French patent superseded Hennebique's ensued. Nonetheless, Hennebique spawned a sizable international building firm based on his "Hennebique system" of concrete construction.

Monier's development of reinforced concrete is one of the most important innovations in construction history. The flowerpot maker had grasped how versatile reinforced concrete was and proposed its use in everything from piping to buildings. It would be nearly 50 years before another milestone in concrete innovation occurred, with the development of prestressed concrete in the early 20th century by Eugène Freyssinet.

From the iron-reinforced cement planting tubs of Monier sprouted a modern world of high-rise buildings and daring bridges.

See also CONCRETE; FREYSSINET, EUGÈNE; HENNEBIQUE, FRANÇOIS; PRESTRESSED CONCRETE; REINFORCED CONCRETE.

Mont Cenis Tunnel (built 1857–1871) *First tunnel through the Alps*   Also known as the Fréjus Tunnel, the Mont Cenis Tunnel was remarkable for the technological innovations it employed and for efforts to protect the health and safety of its 4,000 workers. It was also the first major tunnel to be driven from only two headings. Previous large-scale tunneling efforts were done with multiple intermediate headings accessed by numerous shafts along the length of the tunnel route.

Such access shafts sunk from above would not be possible since the 8.5-mile Mont Cenis tunnel would run 3,937 feet beneath the summit of the Cottic Alps. These shafts not only provide the ability to drive tunnels more quickly since crews can drill and blast through more opposing faces of the tunnel, accelerating the rate of work. Almost as important, access shafts also provide ventilation. This latter purpose is crucial inside the wet environment of tunnels, where heat can be oppressive and the atmosphere unhealthy, if not downright poisonous.

In short, the technical and safety challenges confronting the builders of the tunnel were immense but would be surmounted not only by the innovative brilliance of its chief engineer Germain Sommeiller but by his humanity.

A plan for tunneling between France and Italy through the Alps was first proposed in 1833 to the Kingdom of Sardinia, whose regions included not only the island of Sardinia but the area encompassing the Alps between what are now Italy and France. Engineers sneered at the plan, saying it was impossible to attempt so long a tunnel. Amid the usual challenges of drilling and blasting through nearly nine miles of rock, the major technical challenge was supplying fresh air. The opposing teams of miners would tunnel four miles from their respective portals, which were their only source of air.

The problem of fresh air was not a minor one. The workers would exhaust the oxygen and exhale carbon dioxide. In addition, gaslights for illumination would also consume oxygen. Black powder blasting conducted every day would add more carbon dioxide to the air along with acrid smoke and other gases. This stale air would then be added to heat generated by human exertion and machinery. Mix in the dust, humidity, and human airborne diseases such as tuberculosis and the health issues become a serious concern.

However, technical innovations continued, and the conversion of a steam rock drill invented by an English engineer, Thomas Bartlett, in 1854 provided a key innovation for excavating the Mont Cenis tunnel and for providing fresh air to miners. Originally designed to use steam, Bartlett's drill was modified by Sommeiller to use compressed air instead. The modification simplified the drill's operation and made it lighter, and the miners would welcome its exhaust gas—compressed air.

This innovation, along with the design prepared by Sommeiller and his assistant Severino Grattoni, were reviewed by a Sardinian commission who determined the technical hurdles to the tunnel project had been resolved. The Fréjus Act of June 29, 1857, provided 40 million francs in funding for the tunnel. Blasting began August 18 of the same year. At the time it was estimated that 20 years would be required to complete the project, meaning each heading would have to penetrate a minimum of three feet per day seven days a week. Thanks to technical innovations that Sommeiller himself developed, the project would take only 14 years.

Sommeiller knew the poor history of major European projects in respect to the welfare of workers and took steps to provide adequate housing and other facilities in the alpine villages near each portal. The shameful historical fact was that the arrival of hundreds of workers at most of the 19th century's great construction projects created tent cities and unsanitary conditions.

Outbreaks of cholera and other diseases often did more to kill workers than the hazards of a canal or tunnel project. The villages of Modane and Bardonécchia were at the east and west portals, respectively, and became the recipients of a building program to provide solid housing for the influx of workers. Schools were built, as were bathing facilities and community buildings. Along with his technical innovations and engineering accomplishments in designing and building the tunnel, Sommeiller's actions to create a healthy environment for mid-19th-century workers constituted another first.

Although Sommeiller and the Sardinian government had high hopes for rapid tunneling progress, the compressed air rock drills would not see use during the early phase of the project. The problem was the system for providing compressed air to the drills. The first system used the force of water traveling down a 164-foot pipe to compress air. This compressed air, entrained in the water, was bled off by valves and stored in tanks, to be sent by pipe into the tunnel. It was a centuries-old technology.

An insurmountable problem arose that doomed the system when a valve at the working face of the tunnel proved too weak. When the air-powered rock drills were not in use, the valve was to open and vent the air into the tunnel's atmosphere. During operation, the valve invariably became damaged, and because a durable enough metal had not been developed to remedy the problem, the system was deemed unusable.

Sommeiller stepped in to design a system using the piped water to turn waterwheels to operate piston compressors. This replacement system worked remarkably well and eventually made it possible for Sommeiller to begin using compressed air rock drills, although these problems delayed their use until four years after construction began.

Between 1857 and 1861, tunneling progress had been terribly slow. The tunnel progressed by means of blasting that involved the traditional hand-drilling of shot holes. Progress was painfully slow. During the years before incorporation of mechanical rock drills, the average distance the tunnel was advanced every 24 hours was a mere nine inches.

When Bartlett's drill bearing Sommeiller's changes was put into use on January 12, 1861, there was improvement in the rate of drilling, but not much. During 1861, the tunnel was advanced a disappointing 557 feet. Although it doubled the rate achieved by hand-drilling, it was barely worth the added cost of the equipment. The rates of advance kept improving, however, and by the end of 1862 crews working from the Bardonécchia portal penetrated 1,278.5 feet of rock, nearly five times the rate accomplished by hand-drilling.

Sommeiller kept busy adding new refinements to Bartlett's rock drill, making it even more efficient. By the middle of 1863, this improved drill was placed into the hands of the miners and the rates of advance soared. During 1864, miners excavated 2,038 feet of rock from the Bardonécchia side and 1,529 feet from the Modane portal.

The rates of advance improved as work progressed on the tunnel. Breathing the exhaust air from their drills, the miners moved through 2,917 feet of rock on the Bardonécchia end in 1870 while their colleagues drilled and blasted through 2,447 feet from the Modane side to the north. Whereas miners had recorded an average daily rate of only nine inches while drilling by hand, a record daily rate of 14.9 feet was achieved in 1870. Although these advance rates might horrify modern tunnel engineers, the accomplishments of Sommeiller and his miners were astonishing at the time.

Despite Sommeiller's eagerness to use any technical innovation available, one he chose to avoid was the newly patented explosive known as dynamite. Although far more powerful and safe than black powder, the Swedish chemist Alfred Nobel's 1863 invention would not be used on the Mont Cenis Tunnel. Until 1870, when actual excavation work was completed on the tunnel, black powder was the explosive of choice. The reasons for this were probably numerous. The properties and capabilities of black powder were well understood by Sommeiller's blasting crews, and Nobel's invention was still in its infancy. By the time excavation ended on the tunnel, at least 580 tons of black powder had been used in 2,954,000 shot holes. These explosions tore loose 960,000 cubic yards of rock, which was hauled, or mucked, out of the tunnel.

Sommeiller's other major innovation was to place from four to eight of the improved rock drills on a carriage that was rolled along a track as the mining progressed. This allowed the miners to move the carriage forward, adjust the position of the drills, and then hammer shot holes into the rock in a mass production process. Configuring the drills in this fashion became an innovation used in later decades. Devices of this type would later be known as jumbos.

Directly behind the drillers and blasters were hundreds of other miners widening the pilot headings. Farther back were 90 more men and boys lining and reinforcing the tunnel's roof and walls with masonry. This reinforcement effort was supported by hundreds of others who transported building material to them from outside the tunnel.

As the tunnel advanced, the problems of ventilation became more pronounced. Despite the operation of the compressed air system, which sent air to the working faces of both sides of the tunnel, temperatures had increased inside to nearly 90 degrees Fahrenheit, and it was far more difficult to move the stale air from the tunnel.

To combat this problem, what can only be described as a fireplace was installed above the southern end of the tunnel to create a draft through a hole in the tunnel roof. Since this idea was as bad as it sounds, it failed to improve the air in the tunnel, and a 32-inch fan was then put in place. This was used to blow air into the lower part of the tunnel while the stale air was drawn through the upper half. Exhaust air was ventilated from the northern side by pumps. Conditions were never ideal, but the workers survived.

The tunnel project was considered so important that not even the 1859 Italian War between the Franco-Sardinian alliance and Austria was allowed to stall its progress. Even the peace that followed that divided the tunnel between France and Sardinia was not a major problem. At war's end Sardinia's borders were redrawn, giving Modane to France, which agreed to fund half the tunnel's cost.

Everyone wanted the tunnel. In 1867, a decade after work began, the Italian government contracted with Sommeiller and Grattoni to complete the tunnel at a fixed price for each meter excavated. The work continued as Sommeiller pushed himself incessantly to resolve technical issues while managing the massive project.

On Christmas Day 1870, workers managed to blast through the last bit of rock separating the opposing tunnels to form a single passageway through the Alps. The tunnel, which was by far the longest built up to that time, had tested the ability of the surveyors and engineers to connect the two excavations. On the day of the breakthrough, it was found that the tunnels were out of alignment by only 18 inches, an extraordinary achievement for the time.

As the project neared completion, it would be a triumph Sommeiller would not savor. The engineer died on July 11, 1871. In an odd footnote to this massive engineering accomplishment, Sommeiller became a nonentity during the September 17, 1871, ceremonies inaugurating the tunnel. Sommeiller, whose inventiveness and dedication made the tunnel possible, was not mentioned by any of the French or Italian dignitaries toasting the tunnel.

Sommeiller shared an even grimmer distinction with the other famed alpine tunnel builders of the 19th century who died before the completion of their tunnels. Louis Favre, the chief engineer of the Saint Gotthard Tunnel, and Alfred Brandt, chief engineer of the Simplon Tunnel, died during before their own works were completed.

See also BRANDT, ALFRED; BROOKLYN-BATTERY TUNNEL; FAVRE, LUIS; ROCK DRILL; SAINT GOTTHARD TUNNEL; SIMPLON TUNNEL; SOMMEILLER, GERMAIN.

Moses, Robert M. (1888–1981) *New York's imperious builder* Robert Moses began his public service career in New York City as a brilliant and idealistic reformer dedicated to replacing political patronage with a merit-based hiring system. The naïve Moses saw this effort swatted down by corrupt Tammany Hall politicians, learning an indelible lesson about political power. Moses persisted in his efforts at reforming the city's operation and developed a reputation as the enemy of vested interests and corruption. In 1933, Fiorello La Guardia appointed Moses both chief of the New York City Park Commission and head of the Triborough Bridge Authority.

Unfortunately, Moses learned the lessons of power so well he eventually became a symbol of the type of unbridled power he once opposed, power to which he himself became addicted. He craftily expanded his authority and made the Triborough Bridge and Tunnel Authority a self-funding entity answerable to virtually no one. Moses moved from park development to roadway construction and eventually into the development of massive public housing projects in the 1950s. During the 35 years that Moses held 14 public positions and secured his power, he had overseen the building of more than 600 parks and the city's most impressive bridges. Moses, the champion of the little man and the principles of democracy, was always appointed by politicians and never elected to a single position he held.

He pushed through 16 major expressways and the construction of seven of New York's most important bridges, including the massive Verrazano Narrows Bridge, which was the longest when completed in 1964. The length of the span across the Verrazano Narrows would not be exceeded until the construction of Britain's Humber Bridge 17 years later. Among the 10 bridges he pushed into existence were the Throgs Neck Bridge, Cross Bay Veteran's Memorial, Henry Hudson, Bronx-Whitestone, Marine Parkway Gil Hodges Memorial, and Triborough Bridges. One of these 10 water crossings, the Robert Moses Causeway, carries his name. By the end of the 1990s, the bridges and tunnels that were once under his control were carrying 750,000 cars each day.

Unyielding and unafraid, Moses was masterful at securing political power. He carefully crafted a public persona as an incorruptible public servant and amassed so much influence that his nominal bosses, the governors and mayors of the state and city, were forced to pay homage to Moses, an action that politicians seldom saw reciprocated. Moses, who lived to be 92, is now remembered as one of the most vindictive public officials in New York City's history.

The politically deft Moses made one fatal error, and that was to make an enemy of one of the few men in America who had the power to thwart him. That man was Franklin Delano Roosevelt. Moses had a disdain of the patrician Roosevelt and initially poisoned their relationship by opposing the then New York governor's efforts in the 1920s to establish a wooded park in upstate New York.

In the fourth year of the Great Depression, Moses was solidly in power at what was then the Triborough Bridge Authority, but the election of Roosevelt to the presidency in 1932 left him vulnerable. In 1934, Roosevelt demanded Moses's removal from the Triborough Bridge Authority in exchange for Public Works Administration money that was desperately needed by the city. Moses stood his ground with La Guardia, who was clueless as to how deeply Roosevelt hated the mayor's park commissioner until threatened with the loss of federal money. If fired, Moses coldly announced, he would go public and smear La Guardia with allegations that he was subject to outside pressure. The threat forced La Guardia to stand up to Roosevelt's bid to unseat Moses, but the acrimony between the president and the man known to many as the "Boss" or "RM" lived on.

To the public, Moses crafted an image of himself as a selfless public works hero building parks for youths and providing expressways for growing numbers of automobiles. Moses went to bat for the people as he stood firm against the corrupt and callous political status quo. But Moses had been seduced by power and by the expansiveness of sweeping public works projects, and he clung to power by any method at hand, including character assassination and a cozy relationship with journalists and editors.

Moses did not like vehicular tunnels, which he believed were too expensive and carried too little traffic. The Moses biographer Robert Caro pointed out in his brilliant *The Power Broker* that Moses did not consider tunnels suitable monuments to his own ambition; bridges were far grander symbols. Moses waged bitter and unsuccessful fights against both the Queens-Midtown Tunnel (completed in 1940) and the Brooklyn-Battery Tunnel (completed in 1950).

Moses's fight against the Brooklyn-Battery Tunnel put him once again into Roosevelt's political gun sight, and this time the president craftily made sure his fingerprints were not on the weapon. Moses's

counterproposal of an ugly view-blocking bridge was assassinated by Roosevelt's secretary of war, Harry H. Woodring, who ruled (falsely) that a bridge at the entrance to the East River could pose a potential hazard to navigation in wartime. In the end, however, Moses acquired control of both these tunnels when the New York City Tunnel Authority fell under the Triborough Bridge Authority after World War II.

Moses fell into disfavor when his infrastructure projects, which were conducted with virtually no oversight or control, began to create blight in the very areas Moses claimed he was trying to rejuvenate. Massive expressways and bridge approaches amputated neighborhoods while encouraging a heavy flow of automobile traffic into Manhattan. Oddly enough, Moses never drove a car and therefore never paid a toll on one of his many toll roads. The other sin that soured the public on his legacy was Moses's construction of monolithic housing projects that proved to be not only social failures but eyesores as well.

In 1962, the 73-year-old Moses was stunned when Governor Nelson Rockefeller refused to provide a customary dispensation from the mandatory retirement age of 65. This effectively prevented Moses from retaining his position as chief of the State Council of Parks. An angry Moses then resigned his other municipal park positions in a bid to force Rockefeller to back down, something the millionaire politician did not do. Moses ended up holding only two jobs, as director of the Triborough Bridge and Tunnel Authority and as president of the 1964 World's Fair Corporation.

By 1968, Moses's political clout had waned further, and the newly formed Metropolitan Transit Authority would become the Triborough Bridge and Tunnel Authority's parent agency. Although Moses had acquiesced to the change after being promised a ranking position on the board of the new agency, the appointment did not materialize and the 79-year-old Moses was left without a public position for the first time in 44 years.

See also AMMANN, OTHMAR HERMANN; BROOKLYN-BATTERY TUNNEL; QUEENS-MIDTOWN TUNNEL; SINGSTAD, OLE; VERRAZANO NARROWS BRIDGE.

Mostar Bridge (built 1557–1566) *Victim of the Bosnian civil war* Bridges are perhaps the most symbolic of structures since they unite people by overcoming obstacles to travel, but the Mostar

A symbol of both divisiveness and unity for 427 years, the Mostar Bridge in Bosnia and Herzegovina was vindictively blasted into rubble by Croat troops in 1993 during fighting between Christians and Muslims. The bridge that linked the Christian and Muslim sections of the city of the same name was to be rebuilt by 2004 through an international effort. *(Robert Cortright, Bridge Ink)*

Bridge built by the Ottoman Empire in the Bosnia and Hercegovina town of Mostar may be the most symbolic of them all. This venerable bridge reflects both the unity and the division of the region.

Although the bridge is commonly known as the Mostar Bridge, its name is actually derived from the words *Stari Most,* or "Old Bridge." Over the years the name shrank to *Mostar,* and the town whose east and west sections it united drew its name from the bridge.

The Mostar Bridge was built during the 16th century, when the Ottoman Empire was at the pinnacle of its power under the ruler Suleiman the Magnificent. The Muslim Ottomans demanded not only taxes from the Balkan territories but also non-Muslim Christian children, who became slaves in the service of the empire. This conscription of youths between five and 18 years of age was known as *devshirme* and targeted those in the vassal Balkan states. This human levy figures prominently

in both the construction and the symbology of the bridge.

Under Suleiman's reign architecture and art flourished, and the architect Najrudin was commanded to design a permanent stone bridge to replace a less reliable wooden span at the Neretva River. Najrudin's teacher was Mimar (Architect) Sinan, one of the world's most accomplished civil engineers of the period. Sinan was himself a non-Muslim forced as a child into Ottoman service through the devshirme system, which elevated former Christians into high-ranking positions within the empire's bureaucracy, though its primary purpose was to ensure subservience in its territories.

In 1538, Suleiman appointed Sinan as chief architect of the Ottoman court. Sinan lived to be more than 100 and is credited with designing hundreds of daring mosques, bridges, and public buildings. It is Sinan's architecture that made Istanbul the architecturally beautiful place it is today. And it was

Sinan who taught Najrudin, and it would be Najrudin who would build the Mostar Bridge.

Since Sinan was in his prime during construction of the Mostar Bridge, it is possible that he and Najrudin discussed its design. Even before it was built, the bridge attained a level of symbolism for the Christians in the Balkans, who alternately loathed and welcomed the devshirme as both curse and opportunity. The bridge demonstrated Ottoman power and skill, as well as the knowledge of the former Christian youth Sinan through his student Najrudin.

The traditional bridging location of the Neretva River was only 92 feet wide, but this distance was not easy to bridge with a single arch in the 16th century. Najrudin's solution was to build a steep arch that rose 69 feet from the surface of the river while the springings of the arch rested upon massive stone abutments built into the banks of both shores. Najrudin built the bridge of relatively soft limestone, which was easy to quarry yet strong enough to accomplish the task.

The bridge took nine years to complete. It was rumored that Najrudin became so nervous that the arch would collapse once its supporting falsework was removed that he fled the city. It is equally likely that Najrudin knew exactly what he was doing, despite the daring height of the arch, and calmly watched the falsework removed. Like most bridges built at the time, the Mostar was equipped with a tower to deter enemy forces from crossing the bridge. Because of the eastern orientation of the Ottoman Empire, the tower protected the east bank from the west.

The bridge endured centuries of floodwaters, wars, and strife, and its stark white beauty became a favorite subject for painters and sketch artists. The local population composed of Muslims and Christians came to love the bridge and referred to it for hundreds of years as the "Old Man." One writer has described the bridge as a white necklace surrounding the blue neck of the Neretva.

With the fall of the Soviet Union, the former Yugoslavia disintegrated as its member states sought to form their own nations. Bosnia and Herzegovina seceded in 1991, and the new nation soon fragmented into conflict among its Muslims, Catholic Croats, and Orthodox Serbians. Centuries of anger and recrimination among the groups boiled over with dreadful and deadly consequences that included genocide and indiscriminate killing.

In Mostar, the Muslims formed up on the east of the Neretva and the Croats to its west. As the fighting raged, those who loved the old bridge tried to protect it from small arms fire and shrapnel with scaffolding and a protective shroud of old tires.

On November 9, 1993, Croat troops, claiming Muslim fighters were using the bridge, directed tank cannon fire at the venerable span. After a few rounds of high-velocity antitank ammunition struck the arch, the 427-year-old bridge collapsed into the waters of the Neretva. The bridge that had been not only a symbol but also a physical means of unification was not available to the people it served for the first time in nearly five centuries.

When fighting in the region subsided, outraged historians, officials, and ordinary citizens decided to act. With the arrival of United Nations troops an effort began in 1997 to recover the stones from the riverbed using Hungarian army divers and combat engineers. In the meantime, a temporary suspension span was placed at the site of the original bridge to reconnect sides of the city. The effort to recover the stonework using divers and pontoon-mounted cranes was completed in 1998.

By 2000, a team of experts had been assembled to study how the bridge was originally built in order to direct its accurate reconstruction. The effort to rebuild the bridge with its original materials has been promised financial support from not only the World Bank and the United Nations but also Turkey and private sponsors. The reconstruction of the bridge was expected to take until 2004 and cost at least $5 million.

See also ABUTMENT; ARCH BRIDGE.

movable bridges　Although not the most beautiful or exciting types of bridges, lift, swing, and bascule bridges are arguably among the most interesting. The spans of movable bridges have traditionally moved in one of three ways. In the case of a bascule bridge (often commonly referred to as a drawbridge) each span or leaf is tipped upward to create an opening. Lift bridges use machinery and cables to raise homogenous segments upward in an elevator fashion between towers. The swing bridge merely pivots at its center or on its shore end so that its movable span is placed parallel to the waterway.

There is something a bit awesome, if not slightly unsettling, about seeing a rather large bridge carrying rail or auto traffic one minute, then moved out of the way the next. Movable bridges have existed for centuries, although the earliest types were very modest designs that allowed people to swing a bridge out over a waterway by hand. The swing bridges, like other movable bridges, were cheap alternatives to constructing high arch bridges over traffic-carrying waterways.

Eventually, swing bridges developed into mighty and complex mechanisms. The movable span of a swing bridge is mounted on a massive pin that is turned by a motor-driven gear system. Braking systems are often employed on the larger swing bridges so the bridge tender can slow the bridge span during opening and closing to prevent damage. Although swing bridges open rather slowly, their weight can generate a huge amount of inertia that has to be controlled. The operation requires skill since the larger versions of these spans must be moved with great care. The weight of the leaves can be extraordinary since the leaf is essentially a cantilever span requiring solid and heavy construction to support its own weight and that of the traffic it carries. Each of the two movable leaves of Seattle's Southwest Spokane Street Swing Bridge weighs 7,500 tons, nearly equal to the weight of a modern U.S. Navy destroyer. However, smaller swing bridges are designed for foot traffic alone and can be found crossing small creeks and bayous.

Another type of swing bridge is one supported by a massive pivot at its center, allowing a horizontal span to turn perpendicular to the stationary roadway or rail deck of a bridge. This type of swing bridge is usually located in large shipping channels and allows two-way maritime traffic to pass on each side of the pivoted span.

In this 1944 photograph are a vertical lift bridge with its movable segment in the up position and a railway swing bridge with its movable span turned in the open position. A backdrop for these bridges over New Jersey's Passaic River is a fixed, cantilever truss bridge. *(National Archives, New Jersey Department of Transportation)*

Movable bridges, particularly large ones, are usually manned 24 hours a day, adding to the cost of operation. This expense is usually outweighed by the necessity of keeping a waterway open for commercial shipping traffic and by the elimination of the need for a bridge design that might have cost millions more to build. Some smaller swing bridges that are large enough for automobiles still operate by using the muscle power of a bridge tender. The Snow-Reed Bridge in Fort Lauderdale, Florida, is still opened and closed 20 times daily by a bridge tender using a hand crank. The 48-foot bridge was built in 1925 and is named after a pair of former Fort Lauderdale mayors.

See also SEVEN MILE BRIDGE.

mucking When mining or tunneling, the proper term for the spoil or earth being removed is *muck,* and the term for transporting it from the excavation is *mucking* or *mucking out.*

Although the British term *mucking about* is often used to describe someone who is puttering or otherwise spending time unwisely, mucking is one of the most complicated and expensive parts of tunneling. Although a great deal of excitement if not romance surrounds the process of digging, drilling, and blasting galleries beneath the earth, the banal truth remains that what is excavated must be moved aboveground for disposal. This process requires careful planning and a sizable amount of money and is equal in importance to the actual digging in terms of completing a project.

Traditionally, earth was carried out by horse-drawn cart and later by rail cars pulled by small locomotives. When tunnel-boring machines came into use, a conveyor belt system was used to move the earth to the rear of the massive machine. Muck, also known as spoil, can be mixed with water to form slurry for removal by pumping it out of a tunnel through pipes. This was the solution seized upon by the French while they constructed their portion of the Eurotunnel, also known as the Channel Tunnel.

N

National Bridge Inspection Standards *Response to the Silver Bridge collapse* Sailors say that U.S. Navy regulations are "written in blood," based as they are on fatal lessons; the same might be said of the decision by the federal government to implement a program that would conduct a systematic and detailed assessment of America's bridges. Established by law in 1968, the National Bridge Inspection Standards were mostly remarkable in that they had not existed before.

The deadly catalyst that forced America to take a hard look at its bridges was the shocking 1967 collapse of the Point Pleasant Bridge (commonly known as the Silver Bridge), which killed 46 people near Point Pleasant, West Virginia.

Incredibly, most of the bridges in the United States were not inspected on a regular or detailed basis even by the late 1960s. America's municipalities, which had experienced an unprecedented road and bridge-building boom to coincide with the explosive increase in automobile traffic, built bridges as fast as possible beginning in the early 1900s and did not look back. Bridges were fairly solid structures and could pretty much take care of themselves, or at least that is what everyone believed until the 2,235-foot Silver Bridge collapsed.

The design of the Silver Bridge between Point Pleasant, West Virginia, and Gallipolis, Ohio, relied on trusted technologies used for more than a century. The truss-reinforced bridge was a suspension span supported not by wire cables but by steel eyebars linked by pins and running down each side of the bridge. A pair of massive concrete piers supported two steel towers over which the steel eyebar chains were draped. An identical bridge was built across the Ohio River at Saint Marys, West Virginia.

Completed in 1928, the Silver Bridge served without incident for nearly 40 years. On December 15, 1967, that unblemished history of service ended without warning when it crumpled into the water while occupied by 37 vehicles. Of those cars and trucks, 31 tumbled into the Ohio River, killing 46 occupants, another 18 persons where rescued from the waters around the collapsed span. The incident stands as the deadliest bridge collapse in America's history. It was also the largest bridge to collapse since the Tacoma Narrows Bridge failure 27 years earlier, a catastrophe that claimed only one life, that of a pet dog.

The nation was stunned by the Silver Bridge event, a catastrophe that at first seemed to have no explanation. It was the first bridge incident investigated by the newly formed National Transportation Safety Board, an agency that ordered the components of the bridge retrieved from the river and reassembled on shore. What soon became apparent was that one of the eyebar links had developed a crack where its eye was pinned to that of another eyebar. The fatigue crack, which had grown worse during the intervening years, finally gave way. Since the real strength of a suspension span lies in its cables, the failure of the single eyebar caused the bridge to lose its structural integrity.

The resulting controversy over the tragedy focused America's attention on the terrible fact that America's municipalities were not conducting

These twisted remnants of the Silver Bridge suspension span between Ohio and West Virginia are an eloquent indictment of America's failure to inspect and maintain its bridges properly. The 1967 collapse of this bridge over the Ohio River killed 46 and prompted federal rules requiring states to inspect and report the condition of their bridges. *(Federal Highway Administration)*

regular inventories or assessments of their bridges. Engineers pointed out that America's older bridges were carrying far heavier loads than they were designed for by the 1960s, and that many were in such poor shape that they required load-limit warnings. America, preoccupied as it had been with building bridges, discovered it had roughly a half-million bridges of all types and that a shockingly high percentage were in distressingly poor shape.

Within a year of the disaster, the Federal Highway Administration was authorized by the Federal-Aid Highway Act of 1968 to establish National Bridge Inspection Standards to be applied in a program that would inspect bridges on Federal-Aid

roadways. Federal law broadened the scope of the program in 1978 to include all bridges over 20 feet in length.

The comprehensive standards, to be used by all states, spelled out in minute detail how bridges would be inspected. Under the law, the states would have to provide a complete inventory of their bridges every two years along with an assessment of their condition based on these standards. The program also gave the Federal Highway Administration authority to require that all inspectors would be evaluated and certified as to their competency. This compilation of bridges and their condition is known as the National Bridge Inventory.

The goals of the National Bridge Inspection Standards are to keep track of those bridges that pose a serious hazard to the public and to repair any structural problems quickly to prevent another tragedy like that at The Silver Bridge.

When it became apparent how bad America's bridges were, the Federal-Aid Highway Act of 1970 was enacted to establish the Special Bridge Replacement Program. This program made money available for the emergency replacement of important bridges considered unsafe.

The program has continued with the Federal Highway Administration, within the U.S. Department of Transportation, which promotes research into ways to build better and stronger bridges and testing methods to increase their useful lifespan.

Despite these efforts, America's inventory of bridges has been subjected to increasing levels of use and virtually every state has failed to resolve fully the problem of deficient bridges. As a result of the National Bridge Inspection Standards program, data are now available to reveal the condition of America's bridges. And, despite some progress, the news continues to be bad.

Better Roads, a magazine for roadway contractors and public highway officials, has published the results of the National Bridge Inventory each year since 1979. For 2000, the survey showed that America had 592,091 bridges of which 167,321 were either structurally deficient or obsolete. The deficient bridges amount to 28.3 percent of the nation's spans. However, matters did appear to be getting better, though not much. In 1997, a total of 179,455 bridges were considered unsound and obsolete. In 1998, the total was 178,092, and in 1999 unfit bridges numbered 171,272.

As of 2000, the states with the largest number of deficient and obsolete bridges were Texas with 12,707, Ohio with 11,791, Oklahoma with 9,129, and Pennsylvania with 9,116. However, the numbers can be misleading since only 26 percent of Texas's 48,832 bridges are considered deficient, just below the national average. Ohio, which has the next largest number of bridges with 30,351, has deficient bridges amounting to 39 percent of its total.

Positioned at almost opposite sides of the globe are two of America's states with the highest percentage of deficient and obsolete bridges. A sobering 62 percent of tiny Rhode Island's 750 bridges needed repair or replacement. Of Hawaii's 1,137 bridges, 48 percent were considered substandard. Arizona scored best under the federal bridge inspection standards with only 5 percent of its 6,769 bridges considered deficient.

As for the Silver Bridge, the Silver Memorial Bridge replaced it in 1969. The new bridge's cantilever truss design eliminated the eyebar suspension chains of its predecessor. The Silver Bridge's sister span at Saint Marys was demolished soon after the collapse of the Silver Bridge, as a result of fears that it could also collapse through a similar structural failure. In 1977, it was also replaced with a cantilever truss span.

See also BROWN, SAMUEL; SUSPENSION BRIDGE; TACOMA NARROWS BRIDGE.

National Bridge Inventory *See* National Bridge Inspection Standards.

National Redoubt *Switzerland's underground defense against Hitler* If you think that all the Swiss make are milk chocolate, watches, and distasteful banking deals with Nazis, then come for a walk along the crags and valleys of the Swiss Alps. If your eyesight is sharp enough and you are especially lucky, you may spot one of thousands of Swiss bunkers and fortresses dug into solid rock during World War II. In all likelihood you will not see a thing: These expertly camouflaged strongpoints—which finally became known to the world 50 years after World War II—remain largely secret.

This defensive system was not a single monolithic line of fortifications but a series of small, medium, and large forts from which the Swiss intended to mount defensive operations and counterattacks against an anticipated German invasion. Having watched Adolph Hitler invade Czechoslovakia, Poland, France, Belgium, Holland, and Norway, the Swiss decided they were probably next. The tiny nation embarked in 1940 on a program to fortify the borders of their mountainous nation in a project as secret as it was massive.

Although a geographically small country of only 5 million, Switzerland was superbly equipped to construct underground defenses since its engineers had honed their skills building some of the most impressive railway tunnels in the world.

It took thousands of workers digging and pouring concrete seven days a week, 24 hours a day, to complete the fortifications. Tremendous secrecy surrounded the project during its construction as local Swiss were barred from approaching the bunkers and tunnels. Nearly 10,000 small bunkers, 70 larger fortified strongpoints, and three massive command and control centers were placed underground.

Dug into mountainsides overlooking valleys and other likely avenues of attack, the bunkers and larger fortresses provided protection, rallying points, and supplies for thousands of Swiss troops. Underground kitchens, dining halls, hospitals, and ammunition dumps were concealed deep within the rocky innards of the Alps. It is estimated that the fortifications, if constructed in the 1990s, would cost the equivalent of 10 billion American dollars.

A look at a map reveals the unfortunate geographic situation in which the historically neutral Swiss found themselves. Bellicose Nazi Germany frowned down on Switzerland's northern border, and German-occupied France lay helpless on its western border. To the south was the Axis partner Italy, and on its eastern border sat Austria, which had acquiesced to being part of Hitler's empire.

Switzerland's highest ranking officer, General Henrí Guisan, realized roughly half of Switzerland's lower valleys and plains would have to be abandoned to a Nazi advance since his forces lacked the tanks and aircraft to fight a head-on battle with the Wehrmacht and the Panzer divisions. Guisan decided not to fortify Switzerland's borders but to build a National Redoubt in the heart of the rugged Alps that bisected Switzerland. He gathered his troops in an alpine valley and made them swear an oath to defend Switzerland to the last bullet.

This National Redoubt would seek protection from German air power by burrowing deep into the ground. Multistory subterranean installations would protect the troops until they could decimate the feared Nazi invasion with artillery, mortar, and sniper fire. Aircraft were secreted inside caves and tunnels and artillery positions were hidden inside phony chalets.

Although modern commanders with helicopters, paratroopers, and mobile armored units generally sneer at fixed fortifications, fighting across the mountainous terrain of the Alps to dig out determined troops protected by tunnels and underground bunkers would have been a costly exercise for the Germans.

Nearly five decades after the end of World War II, some of the underground forts have been opened to the public, who can see the water squeezing through the rock walls of the underground facilities into drainage channels along the concrete floors.

Fully equipped kitchens and hospitals in some underground strongpoints still await use in time of war. However, the Swiss government is in the process of selling some of the complexes as it constructs newer underground facilities elsewhere and concentrates on rebuilding its military into a mobile armored force that is less dependent on fixed fortifications.

Although Germany and Italy both developed war plans to invade and carve up Switzerland, this never took place. Germany's hesitation to invade a nation despised by Hitler has historians speculating whether the oldest forms of shelter—tunnels and caves—may have played a role in preventing an invasion by one of the world's most modern and ruthless war machines.

See also MAGINOT LINE.

new Austrian tunneling method *Building tunnels with sprayed concrete* Time is money in construction, and the primary chores in major tunneling efforts involve excavation and reinforcement, both monumentally expensive undertakings. The history of tunneling is the story of finding faster ways to dig and reinforce the excavated cavities. Traditional tunneling evolved from hand excavation to blasting and then progressed to machine excavation with efficient tunnel-boring machines. Despite these advances in excavation technology, reinforcing the shafts and galleries with timber followed by steel continued into the first half of the 20th century.

Soft ground tunneling bedeviled underground workers because tunnels tended to deform under the weight of the ground above and were so unstable that they required immediate reinforcement with timber or metal supports. Hard rock tunnels, seemingly solid, can also be prone to deformation and collapse. To preserve an excavated chamber, not to mention the lives of the tunnelers, reinforcement was often required to ensure the stability of the rock. Although the roof and walls of a rock tunnel might not completely collapse, the movement of surrounding rock could result in rock bursts or rock falls that could kill and cripple workers.

One of the most important and controversial innovations in tunneling was the use of shotcrete, concrete pumped into steel reinforcement, on the walls and roof of tunnels to create immediate reinforcement.

Carl Ethan Akeley's 1907 invention, the cement gun for helping the taxidermist create better animal models, made available a tool for rapidly applying the relatively inexpensive but powerfully strong cement mixture known as concrete using air pressure. The cement mixture originally sprayed with Akeley's invention became known by the trade name *Gunite* and was recognized as an excellent way of stabilizing soil above ground on hillsides. Swimming pool construction was another beneficiary of the technology.

Shotcrete was used in underground tunneling as early as the 1920s, but its full potential as a tool for structural reinforcement was not realized until 1954. This took place when the Austrian tunneling engineer Anton Brunner proposed its use to control the squeezing ground closing in on a 26-foot-diameter water diversion tunnel. Brunner believed that sprayed concrete could rapidly form a supportive lining strong enough to tame the movement of the ground.

Although the process was approved for use in the Runserau hydroelectric plant's diversion tunnel, Brunner's own firm was so uninterested in the process that it left the matter of submitting the process for a patent in Brunner's hands. Since Brunner's shotcrete lining technique had proved successful on the Runserau project, several other tunnels were subsequently reinforced in the same way. At the time the process of rapidly reinforcing tunnels with sprayed concrete became known as the shotcrete method. Brunner tirelessly pushed the use of shotcrete as a tunnel support method, but it would take another decade before it was fully accepted.

By 1962, the Austrian engineering professor Ladislaus von Rabcewicz had coined the term *new Austrian tunneling method* (NATM) in a professional paper on the technique, and by 1964 he and a colleague, Leopold Müller, were overseeing use of this method in the Schwaikheim Tunnel in Germany. Its use in that tunnel generated considerable interest in the technique, which had been refined to include several principles.

The new Austrian tunneling method requires a careful analysis of the ground through which the tunnel is being driven, followed by the application of immediate reinforcement, which may consist of shotcrete, rock bolts, or iron lining rings. A final lining, which often consists of curved prefabricated concrete segments, follows this initial reinforcement.

The soft ground subaqueous tunnel excavated beneath the Thames in London by Marc Brunel during the early 19th century was the first successful tunneling effort in such unstable ground. Using an iron tunneling shield that protected digging works at its front, Brunel's system also allowed brickmasons at the rear to build up a reinforced lining immediately. Although more highly refined, the new Austrian tunneling method follows the same principles of rapidly reinforcing an excavation.

NATM provided an excellent means of tunneling beneath soft ground in urban areas, where engineers needed a way of burrowing subway tunnels beneath busy city streets without disrupting traffic using the cut-and-cover method of tunneling. NATM was chosen for the construction of Frankfurt's subway system expansion in soft clay in 1968 and proved successful. At least 120 tunnels using NATM were constructed in West Germany after 1980 for use by high-speed trains. One of the first large-scale uses of NATM in soft ground conditions in the United States took place during the construction of an extension of the Washington Metropolitan Area Transit Authority subway line beginning in 1983.

To keep the technology economical, careful analysis must be done to determine the types of initial and secondary reinforcement required by a tunnel. Overestimating the amount of material and the type of reinforcement needed could result in unnecessary costs. Failure to appreciate fully the weakness of a section of ground could result in the application of inadequate preliminary reinforcement. The result of the latter miscalculation could be a collapse. If the ground is extremely unstable, it may be necessary to pump cement grout into the surrounding ground or even to freeze the soil until the shotcrete lining can be applied.

NATM takes not only expertise but also a substantial amount of engineering confidence, something that not all engineers have. Dr. Gerhard Sauer, a leading consultant on NATM, points out that some engineers become unnerved by soft ground conditions and abandon the new Austrian tunneling method for more conventional techniques.

Even more dangerous is the improper use of NATM, which can expose workers to sudden collapse of a tunnel in soft ground conditions. The 1994 collapse of a trio of subterranean rail tunnels at Heathrow Airport is an example of a failure to assess ground conditions properly to allow the use of adequate reinforcement.

See also AKELEY, CARL ETHAN; HEATHROW TUNNEL COLLAPSE; EUROTUNNEL; SHOTCRETE.

New London Bridge (built 1823–1831) *Sinking bridge moved to Arizona* Completed in 1209, Old London Bridge with its numerous narrow arches constricted the flow of the Thames, creating a hazard to boats. Incredibly, the bridge continued to carry London's traffic for more than 620 years. A new span had been needed for centuries, but not until the 19th century were officials willing to spend the money to build one.

London turned to John Rennie (1761–1821), an engineer of Scottish descent, who was orphaned at the age of five. He became a millwright and mechanic at a young age because of his penchant for playing hooky and was finally persuaded to attend high school, where his mathematical brilliance was discovered. He subsequently graduated from the University of Edinburgh and became a civil engineer directing efforts to drain marshlands, designing canals, and constructing harbors. He then turned his attention to the building of bridges, designing the Waterloo Bridge (built 1810–1817) and Southwark Bridge (built 1814–1819).

The Southwark Bridge was a tremendous accomplishment since a trio of iron arches with a central span measuring 240 feet supported it. By the early 19th century Rennie was considered one of Britain's preeminent civil engineers and an obvious choice for determining whether the Old London Bridge should be renovated or replaced. After studying the medieval bridge, Rennie told city officials it could be made usable but at a very high price and strongly recommended construction of a new bridge. His findings were debated at length, but both the officials and the public were eventually swayed by his arguments and Rennie was commissioned to design a new bridge.

A bill was passed to allow construction of the bridge in 1823, but this approval came two years after Rennie's death and Rennie's son, John Rennie II, took on the task of completing the project.

The bridge was to be supported by five semielliptical arches spanning 650 feet across the river. The central arch measured 150 feet and the adjacent arches 140 feet each. The shoreward side spans would each measure 130 feet. Combined with its approach spans, the bridge would measure 928 feet.

The bridge began carrying traffic in 1831, and soon afterward the massive stone bulk of Old London Bridge adjacent to the new bridge was dismantled. Soon what became known as the "New London Bridge" was as much a fixture of London as the old one, but it would not last nearly as long. By the 1920s, it became apparent that the bridge was inadequately supported by its foundations and was sinking into the soft clay of the riverbed.

Nonetheless, the heavily used bridge remained in service for another four decades. It survived the German bombs that rained upon London during World War II as well as an increasing amount of traffic. But the settling of the bridge into the Thames riverbed continued. By the mid-1960s the decision was made to replace the bridge with a modern span of concrete.

It was then that the bridge's fate took a very strange turn. An Arizona developer, Robert P. McCulloch, was looking for a gimmick to make people notice his Lake Havasu City development and decided the New London Bridge was the answer. Although transporting the entire bridge—foundations and all—would have been impossible, McCulloch paid $2.5 million for its granite facing. McCulloch would spend an additional $7 million to ship it to America, build an actual bridge superstructure, and attach the granite.

McCulloch's development is located on the Arizona shore of Lake Havasu, which was formed when the Colorado River was dammed by the federal government. The bridge was built from the Arizona shore of Lake Havasu to an island containing Lake Havasu State Park. The New London Bridge, which spent 131 years in the fog and damp of London, now basks in the perennial sunshine of Arizona, a two-hour drive from Death Valley National Park.

As a promotional move, McCulloch's reconstruction of the New London Bridge struck pay dirt in 1972, when a network television special featuring the bridge as its main cast member was aired. Hosted by the Welsh singer Tom Jones and such singing stars as the Carpenters, the "London Bridge Special" ensured that just about everyone heard of the desert development and its refugee bridge.

See also LONDON BRIDGE; OLD LONDON BRIDGE.

New Madrid earthquakes *See* seismic retrofit.

New River Gorge Bridge *See* Bayonne Bridge.

New York Tunnel Authority *See* Queens-Midtown Tunnel.

Niagara Gorge Suspension Bridge (built 1854–1855) *Dramatic and rare railway suspension bridge* Despite a handful of successful suspension bridges in America and Europe by the middle of the 19th century, the public was still hesitant to trust the design type. Suspension spans, including the more successful ones, had proved themselves less than trustworthy. Some collapsed under marching troops; the road decks of others were shaken into wreckage by strong winds.

Nonetheless, the promise of the suspension span type—a stronger bridge requiring fewer piers and far less structural material—was a concept proved by calculations. Even those willing to gamble on a suspension bridge as a roadway for pedestrians or horse-drawn wagons were loath to believe a suspension span, which was flexible by its nature, could not only support the weight of a train but also endure the movement and vibration it would transmit.

But it was a railroad span that was desired over the Niagara Gorge between Canada and New York, and a suspension bridge would be the type chosen for the mission. The 250-foot-deep gorge just down from Niagara Falls made it virtually impossible to construct intermediate piers in the river for a traditional arch bridge or a timber bridge. A single span supported by suspension cables was the answer, and in 1847 the tempestuous and brilliant engineer Charles Ellet, Jr., was hired to design and build the bridge.

The temperamental Ellet became embroiled in fiscal disputes with the New York Bridge Company and the Niagara Falls Suspension Bridge Company. A lawsuit followed, and Ellet abandoned the effort after receiving a court-ordered cash settlement. Ellet's departure in December 1848 left behind a rudimentary pedestrian bridge and a pair of bridge companies still eager for a railway span across the breathtaking gorge. The bridge companies turned to John Augustus Roebling, an engineer whose design had initially been rejected in favor of Ellet's.

Fortunately for the project, Ellet's departure opened the door for a man who was probably far more qualified to build the bridge. Though far less famous than Ellet during the 1840s, Roebling had been not only building suspension bridges but perfecting engineering remedies to make them resist aerodynamic forces, the Achilles heel of the spans that were far lighter than their stone and iron arch counterparts.

This late 19th-century view of John Augustus Roebling's Niagara Gorge Suspension Bridge belies the wild surroundings over which the daring railway span was built. Below the twin-track upper rail deck is a roadway for carriages and pedestrians, and 300 feet below the lower deck is the turbulent Niagara River. *(Author's Collection)*

In 1851, Roebling began building a double-deck roadway designed to carry rail traffic above and a lower level for carriage and pedestrian traffic. Though he originally wanted to have a single roadway, this design was deemed impractical and the double-deck model was adopted.

The span would be the first suspension bridge to incorporate truss reinforcement for its decking. The triangular-shaped reinforcement running down both sides of the timber bridge would provide both stiffness and weight to the span, which would measure 820 feet between the walls of the gorge. Besides weight and stiffness, Roebling's third essential element would be the cabling in the form of suspension cables and diagonal cable stays. A pair of towers erected on each side of the gorge would each be crowned with saddles to accommodate four suspension cables measuring 10 inches apiece. Each of these cables was made up of 3,640 wires.

The suspension cables would then be connected to the deck of the bridge by iron suspenders and diagonal stays that would radiate from the towers at an angle and attach to the deck to brace the whole arrangement. Another 56 cable stays would extend from the lower part of the deck to be attached to the stone on the opposing cliffs. The Niagara Bridge was so heavily reinforced with cable stays that some considered it a hybrid between a suspension span and a cable-stayed bridge. Roebling, respectful of the ability of wind to lift and destroy a bridge deck, was taking no chances.

Roebling, ever the thrifty German, salvaged the high-quality iron wire originally used by Ellet's adjacent walkway and incorporated it into his suspension cables while noting that the oil coating had ably preserved the wires in the damp environment. The stone towers were erected, and massive masonry anchorages were built on each shore. With the completion of these, Roebling began spinning the great lengths of iron wire that were first dipped in hot linseed oil. With the cables in place the massive timbers of the bridge decks were constructed and then hung from the suspension cables by suspenders.

Although the work progressed remarkably well, the job was not to be without its mishaps. In July 1854, cholera struck the area, killing 60 people during the first seven days of the outbreak. When Roebling, who took an active role in treating the sick, began to experience symptoms of the illness, he remained up all night willing away the infection.

The second tragedy to strike was the October 10, 1854, collapse of scaffolding at the bridge site, causing the deaths of two men. A terrible loss, these were the only deaths recorded on the job site, a total that is fairly miraculous, considering that workers spent four years building the bridge, nearly 300 feet above the Niagara River thundering below.

After the completion of the bridge in March 1855, a test crossing was made with a train. Limited to 8 mph, the train made its way across Roebling's unyielding suspension bridge, silencing critics of his stance that suspension bridges were capable of carrying heavy rail traffic. Soon afterward, trains were huffing on the upper deck, and carriages and pedestrians crossing on the lower deck.

Roebling wrote extensively about his bridge at Niagara and was so committed to the suspension design that he encouraged its use for bridges of all types, including railway spans. Many, however, remained cautious about utilizing suspension bridges for rail traffic since the loads and vibration involved are so much greater those generated by wagons and carriages. In 1856, a Roebling article was published in Britain encouraging the construction of railway suspension bridges measuring a mile in length using high-quality steel wire in lieu of iron.

Roebling pressed for a pure suspension span that would be far cheaper and lighter to build. Robert Stephenson, designer of the Britannia Bridge who took Roebling and his reputation seriously, replied: "If your bridge succeeds, then mine have been magnificent blunders." Stephenson did not imply that his bridges would collapse but was admitting his designs might be proved unnecessarily expensive in terms of cost and materials in comparison with suspension span designs.

The popularity of the Niagara span as a rail and wagon crossing had a progressive effect on the bridge's structural integrity. Vibration, heavier loading, and the temporary nature of a timber structure meant repairs became more common with the passage of time. The cabling was refurbished and the anchorages reinforced in 1877 to accommodate the additional stress placed upon the bridge. In 1886, iron suspension towers designed by the famed American bridge engineer Leffert Lefferts Buck were built to replace the stone ones. By 1897, it appeared that those critical of a suspension bridge for train use were right when the bridge was declared unfit for service and demolished.

In defense of Roebling, his biographer David Steinman pointed out that the Niagara Gorge Suspension Bridge had outlasted many of the other spans of the period. Today, nothing remains of Roebling's Niagara bridge but vintage photographs. And, despite the success of the much-criticized bridge, only a small number of suspension bridges today carry rail traffic.

See also BRITANNIA BRIDGE; BUCK, LEFFERT LEFFERTS; CABLE-STAYED BRIDGE; ELLET, CHARLES, JR.; HOLLOW-BOX GIRDER; ROEBLING, JOHN AUGUSTUS; STEPHENSON, ROBERT; SUSPENSION BRIDGE; TACOMA NARROWS BRIDGE; TIMBER BRIDGE.

nitroglycerin   *See* dynamite.

Nobel, Alfred (1833–1896) *Inventor of dynamite*
Dynamite and its inventor, Alfred Nobel, are inextricably linked to tunneling and mining. Born to a self-taught engineer named Immanuel Nobel and his wife, Andrietta, in Stockholm, Nobel was reared and educated in Saint Petersburg, Russia, where his father invented antiship mines and participated in numerous enterprises including a tunneling project.

Nobel, who spoke four languages besides Swedish, was later sent to Paris to study by his tough, self-made father, who lacked any university training. The Nobel family, beset by financial reversals in Russia, later returned to Sweden. While in Paris studying chemistry, Nobel met Ascanio Sobrero, the Italian chemist who in 1846 invented the explosive nitroglycerin. The explosive liquid was the most potent explosive of its time but prone to accidental detonation if exposed to heat or vibration. Sobrero had been disfigured by an explosion of his chemical creation during the late 1840s. Sobrero often lamented the hazardous nature of the chemical he introduced, which was nonetheless put to use despite the risks it posed.

After this introduction to Sobrero and nitroglycerin, Nobel was sent to work for four years in America with the Swedish-American mechanical and naval engineer John Ericcson, in a sort of apprenticeship that marked the training of European engineers during the 19th century. Ericcson had much to teach. The brilliant engineer's feats were numerous, including his innovative creation of the ironclad warship U.S.S. *Monitor* with its futuristic revolving turret. Ericcson's other contributions included development of the propeller, which revolutionized ship propulsion.

However, the Nobel family had become intrigued with the promise of nitroglycerin as an industrial explosive and began to manufacture the temperamental substance. In 1863, Nobel returned to Stockholm to help in this effort. Sadly, the cantankerous amber liquid had no friends, and in 1864 Nobel's brother, Emil, and four workers died in a nitroglycerin explosion. The blast horrified Stockholm officials, who banished the operation to a barge in the middle of a lake.

Nobel redoubled his efforts to find a way to make nitroglycerin safe. In 1866, the resourceful Nobel discovered that by mixing the explosive with a form of clay he could make nitroglycerin insensitive to shock and heat, thus moderating its instability. By 1875, he created a gelatinous form of dynamite that could be detonated underwater. His equally important invention of the blasting cap in 1865 made it possible to detonate the dynamite reliably under virtually any conditions.

Dynamite was rapidly incorporated into mining and tunneling projects since it provided a potent explosive excavation tool with a minimum of hazards. Dynamite was widely used in the construction of Transcontinental Railroad tunnels during the 1870s through the rugged Sierra Nevada mountain range.

Nobel wandered throughout Europe as something of a scientific nomad working at the various laboratories of his far-flung business empire and contributing to the development of modern firearm propellants and military munitions. Before his death in 1896 he had acquired 355 patents for inventions and innovations. Although dynamite was nearly a panacea for excavation work demanded by the Industrial Revolution, Nobel became distressed when he realized he might be remembered only for his explosives work. In the year before his death in Italy, Nobel drafted a will bequeathing his sizable wealth to the formation of a committee that would award prizes for various forms of scientific and literary accomplishment as well as for efforts promoting world peace.

See also DYNAMITE; HOOSAC TUNNEL; MONT CENIS TUNNEL.

Old London Bridge (built 1176–1209) By the 12th century it was decided that a permanent stone bridge should be built across the Thames, and the man for the task was a Catholic priest and bridge builder known as Peter de Colechurch. Father Peter had spent his life rebuilding an existing wooden span known as London Bridge that had spanned the Thames off and on since A.D. 43.

In 1176, Father Peter was asked to build a stone bridge across the river capable of resisting tides, floating ice, or anything else that had damaged or destroyed previous bridges. The task was daunting, considering the technology available in 12th-century Britain. Nonetheless, Father Peter planned what turned out to be one of history's most successful bridges, although its disruption of the Thames's tidal flow menaced river traffic until it was demolished after an astounding 622 years of use.

Father Peter's education as a bridge designer came about with the establishment of the Saint Mary Overy Church, which supported an order of priests who studied the art of bridge-building. One of the first bridges constructed by Father Peter's predecessors was a timber span across the Thames, to replace another that had undoubtedly been lost for one reason or another. Built in 993, the bridge may have been destroyed by the Viking warrior king Olaf Haraldson in 1014 and was made famous in the nursery rhyme "London Bridge Is Falling Down."

The origin of the church, and the order of bridge-building priests it supported, is rooted partly in humorous myth. As the story goes, the penny-pinching owner of a Thames ferry operation named John Audrey pretended to be dead to invoke a ritual of fasting among his gluttonous relatives and employees. Audrey feigned death in an attempt to save on a day's worth of groceries but silently fumed when instead of solemn fasting he heard his family gorging on a bounteous funeral wake. No longer able to endure his family's profligacy, he arose from his bier to protest. Believing the resurrected Audrey to be a demon, a stunned reveler grabbed a nearby oar and delivered a fatal blow to the conniving tightwad, ensuring that Audrey was not late for his own funeral. Audrey's daughter, Mary, is said to have suffered not only the loss of her father but the death of her fiancé, who died in a fall from his horse while riding to the side of his grieving bride-to-be. Distraught, Mary Audrey is said to have made a gift of her father's estate to found the Saint Mary Overy priory and the order Father Peter later joined. It is said she then entered a convent. However it happened, the church (which today is the Southwark Cathedral) was founded and Father Peter and his fellow priests were Britain's premier bridge experts.

What Father Peter faced was construction of a bridge across nearly 1,000 feet of a river that rose and fell up to 19 feet with the North Sea tides. Since the Thames was an important waterway for transportation, the arches of the bridge would have to be tall and wide enough for boat traffic, and a drawbridge segment would be required for bigger ships.

With donations from various sources, including the Catholic church in England, work on the bridge began and would continue haltingly for 33 years.

Father Peter's engineering plan was to construct a stone arch bridge, arch by arch across the river.

The method of creating foundations, or piers, for the bridge was an ancient one. Wooden pilings were driven into the muck of the Thames in an ovoid shape until their tops were even with low-tide level of the river. These enclosures were then filled with stone, creating a foundation for the bridge piers built with masonry above the waterline. When a pair of foundations were completed, workers could then construct a single arch. With fair weather and adequate funds, a single arch could be constructed in a year's time. Unfortunately for the struggle to build the bridge, wars, a lack of money, and uneven official interest in the project delayed its completion.

As the bridge progressed slowly and not always surely, its design became more and more interesting. For one thing, the piers were far from uniform in shape and size. Although most are roughly oval, similar to an overhead view of a boat hull, some are as short as 36 feet and others are as long as 54 feet. A massive pier that was the ninth one from the London side of the Thames was a prodigious piece of work measuring 95 feet long and 36 feet wide. Eventually, a total of 19 arches were built.

This pier—nearly twice the size of the others—would be the foundation for a two-story chapel dedicated to the murdered archbishop of Canterbury, Thomas à Becket, a Catholic clergyman who had opposed King Henry II's attempts to reduce the political influence of his church. Before the death of Henry II in 1180, the monarch instituted a tax on wool to assist the financing of the bridge. This led to the public jest that the bridge was built on foundations of wool. Father Peter, who labored tirelessly to see the bridge completed despite the length of the project, died in 1205, an event that threatened to throw the project into disarray. Father Peter's body was interred within the chapel on the unfinished bridge to which he had dedicated most of his adult life.

King John then recommended that a French bridge expert named Isembert take over the task of completing the job. The record is unclear on exactly who did what, but a trio of businessmen, William Almaine, Serle Mercer, and Benedict Botewrite, were also recommended as comanagers of the project. It is entirely possible the work was completed with the businessmen at the helm as they sought technical advice from Isembert.

When all 19 arches were in place, the effect on the waterway was immense. The too-narrow openings of the arches—ranging from 34 feet to 15 feet along a 936-foot distance—served as nozzles through which water jetted to create a set of rapids that may have drowned as many as 50 boatmen yearly. An old English saying described the situation succinctly: "London Bridge is intended for wise men to go over and fools to go under."

When the bridge was completed, it also contained battlements on both its entrances since enemies and rebels could use it to reach the heart of London. Towers stood high above the bridge along with gates that could be closed at a moment's notice. Like other bridges during this period, the London Bridge became more than a mere way of getting across the Thames. Its wide roadway, besides providing space for a church, also became the location of businesses and homes. This made the bridge a destination as well as a conduit of transport.

Although the bridge was made of stone and was essentially fireproof, the buildings on its roadway were not. In 1212, only three years after its completion, a fire broke out in London and spread to the bridge's buildings. Many drowned when they leaped from the bridge into the Thames to escape incineration.

Maintenance became a problem soon after the bridge was completed, mainly because of the bizarre way governance by royal decree worked. In 1269, King Henry III assigned rent money collected from bridge tenants to his wife, Eleanor, who squandered it without spending a farthing on repairs. Not wanting to complain about his queen while he was alive, London's city officials raised the issue of the bridge's condition after his death in 1272 to King Edward I, who assigned the city the task of keeping the bridge maintained. Maintenance costs mounted, and the first tolls ever charged on the bridge were levied in 1281: Pedestrians paid a farthing (one fourth of a penny), and horsemen had to plunk down a full penny.

Disaster struck the bridge's buildings during the winter of 1281, when a massive amount of ice caused five of the 19 arches to collapse. Badly needed repairs to the bridge were financed by a complex toll formula levied against goods crossing the bridge. Even the poor boatmen, who had spent 72 years fighting their way against the bridge-induced rapids, must have blinked (or worse) when they were also assessed a toll for merely passing under the bridge that had killed so many of their colleagues.

Because the bridge was so heavily traveled, it also served as something of a brutal advertising medium for the government, which made it a policy to mount the heads and other body parts of traitors on the north gate leading to the drawbridge, gaining it the name of "Traitor's Gate." One of the more notable heads to occupy the gate was that of Sir Thomas More. His head was impaled on the gate July 6, 1535, and later recovered by his daughter, who was buried with it.

Despite living down the street from a display of severed and decomposing human body parts—a practice halted in 1678—the bridge was considered a top-drawer address mostly because running water was nearby. (One needed only to lower a bucket to the river.) The proximity of the river also made the nettlesome problem of garbage and sewage disposal a snap. In true medieval fashion, everything simply went straight into the river. In the days before running water and indoor plumbing, life on the bridge was a veritable breath of fresh air. For those bobbing below in a boat trying to dodge garbage and sewage, this method of disposal may have been somewhat bothersome. (The river would remain little more than an open sewer until the late 19th century.)

Despite the condition of the water, the bridge became something of a public utility in 1580 when a Dutch entrepreneur, Peter Moris, obtained a 500-year lease to install water wheels at one end of the bridge to pump this delicious water to nearby homes. The business thrived and was pumping water to 10,000 homes and businesses by 1701.

Moris's enterprise triggered an era in which the Old London Bridge became an industrial location with small water-wheel-powered mills based on the foundations, and during the 1600s houses gave way to businesses and shops. For the ensuing 100 years the miniature world of Old London Bridge bustled along.

However, the sun was setting on Father Peter's magnificent bridge. Time had taken its toll, as had efforts to widen and renovate the span. After 500 years of operation the bridge had been widened twice, once in the late 1600s and again in the 1750s. The waters of the Thames were made so wild by the constrictive arches that some boat passengers disembarked and walked along the bank before reboarding to complete their journey. In an effort to reduce the turbulence caused by the bridge, two of its center arches were consolidated into one, but this only served to create a scouring effect on the riverbed that began to undermine the piers. The bridge, which had stood as an unshakable monolith, was now threatened with collapse. To remedy the scour problem, so many tons of protective stone or riprap were dumped around the submerged foundations that they posed a hazard to boats navigating the river.

In addition, what was a silk-stocking address for nearly 600 years had, by the 18th century, become what today's realtors call a "distressed property." London's expansion meant the bridge's space was no longer essential, and what homes and shops remained were rundown and unprofitable. The process of demolishing the buildings on the bridge began in 1757. By this time, of the 19 original arches, seven had been completely rebuilt and three had been removed.

By 1821, all were agreed the Old London Bridge had to go to make way for a new bridge designed by John Rennie, Sr. The new bridge was completed in 1831, and the demolition of the existing bridge was completed that year as well. During the removal of the 600-year-old bridge, workers discovered that many of the elm piles driven into the mud for construction of the foundations were as solid as the day they were cut down. Father Peter's bridge had stood the test of time and just about everything else.

See also ARCH BRIDGE; NEW LONDON BRIDGE; PILE; RIPRAP; SCOUR; STIRLING BRIDGE.

O'Shaughnessy, Michael Maurice *See* Golden Gate Bridge.

P

Paris Métro   *See* cut-and-cover tunneling.

Pauli truss   *See* Smithfield Street Bridge.

pier   The term *pier* is most commonly associated with a horizontal platform extending out over the water to be used for fishing or the docking of boats. For bridge engineers, however, the term refers to a vertical support structure that bears the weight of a bridge's superstructure and roadway. A pier can be constructed of timber, masonry, steel, or concrete.

Although the term is often used to describe the entire vertical support of a bridge, a pier is actually composed of three major elements: the pier itself, the foundation upon which it rests, and the footing upon which the foundation sits beneath the ground or water. Such footings, made of rock or timber in early times, gave way to masonry structures and poured concrete. These footings were usually placed against bedrock or extremely stable soil when possible and were made far larger than the footprint of the pier itself to distribute the weight of the bridge over a larger area. The foundation was then built upon the footing, which then accepted the vertical column, or pier, that upheld the roadway. The entire structure is usually referred to as the pier.

A steel pier extending from the foundation might be bolted to fittings in concrete or masonry. Masonry piers might be laid atop the foundation until they reached the appropriate height. A concrete pier might be connected to the foundation and the footing with steel reinforcement and then poured as a homogenous unit atop the foundation.

A lone Canadian soldier standing guard at the Quebec Bridge during World War I offers an excellent indication of the massiveness of the masonry piers supporting the cantilevers of this gargantuan bridge. *(National Archives of Canada)*

229

pile   A pile may be a column of timber, concrete, or steel driven hammerlike into the earth beneath the water or on dry land. Protruding from the surface of the water or ground, piles serve as piers directly supporting a bridge span. This is an extremely rapid method of erection and was used in the construction of Louisiana's Lake Pontchartrain Causeway, the world's longest bridge. Piles used in this way are referred to as bearing piles. The tops of bearing piles themselves are fitted with a tee, or a common cap, to allow distribution of the weight of a bridge segment. A pile driven at an angle to provide lateral support for a row of vertical bearing piles is referred to as a batter pile.

Piles are frequently used to form the framework of a cofferdam submerged in water during the construction of bridge piers or foundations. This box of sheet steel or timber may or may not be pumped dry before it is filled with concrete or rock to serve as the foundation for a pier.

End-bearing piles are those driven completely into ground that is relatively soft until they reach bedrock, or at least a type of soil firm enough to support the weight of a bridge. These buried piles then serve as the base, or footing, of a foundation that is built overhead. Although it seems incongruous, wooden piles can actually last hundreds of years if fully buried. Wooden piles can even be driven into rock if a steel punch has generated a pilot hole for their entry. Wooden pilings were used to lock concrete foundations to the submerged bedrock during construction of the Seven Mile Bridge in the Florida Keys. This was done by driving the pilings into the limestone so they protruded into the forms of the poured concrete piers.

Engineers have used piles for thousands of years to assist in bridge construction, but the method of driving them into the earth has changed. As man devised clever methods to harness human and animal muscle power, they went from hammering relatively small piles in by hand to doing the same chore with pulleys that lifted and dropped stones onto the piles. When done the old way, pile-driving was an agonizingly slow process.

Mechanization, including the use of pneumatic pile drivers that lift and drop massive weights 50 times a minute, has made it possible for piles dozens of feet in length to be pounded into place in hours instead of days. Pile-driving is a construction art in itself and is usually practiced by firms that specialize in the craft.

pipe-jacking   Microtunneling technology, or the digging of small tunnels less than six feet in diameter, has become a major industry as firms place sewer, water distribution, and telecommunications lines beneath heavily developed urban areas. Once the hole is dug, lining the miniature tunnel is often accomplished by the technique of pipe-jacking.

The process involves the construction of a jacking pit where the pipe sections enter the tunnel and hydraulic jacks push the pipe through the ground. New sections are added, and the jacking continues until the line is completed or until the piping is connected to another segment already in place.

One of the enemies of the pipe-jacking process is friction, which can limit the distance pipe can be pushed through the soil from a jacking station. Remedies for this problem include the addition of bentonite slurry or other types of lubricants to reduce friction. If friction is too great or the pipe sections are too long, intermediate jacking stations are excavated and the pipe is pushed through sequentially.

This technology is similar to tunnel-jacking, or the shoving of an entire, large-bore tunnel through the ground. Although potentially more expensive, the jacking of pipe and tunnels is a cost-effective way of tunneling without disturbing aboveground activities, offsetting the cheaper but disruptive method of laying pipe in an open trench.

See also CENTRAL ARTERY/TUNNEL PROJECT; MICROTUNNELING; TUNNEL-BORING MACHINE.

pneumatic caisson   *See* caisson.

pneumatic transit   *See* Beach, Alfred Ely.

Pont du Gard   The passion of the Romans for conquest was equaled only by their love of running water and bathing, a pastime that prompted some of the Roman Empire's most impressive engineering and construction feats. The Roman addiction to running water made Rome not only the masters of the "known world" but master plumbers as well. The Pont du Gard, an aqueduct project considered a milestone in Roman Empire engineering, continues to awe those who visit it in southern France more than 20 centuries after it was built.

Although most Roman water supply engineering took place at ground level in the form of canals or below ground with subterranean pipes and tunnels, the need to carry water across deep valleys required that Rome's accomplished engineers build aqueducts. These water-carrying bridges transported water for drinking and sanitation as well as for the luxurious public baths and the famed fountains of Roman cities.

One of the most beautiful and well preserved of the Roman aqueducts is that built over the Gard River in the Provence region of southeast France. This picturesque locale is packed with history and was originally inhabited by the Celts. It was occupied by the Greeks in 600 B.C. and later by Phoenicians. The Romans arrived in the region more than a century before the birth of Christ. The city of Namausus, 65 miles north of Marseilles, fell under Roman control in 121 B.C., and its inhabitants were granted Roman citizenship in 28 B.C. The city, which is today called Nîmes, was among the wealthiest of the Roman settlements in the south of France.

During the Roman era, Nîmes became a favored retirement area for soldiers who had fulfilled their duty in the famed legions of the Roman Army. A major reason the city was a bustling center of commerce was its location astride a major land route between northern Italy and Rome's colonies in Spain. It was only a matter of time before the city's growth, and the Roman penchant for fresh water, outstripped the existing source of spring water inside the city. Roman engineers were soon seeking an additional source of water and a way of transporting it to what is now Nîmes.

This project called for the city to be connected with a freshwater supply that, unfortunately, existed some 30 miles to the northeast. The water percolated through a limestone spring at a locale known as the Fountain of the Eure near the village of Uzés. Although canals could carry the water much of the distance, the deep Gard River gorge presented an

The magnificent Pont du Gard aqueduct near Nîmes, France, has endured for nearly 2,000 years. Built with Roman engineering and French limestone, it carried water through its flume, or water channel, for hundreds of years. After centuries of abuse and neglect, the aqueduct is now protected by the French government and has been restored to its original configuration. *(Robert Cortright, Bridge Ink)*

unavoidable obstacle. Although the water could be easily channeled down the gorge, it would have been impractical to pump it back up the other side. The answer would be an aqueduct, the longest and tallest ever built by Roman engineers. It would have to rise 160 feet above the floor of the valley and span a distance of more than 900 feet.

The decision to span the Gard River gorge with an aqueduct was made more than 20 years before the birth of Jesus Christ. The idea of building the massive aqueduct was that of Marcus Vipsanius Agrippa, who was one of Rome's most able soldiers and administrators. It was Agrippa who had consolidated Rome's hold on Gaul, as France, Belgium, and parts of Germany were then known.

Agrippa, like the aqueduct whose construction he ordered, was a fascinating bit of Roman history on his own, having served as both general and admiral. He backed construction projects throughout the empire and was a lifelong friend and supporter of Emperor Augustus, the adopted son of Julius Caesar. Although of common birth, Agrippa rose to a level of considerable influence thanks in no small part to his closeness with Augustus and his own skill as a soldier and administrator. Agrippa used his vast personal fortune to help finance construction of Rome's original Pantheon (later rebuilt by Emperor Hadrian and dedicated to the popular Agrippa) and promoted infrastructure projects, including roads and Rome's famed public baths. His other notable achievement was that his military expertise had been instrumental in the 31 B.C. naval defeat of the rebel forces of Mark Antony and Cleopatra in the political turmoil that followed the murder of Julius Caesar.

The Pont du Gard (pont is the French word for "bridge") was a crucial element of a water transport system that wound from the artesian springs near the mountain village of Uzés to Nîmes. Roman engineers used canals, tunnels, slightly elevated water troughs, and several smaller aqueducts to traverse the Massif Central, a mountainous region in southern France. Roman engineers depended on gravity to assist the flow of the water to its destination, and their task was made exceedingly difficult by the source of the water, only 55 feet above the elevation of Nîmes. This required the engineers to plan every foot of the aqueduct carefully to ensure it had a constant gradient leading to its destination.

Although the name of the aqueduct's designer has been lost to history, only an administrator with the clout and ability of Agrippa could jump-start the concept into a reality. However, it is not exactly clear when the Pont du Gard was built, although its construction may have been completed around 19 B.C. The entire water project may have remained under construction for nearly 100 years until the reign of Emperor Trajan, which lasted from A.D. 98 to 117.

The Pont du Gard is a superb model of Roman engineering and is considered the finest remaining example of a Roman aqueduct not only because its construction is extraordinary but because it has survived the centuries in relatively good condition. Designed in three tiers supporting three levels of arches, the bridge is constructed of locally quarried limestone blocks, some weighing as much as six tons. Visitors to the area can still see the two nearby quarries where limestone was cut and shaped for the aqueduct.

The dimensions of the aqueduct are impressive even by today's standards. As a result of the natural V shape of the Gardon Valley that it spans, the tiers of the Pont du Gard are of three different lengths. The first tier in 465 feet long and supported by six arches measuring 72 feet in height. The second tier, which also contains a roadway, is 794 feet in length and is supported by 11 arches that are 65 1/2 feet high. The third and highest tier of the Pont du Gard reaches 902 feet across the Gardon Valley. This span rides upon 35 smaller arches measuring 23 feet in height. This uppermost tier contains the raison d'être of the structure—a cement-lined water trough that rises just over 160 feet above the valley's floor. This conduit has six-foot-high walls and is four feet wide.

The piers supporting each arch of the aqueduct are singularly slender. This design is in defiance of Roman engineering tradition but is completely adequate structurally and gives Pont du Gard a exceedingly thin appearance. Ordinary Roman design dictated that piers be at least one third as thick as the length of the arch span. The piers of Pont du Gard have an almost airy 1/5 ratio. The designers of the bridge apparently knew that discarding the traditional and heavier pier design would not affect strength. A 1958 flood raised the level of the Gard River to the top of the first tier of arches, destroyed nearby modern bridges, and killed 35 locals, but the Pont du Gard was unscathed.

The workmanship of the Pont du Gard is superb, with the stone cut and placed with such tight tolerances that it was unnecessary to use cement between their joints. Placement instructions can still be seen carved into the rough face of the limestone blocks along with messages and graffiti left by those who built the bridge. Stone blocks jutting from the sides of the arches might appear to a passerby to be mistakes in the stonework, but this feature is a common element of Roman bridge construction. These appendages were intentionally installed to support wooden scaffolding and falseworks for assembling the arches that would be completed only when the voussoirs, or the wedge-shaped stones of the arch and the final keystone, were in place. As an integral part of the aqueduct, these jutting stones served to reduce the requirement for timber during construction and provided handy supports for scaffolding during subsequent repairs. When operational during the Roman period, the Pont du Gard's trough carried 44 million gallons of water daily to Nîmes. This amounted to roughly 105 gallons for each person in Nîmes, a figure that is 25 gallons higher than average daily water consumption in a modern American home.

A major reason for the longevity of the Pont du Gard is that Roman engineers were able to place the bridge's foundations directly on bedrock. However, after 2,000 years of service some of the foundations were undermined by the hydrologic action of the Gard River in a process known as scour. This foundation damage has been repaired, and engineers are monitoring the condition of the foundations on a regular basis.

Although it is in remarkable condition, Pont du Gard has suffered its share of physical insults during the intervening centuries. Great sections of the 30-mile water transport system—which fell into disuse with the disintegration of the Roman Empire—were dismantled and the stone components converted into local building materials. Pont du Gard was kept intact since it was also pressed into service as a pedestrian crossing over the Gard River at least since A.D. 1295. Locals chiseled away at the piers of the second tier around this period to make it easier for traffic to pass. This and other ill-considered renovations almost caused the aqueduct to collapse.

To provide more room for traffic, the arches of the first level were widened between 1743 and 1745 by the engineer Henri Pitot and some of the stonework damaged in the previous centuries was repaired. Beloved by the local population and revered by engineers and historians, the Pont du Gard was added to a list of historically significant structures by the French government in 1840. Between 1843–1845 and 1855–1857, additional work was done to restore the aqueduct's structural integrity and original appearance.

Provence, a region of France already popular to sightseers both local and foreign, counts the Pont du Gard among its most famous attractions. In recent years an estimated 2 million people annually have seen the remarkable structure. Traditionally, tourists were allowed to clamber to the top of the open water conduit for a closer look. This practice was halted temporarily in 1992, when it was decided to conduct a massive renovation of the bridge that continued in several stages until completed in the summer of 2000.

During this work, masons replaced 800 tons of limestone in the bridge, chiseled in the same fashion as the original stone, and will renew the grout in the flagstone lining of the elevated channel to prevent rainwater from streaming down through the arches below. Despite the massive amount of stone replaced in the bridge, it amounts to only one twentieth of the entire volume of stone in the imposing span.

The aqueduct is considered so important historically that in 1985 the United Nations Education, Scientific, and Cultural Organization (UNESCO) also listed it as a World Heritage Site.

See also ARCH BRIDGE; FALSEWORK; SCOUR; VIADUCT.

pontoon bridge For nearly 3,000 years soldiers and sometimes civilians in need of an efficient crossing that could be built in great haste have depended upon the pontoon bridge as a way of keeping their feet and their war machines dry.

The pontoon bridge concept combines boats or floats known as pontoons to support a roadway, making the bridge as much a boat as it is a roadway. As most for ancient bridging technologies, no one is quite sure who built the first pontoon bridge, but its initial use is often credited to the Iraq-Persia region perhaps as far back as 800 years before the birth of Jesus.

The Assyrians are credited with building some of the earliest pontoon bridges as they conducted successful campaigns to build their empire in what was then known as Mesopotamia. Accounts have the Assyrians using inflated animal hides and small boats to provide flotation to their pontoon bridges. The neighboring Persians, with their own military

U.S. Army vehicles roll across a pontoon bridge erected over a fast-moving river near Branau, Austria, in the spring of 1945. Buoyed by inflatable pontoons, the prefabricated metal decking serves as a roadway. To the left is a truss arch bridge probably wrecked by retreating German troops. *(National Archives)*

ambitions and hordes of cavalry, also developed pontoon bridges and built an extraordinary one around 480 B.C. To move the Persian leader Xerxes' army across the Dardanelles for the invasion of Greece, military engineers are said to have lashed a roadway across the decks of nearly 700 boats.

All this was happening while Rome was just organizing itself into a powerful city-state. By the time Jesus was born, the Romans were the preeminent military force in the Mediterranean and had long since mastered pontoon bridge technology. It is believed that one of the original bridges across the Thames River in London was a pontoon span thrown across the tidal river by the Romans around A.D. 43, when they arrived to establish their garrison city of Londinium.

The ability to bridge terrific distances made pontoon bridges a tool of military commanders for centuries. Napoleon employed the floating bridges in his far flung campaigns, and during the Civil War both the Union and Confederate armies were busy erecting pontoon bridges where permanent spans had been destroyed or where no bridge had ever existed. Although there is no way of knowing precisely how many pontoon bridges were erected during the Civil War, the number probably ran into the thousands.

After the war some pontoon bridges that had been left in place were allowed to remain and were maintained for many years until more permanent crossings could be erected. The obvious problem posed by pontoon bridges was that they blocked river traffic, although this was sometimes advantageous in war. Pontoon bridges could also be vulnerable to flood conditions when a river's level rose too quickly and its flow was too rapid. Nonetheless, by World War I, Allied and German armies were both using specialized pontoon-bridging equipment.

By World War II, pontoon bridges that used not only rigid boats as the pontoons but also inflatable rubber boats specially designed for the task were developed.

Deflated for transport, the rubber semirigid boats could be quickly inflated and floated into position, sometimes by using a small outboard motor. During the Allied drive into Germany these bridges were often erected under intense enemy fire but were capable of supporting the weight of tanks and fully loaded supply trucks. A cabling system was used to connect the decking and provide extra stability to the bridge. The deck also distributed the weight of the crossing vehicles among several pontoons so one section of the bridge would not sink too deeply.

When U.S. Army tank forces were ordered into Bosnia in 1995 for peacekeeping duty, they had to await the construction of a pontoon bridge across the Sava River by combat engineers since the original permanent bridge had been destroyed. The 2,043-foot pontoon bridge was the longest pontoon span built by the Army since World War II and accepted the weight of 70-ton M1A1 tanks that moved at 3 mph across the bridge. To allow the movement of river traffic, sections of the pontoon bridge were removed during brief periods to create an opening.

There are a handful of full-time pontoon bridges still in service in the United States, including several notable examples in Washington state. Although far more substantial than military pontoon bridges and designed to carry ordinary vehicular traffic, a pair of parallel floating or pontoon bridges cross Lake Washington in Washington state, the Lacey V. Murrow Bridge and the Homer M. Hadley Bridge.

See also ARMORED VEHICLE LAUNCHED BRIDGE.

Pontypridd Bridge (built 1750–1756) When you hear the story of the 18th-century Pontypridd Bridge it is actually a bit comical—it is the one about the bridge built by a self-taught stonemason in South Wales that may have collapsed twice before being completed. And, when completed, it was too steep for use by wagons or carriages. Seeing the gracefully proud bridge and learning it was once Britain's longest unsupported stone span give the bridge and its design a bit more gravity, something that this bridge has been resisting for nearly 250 years.

The bridge is a beautiful structure designed and built by the stonemason William Edwards over the Taff River. The bridge is known both as the William Edwards Bridge and as the Pontypridd Bridge, the latter name derived for its location at a shallow, fording point on the Taff River. The Welsh phrase *Pont-y-ty-Pridd* means "bridge of earth."

Although historians and engineers are still researching the history of the bridge, what is known is that local officials paid Edwards, the local stonemason, to bridge the Taff River with a stone structure to replace a less permanent wooden bridge. Floodwaters rushing down the Taff had—without bothering to notify local officials—removed the wooden span.

To avoid placing supporting piers in the unfriendly river, Edwards opted for a single arch bridge that would touch solid ground only where it met the banks of the Taff, 140 feet apart. Edwards was nothing less than gutsy since his unsupported arch would be 30 feet longer than any other in the British Isles.

Construction on the bridge probably began around 1750, but the project had some problems. On removal of the falsework that supported it during construction, the bridge collapsed and was rebuilt by Edwards, who, some say, was exasperated to see it collapse a second time. Nevertheless, Edwards persevered, and in 1756 his task was accomplished.

To make the bridge work, Edwards resorted to some tricks of his own that engineers are still trying to unravel. To reduce the weight of the bridge, Edwards incorporated a trio of holes on each of its sidewalls where they touch the shore. The holes range from large to small with the largest on the shore side. Some documentation suggests Edwards also incorporated other holes within the structure, with those hidden behind a masonry façade.

To make the arch bridge support its own weight, Edwards gave it a rather sharp angle that works in terms of structural integrity. Designing it as a pedestrian bridge, Edwards incorporated stone steps in the roadway to assist those walking up the steep incline to its apex. A second but more conventional bridge employing piers was built next to it in 1857, though Edwards's work of art remains. That Edwards made his bridge finally work is a spectacular feat in the world of bridge engineering, in which arch depth to span ratios matter a lot. Edwards's little bridge in Wales had an arch thickness of 2.49 feet and a length of 140 feet, giving his bridge a thickness to span ratio of a respectable 1/56.

The bridge continues to interest engineers, as do other stone arch bridges in use around the world that demonstrate tremendous resilience and a capacity to endure loads exceeding those originally anticipated. Cardiff University in Wales is conducting research on the Pontypridd Bridge by crafting miniature models of the span and subjecting them to heavy gravitational forces in a centrifuge.

See also ARCH BRIDGE.

popping rock *See* rock burst.

powder monkey Perhaps the silliest name in the world for one of the most dangerous and serious jobs, the term *powder monkey* today refers to anyone who handles explosives, and these skilled explosives experts were responsible for advancing tunnels at a phenomenal rate. The origins of the term are cloaked in history, but sometime before the 18th century it emerged to refer to preteen boys employed to fetch black powder from the magazines of wooden-hulled warships. Small, nimble, and fearless, these powder monkeys would clamber up the decks with the ingredient necessary to reload cannons during naval combat.

Children's rights activists, and just about everyone else, would today be appalled by this use of children. However, powder monkeys were just the right size for maneuvering in the crowded magazines and then—carrying the precious gunpowder—moving through a crew of sweating tars working to load and fire a cannon. They were often no more than 12 years of age and saw service in the world's navy's for centuries, including those of the Union and the Confederacy during the Civil War.

With the advent of larger guns and modern propellants encased in bags or brass casings, the romantic, tough, and dangerous career of the powder monkeys at sea ended, but the colorful name lived on. At some point, certainly by the 1800s, the term was applied to anyone who handled explosives such as black powder and, by the late 1800s, nitroglycerin and dynamite.

Other names for those who transported and used explosives included *shooter, blaster,* and *powderman.* The term *blaster* reemerged in the 1980s as a street slang for someone who smoked crack cocaine, undoubtedly referring to the explosive high obtained by the addicts.

Landlocked powder monkeys working inside tunnels perhaps had the most dangerous job. Inside poorly ventilated tunnels and using crude technology, they would carefully tamp a measured amount of black powder into shot holes drilled into the face of a tunnel. Sometimes the powder might detonate during its placement. Designed to shatter rock, a blast could generate an untold number of death-dealing fragments. Even if the black powder were ignited at a safe moment, the gases emitted by its detonation still produced clouds of toxic gases that had to be ventilated lest they sicken or incapacitate the miners.

At the arrival of the famously unstable liquid explosive nitroglycerin, the powder monkeys learned the wiles of this new blasting agent. With nitroglycerin, the job of the powder monkeys was perhaps even more dangerous, although the results obtained with the liquid were impressive. When Alfred Nobel patented a way to stabilize nitroglycerin by using a clay mixture, the powder monkeys were given the explosive gift of dynamite that was so stable it would not detonate if exposed to a stray spark or subjected to sudden vibration.

The job of the powder monkey became a complex and highly refined one. With careful calculations indicating the precise depth and placement of the shot holes filled with the proper amount of explosives, skilled powder monkeys could remove rock and shape tunnels with great precision.

Although powder monkeys once crudely detonated their charges with trails of black powder or lighted fuses; the introduction of electrical ignition systems and improved blasting caps, also known as detonators, meant they could move well clear of the blast site and trigger the explosives through remote control. This electrical method of detonation entered American use in 1868, in the Hoosac Tunnel, where nitroglycerin was employed.

The dangers faced by powder monkeys and the drillers of the shot holes were sometimes as ironic as they were deadly. During Australia's construction of massive tunnels for the Snowy River Scheme —a hydroelectric project built between 1949 and 1974—some of the drillers died when they attempted to deepen existing shot holes after a blast. Unexploded dynamite would sometimes detonate when drill bits struck the remaining explosives. One who died in this way was Ernest Vecchiato, 29. The blast that resulted on October 26, 1959, perforated Vecchiato from head to toe with hundreds of granite splinters. Although Vecchiato did not die immediately, his numerous wounds proved fatal.

See also DYNAMITE; HENRY, JOHN; HOOSAC TUNNEL; NOBEL, ALFRED.

prestressed and post-tensioned concrete Although concrete possesses massive compressive strength, or the ability to resist crumbling under great weight when used as a vertical column, its ability to carry tension when used in a horizontal fashion is poor. Although steel-reinforced concrete beams acquired additional tensile strength through embedded steel reinforcement, their brittleness when accepting a load resulted in cracking. The cracking not only weakened the strength of the beam but allowed the infiltration of water and road salts corrosive to the steel reinforcement.

The solution to this problem is the installation of steel tendons locked under tension into the ends of a concrete beam. A simplified way of understanding the concept of prestressed or post-tensioned concrete is to imagine a stack of poker chips lying on a table. Vertically stacked, they can accept massive pressure from above, but held horizontally between two fingers, they cannot withstand the slightest downward force without clattering to the floor. Drilling a hole through the center of the plastic chips and threading a string through them that is then powerfully tightened creates a cohesive unit that can barely be budged.

A tendon stressed and then tightened through the center of a concrete beam adds tensile strength by transferring that strength from the ends along the horizontal length of the beam. When a beam is stressed, the concrete is slightly cambered, with the hump of the beam ready to accept the load and the curved underside capable of flexing downward without cracking. When the force of the tendon is locked into the concrete beam and external loads are applied, the top surface flattens and goes into more compression, whereas the compression of the bottom surface is reduced as it flattens. As long as the forces do not lead to sufficient tension in the bottom surface to overcome the initial compression, the concrete does not crack.

There are two ways of locking this compression into the concrete: prestressing and post-tensioning. In the case of prestressing, which is performed on long concrete beams fabricated in a factory, the concrete is put into compression by initially stressing steel cables in tension inside the beam forms. After the concrete is cast and cured, the cables under tension are locked into place. When the force stretching the cable is released, the cable unsuccessfully attempts to resume its original, shorter length but succeeds only in transferring compressive force into the concrete.

The concept is similar for post-tensioning except that pathways for the cables are cast into the concrete beam. Once the concrete hardens, cables can be threaded through these ducts, anchored at one end, and then pulled from the other end with powerful jacks to a specific tension before being locked in place. This process allows crews to adjust the tension for specific structural needs. As in prestressing, the cables transfer a

A coal miner lies crushed beneath a large rock that has fallen from the roof of a mine sometime during the early 1900s. Incidents like these, known as rock falls, can be precipitated by progressive failure of surrounding rock after excavation. *(National Archives)*

compressive force that enables beams to endure far greater tensile forces.

The concept of prestressing and post-tensioning concrete occurred to the French bridge engineer Eugène Freyssinet in 1910 after he observed that concrete had the ability to bend and creep while under stress. Freyssinet honed the technology that is now employed worldwide in bridge construction. Like most superb engineering innovations, it is elegantly simple and logical.

See also CONCRETE; FREYSSINET, EUGÈNE; REINFORCED CONCRETE.

progressive failure During tunneling operations in hard rock the excavation process removes supporting masses of rock that have been in place millions of years. Whereas some rock is extremely stable, or competent, other rock is fractured and loose and unable to support itself or the rock sur-

rounding it. This type of rock is referred to as incompetent.

Should a loosened rock fall, its loss tends to precipitate the loss of other rocks since all are part of a structural jigsaw puzzle. As each rock is lost, the cooperative support shared is diminished, causing more rocks to fall and leading to what experts in rock mechanics call progressive failure.

This problem is worse in rock that is naturally fractured or is damaged during blasting or cutting and can pose dangerous conditions for workers. Rapid reinforcement of the roof and walls of a tunnel with shotcrete, or rock bolts with steel mesh, can reduce or eliminate these problems.

See also RUNNING GROUND; SHOTCRETE; SLURRY SHIELD MACHINE.

Purcell, Charles H. *See* San Francisco–Oakland Bay Bridge.

Quebec Bridge (built 1904–1917) *Record-breaking bridge that collapsed twice, killing 88* Beginning in 1851, business and civic leaders of Quebec dreamed of a railroad bridge that would link their French Canadian city to the southern shore of the Saint Lawrence River and thus to eastern Canada and America's New England states. It would be a mammoth undertaking since the bridge would have to span a river more than 3,000 feet wide.

No less than 65 years was required to convert the dream into a reality, but the process would see the bridge collapse twice during separate construction attempts, killing a total of 88 men. The first collapse would also crush the career of one of American's most esteemed bridge engineers.

Despite the engineering mistakes that caused the first collapse and the freak failure of a component that caused the second, the Quebec Bridge still reigns as the longest cantilever span in the world.

The Gulf of Saint Lawrence is a massive body of water to the north of Quebec that narrows into the Gaspé Passage and continues southward to become the mouth of the Saint Lawrence River. Not until the mighty Saint Lawrence reaches a point just south of Quebec does it narrow enough to make bridging the waterway realistic. Although plans were bandied about, little but talk took place until the Quebec Bridge Company was formed by the Canadian Parliament in 1887 with the mandate to build a bridge within three years.

The deadline came and went not once but three times with Canada's Parliament granting extensions each time. In 1898, the Quebec Bridge Company finally advertised for bids for a bridge over the Saint Lawrence. Of the seven designs submitted, four were of the cantilever type and three were suspension bridges.

Suspended rail bridges had never really caught on, though the success of Scotland's Forth Bridge—a cantilever span completed in 1890—had impressed the Quebec Bridge Company's officials. The Forth Bridge had been heavily overbuilt in response to the deadly 1879 collapse of Scotland's Tay Bridge, which killed 75.

The Quebec Bridge Company asked the American bridge engineer Theodore Cooper to assess the plans as a consulting engineer, and he selected a cantilever design presented by the Phoenix Bridge Company. Cooper had built his reputation as one of the key engineers on the historic Eads Bridge over the Mississippi River and had designed dozens of railway bridges during a long and distinguished career. Nearing 60 in 1898, Cooper was considered an unassailable authority on bridge design. He accepted the job for so small a fee it precluded his hiring assistants.

The New York–based Cooper also recommended that the foundations for the bridge's two piers be placed farther apart to make their construction easier in shallower water nearer shore. This recommendation was followed, and the length of the central span was widened from 1,600 feet to just over 1,800 feet. Although some have made the case that Cooper widened the central span of the bridge to exceed the record-setting 1,700-foot main span of the Forth Bridge, Cooper's original intent was prob-

The center span being raised into position between the two trussed cantilevers of the Quebec Bridge crashed into the Saint Lawrence River at the very moment this photograph was snapped September 11, 1916. This mishap killed 13 bridge workers and was the second collapse during the construction of the world's longest cantilever bridge. *(National Archives of Canada)*

ably to reduce the staggering costs associated with the foundation work.

When the Quebec Bridge Company decided to analyze the increased stresses on the design caused by widening of the central span, Cooper was insulted by this proposal and made it clear his design work on the bridge needed no other approval than his own. Incredibly, as Henry Petroski pointed out in *Engineers of Dreams,* Cooper had no staff in New York to assist him in the complex calculations that in those days had to be done with pencil and paper. By imperiously brushing aside double-checking of his work by others, Cooper rejected a method of properly calculating the new stresses wrought by the changes he instituted.

Since the Quebec Bridge Company was to raise money for the bridge by selling stock to investors, its funding was far from guaranteed, and the enthusiasm of the Phoenix Bridge Company wavered accordingly. As a result of money concerns, engi-

neering and research were meager until 1903, when the Canadian government passed the National Transcontinental Railway Act, which guaranteed the bridge's financing as part of a coast-to-coast railway.

The Phoenix Bridge Company engineer overseeing the project was Peter L. Szlapka, who had virtually no experience as an onsite construction engineer. Cooper, once famous as a hands-on designer, now detested visiting the bridges that he designed and visited the Quebec work site infrequently, claiming ill health.

After the completion of the foundation work, construction began on the side spans from shore. By 1907, work was under way on the cantilever arms outward from the massive piers. Szlapka designed the lower chords of the cantilever but failed to calculate the weight of the steel required properly and therefore erred in computing the resulting stresses on the structure. Cooper, essentially working alone, accepted Szlapka's calculations without question. It was later reckoned that the additional stresses involved might have been 20 percent greater than Szlapka had estimated, far exceeding what the cantilever truss could hold.

By early August 1907, steel beams had been deformed on the cantilever, alarming Cooper, who sent telegrams seeking reasons from engineers at the site for this buckling. Quebec Bridge Company officials confused the issue by claiming some of the steel was bent three quarters of an inch before installation. This claim did not placate the engineer Norman R. McClure, who realized the steel had bent a full two inches by August 27, prompting a cessation of work on the bridge. McClure rushed to New York to discuss the matter with Cooper, who had not visited the bridge during the previous two years. Incredibly, the bending of the steel components of the cantilever truss segments did not alarm the engineers at the site, who ordered the work resumed in McClure's absence.

After Cooper and McClure conferred, the elder engineer dashed off a telegram on August 28 to the Phoenix Bridge Company headquarters in Pennsylvania warning against adding any more components to the south cantilever. As if Cooper had been opposed by the Fates, a telegrapher's strike may have played a role in delaying the message. Whatever the cause, the telegraph was not read in time to be relayed to the work site before the south cantilever, built to a length of 600 feet, had failed.

Just before the 5:30 P.M. quitting time on August 29, 1907, 92 workers were on the bridge when the explosive sounds of huge girders collapsing under their own weight shook adjacent houses and rumbled like thunder in nearby Quebec. Seventy-five men died in the collapse, including 33 Canadian Kahnawake Mohawks.

A Canadian government investigation followed, and Cooper and Szlapka were held responsible for the collapse. Cooper, who was by then 68, revealed he had trusted Szlapka's work and admitted he had not visited the bridge in person during the preceding 24 months. It became apparent that Cooper's reputation as one of North America's foremost bridge designers had negated a review of the bridge's design that would have revealed its flaws.

The American bridge designer Gustav Lindenthal was highly critical both of Cooper's role in the disaster and of a culture that seemed to place little value on the all-important contribution of the engineer. Lindenthal blistered Cooper's willingness to accept the pittance of $3,570 a year for his engineering services since it crippled Cooper's ability to hire a staff with which to calculate the stresses upon the structure properly. This was an example, Lindenthal claimed, of the low esteem in which engineers and their crucial contribution were held.

In the aftermath of the investigation, a new three-man engineering team was recruited, including Polish-American bridge designer Ralph Modjeski, who would be the only one of the trio to remain with the project until its completion in 1917. Another cantilever bridge was designed, this one possessing a wider bridge deck than the last. After the collapsed steel was removed from the site of the south cantilever's collapse, an unsuccessful attempt was made to widen the existing piers to accept the new bridge. This idea was abandoned in favor of having an entirely new bridge built nearby.

New foundation work began with pneumatic caissons to allow men to excavate the foundations to bedrock in a compressed air environment. On the north side of the Saint Lawrence, workers descended to 55 feet, where they struck bedrock. On the south side, workers sank the excavation to 80 feet, where they reached a hard layer of acceptably firm compacted gravel. Concrete then filled the interiors of the caissons, whose roofs had been built up above the waterline with courses of masonry as they sank to their required depth.

The work proceeded smoothly with the benefit of hindsight that encouraged the engineers to calculate the stresses of the cantilever truss carefully to preclude the type of failure that had occurred in 1907. Massive box girders of nickel-steel alloy, stronger than the steel in the original bridge, were used in the components of the new Quebec Bridge. With the side spans and the cantilever truss segments fully built by 1916, all that remained was to close the center gap between the opposing cantilevers with a 640-foot segment.

On September 11, 1916, boats towed the 5,000-ton segment to the middle of the river, where jacks slowly raised it 130 feet to the cantilevers. A complicated steel casting that constituted a corner of the segment fractured, and the span twisted itself into wreckage as it plunged to the water below, an unlucky moment captured in an extraordinary photograph. The mishap killed 13 bridge workers and injured 14 more, raising the bridge's total of worker fatalities to 88, a number that varies slightly with different accounts. Another segment was fabricated and set in place within a year, and the first train crossed the bridge on October 17, 1917, an event then overshadowed by World War I.

The bridge's main span measures 1,810 feet between the foundations, and its total length is 3,239 feet, including its side and approach spans. Its deck width is 88 feet—21 feet wider than the deck of the original ill-fated span. The bridge cost $25 million in 1918, roughly equivalent to $350 million in today's dollars. The span was turned over to its owner, Canadian National Railways, in August 1918. In 1999, a multiyear rehabilitation process that included removal of lead paint and application of new rust-resistant coatings began. The cost of this work will total $60 million.

The Quebec Bridge today contains a single rail line, a trio of auto lanes, and a pair of walkways. In 1996, it was named a National Historic Site by the Department of Canadian Heritage.

See also CANTILEVER BRIDGE; COOPER, THEODORE; FIRTH OF FORTH BRIDGE; LINDENTHAL, GUSTAV; MODJESKI, RALPH; MOHAWK IRONWORKERS.

Queens-Midtown Tunnel (built 1936–1940) *The East River's first vehicular tunnel*

The success of the Holland Tunnel connecting New Jersey with Manhattan beneath the Hudson River paved the way for the Queens-Midtown Tunnel, which would eventually link Manhattan with Queens across the East River. Completed in 1927, the Holland Tunnel was the world's first long vehicular tunnel and demonstrated that bridges were not the only means of allowing automobiles to negotiate water obstacles. Before the Holland Tunnel was opened, officials were already talking about building a tunnel across the East River, but the project was to face daunting obstacles.

Like many infrastructure projects conjured during the 1920s and slated for construction in the following decade, the Queens-Midtown Tunnel was delayed by the Great Depression when the high hopes of the Roaring Twenties were choked by the 1929 stock market crash. Tax revenues withered and municipal officials were more concerned with staving off mass hunger than with pouring millions of nonexistent dollars into massive infrastructure projects.

However, election of Franklin Delano Roosevelt as president in 1932 ushered in his New Deal program, which committed billions of federal dollars to infrastructure projects with the dual purpose of not only modernizing America but also employing its idle workers. One of Roosevelt's "alphabet soup" of agencies designed to breathe life into the economy was the Public Works Administration (PWA), headed by the liberal secretary of the interior Harold Ickes. Fortunately for New Yorkers, the proposed Queens-Midtown Tunnel was precisely the sort of project the Roosevelt administration wanted to promote.

The need for the tunnel was obvious since the four bridges spanning the East River were already becoming congested, carrying as they were automobile, trolley, and subway traffic during the 1920s. It was decided that a vehicular tunnel would be the traffic artery of choice to complement the Brooklyn, Manhattan, Williamsburg, and Queensboro Bridges. Thus, the Queens-Midtown Tunnel became a project touted by officials as the answer.

The Port Authority of New York and New Jersey had built the Holland Tunnel, but no municipal agency existed to oversee the construction of a tunnel within New York City itself. However, the establishment of the New York Tunnel Authority remedied this dilemma when it was created by the New York State Legislature in 1936. The New York mayor, Fiorello La Guardia, pushed the authority into existence in his eagerness to see the $58 million project accomplished.

La Guardia lobbied Washington heavily for the federal loans and grants that became available under President Franklin Roosevelt's New Deal programs and enthusiastically pushed for the construction of the tunnel that would provide jobs to unemployed New York construction workers.

New York had not a cent with which to support the project, as La Guardia made clear when he appointed the Tunnel Authority's three board members, saying, "You are starting from scratch, with no appropriation and nothing but an idea and a law." The appointees, Albert B. Jones, Albert T. Johnson, and William H. Friedman, soon obtained a $41,000 loan from the city with which to develop tunnel plans that would be presented to the PWA for review.

If the economic climate were not tough enough on an expensive project like the tunnel project, the unique political problems in New York promised to be even more daunting. Robert M. Moses, who was nearly invulnerable politically as kingpin of park development and public works projects in New York, found he was excluded as a board member of the newly formed authority. Moses, whose lust for power had long been out of check, had unsuccessfully attempted to torpedo the legislation creating the Tunnel Authority.

The decision to exclude Moses from the Tunnel Authority was made by La Guardia, who knew full well that a deep and longstanding hatred existed between Moses and President Roosevelt, the man who held sway over the federal dollars available for infrastructure projects.

Unable to control the tunnel project, Moses attempted to assassinate it with delays. Ickes had indicated that unless construction on the project began during Roosevelt's 1936 campaign for reelection, the $58 million earmarked for the tunnel would be used elsewhere.

Knowing this, Moses sought help from allies in the state legislature to sponsor legislation that would erase the Tunnel Authority as an agency while calling for a year-long delay to study the project. Both proposals would have dealt a death blow to the tunnel. The bill sought by Moses to extinguish the Tunnel Authority was eventually defeated and the project moved forward.

In 1936, the Public Works Administration provided a loan of $4,130,000 and a federal grant of $11,235,000. Since the federal government required that the tunnel retire its own debt, the Queens-Midtown Tunnel would charge a toll at its entry plazas on both sides of the East River. The experienced tunnel engineer Ole Singstad and his staff had labored to come up with a design for the Queens-Midtown Tunnel and its crucial ventilation system, a design that would have to obtain approval from the PWA. The project would include a pair of parallel bores 31 feet in diameter, each providing room for a pair of lanes on a 21-foot-wide roadway. With the defeat of the legislation sought by Moses and the approval of Singstad's engineering plans, the tunnel project was close to becoming a reality.

With millions out of work and some of those hardest hit by the depression being African Americans, the PWA chief, Harold Ickes, worked out quotas with representatives of the National Association for the Advancement of Colored People to ensure representative numbers of blacks were employed on federally funded projects, so that jobs would be distributed more equitably. Although far from the affirmative action programs that would evolve during the 1960s, it was an effort to prevent outright exclu-

sion of some of America's workers hardest hit by the shattered job market of the 1930s.

Work on the twin tubes of the tunnel began October 2, 1936, when President Franklin Roosevelt pressed a button in nearby Queens. After workers excavated vertical access shafts on each side of the river, tunneling shields were assembled on each side. These shields were equipped with cutting edges to slice through soft earth at the face of the tunnel as well as air locks to allow workers to enter the compressed air environment of the working area. The four shields would progress toward each other from opposite sides of the river until they met.

As the shields moved forward an erector arm lifted massive iron segments into place, where they were bolted together by workers. Each time the shield progressed 32 inches a new segment of lining was put into place. A series of 28 hydraulic jacks then pushed against the 31-foot-diameter lining, moving the huge shield forward. As the lining rings were assembled and bolted together, workers were removing earth and rock from the pressurized work area using mucking cars running on small railway tracks. Other workers were excavating through the rocky face of the tunnel by drilling blast holes and then filling them with precisely measured charges of dynamite. Each shield was able to progress at an average of 18 feet per week.

As the four bores approached each other, careful measurements were taken to ensure they would meet head-on beneath the East River. The bore that would carry the Queens-bound traffic would measure 6,272 feet in length, the one carrying traffic to Manhattan would measure 6,414 feet. The tubes would curve to 103.5 feet beneath the surface of the river.

Because of the depth of the tunnel beneath the river, it was essential that the sandhogs work in a pressurized environment to prevent water from infiltrating the excavations. Working "under pressure" meant the sandhogs had to limit their time underground to four hours at 37.5 pounds of air pressure, 2 1/2 times normal air pressure.

The tunnel engineer, Singstad, designed the ventilation system to exchange 3 million cubic feet of air in the tunnel each minute. Huge ventilation buildings located on each side of the East River are equipped with a total of 46 massive fans for this task. A ventilation channel below the roadway carries fresh air into the tunnel, and ducts in the roof of the tunnel bores carry away the automobile exhaust.

When the shields met at the midpoints of the two tunnel bores, the steel cylinders were sliced to pieces by acetylene torches and hauled out piecemeal since removing them the same way they were pushed through the bores would have been impossible. A concrete lining was installed inside the iron reinforcing rings and the roadway and suspended ceiling were also installed.

The Tunnel Authority deftly marketed the Queens-Midtown Tunnel, saying the artery would allow motorists to "Cross in 3 minutes. Save in 3 ways: Time, Miles, Gas." As a result of advertising, coupled with the convenience offered by the tunnel, 4.4 million vehicles passed through it during its first full year of operation in 1941. The original toll charged for the tunnel was 25 cents. However, the lingering effects of the depression accompanied by fuel rationing and the entry of a sizable portion of the population into military service between 1941 and 1945 prevented the tunnel's potential from being realized fully until after World War II.

By the end of 60 years of operation, the tunnel was counting 25 million toll-paying vehicles passing through its twin tubes yearly. It remains in operation, although it has undergone extensive maintenance during the ensuing decades. In 1993, the tunnel's iron and steel components were provided a $17 million cathodic protection system designed to halt corrosion. The system counteracts the normal electrical current that causes the tunnel's iron-based components to break down into iron oxide.

See also BROOKLYN-BATTERY TUNNEL; CAISSON DISEASE; CATHODIC PROTECTION; EAST RIVER GAS COMPANY TUNNEL; GREATHEAD, JAMES HENRY; HOLLAND TUNNEL; MOSES, ROBERT M.; SANDHOGS; SINGSTAD, OLE; TUNNELING SHIELD; TUNNEL LINING.

R

recompression *Treatment for caisson disease* The cause of the bends, or caisson disease, in compressed air workers is that the increased air pressure of their work environment compresses nitrogen bubbles into their tissues. When workers emerge too quickly to normal air pressure without slow decompression, a variety of medical problems, ranging from embolisms and paralysis to a horribly painful death, occur.

Compressed air work began in caissons where bridge piers were excavated in 19th-century England and France and arrived in America during work on the Eads Bridge and the Brooklyn Bridge. Physicians and engineers on both sides of the Atlantic were puzzled by the illness, and numerous theories as to how best to prevent the disease and, failing that, how to treat it, were put forward. Some theories held that men who led dissolute lifestyles were at greater risk; others declared that hot showers and warm surroundings would prevent caisson disease.

By the late 1800s, it was clear that by recompressing sick workers in the airlocks and slowly bleeding off the pressure, many if not all of the effects of caisson disease would be reduced or eliminated.

One of the first to employ this technique was the engineer E. E. Muir, who was part of the British-led second effort in 1889–1891 to complete the Hudson Tubes between Manhattan and New Jersey. Muir set up a separate airlock that came to be known as a "medical airlock," where compressed air workers were returned to the pressure that they experienced underground. They were then decompressed much more slowly to allow the gas to dissipate at a safe rate.

See also CAISSON DISEASE; HUDSON TUBES.

reinforced concrete Although capable of supporting massive weight because of its compressive strength, concrete has a major flaw in its brittleness, or lack of tensile strength. If a piece of concrete is stretched (or put in tension) rather than squeezed (or put into compression), its resistance to cracking and failure is small. It would be up to a French flowerpot maker to solve a problem that had confounded trained engineers.

A French gardener, Joseph Monier, manufactured planting pots using cement but was dismayed by the inability of the cement to withstand cracking. Taking a cue from the centuries-old technique of adding straw and twigs to mud bricks, Monier added iron wire to his concrete pots. Monier's concrete planting tubs were made many times more resilient.

Monier's iron mesh eventually became a steel-reinforcing bar that made concrete a far more useful structural material. The reinforcement of concrete is a straightforward process. Steel bar of the required thickness is simply placed inside the bridge components or foundations, and concrete cast around them.

Reinforced concrete was on its way to becoming a superb material in the construction of bridges and tunnels by the end of the 19th century. Reinforced concrete allowed bridges to take on a thinner, more graceful appearance since the material could be shaped into slender arches with the steel reinforcement providing additional strength. Bridges made with reinforced concrete took on a sleek modernistic and homogenous appearance. Another Frenchman, François Hennebique, began building some of the most beautiful bridges in Europe. Hennebique's Vienne River Bridge at Châtellerault, France, is considered a fine example of an early reinforced concrete span. Completed in 1899, it had a central span measuring 172 feet. Its total length of 434 feet made it the longest reinforced concrete bridge built in the 19th century.

Reinforced concrete remained an integral part of virtually every large bridge constructed during the 20th century, if not in the bridge deck or towers, then certainly in the foundations for the piers. Although reinforced concrete was not put to large-scale use in tunneling during the 19th century, it became the material of choice for lining water-carrying tunnels by the 1930s and by the middle of the 20th century had become an important means of providing reinforcement for tunnels in soft earth.

One of the first major uses of reinforced concrete in tunnel construction was during the Hoover Dam project, when four 50-foot diversion tunnels were bored through solid rock. The reinforced concrete linings were a massive 36 inches thick.

By the 1950s, it was determined that concrete sprayed against tunnel walls under pressure into a meshwork of reinforcing steel could provide extraordinary strength and safety in soft earth conditions. This type of reinforcement became a cornerstone of a tunneling method known as the new Austrian tunneling method, or NATM.

Research into methods of reinforcing concrete has provided additional ways of making concrete able to withstand cracking. Small steel and plastic fibers are now mixed with concrete to give it added resistance to cracking. The renovation and strengthening of Sir Marc Brunel's 19th-century Thames Tunnel during the 1990s used sprayed concrete that was reinforced with slivers of plastic and steel.

See also CONCRETE; HENNEBIQUE, FRANÇOIS; MONIER, JOSEPH; NEW AUSTRIAN TUNNELING METHOD; PRESTRESSED AND POST-TENSIONED CONCRETE.

Rennie, John *See* New London Bridge; Old London Bridge.

Rialto Bridge (built 1587–1591) *Venice's enduring landmark* Crossing the narrowest part of Venice's Grand Canal is one of the world's most unique and beautiful spans, the Rialto Bridge. The bridge was a marvel of Renaissance engineering and is today one of the most popular tourist attractions in the city. And although the bridge is barely 83 feet long, its design and construction were surrounded by as much controversy as the building of the Brooklyn Bridge.

A wooden toll bridge originally spanned the Grand Canal. Built in 1252, this first bridge came to be known as the Money Bridge. Because of its wooden construction the bridge needed constant repair and replacement. In 1450, the bridge's railings collapsed as a result of either overloading or deterioration and the bridge was rebuilt. By 1458, the bridge was renovated to provide room for shops on its roadway. The fees paid by the shop owners covered the cost of maintaining the bridge so the tolls were eliminated.

After the elimination of the tolls, the bridge acquired the name *Rialto,* a contraction of the Italian term meaning "high bank," and the name of the district in which the bridge was located.

The bridge was an essential crossing in Venice, but the wood span was susceptible to rot and fire. When a blaze swept through the neighborhood surrounding the Rialto Bridge in 1513, the bridge narrowly escaped destruction. The lesson was not lost on those in power and it was decided that a stone bridge would have to be built. This idea entered the discussion stage, where it remained for the next 74 years. A design was selected in 1570, but Italy's war with the Ottoman Empire delayed the project another 17 years.

In 1587, an arch bridge design by Antonio Da Ponte (1512–1597) was accepted. That year the architect-engineer began driving 1,200 timber piles into the muck to provide foundations on each side of the canal for the span's heavy stone abutments. The 83-foot bridge was built not only for pedestrians but also for shoppers, with a 66-foot wide roadway to accommodate two rows of shops and a trio of walkways.

Although Da Ponte designed the bridge and oversaw its construction, a myth has persisted for centuries that Michelangelo (1475–1564) or another eminent architect was the designer. Da Ponte, a renowned architect of his time, did have help on the project from his nephew, Antonio Contino, an archi-

Its delicate, almost ornamental appearance belies the strength inherent in the Rialto Bridge spanning the Grand Canal in Venice. The stone arch bridge was built in 1591 to replace timber bridges that required constant replacement. *(Robert Cortright, Bridge Ink)*

tect who studied under Da Ponte. Curiously, *ponte,* the Italian word for "bridge," makes Da Ponte's name "Antonio of the Bridge," begging the question as to whether his name derived from an ancestral home near a bridge or from his own fame as a designer of such spans.

As the work progressed, Da Ponte's bold design sparked a chorus of criticism from jealous architects who had unsuccessfully sought the commission to build the bridge. Detractors claimed that the bridge was unsafe because of faulty design and that its abutments were incapable of supporting the arch. A commission of architects was appointed to review both the design and the construction of the unfinished bridge. The commission gave the bridge a clean bill of health, reporting that Da Ponte's span was safely designed and well constructed. The bridge was completed in 1591.

Sustained by a single stone arch, the highest clearance beneath the bridge's arch is 24 feet to make room for the copious amount of river traffic.

The bridge has become a must-see attraction for tourists visiting the city. Although the presence of shops on the bridge was nothing unusual in medieval and Renaissance times, the Rialto Bridge provides a stunning sight to gondola-borne visitors, standing as it does over a bend in the canal framed by centuries-old ornate buildings.

Although criticized by some through the years as unattractive or as flawed in design, the bridge has steadfastly remained in place for more than four centuries. Some critics compare the Rialto Bridge to the Eiffel Tower in that its design is so unique as to place it in a class of its own.

Da Ponte's nephew learned his craft well, for in 1600 he completed work on a bridge of his own design, the famed Bridge of Sighs over the Rio di Palazzo. The span connected the second stories of an infamous prison—designed by Da Ponte—with the torture chamber of the Doge's Palace across the waterway. Although the proper name of the span is the *Sospiri Bridge,* the commonly used name is derived from the legend surrounding the hopeless sighs of prisoners crossing the bridge from their cells to undergo torturous interrogations.

See also ARCH BRIDGE.

riprap *See* scour.

rock bolt *Putting the screws to incompetent rock*
It is commonly believed that tunneling through rock provides an excavation in material that is cohesive enough to be self-supporting. Despite the hardness of rock, seldom do tunnelers find circumstances in which some type of tunnel support is not eventually required.

Among the tunneler's arsenal of roof and wall support tools is the rock bolt, with which massive bolts are screwed into predrilled holes to tie loose rock together to prevent collapse. A bearing plate is then screwed on the outside end to bear the weight of the rock.

To provide additional protection from falling rock, wire mesh similar to chain link fencing can also be secured beneath these plates. This mesh secures loosened rock that might later fall. By preventing rocks from falling, this halts what is known as progressive failure, since the loss of rocks from a tunnel roof precipitates the shedding of additional rock.

As is the case with much tunneling technology, it is not always certain where the rock bolt originated since many innovations have been hatched on the spot or borrowed from related enterprises. Rock bolt technology may owe its beginnings to a stone quarry in Wales, where rock bolts were used to stabilize rock faces in 1862. Rock-bolting did not exactly catch fire as a method of tunnel reinforcement: It had been used on only a handful of American hard rock tunnel projects by the 1950s.

One of the first major uses of rock bolts was in the two underground power stations of Australia's massive Snowy Mountains Hydroelectric Scheme. The project, which recruited mining engineers from around the world, wanted to ensure that the roofs of the football-field-sized galleries were secure.

The rock bolts originally used in the project were mechanically anchored, meaning they were designed to secure themselves to rock at the very end of their reach. The bearing plates were then attached and tightened to a specific torque to keep rock masses secured. Originally thought of as only a form of reinforcement, rock bolts were considered by engineers on the Snowy Mountains project an integral part of the tunnel support system.

Rock bolts installed on the second underground power station were held in place by a grout mixture pumped into their hollow shafts. The grout oozed through holes machined into the bolt to fill surrounding rock cavities before hardening. This was one of the first if not the first instance of rock bolts' being secured with grout, a method that essentially worked to stabilize the rock and simultaneously glued the bolt in place.

Rock bolts are not placed willy-nilly but are carefully sited by engineers on the basis of the characteristics, or rock mechanics, within the space being excavated. Rock bolts can be six to 16.5 feet in length or longer as required.

Rock bolts today boast highly technical designs that allow them to anchor themselves once they are seated. Since many are grouted with various types of compounds, including epoxy-type resins, to secure them in place and bond them to the surrounding rock, they are hollow to allow the pumping of the grout. Perforations along the bolt allow the grout mixtures to ooze into the surrounding rock. Other rock bolts are less exotic and are manufactured from Rebar material lacking a hollow center.

Another major use of rock bolts was during construction of the North American Aerospace Defense Command's (NORAD's) Cheyenne Mountain complex completed in 1965. During this project, gigantic caverns were carved out of the inside of the mountain by explosives. To reduce the hazard of falling rock in case of a nuclear attack, 115,000 rock bolts were emplaced along with steel mesh to consolidate the roof and walls.

rock burst *Nature's deadly underground missiles*
Tunneling in hard rock entails a hazard that terrorizes miners and is little known to those who work above ground. It is a lurking killer whose threat intensifies as a tunnel is driven deeper and surrounding rock becomes harder. The danger is known as rock burst, in which the stresses caused by the settling of the surrounding cause an explosive fracturing of the rocks.

The energy released in rock bursts can be immense and is capable of hurling slabs of rock across an underground space at terrific velocity. So powerful is the force that the blast of air created by the passing rock can knock miners down. Throughout the centuries miners have been killed by this natural phenomenon, which cannot always be prevented. Deaths from rock bursts still occur regularly throughout the world.

Experience has revealed that shallow mines have less of a problem with rock bursts than deeper mines that extend several thousand feet below the surface. The obvious explanation is that the stresses encountered deeper underground are far greater because there is so much more overlying rock. The second aspect of this hazard is that rock bursts tend to occur in rock that is extremely hard and less flexible than softer rock.

When rock is blasted or dug by a tunnel-boring machine, even the hardest rock reacts to the loss of structural support. This effect, called strain by engineers, is most pronounced in rock that is weakened by faults or interspersed with weaker rock. As the rock settles, weaker, elastic rock bends and absorbs the strain, forcing more stress on the harder rock, which may release this energy by bursting free of the tunnel's surfaces.

The problem of rock burst is surprisingly complex since some extremely deep tunnels, with rock conditions ordinarily conducive to such bursts, experience no problems. This has led engineers to surmise that factors in addition to strain caused by the weight of overhead rock are involved. Seismic movement and regional geologic characteristics may also play roles in this hazard.

Rock bursts are generally localized, meaning that they occur along the roof, side, or floor (the invert) of a tunnel. The presence of weak rock overlaid on strong rock may direct these bursts in a localized, directional fashion. In simple terms, the weak layer may experience considerable movement, placing strain on a specific location inside the tunnel lined with hard, unyielding rock. When those forces become too great for the stiff rock, they are released by an explosive failure.

Immediate reinforcement of the excavation by supports is crucial in preventing injuries caused by rock burst. Such supports can include rock bolts drilled deep into the tunnel walls and crown to support steel mesh. Another method is the erection of steel ribs, prefabricated concrete linings, or sprayed concrete known as shotcrete. Before the 20th century, timber supports were often used, but this type of support has gone the way of the prospector's burro. The frightening thing about rock bursts is their ability to wreck some supports after their installation and cause injuries nonetheless.

Rock goes through several phases when it is excavated, and the problem of bursts is greatest during what is known as the transient phase in the minutes and hours after blasting. Unfortunately for miners in deep shafts, rock bursts are not a hazard that necessarily disappears with time and may occur years after excavation.

If this hazard were not bad enough, the nerves of the miners are tested even more by the loud but usually harmless false rock bursts that occur shortly after blasting. Although their sharp report is similar to that of an actual rock burst, they are little more than noise. This harmless phenomenon is known as popping rock to miners.

Despite modern safety techniques and a growing understanding of rock mechanics, rock burst events still occur, sometimes with fatal results. In September 2000, Rodney Criddle, 42, was killed when a rock burst sent rock crashing into an excavation machine he was operating in a subterranean gold mine in central West Australia. At the time of the accident, Criddle was nearly 1,700 feet underground.

The federal Mine Safety Health Administration (MSHA), which regulates safe operations of mines and investigates mishaps and fatalities in underground excavations, considers rock bursts a serious issue. The Code of Federal Regulations requires that any mine operation experiencing a rock burst serious enough to disrupt work or cause injury be reported to the MSHA within 24 hours. The regulation also states that companies must prepare a plan for dealing with future rock bursts within 90 days.

See also POPPING ROCK; SPALLING.

rock drill The excavation of hard rock tunnels with the use of explosives is a two-stage process involving the primary step of drilling shot holes to a certain depth and in a specific pattern and then filling them with explosives.

It is crucial to place the explosives inside the rock to crumble the material so it can be removed. Merely placing explosives beside a rock face would produce far less rubble since the shock waves of the explosives are not transmitted efficiently into the surrounding rock. It should be no surprise that drilling shot holes is the most time-consuming portion of this process. When it was done by hand, it was excruciatingly slow but the development of rock drills during the 1840s revolutionized tunneling.

One of the first was the steam-driven rock drill developed by the Philadelphian J. J. Couch in 1848, which propelled a metal rod into a rock face like an extremely heavy spear. Although a potentially good siege weapon for medieval warfare, it was not really what tunnelers needed. Couch was granted a patent for his device in 1849. His machine sparked little interest, and steel-driving men like the mythical John Henry continued to pound drills with sledge-hammers.

Joseph W. Fowle, a mechanic who assisted Couch in building Couch's drill, went his former collaborator one better when he built and patented a more practical machine that actually drilled holes into hard rock with a drill steel using a steam-driven piston.

Couch's machine could throw only a rod whose striking force was limited by the rod's own weight. Fowle's machine slammed the drill into the rock and rotated it as a human "turner" did. Fowle obtained a patent for his drill in 1851.

Charles Burleigh purchased Fowle's patent in 1865 and marketed the drill in 1866 to those building the Hoosac Tunnel in Massachusetts. The drill lacked robustness because of its light construction and frequently broke down. Known as the Brooks, Gates, and Burleigh drill (or more commonly as the Burleigh drill), it was redesigned and returned to service in the Hoosac Tunnel. The improved drill worked well and was about the only bright spot in the tortured history of the Hoosac project.

Tunnels and the tools to build them with were a growth industry during the latter 1800s as a result of worldwide railroad expansion. The issuance of 110 American rock drill patents between 1850 and 1875 is ample proof of this. Dozens of patents for rock drill innovations were also granted in Europe at around the same time.

Although some rock drills were driven by steam, the Brandt rock drill utilized water as a hydraulic power source and cored out cylinders of rock to create shot holes. Alfred Brandt's machine saw service in several major alpine tunnel projects, including the Simplon Tunnel. Although Brandt's was a good rock drill that was more efficient since it used water pressure instead of compressed air, a superior compressed air rock drill built by the American Simon Ingersoll in 1871 displaced Brandt's drill. The Ingersoll drill is widely considered to be the finest rock-drilling machine of its time.

Miners who were part of the 25-year effort to complete Massachusetts's 4.75-mile Hoosac Tunnel stand in an elevator alongside a Burleigh rock drill sometime before the completion of the tunnel in 1875. The long-suffering project would not have been possible without the drill invented by Charles Burleigh, who improved upon the Sommeiller drill. *(Author's Collection)*

The lineage of the Ingersoll machine is notable since it was based on the Fowle-Burleigh drill patents that were purchased by Ingersoll, as was the Brooks, Gates, and Burleigh Company.

The Denver machinist C. H. Shaw developed one of the most successful early hammer drills in 1890. This drill allowed shot holes to be drilled in the roof of excavations. It was also equipped with a leg known as an airleg that allowed the drill to be pushed forward with air pressure as it penetrated rock. This capability of keeping a drill moving into the rock face is known as a feed.

One of the main problems with drilling downward vertical holes was the accumulation of debris that impeded progress. J. G. Leyner, a Denver rock drill repairman, developed a hollow drill that blew air into the hole to remove loosened debris. This innovation made downward-drilling rock drills practical, and his invention was patented in 1897. The only major redesign required by Leyner's machine (after coughing miners returned his first production lot because of the dust they stirred) was the addition of a system that injected water into the hole during drilling.

The development of superhard tungsten carbide cutting bits for rock drill steels during the 1920s set the stage for rock drill cutters that would last longer and cut more efficiently. The Swedish firm Atlas Diesel—now Atlas Copco—designed a lightweight drill with an airleg and equipped its drill steels with superbly designed tungsten carbide cutters. The RH-656 rock drill required only one worker for its operation and became the basis for a method of efficient, inexpensive rock drilling known as the Swedish method. The drill was marketed to a world eager to resume commercial pursuits in 1946. Updated versions of this rock drill are still sold worldwide.

Germain Sommeiller was the first to mount multiple rock drills on a rail carriage during the Italian-French Mont Cenis Tunnel project in 1863. This innovation allowed the drilling of multiple shot holes using rock drills designed by the English engineer Thomas Bartlett that were modified by Sommeiller. The concept of mounted multiple drills on a single platform was later adapted to vehicles. It allowed workers to drive to a rock face and begin drilling shot holes in a mass production process.

See also BRANDT, ALFRED; DRILL-BLASTING; HOOSAC TUNNEL; SOMMEILLER, GERMAIN.

rock fall *See* progressive failure.

Roebling, Emily Warren (1843–1903) *The woman who helped build the Brooklyn Bridge* Although not trained as an engineer and living in a time when women were supposed to be seen and not heard, Emily Warren Roebling had intellect, tact, and courageousness that were instrumental in the completion of the famed Brooklyn Bridge.

Although there is some controversy as to how much of an engineering role she played in the construction of the bridge, it is obvious that her keen understanding of suspension bridge technology allowed her to supervise its progress as a representative of her husband, Washington Augustus Roebling. Crippled and made partially blind and mute by caisson disease in 1872, Roebling faced expulsion as chief engineer on the bridge designed by his father,

John Augustus Roebling. So debilitated was the younger Roebling that he would not visit the bridge during the final 11 years of its construction.

With what strength he could summon, Roebling began to dictate engineering details to Emily Roebling but remained a steadfast recluse in his home overlooking the site of the bridge. In conversations with her husband, Emily Roebling learned the principles behind the design of the bridge and was soon making daily trips to the work site to instruct the contractors on how to complete the work.

As Washington Roebling and his father made plans to build the Brooklyn Bridge in 1867, Emily and her husband traveled to Europe to review European use of pneumatic caissons that were necessary for construction of submerged pier foundations required for the Brooklyn Bridge.

When cost increases and delays coupled with concerns over Washington Roebling's illness created extensive criticism in the newspapers in 1882, the New York Bridge Company trustees demanded that Roebling attend a board meeting to answer the charges. When Roebling declined to do so, trustees said they intended to dismiss him as chief engineer.

At about this time Emily Roebling delivered a talk to the American Society of Civil Engineers that effectively and eloquently answered the criticisms of her husband's handling of the project. As a result of her presentation, Emily Roebling gained enough support for her husband to convince the bridge company trustees to table their proposal to fire him.

When the bridge was officially opened on May 24, 1883, U.S. Representative Abraham Stevens Hewitt recognized Emily Roebling's contributions to the completion of the bridge. The historian David McCullough, in his exhaustive *The Great Bridge*, pointed out that Emily's almost constant presence at the bridge and deep participation in the project led to false gossip that she was the true designer of the span. It does appear that Emily Roebling's role in building the bridge is not mythical. She rose not only to the defense of her husband but to the task of acting as an engineering intermediary.

Emily Roebling later became an active participant in numerous organizations and obtained a law degree from New York University. She was also inducted into the Rensselaer Polytechnic Institute Alumni Hall of Fame, an honor she shared with her husband, who was a graduate of the institute's class of 1857. Stricken with stomach cancer, she died in 1903.

See also BROOKLYN BRIDGE; ROEBLING, WASH-
INGTON AUGUSTUS.

Roebling, John Augustus (1806–1869) *Designer
of the Brooklyn Bridge*   In itself, John Augustus
Roebling's design of the Brooklyn Bridge is enough
to secure his fame as one of the world's great civil
engineers. His intuitive brilliance allowed him to
construct what is widely considered the world's
greatest bridge, a span that was a quantum leap
ahead of all the bridges that had been built before.
To do so, Roebling had to overcome engineering
obstacles that eluded not only the best engineers of
his own century but many of those in the first half of
the 20th century as well. The Brooklyn Bridge still
looks down on the East River as a monument to his
abilities and as a powerful symbol of technical innovation.

Less well known is the extraordinary path that
led Roebling to America from relatively humble
beginnings in Germany. The son of a tobacco shop
owner in Mühlhausen, 150 miles southwest of
Berlin, Roebling was a superb mathematician and
artist who at 14 passed an examination that certified
him as a "master builder."

Roebling was pressed to obtain a university education by his mother, who scrimped and saved so he
could attend Berlin's Royal Polytechnic Institute. At
the institute between 1823 and 1826, Roebling
became fascinated by discussions of the precarious
suspension bridges of the day. Although these spans
frequently collapsed, the potential of the suspension
design to allow the construction of lightweight
bridges of great length captured the imagination of
Roebling. At the institute Roebling became familiar
with two Pennsylvania suspension bridges, the first
built by the American James Finley in 1801 and the
other constructed by Erskine Hazard and Josiah
White in 1816.

While at the Polytechnic Institute, Roebling also
befriended Georg Wilhelm Friedrich Hegel, the
renowned philosopher, who was lecturing at the university during Roebling's studies. Hegel, a deep
believer in an orderly society in which freedom is
paramount, made a lasting impression on Roebling
and convinced him to consider moving to America.

After graduating from the Polytechnic Institute
and entering government service as a road builder,
Roebling decided that freedom meant being able to
seek success away from the stultifying confines of a
Germany hidebound by tradition. What Roebling

The Brooklyn Bridge designer John Augustus Roebling did not
invent the suspension bridge but made it reliable while similar
bridges collapsed in America and Europe. The brilliant engineering triumphs of this autocratic German immigrant occurred after
he had flopped as a farmer and a canary breeder. *(Historic American Engineering Record, National Park Service)*

wanted most of all was the freedom to design and
build bridges, an opportunity a twenty-something
engineer would be denied in Germany.

After working several years as a road builder for
the government, Roebling and his brother, Carl,
decided they would organize a group of fellow Germans and emigrate to America. This decision was
dangerous since the Prussian authorities prohibited
skilled workers from leaving German states under
their control. Roebling made his plans, and by May
1831 he and his party left the port city of Bremen
aboard the merchant ship *August Eduard*. The
group of 92 settlers eventually made their way to
Pennsylvania, where land was purchased to found a
farming community they named Saxonburg.

Although free of Prussian authorities and suffocating German tradition, Roebling now found himself under the yoke of a plow and discovered his
impatience made him ill suited for farming. Roebling
then decided to breed canaries, an enterprise that

seemed to move him even further from his dream of building bridges. At this he was worse that he was at farming. Roebling's unusual entrepreneurial venture was a disaster as a result of circumstances beyond his control. Since canaries are valued for their singing and since male canaries sing more strongly than females, Roebling needed male birds, but virtually all his hatchlings turned out to be silent females.

After five years of farming and an abortive attempt at canary ranching, Roebling married Johanna Herting in 1836. A year later Roebling experienced two momentous events: the birth of his first child, Washington Augustus Roebling, and news that he had been granted American citizenship. By this time, Roebling had taken to designing homes and digging water wells as ways to exercise his engineering skills and escape the drudgery of agriculture.

He also began to invent and patent gadgets of all types. One device was a gauge that sounded an alarm when the water level in a steam boiler was dangerously low. Another was a propeller for moving ships through the water. As steam locomotives became prevalent and embers from their smokestacks set the surrounding countryside ablaze, Roebling invented a device for arresting sparks that was used on at least one locomotive. Despite these successes, Roebling needed full-time employment and in 1837 began work as a canal engineer. He later became a railroad engineer designing routes for the Harrisburg and Pittsburgh Railroad.

Roebling still yearned to build bridges, and despite a steady stream of canal and railway engineering jobs, he had yet to build a single span. Despite certification in Germany as a civil engineer and the scarcity of trained engineers in early 19th-century America, Roebling needed actual experience on a bridge project to kick off his career. His desire to assist on a bridge project led him to seek work from Charles Ellet, Jr., a man who would later become both nemesis and rival to the German immigrant, though never his equal as a bridge engineer.

Ellet, learned Roebling, was vying for a contract to build a bridge over the Schuylkill River at Fairmount, Pennsylvania. In 1840, Roebling penned a respectful letter seeking the opportunity to be Ellet's assistant on the project. Ellet's reply was a tad haughty, and Roebling's subsequent and diplomatic attempts to give Ellet ideas and suggestions were answered with silence, according to the Roebling biographer David Steinman. However, the Ellet biographer Gene Lewis disputes this, asserting that the two were exchanging cordial letters about bridge theory in 1841. Nonetheless, Ellet apparently believed he had no use for Roebling's services at Fairmount and was jarred to learn soon afterward that a competitor named Andrew Young had been granted the contract and had hired Roebling as an assistant.

Ellet, a shrewd self-promoter, stepped in and snatched the contract from Andrews by offering to finance part of the project. There would be future encounters with Ellet, but, in the meantime, Roebling would invent two of the most important devices of his life—iron rope and wire suspension cable.

Although some suspension bridges had utilized iron bars as chain suspension cables, Roebling and other engineers knew that iron wire would provide greater strength at a fraction of the weight of iron bars. As a result, the use of wire made the highly efficient concept of suspension bridges even more efficient. However, there were several problems with using wire. When exposed to moisture, the relatively slender wires would be prone to rusting. In addition, a way had to be found to ensure each wire sustained uniform loading. Roebling's solution was to drape the wires parallel to each other to form a cylindrical shape. To weatherproof the bundle of wires, Roebling recommended oiling the wires and then tightly wrapping a protective sheath of wire around the cylindrical bundle then painting it. He sought a patent for this remarkably logical idea in 1841. This technique would form the basis for suspension cables on all major suspension bridges for the next 150 years.

His other innovation was even more brilliant in its simplicity. Having watched the use of plant fiber ropes in the pulling of rail cars up hills and mountains on America's portage railways, Roebling saw how often the ropes failed. It was then that he hit upon the idea of braiding iron wire into rope and using it for hauling the boats to the next canal. Such a rope would be stronger and lighter, last longer, and be safer for all involved. Iron rope also promised to be no more expensive than the huge hemp ropes that sometimes measured nine inches in diameter.

Roebling, whose home remained near the settlement he helped found, hired his neighbors to help him braid iron wires into ropes. The iron was woven together and then twisted by hand-powered machines in the same way hemp rope was made. The result was cabling equal to the needs of the

Industrial Revolution. Roebling, who later developed machinery to help with the process of creating wire rope, wasted no time in applying for a patent on the process, which was granted in 1842. After meeting resistance and even instances of sabotage, the iron rope was put into use on portage railways and later found acceptance as rigging on sailing ships and as elevator cable. Roebling later promoted the use of iron cable to replace belts in the transmission of power from waterwheels.

Roebling's first bridge project would carry neither wagons nor trains but a canal. An aqueduct carrying a canal across the Allegheny River at Pittsburgh had suffered so much ice damage during the winter of 1843–1844 that it required replacement. When Roebling learned of this, he immediately went to work designing a bridge as replacement.

Employing his method of wire cable construction, Roebling designed a suspension bridge of unparalleled strength and successfully sold his concept to the canal company, despite competition from 43 other designs. Roebling accomplished the task within the $62,000 budget allocated, proving himself not only as an engineer but as an administrator as well.

Even as he worked to complete the bridge, an April 10, 1845, fire that razed much of Pittsburgh also consumed a wooden bridge over the Monongahela River at Pittsburgh's Smithfield Street. This calamity provided Roebling another bridge-building opportunity since the bridge was crucial to the devastated city. He proposed a suspension bridge as a replacement and his design was accepted.

Armed with a $55,000 budget, Roebling constructed the bridge using seven piers (remnants of the destroyed bridge) that carried eight spans measuring 188 feet apiece for a total length of 1,504 feet. As in the Allegheny Aqueduct, suspension cables also supported Roebling's Monongahela Bridge. Roebling had economically converted the old abutments for the destroyed bridge into anchorages for the suspension cables. The bridge carried traffic across its two streetcar tracks, two-way carriage lanes, and double walkways from 1846 until 1881, when it was demolished to make way for a larger bridge.

By 1848, Roebling's wire and cable operation had expanded so greatly that he decided to move the enterprise to Trenton, New Jersey, where he constructed a new factory site that was operational by 1850. By the first half of the 20th century what eventually became John A. Roebling's Sons Company flourished, employing 10,000 workers.

Regardless of his success as a manufacturer of iron rope, Roebling was still consumed with a desire to build suspension bridges, and in 1845 he found himself competing directly with his old rival Ellet for another bridge contract. This one involved a suspension span across the massive Niagara Gorge, just down the river from the massive falls for which it was famous. Ellet's salesmanship gained him the contract by 1847, and while he managed to construct a narrow walkway over the Niagara, his relationship with the Canadian and American bridge companies financing the product soured. Ellet eventually left the scene, to be replaced by Roebling, who undertook the job of building the bridge in 1851.

Although Roebling would later build bigger bridges, none was to be as visually dramatic as this one across the 244-foot deep Niagara Gorge, spanning an 820-foot-wide chasm. Roebling proposed a suspension bridge capable of carrying an upper deck for railroad use and a lower deck for carriages. Roebling's successful completion of the Niagara project stunned engineers around the world, who were disinclined to trust suspension bridge designs.

During this period Roebling established his method of supervising his rope and cable business by correspondence during extended absences from the Trenton home where his wife was raising a growing family by 1851. Although the couple had a total of eight children, Roebling stayed busy with his engineering projects and gave little heed to his wife's responsibilities, in line with prevailing attitudes of the day.

Despite his liberal political views, fair treatment of others in business, and basic decency that included his hate of slavery, Roebling was austere, remote, and authoritarian. He paid little attention to his wife and family as he pursued his single-minded goal of earning a living and, most importantly, building bridges. So removed was Roebling from his family life that he learned that his wife had given birth to his fourth son in 1854 in a letter from his trusted factory manager Charles Swan. Correspondence indicates that Swan had become something of an assistant to the entire family. It was Swan to whom the son, Washington Roebling, had to write for books and postage stamps while away at school, and it was Swan who would write to the traveling Roebling of Roebling's wife's failing health.

The Niagara Falls Bridge was essentially completed in 1854, but the first locomotive to cross it did not do so until March 8, 1855. For 41 years the bridge endured passing trains and wagon traffic, prompting Roebling's biographer Steinman to point out that the lifespan of less exotic bridges during the same period was around 20 years. The bridge is considered to be of a hybrid design since it employs not only suspension cables but also cable stays that run diagonally from the bridge's towers across to the road decks.

Roebling authored a paper on the principles behind the construction of the span, explaining that suspension bridges could be made stable through the use of "weight, girders, trusses and stays." Roebling also issued a warning:

And I will here observe that no suspension bridge is safe without some of these appliances. The catalogue of disastrous failures is now large enough to warn against light fabrics, suspended to be blown down, as it were, in defiance of the elements. A number of such fairy creations are still hovering about the country, only waiting for a rough blow to be demolished.

Roebling also referred to the May 17, 1854, collapse of Ellet's Wheeling Bridge, whose deck was smashed by winds that turned it into a writhing and undulating mass.

However, Roebling's design and construction of the Niagara Falls Bridge proved the worth of a true suspension span without intermediate piers. Its successful completion propelled Roebling to the forefront of the community of bridge engineers.

Roebling's reputation was such that he was able to acquire the contract to design and build the Cincinnati-Covington Bridge over the Ohio River. Work on the bridge's pier foundations began in 1856, but winter weather halted the work until 1857, when a financial panic crippled efforts to fund the project. When money did become available for construction, the 1861 outbreak of the Civil War again delayed work on the bridge.

Roebling, who had been appalled by slavery since his arrival in America in 1831, was dining with his family in their Trenton home soon after the April 1861 Confederate bombardment of Fort Sumter. Roebling family lore tells of the senior Roebling's barking at his son, Washington, who was attempting to eat a potato, "Don't you think you have stretched your legs under my mahogany long enough?" The younger Roebling, who was no less tough than his blunt father, walked away from his family's home without a word, enlisted in the army, and did not communicate with his father for another four years.

In November 1864, Roebling was notified of the death of his wife and traveled home for the funeral. The stern and frequently absent Roebling wrote in the family's Bible of his wife, "My only regret is that such pure unselfishness was not sufficiently appreciated by myself." The Cincinnati-Covington project, one of Roebling's most magnificent, was finally completed in 1867.

However, one final suspension bridge remained in Roebling's career. Roebling was well familiar with the time-consuming ferry ride across the East River between Manhattan Island and Brooklyn, and as early as 1857 he had mounted his own campaign to convince the public that a suspension bridge could span the divide. The publisher Horace Greeley printed Roebling's proposal in a newspaper, but the daring plan attracted more doubters than support, as Greeley himself expressed leeriness about the plan. Although Roebling had successfully built a suspension bridge across the 820-foot Niagara Gorge, the distance between Manhattan and Brooklyn was seven times that, a daunting 5,989 feet.

Roebling's reputation, however, had only grown during the intervening years. In May 1867, the New York Bridge Company unanimously approved Roebling as the designer and builder of the Brooklyn Bridge.

The span he proposed would consist of two towers on each side of the river from which cables would support a center span of 1,595.5 feet. Had he offered this bold design before completing his Niagara and Cincinnati-Covington bridges, Roebling would have been literally laughed out of town. However, by 1867, Roebling's daring plan was being taken seriously, although some wondered whether the central span was too long to be practical.

Roebling was on the Brooklyn shore conducting a final survey of the bridge site on July 6, 1869, when a ferry slammed into the dock, compressing the pilings on which he stood, smashing one of Roebling's feet. Roebling was taken to his home, where a surgeon amputated the crushed toes of his right foot. Characteristically, Roebling told the physician to get on with his grisly chore and endured the surgery without anaesthetic. Unfortunately, the injury led to an infection with tetanus, a bacteria that finds a friendly habitat in deep wounds.

Roebling, like many self-made people with overpowering personalities, believed one could ward off sickness through sheer will. (Roebling was convinced he had survived a cholera epidemic during construction of the Niagara Falls Bridge by pacing in his room and concentrating on remaining well.) Roebling would be less lucky with tetanus, an infection that attacks the nervous system. For weeks, Roebling held on, but the horrible symptoms, which include severe muscular spasms and convulsions, wracked his body. The German immigrant's great journey ended July 22, 1869. Roebling was 63 years old.

After his death, Roebling's title of chief engineer would be passed to his son, Washington, who would oversee its construction for the next 14 years. When finally built, it would awe the world and become a symbol for New York City. The John A. Roebling's Sons Company became a premier bridge subcontractor by specializing in spinning suspension cables with wire produced at its Trenton factory. The firm spun the suspension cables of the Golden Gate Bridge and the George Washington Bridge.

Roebling's genius can still be seen in the Brooklyn Bridge and the Delaware Aqueduct built in 1847, which remains as the oldest American example of a suspension bridge.

See also BRIDGE AERODYNAMICS; BROOKLYN BRIDGE; CINCINNATI-COVINGTON BRIDGE; DELAWARE AQUEDUCT; ELLET, CHARLES, JR.; GOLDEN GATE BRIDGE; NIAGARA GORGE SUSPENSION BRIDGE; ROEBLING, EMILY WARREN; ROEBLING, WASHINGTON AUGUSTUS; WHEELING BRIDGE.

Roebling, Washington Augustus (1837–1926) *Chief engineer of the Brooklyn Bridge* The first child of the imperious, stern, and brilliant engineer John Augustus Roebling, Washington Augustus Roebling cemented his place in engineering history by supervising the completion of the Brooklyn Bridge after the death of his father.

Born in Saxonburg, the settlement founded by his father, Roebling was the recipient of his mother's kind affection and his father's stern discipline and volcanic anger. In 1849, John Roebling moved the family's wire rope business to Trenton, New Jersey, where the 12-year-old Roebling attended a private school and worked in his father's wire rope mill.

At 17, Roebling was then sent to the Rensselaer Polytechnic Institute in Troy, New York, to acquire an education in engineering. While there he rubbed shoulders with classmates such as Theodore Cooper, later an assistant engineer on Saint Louis's Eads Bridge and consulting engineer on the ill-fated Quebec Bridge.

On graduation in 1857, Roebling entered the employ of his father, manufacturing wire rope in Trenton, and assisted him in the construction of bridges, including one over the Allegheny River at Pittsburgh. All this changed in 1861, when the senior Roebling barked at his son Washington, who was attempting to eat a potato, "Don't you think you have stretched your legs under my mahogany long enough?" Without a word, the younger Roebling rose from the table, left home, and enlisted in the army. He did not speak with his father during his four years of military service.

Unflappable and capable, Washington Augustus Roebling oversaw the construction of the Brooklyn Bridge. After being severely disabled by caisson disease, Roebling was unable to visit the project during its last 11 years and his wife, Emily, acted as his engineering emissary. *(Library of Congress)*

Roebling's military career was nothing less than extraordinary. The 24-year-old Roebling enlisted as a private in the New Jersey State Militia when he could have immediately obtained an officer's commission or avoided military duty entirely. He saw action as an infantryman and within four months was promoted to sergeant when his unit became an artillery battery. Four months later he was elevated to the rank of lieutenant.

Because of the unusual organization of the Roebling family, most of the young man's letters home were written not to his father but to Charles Swan, John Roebling's trusted factory manager. These letters chronicle Roebling's exploits with his artillery unit in Virginia and his subsequent transfer to the staff of a general, when Roebling became an engineering officer assigned to build dozens of bridges during the next three years. Roebling also designed a military suspension bridge that could be erected by troops in a short period, with the John A. Roebling Wire Rope Company manufacturing the cabling for the bridges.

As a result of his engineering duties and the importance of bridges in the wartime movement of troops and equipment, Roebling found himself dispatched throughout Virginia, where he struggled to build bridges to replace those destroyed by floods or Confederate activity. A man of few words with a laconic sense of humor, Roebling was no dashing George Armstrong Custer but an educated civil engineer who worked shoulder to shoulder with his men in building bridges. During one such assignment Roebling managed to erect a 1,000-foot suspension bridge across the Rappahannock River using the existing piers.

After the battle at Chancellorsville, Roebling was ordered aloft in a captured Confederate hot air reconnaissance balloon to observe the disposition of the enemy forces. His precise and detailed reports provided the first warning that General Robert E. Lee was heading to Pennsylvania in a move that would result in the Battle of Gettysburg. Throughout the war the engineering officer would turn up precisely where he was needed.

When it was learned that the Union army had no worthwhile topographical maps of the Gettysburg, Pennsylvania, area, Roebling realized his father possessed one and rushed to Trenton, New Jersey, to retrieve it for military planners, a visit about which his father knew nothing. After eluding Confederate troops for days, Roebling arrived with the map on the first day of the historic three-day battle.

Roebling was with General Gouverneur Kemble Warren, the Army's chief of engineers at Little Round Top. Warren had ordered the important position seized with a small number of troops when he realized the Union forces were about to be outflanked by the troops of the Confederate general John Bell Hood. Then Roebling and Warren spotted another Union brigade and directed them to Little Round Top, where Roebling helped to drag artillery into position to thwart Hood's furious assault. The defense of Little Round Top is credited with preventing a devastating rout of Union forces.

As the war continued so did Roebling's promotions. He would rise to the rank of colonel, attaining the position of aide-de-camp to Warren, and for the remainder of his life he would be known as "Colonel Roebling." A fellow officer described Roebling's personality as that of a man who appeared unexcited if not completely bored by the dramatic activity around him. Roebling was also described as having tattered trousers cuffs, caused by his propensity for wearing low-quarter shoes instead of boots in the field. During the last month of 1864, Roebling was a member of a raiding party that plunged behind enemy lines to destroy a railroad under Confederate control.

It was during his duty with Warren that Roebling met and fell in love with Warren's younger sister, Emily, an educated and capable woman who would later play a crucial role in the building of the Brooklyn Bridge. The two were married January 18, 1865. Soon afterward Roebling was released from duty, whereupon he went to Cincinnati to help his father complete the Cincinnati-Covington Bridge, a project delayed by the Civil War. During this two-year period Roebling was able to familiarize himself with large suspension bridge design.

The next major project the two would cooperate on was perhaps the most stupendous civil engineering project of the 19th century—construction of the Brooklyn Bridge, with a central span of more than 1,500 feet. After John Roebling was granted the contract to build the bridge in 1867, Washington Roebling was dispatched to Europe, where he and his wife traveled for a year observing pneumatic caisson work and bridge construction techniques.

A month after his father's July 22, 1869, death of tetanus, Roebling was appointed chief engineer of the bridge project. During the dangerous and

Herculean efforts to sink the pneumatic caissons to bedrock for the piers, Roebling demonstrated the same cool courage he had exhibited on the battlefields of the Civil War.

During the excavation of the Brooklyn caisson in 1870, a smoldering fire ate its way into the timbers of the caisson's roof, prompting Roebling to order it evacuated and flooded. When man-sized boulders began to block excavation work in the caissons, Roebling decided that blasting was the only way to expedite the work, and it was Roebling who entered the caisson to conduct the dangerous experiments with black powder before going ahead.

Roebling and a group of workers almost lost their lives in mid-December 1870, when the upper door of a caisson's supply shaft was left open at the same time the lower door was unlocked to allow a load of gravel into the pressurized work area. This caused a loss of air pressure that allowed water to flood into the submerged excavation chamber. As the water rose, Roebling realized the upper door was allowing the air to escape and managed to slam shut the lower door, saving all from drowning.

But the bridge project that indirectly led to his father's death would also come dangerously close to killing the son as well. Roebling, who led by example, was frequently in the deeper New York caisson for longer hours than the average worker judging the progress and making recommendations. Roebling, as were at least as many as 100 workers on the project, was stricken by the bends. An incoherent Roebling was carried from the caisson to his home in the summer of 1872, when all believed his death imminent.

Roebling bounced back, but when the 35-year-old engineer attempted to return to work he experienced a relapse at the caisson. This time Roebling suffered a full range of caisson disease symptoms as the expanding nitrogen bubbles paralyzed much of his body, impaired his vision, and created an unbearable nervousness in the usually unflappable Roebling. Even his vocal chords were affected. Horrified he would die before the bridge was completed, Roebling began dictating engineering instructions to his wife and spent his days propped up in a chair watching the construction from the window of his home in view of the bridge site. Incredibly, from 1872 until the bridge was completed in 1883, Roebling did not set foot on the work site, instead eyeing the progress through binoculars or a telescope from his home.

In 1882, newspaper criticism of Roebling as an absentee engineer rose to a crescendo, and the trustees of the Bridge Company demanded he attend a board meeting. When he declined to do so on health grounds, a resolution was passed to strip him of his job as chief engineer. Emily presented his case to the American Society of Civil Engineers with such eloquence that the American engineering community agreed that Roebling should remain in his position, silencing the critics and preserving his role in building the bridge.

The completed bridge was officially opened to tremendous fanfare on May 24, 1883, and Roebling, still heavily afflicted by caisson disease, watched the celebration from his window.

See also BROOKLYN BRIDGE; CINCINNATI-COVINGTON BRIDGE; ELLET, CHARLES, JR.; ROEBLING, EMILY WARREN; ROEBLING, JOHN AUGUSTUS; SUSPENSION BRIDGE; WHEELING BRIDGE.

Royal Albert Bridge at Saltash (built 1855–1859)

It is too bad the British civil engineer Isambard Kingdom Brunel's creative interests were so far-flung. Had he not fragmented his genius in so many directions he might have produced more extraordinary bridges like the Royal Albert Bridge near Saltash, England.

Designed by Brunel for the Great Western Railway for a crossing of the Tamar River in England's Cornwall region, it resembles no other bridge. Completed two years before the start of the American Civil War, the bridge has an eerily modern, if not alien, look. The huge, lenticular tube arches curving high above the river on slender columns are visually jarring.

By the mid-1800s, British railways were still working to link Great Britain by rail and still finding rivers, estuaries, and straits among their biggest obstacles. By the mid-1840s, Great Western Railway had extended to the Tamar River, a 1,100-foot-wide waterway west of Plymouth, England. Designs were called for and all were found wanting for one reason or another. Brunel had been the chief civil engineer for the railway and had built scores of its railway bridges, although, to accelerate their construction, most had been of timber. He proffered several timber bridge designs but was turned down since they would not be tall enough to allow the passage of ships. He then proposed a four-span bridge, which was vetoed since it required too many submerged piers, and then a

One of the world's most unique bridges, Isambard Kingdom Brunel's Royal Bridge at Saltash is supported by a pair of lenticular arches composed of hollow iron tubes at the top and eyebar chains below. An ailing Brunel died the same year that this bridge was completed. *(Godden Collection, Earthquake Engineering Library, University of California, Berkeley)*

single-span bridge, which would have been unacceptably expensive.

At last, Brunel conjured up the design that was accepted. It would contain a pair of 465-foot elliptical tube arches of iron supporting two spans. The arches, with their humps reaching skyward, would be ribbed, or connected to the rail deck by vertical iron beams. Suspension chains connected at the ends of the two arch tubes would provide suspension bridge–type support. The downward-arching iron chains would also visually counterbalance the upward arch of the nine-foot-wide tubes. Brunel's practical experiments demonstrated that the semicircular tube arches were the best means of distributing the stresses of supporting the bridge.

In a strange twist of fate, the iron chains obtained for the bridge were originally forged for the Clifton Bridge designed by Brunel in 1829—his first bridge construction effort. Because of problems in raising adequate money, the Clifton Bridge project had languished for two decades. Since the chains

were unused, Brunel purchased them for the bridge over the Tamar River.

The design required only one deep foundation set on rock excavated below the river's 70-foot depth and two shallow foundations near each bank. The bridge's approaches would be supported not by arches but by tall, slender masonry columns. Brunel began experimenting with the most practical ways of excavating down to bedrock, and in 1848 he initially designed an open-ended iron tube to serve as a cofferdam. The tube was pumped dry and workmen entered to make test borings of the rock. As financial difficulties delayed full-scale work on the bridge, Brunel decided that the best way to allow workers to excavate to the submerged bedrock was with the use of a pneumatic caisson. Brunel designed a unique pressurized caisson that consisted of a pair of concentric tubes in which only the outer tube was pressurized, where workers would be excavating to allow the tube to cut toward the bedrock. The interior tube would be open to normal air pressure where other workers would labor to remove the rest of the spoil from the riverbed.

The effort to reach the bedrock required excavating to a depth of 16 feet beneath the muck. Once seated, masonry would fill the caisson, providing the bridge its center pier foundation. When the two shallow water foundations and their masonry piers were completed, the cast iron caisson used for the deep, central foundation was unbolted and towed away, ending the foundation work in 1855.

The innovative construction techniques used on the bridge did not end with the caisson work. The truss-reinforced arches with their suspension chains were assembled on shore and floated into position on specially designed pontoons. These pontoons could be filled with or emptied of water to allow them to rise and fall while afloat. Each of the two 465-foot sections weighed 1,000 tons. Once completed, the first span was loaded on its pontoons and towed between its not-yet-completed supporting columns on September 1, 1857. In a careful choreography, the water was pumped from the pontoons and the huge bridge section was elevated until it could be placed atop the column. Although this modular form of bridge construction is now commonly known as the large block method, its use in Brunel's time was unprecedented.

Much of the construction work on the bridge was either unseen inside the submerged caisson or rather mundane, but the lifting of the prefabricated arch section was a local event. Brunel predicted this situation and issued pleas for absolute silence as five navy ships helped maneuver the floating bridge section into place directed by flag signals. When the job was completed without a hitch, the spectators let out a rousing cheer and a Royal Marine band struck up a tune. As masonry work progressed, hydraulic jacks would raise the section higher until the supporting columns were completed. The second section was raised in the same way in July 1858. Once the truss-reinforced arches and their suspension chains were in place, iron girders were attached to form the rail deck of the bridge.

Sadly, this was Brunel's last great endeavor. By the time the bridge was nearing completion the engineer was extremely ill, suffering with Bright's disease, a serious kidney ailment. Before his death on September 15, 1859, an ailing Brunel was taken by rail car to the bridge, when he had his only look at the completed project. England's Prince Albert, the husband of Queen Victoria, officially opened the bridge May 3, 1859.

A tribute was paid to Brunel by the firm that produced much of the iron work for the bridge by manufacturing raised letters spelling out *I. K. Brunel, Engineer* on the faces of the columns at each end of the bridge. Coming and going, those aboard trains crossing the river are left with no doubt as to the identity of the bridge's designer.

The bridge, which was to have accommodated a double-track rail line, was reduced in scope to a single track to save 100,000 British pounds at the time of its construction. Looking virtually new, the bridge carries rail traffic east and west over the Tamar River. By 2000, it had been in service for 141 years.

See also BRUNEL, ISAMBARD KINGDOM; SMITHFIELD STREET BRIDGE.

running ground   One of the many unpleasant surprises awaiting miners is the discovery that their seemingly solid tunnel has been invaded by a flow of wet, jumbled rock fragments. Known as running ground, this debris can be pushed into a tunnel with extreme force capable of damaging or collapsing steel or concrete supports. Once it breaks into a tunnel, it resembles a landslide.

The problem is usually the result of fractured rock surrounding the tunnel that not only provides the source of the debris but also is no longer capable of providing structural support to its own mass.

One of the major problems associated with running ground, so named because it moves into and through the tunnel, is that it usually carries with it a considerable amount of water. This must be drained and the debris removed.

Although grouting, steel ribs, and concrete supports can provide structural strength against running ground, an even better tactic is to find ways of draining off subterranean water, which heightens the ability of the crushed rock to move. Waiting until the last safe moment to reinforce the tunnel allows the remainder of the running ground to make its way into the excavation.

See also PROGRESSIVE FAILURE; SQUEEZING GROUND.

safety lamp (1815) *Lighting the way safely for miners* Miners working deep beneath the ground to extract coal live in an environment rife with potentially fatal hazards, the worst of which are the explosive coal dust and methane that they frequently encounter. Because miners needed light to work and because flaming candles and lanterns were the only source of light for centuries, miners often triggered the massive explosions that claimed their lives.

Ironically, two of England's most famous inventors independently and simultaneously came up with deceptively simple ways of providing a flame-based light for miners that not only eliminated the chance of combustion but also provided a means of detecting flammable methane gas.

Although the lamps were designed not to trigger mine explosions, the resulting controversy over who should have been credited with inventing them detonated a dispute that continues until the present.

Generally credited as the first to come up with a safe lighting solution for coal miners was George Stephenson, a self-educated engineer who began his working life as the operator of a steam engine that ran the water pumps in a coal mine. By 1815, Stephenson was already conducting experiments with a safety lamp after determining that a small enough opening could allow air to feed an illumination flame and still prevent the flame from reaching the potentially explosive atmosphere outside the lamp.

Stephenson, who had proved his physical courage and coolness in the face of danger when he once successfully directed efforts to extinguish a sub-terranean mine fire, carried his safety lamp in a mine filled with methane to prove it worked. Stephenson, who became something of a working-class hero among local miners, was given a gift of 1,000 pounds raised among the impoverished miners with whom he worked. Although Stephenson would later go on to design the first practical locomotive, the safety lamp was his first notable invention.

At virtually the same moment, the famed English chemist and scientist Sir Humphry Davy also grasped the concept of allowing the entry of oxygen to a lamp while preventing the escape of flame by using an extremely fine metal grill. When Humphry learned that Stephenson had actually tested a lamp based on essentially the same principle, the brilliant Davy waged a bitter and long campaign to establish himself as the inventor of the safety lamp.

Stephenson, true to his kind and generous nature, did not reciprocate in this affair, choosing instead merely to continue in his own pursuits. Although miners used Davy's lamp, some complained that the fine metal grill could sometimes grow hot enough to glow red and trigger the combustion of methane gas. Stephenson's lamp would merely sputter out, warning miners that a dangerous level of methane was present.

These safety lamps and later versions based on Davy and Stephenson's designs were used until the beginning of the 20th century, when lamps using acetylene generated by the combination of carbide and water came into use. These newer lamps provided a considerable amount of light, although they still presented the danger of an open flame.

However, by the 20th century, greater efforts were being taken to reduce the levels of methane and coal dust in mines through ventilation.

See also STEPHENSON, GEORGE; VENTILATION.

Saint Barbara *Patron saint of miners* If a group of workers ever needed a patron saint, it is the fraternity of miners and tunnel builders, who have traditionally labored under some of the cruelest and most dangerous conditions in the world. The patron saint of these underground workers is a woman who is said to have sought refuge in a cave after her conversion to Christianity threw her father into a murderous rage.

The only problem with Saint Barbara, according to the Catholic church, is that she may be nothing more than an extremely interesting myth. The saint, in short, is officially no saint. But this small detail is not about to erase the devotion of miners around the world, who have prayed and built subterranean altars to Saint Barbara for nearly 1,300 years.

The stories of Saint Barbara are epic tales of a woman subjugated by a cruel father around A.D. 300. Generally, these stories feature a pagan father fearful of her interest in Christian theology who imprisons her in a tower. While dad tries unsuccessfully to force her into arranged marriages, Barbara spends her time studying Christian theology. When he learns that she is a devout Christian, her father becomes homicidal.

As the story goes, Barbara temporarily escapes her father's clutches and hides inside a mine until betrayed by a shepherd. Because her conversion is the last straw, authorities order that Barbara endure a series of hideous tortures, each of which is miraculously thwarted. Dear old dad is then authorized to decapitate his own daughter and happily complies, since nothing else seemed to help. No sooner does he lop off his daughter's head than a bolt of lighting thrown down from Heaven kills him on the spot.

The tale of Saint Barbara has so many versions that she has been considered a patron saint by everyone from architects to artillerymen and is believed to provide protection from all forms of violent and sudden death. In the case of miners, who face floods, cave-ins, and suffocation, she seems to be a most appropriate saint, especially since a mine provided her temporary refuge.

The Catholic church, after years of study, ruled that Saint Barbara was nothing more than a good yarn and could not be considered a true Christian martyr. Nonetheless, miners still erect shrines to Saint Barbara at the entrances to tunnels, believing she shelters them from harm. Like soldiers who say there are no atheists in a foxhole, miners apparently believe none exists in tunnels, either.

The tradition of paying homage to Saint Barbara is not considered foolishness even by 21st-century miners engaged in the most impressive and modern projects. Swiss miners working on a new 35-mile Saint Gotthard high-speed rail tunnel have erected shrines to Saint Barbara at the entrances to the tunnel, a project whose completion date is 2010.

The Saint Barbara tradition is so strong that Saint Barbara Day, which is December 4, is often a holiday for miners and tunnelers. Festivities sponsored by mining companies often occur on this day, particularly in Austria.

Saint Gotthard Tunnel (built 1872–1882) *Alpine passage to death and financial ruin* By the mid-1800s every town in Europe and America wanted a rail line, a connection to the outside world, a swift, reliable link to goods and customers. But most of all, no one wanted to be left out or left alone while the rest of the world was connected by the locomotive. And the Swiss knew that a rail line would someday cross the rugged Alps at Saint Gotthard Pass.

For centuries, north-south travel between Italy and Germany moved over the Alps at Saint Gotthard Pass, whose 6,915-foot elevation had been crossed by a pack trail since at least the 6th century. The pass was the gateway through the mountains connecting Italy to the south with northern Europe. The trail had been widened into a carriage road in a 12-year effort beginning in 1818, but a railroad would be far better. For 18 years politicians orated, businessmen lobbied, and average persons thought of the opportunities that would trickle down to them.

The options were plain: Lay switchback tracks near the summit of the pass and then drive a short tunnel through the rock, or punch a far longer tunnel through the base of the mountain range to eliminate the need for steeply graded tracks. The second plan was bold, but the first would expose the high-altitude rail line to impassable winter snows. In 1871, a conference selected a plan for a long tunnel prepared by two engineers, R. Gerwig and A. Bech.

The plan proposed a nine-mile tunnel through the Alps north of Airolo and south of Göschenen. Although the plan was daring, the completion that year of the 8.5-mile Mont Cenis Tunnel linking France and Italy proved that the Saint Gotthard Tunnel was possible. However, it would not be easy.

Money for the effort was scarce, but the governments of Germany and Switzerland both pledged 20 million francs, and the Italian government agreed to pay 45 million francs. The remainder of the money would be generated by revenue bonds issued by the Saint Gotthard Railway Company, which would build and operate the rail line. The railway company, made leery by horror stories of tunnel projects beset by technical problems and cost overruns, drew up a severe construction agreement that horrified every contractor but one.

Louis Favre, a Genovese contractor, was the only one confident enough (or insane enough) to agree to the impossible terms of the contract: Favre obligated himself to finishing the tunnel within eight years and put up a forfeiture bond that today would be equal to $9 million. Alone, either of these terms could destroy him, but Favre also agreed to pay a penalty equivalent to $5,500 for each day he overshot the October 1, 1880, deadline. After six months the penalty would rise to $11,000 a day. Should he finish the job ahead of schedule—truly a fantasy—he would be paid the $11,000 amount for each day left till the deadline. Missing the completion deadline by a full 12 months meant the entire amount of the forfeiture bond would revert to the Saint Gotthard Railway Company.

Favre was no fool but rather a carpenter who had struggled to become a wealthy construction entrepreneur. He was willing to gamble that he could bring the project in on time and within budget. Favre made two fatal errors in his approach to the tunnel project: First, he accepted the validity of a geological survey of the rock conditions that erroneously anticipated solid, self-supporting rock for the entirety of the tunnel route. Second, Favre chose a slow tunneling method.

The rock was far from solid and was honeycombed with fissures carrying millions of gallons of water ranging from scalding to cold. The enormous task and the conditions under which nearly 400 workers would die generated labor unrest and strikes as well as official inquiries into the tunneling operations. This epic struggle on the tunnel began on October 9, 1872.

Preliminary investments had to be made to provide housing for nearly 2,000 men in villages on each side of Saint Gotthard Pass and for the construction of water-powered air compressors and attendant workshops. Compressed air was essential to power pneumatic drills whose hoses and exhaust would pump fresh air to the miners. By the time Favre's men had built the portals on both sides of the pass, Favre had already consumed two years of the eight he had for the job.

Favre had advantages that the Mont Cenis tunnelers lacked or decided not to use—dynamite and the Ferroux rock drill. Dynamite, invented in 1866, was a safer, more powerful explosive than black powder, and the Ferroux innovation was a capable pneumatic drill that was employed on the northern heading at Göschenen.

After blasting less than 500 feet into the tunnel from the north, Favre's men encountered high-pressure water jetting from the "solid" rock at a rate of 2,600 gallons per minute. The southern advance encountered conditions that were even worse, and tunneling operations there took place in a wet hell of high-pressure water. At 2,000 feet miners in the southern heading were inundated by water roaring through the rock at 3,000 gallons per minute, a travail that did not end for two full years.

Far from a mere nuisance, the water required the miners to install drainpipes and to work in unhealthy conditions. Respiratory infections and diseases were rampant. Dynamite, which itself was a mixture of clay and nitroglycerin, sometimes melted and poured out of the shot holes before it could be detonated.

Saint Gotthard Pass was to provide an endless supply of discomfort, death, and disease to those who tried to circumvent its heights. A parasitic intestinal infection known as "miner's anemia" felled thousands who contracted it in the unclean tunnel and the poorly organized mining camps. When the tunneling was dry, dust choked the air.

Many workers, according to Göstra Sandström in *Tunnels*, could not last more than four months before they were physically broken. Since animals were widely used in the tunnels to transport materials, they also suffered a fearsome toll, as roughly 30 horses and mules died or were killed monthly. All this occurred in an environment heated to nearly 122 degrees Fahrenheit. In the 10 years of building the tunnel an average of 37 workers died each year.

Despite the obstacles, Favre made extraordinary gains with his workers. By May 1875, the combined progress of both headings totaled 2.3 miles. At this point it appeared to Favre that he now had an even chance of complying with his impossible construction agreement. But costs on the project had soared as the difficulties mounted, and the original $205 million railroad and tunnel project had ballooned to $315 million in today's dollars, of which $52 million was for the tunnel alone. The stock lost value and the German government threatened to withdraw support, but Favre's accomplishments in the face of horrible conditions resolved the crisis by reassuring investors.

The multinational work force from Germany, Switzerland, and Italy was also unhappy with living and working conditions. A local physician reported to the Swiss Parliament that the wet roads in Airolo and Göschenen were sewage mingled with mud. Hearings were held and Favre defended his treatment. Asked whether he was endangering workers for the sake of profits, Favre denied the charge but qualified his answer by saying, "Such a task is not possible without victims." Although brutal by today's standards, Favre's attitudes were in line with those of other 19th-century managers who believed that the war of progress required casualties.

Despite Favre's defense of his methods, historians have noted that Favre's men were driven beyond safe limits by the demands of the contract. And although Favre may not have been pushing his workers for the sake of profit, he was certainly pushing them to avoid incurring a crushing financial loss.

When a group of Italian workers decided to lay down their tools and strike on October 27–29, 1875, local police were called and brutally put down the work stoppage. Three of the workers were killed by police gunfire, and others were wounded. This incident remains a black mark in Switzerland's history of labor relations. It was also made clear that police and military personnel would readily back business when labor problems threatened the large tunnel projects of the 19th and early 20th centuries.

The cloistered nation of Switzerland was also transformed by projects like the Saint Gotthard Tunnel. Since Switzerland had too few engineers and craftsmen for the projects, Italians, Germans, and Austrians rushed to take the jobs. It is estimated that only 10 percent of those who worked on the Saint Gotthard Tunnel were Swiss since local people would not work for the low wages accepted by the foreigners. These projects are what converted Switzerland into something approximating an immigrant nation during the 19th and early 20th centuries.

By 1877, Favre's hopes began to dim. Progress slowed, and by the middle of that year his miners were 500 feet behind schedule as a result of the presence of extremely hard rock. Worse yet, brickmasons and their helpers were unable to keep pace with the massive job of lining the tunnel.

Then a series of hammer blows befell the project. An unexpected decrease in the amount of water to power the southern air compressors forced workers to drill by hand, a method 15 times slower than the use of pneumatic rock drills. A higher rate of drilling could not resume until an aqueduct was built to carry in additional water. Nothing was coming easily for the Saint Gotthard Tunnel.

The deeper the miners went beneath the pass, the poorer the air became and the higher the temperature rose. The heat was partly the result of the naturally occurring geothermal gradient, the rate at which higher temperatures are encountered while descending into the Earth. This, the body heat of men and animals, along with the heat generated by machinery and lighting, raised the tunnel's interior to 122 degrees Fahrenheit near the headings.

The air quality, which was never good, worsened as the headings advanced farther from the portals. Since the depth of the tunnel precluded intermediate ventilation shafts, it became more difficult to provide adequate ventilation. Men and animals both consumed what oxygen there was while exhaling carbon dioxide. The fetid, dank tunnel was often flooded and filled with sewage. In the middle of this maelstrom was Favre, pushing his way through the water, shouting directions, and urging his men to work as rapidly as possible.

On July 19, 1879, in the seventh year of his struggle, Favre was inside the tunnel when a massive heart attack took his life. The 53-year-old Favre was buried at Göschenen's Gotthard Cemetery with the workers he had pushed so hard. Although he had presided over a workplace health disaster of immense proportions, he had not done so from a distant office.

Only 15 months remained on the contract and the work struggled on under Favre's assistants. The crews in the opposing headings encountered each other February 29, 1880, when a final dynamite blast made both headings a single tunnel. What tempered the exhilaration of the moment was that the contract allowed only another seven months to widen the tunnel to its full dimensions and complete its masonry lining, tasks that were woefully behind schedule.

As if the geologic characteristics of Saint Gotthard were having one last laugh, a 300-foot section completed two years earlier began to ooze like a plastic blob into the tunnel. Its force was so great it collapsed the heavy masonry lining as well as subsequent repairs. This swelling ground, as tunnelers call it, is rock and clay whose volume increases exponentially after absorbing water or even moisture from the air. Workers had to be diverted to rebuild this section of the tunnel, using a massively thick floor, sidewalls, and arch to control the swelling ground.

After the tunnel segments were connected by the breakthrough, it took another 21 months to complete the tunnel. By the time the tunnel was finished in 1882, the project had exceeded the one-year overrun limit by 14 months. The dead Favre's unforgiving contract meant that the Saint Gotthard Railway Company would take his $9 million completion bond, destroying his company.

On May 23, 1882, the railway and the tunnel were completed and the obligatory celebration took place. Favre and hundreds of his workers missed the festivities as they slumbered on at the Saint Gotthard Cemetery beneath wilted sprays of edelweiss. Of those who labored in the tunnel, 177 died accidentally inside its confines and another 133 were killed outside. Another 877 were seriously injured and only 30 percent of the thousands who toiled in the tunnel escaped the parasite-induced miner's anemia.

The death toll among the tunnel workers became a national scandal. The tunnel, on completion, was the longest in the world, exceeding the length of the 8.5-mile Mont Cenis Tunnel by just over half a mile.

Favre's company was ruined by a grossly unfair contract enforced by the pinchpenny Saint Gotthard Railway Company. As for Favre, who essentially bankrolled much of the tunnel that was literally a nine-mile vale of sorrows, he eventually became something of a hero in Switzerland. In 1932, a 50th-anniversary postage stamp recognizing his role in completing one of Switzerland's most important tunnels was issued.

See also GEOTHERMAL GRADIENT; MONT CENIS TUNNEL; ROCK DRILL; SWELLING GROUND.

Saint Johns Bridge   *See* Steinman, David Barnard.

Salginatobel Bridge (1930) The Salginatobel Bridge of reinforced concrete in western Switzerland seems to spring from one side of the Salgina Gorge to the other. Painted a startling white, this bridge by the brilliant Swiss engineer Robert Maillart appears to be the essence of movement frozen in time. It has become an enduring symbol of modern structural design and represents a body of work by a designer little appreciated while he was alive. He would later be praised as much for his prowess as an artistic experimenter with visual design as his mastery of the structural realities of engineering, and this bridge would become the symbol of that creativity.

The Salginatobel Bridge is of a hollow-box arch design and a good example of the ingenious engineering and stylistic elements Maillart favored in his spans. The floor of the arch is a stupendously thin curve, and the sidewalls of the hollow-box arch run along its top and taper to the abutments on each side of the gorge. The bridge connects the Swiss town of Schiers to Schuders on the east side of the Salgina Gorge.

The Salginatobel Bridge provides an example of an efficient, cost-effective, and safe design that is thrilling to see—hallmarks of Maillart's philosophy. Maillart's concept of construction allowed his fascinating bridge designs to compete against established, old-fashioned bridge designs because his were more economical. Maillart's secret was to construct a thin but efficient arch of reinforced concrete, using relatively light scaffolding or falseworks to reduce construction expenses, a technique he followed in construction of the Salginatobel Bridge. Once the arch was in place, the scaffolding could be removed because the arch could assume the weight of additional construction. Often, falsework and scaffolding consumed a considerable amount of the construction budget and had to remain in place until a bridge was completed.

Maillart's bridge is almost skeletal in appearance as Maillart eliminated the spandrel walls, which were traditionally built to cover the empty spaces between the arch and the road deck at the abutments. He also stripped the bridge and its approaches of a traditional façade of stone to allow the world a gander at the unadulterated musculature of the bridge. The design is mechanistic, sleek, and spare.

Maillart's bridge over the Salginatobel is only barely an arch bridge since its curve has been severely reduced by its designer. But it is an arch

The Swiss civil engineer Robert Maillart's most famous visual statement is the Salginatobel Bridge. His masterful use of reinforced concrete resulted in a span that expressed energetic motion. Images of the Salginatobel Bridge, along with other Maillart creations, were exhibited at New York's Museum of Modern Art in 1947. *(Godden Collection, Earthquake Engineering Library, University of California, Berkeley)*

bridge nonetheless and contains movement-allowing hinges at the center and both ends of the arch. Concrete arches of all types are prone to failure at their apex and where they terminate at their foundations or springings. Maillart's employment of hinges remedied the problem. These hinges allowed the concrete segments to rotate without cracking.

Although the reduced arch is superior in terms of bearing the weight of the span across the football-field-length gorge, it also provides a sense of movement. The severe arch of a traditional bridge is bent almost to the point of becoming a semicircle and appears like an archer's bow pointed upward, awaiting the launch of an arrow. Maillart's shallow arch looks more like a bow with its stored energy in the process of being released.

The hollow-box arch enclosed by its own walls and floor and "roofed" by the road deck provides massive strength and stability to the bridge. The material used in the bridge is reinforced concrete, concrete with a skeleton of steel. The steel gives the Salginatobel Bridge's concrete the ability to endure heavier loads than the brittle concrete could handle alone. Built in 1930, the bridge continues to carry traffic and serve as an attraction for designers, bridge enthusiasts, and the public since it has become an icon of modern design.

The Salginatobel Bridge is derived from the experience and research gained by Maillart in his design of the Tavanasa Bridge 33 miles through Switzerland's rugged alpine terrain to the east. Maillart's designs were daring and emotionally jarring but perhaps too much so for staid Swiss authorities. As a result, they were often of medium size and consigned to rural regions, where their economy and reliability made them desirable. Their beauty was thrown in for free by Maillart.

See also ARCH BRIDGE; CONCRETE; HINGE; HOLLOW-BOX ARCH; MAILLART, ROBERT; REINFORCED CONCRETE; TAVANASA BRIDGE.

sandhogs *America's tunnel workers* Their nickname has a humorous ring to it, but those who excavated subaqueous tunnels and bridge foundations in compressed air burrowed a place in history as some of America's toughest and most courageous workers. Their existence in what is arguably one of the most dangerous work environments was complicated for decades by a complete misunderstanding of the causes of decompression illness, also known as caisson disease, a side effect of working in compressed air.

So high were the death and injury rates that America's sandhogs moved early in their history to organize and protest not only their wages, but also their monumentally hazardous working conditions.

Although *miner* and *sandhog* are essentially interchangeable terms, *sandhog* was the name initially reserved for those tunneling and digging underwater in New York City during the 19th century. They gained early fame when they labored in the submerged caissons of the Brooklyn Bridge's piers during the 1870s and in the train tunnels and subway tubes of New York, where a hazardous compressed air environment was often necessary to prevent water infiltration.

Aside from building America's bridges and tunnels, sandhogs are most famous for becoming horribly ill. With the advent of compressed air work in the 19th century, workers fell victim to a mysterious malady that soon became known as caisson disease.

Not only did sandhogs labor in wretched and sometimes horrifying conditions: The very atmosphere they worked in was life-threatening. The high-pressure air used in many subterranean projects forced ordinarily harmless nitrogen gas into the tissues. When sandhogs emerged into normal air pressure, they could be wracked by terrible symptoms as the compressed nitrogen bubbled back to its normal volume.

For some caisson workers the results were crippling if not fatal, and for decades no one understood that the pressure caused a deadly effervescence

A crew of tunnel workers, known in America as sandhogs, tightens the bolts of iron lining rings inside one of the Lincoln Tunnel's three tubes during the early 1930s. Sandhogs trace their lineage of dangerous subaqueous work in compressed air to the foundations dug for the Eads Bridge and the Brooklyn Bridge. *(National Archives)*

inside the body. Other symptoms could include life-threatening pulmonary embolisms and a range of neurological disorders including paralysis.

The mysteries of caisson disease persisted for nearly 50 years because physicians were unable to make a definite connection between incomplete decompression and illness. During the 19th century, the sandhogs were given little advice other than to abstain from alcohol consumption and to have plenty of sleep between work shifts. Decades would pass before physicians fully understood the consequences of compressed air work.

Aside from the immediate effects of the bends, it was eventually learned that compressed air workers fell victim to another serious condition, osteonecrosis, also caused by improper decompression. The medical condition, whose Latin name translates into "dead bone," is more likely to strike compressed air workers than commercial divers, according to medical studies.

Osteonecrosis can occur years after what physicians term the insult to the worker's body: an incomplete period of decompression. The condition essentially results in the death of a segment of bone, most often near a hip or shoulder joint. Sometimes that bone rots into a wet pulp or decays into a leathery mass. If untreated, as it was for many sandhogs over the years, the condition can be disabling and viciously painful.

Although sandhogs are most famous for their work within the pressurized caissons of the Brooklyn Bridge, their history actually began with the construction of the Eads Bridge in Saint Louis. This project required pneumatic caisson projects work, including one project in which workers had to dig a foundation 127 feet beneath the surface of the Mississippi River. Those who labored far beneath the water building the Eads Bridge during the late 1860s initially became known as submarines rather than sandhogs. At least 14 died during efforts to sink the two caissons of the Eads Bridge.

The turnover among the work force inside the Eads Bridge caissons was staggering, and the dirty, frightening, and dangerous job was often grabbed by the hungriest of American workers, immigrants, in a tradition that would persist into the 20th century.

The same hazards faced workers sinking the piers of New York's Brooklyn Bridge in 1870, as dank and dangerous conditions were coupled with a humiliating lack of toilet facilities. The risk of a "blowout," or sudden loss of air pressure, haunted the men as they worked. They simultaneously risked drowning, caisson disease, myriad infectious illnesses, ordinary workplace injuries, and death, being blown to bits by explosive charges. Still, the lure of steady work kept men inside the caissons.

Conditions for the sandhogs became unbearable as the depth of the caisson on the New York side sank to 70 feet below the surface of the water, the point at which the increase in air pressure began to cause caisson disease. The chief Engineer, Washington Roebling, increased the daily wage to $2.25 in an effort to keep the workers on the job.

But the April 22, 1872, death of a German immigrant, John Meyers, as a result of caisson disease prompted the workers to organize a strike on May 8. They demanded $3 per day and a four-day workweek. Management countered this demand with an offer of $2.75 and a five-day week. The sandhogs walked off the job but returned after being threatened with replacement, accepting management's pay raise and workweek offer.

At the time, the sandhogs also made their case to the newspapers, enlisting public support by revealing the Goyaesque conditions inside the caissons. This move helped to gain public sympathy. The decision to turn to the journalists also added to awareness of the challenges they faced. This notoriety also helped boost their ascendancy in the folklore of the American workingman.

Despite the strike and its settlement, the Brooklyn caisson job was one of the toughest during a century when the lot of workers was incredibly harsh. The turnover rate among workers at the Brooklyn caisson was staggering, as roughly 100 men walked away from the job each week, many of them impecunious immigrants ordinarily desperate for cash. This meant that the entire complement of 2,700 who worked at or in the Brooklyn caisson would be replaced nearly every six months.

The mysterious physical maladies that accompanied compressed air work were numerous, but one ominous peril ranked above all others—the danger of a blowout in a tunnel or a caisson, when a rupture allowed a sudden loss of air pressure followed by uncontrollable flooding.

At one point during the excavation of the Brooklyn caisson for the Brooklyn Bridge, a blowout did occur. The incident took place on a Sunday morning when the site was fortunately unoccupied. Although no lives were lost when air pressure spewed from the inside of the massive caisson, a geyser rocketed 500 feet above the East River, pelting surrounding buildings with rocks and an unpleasant coating of reeking riverbed muck. No one was injured, but the incident served as an example of the hazards of caisson work.

By the end of the 19th century, sandhogs were laboring in tunnels in Boston, Chicago, and other major American cities. Modern infrastructure was needed for these growing urban centers. Tunnels could transport clean water into a city and could transport sewage away while providing conduits for gas service and transportation. The efforts of the sandhogs that turned the plans of America's great engineers into reality were a national necessity.

New York's sandhogs, unemployed after completion of the Brooklyn Bridge's final caisson work in 1874, would soon be put to work digging rail and subway tunnels. Instead of simply digging straight down, as in the caissons of bridges, they would now be digging horizontally through the muck and rock of their city.

When the Hudson Tunnel Railroad Company began work on a rail tunnel beneath the Hudson River between New Jersey and Manhattan in 1874, it issued a call for sandhogs experienced in compressed air work. The tunnel project, conceived and designed by the railroad engineer DeWitt Clinton Haskin, meant tunneling beneath the riverbed of the Hudson in a compressed air environment. The plan was insanely risky because it did not involve the use of a protective tunneling shield. Haskin would rely only on compressed air to prevent the tunnel from collapsing as workers raced to reinforce it with iron plates and brick masonry.

Haskin's gamble was a bad one that the sandhogs paid for with their lives. At 4:30 A.M. on July 21, 1880, a blowout in a section where weak and porous gravel formed the riverbed flooded the tunnel, killing 20 sandhogs. The man in charge of the crew at the time was the assistant superintendent, Peter Woodland. As water rushed into the tunnel, Woodland realized what had happened and yelled to his men, "Make for the lock!" Nine sandhogs were able to escape, and one was killed by a falling timber during the mad dash.

After the survivors slammed the airlock door with the help of Woodland, who remained outside, they watched through a glass porthole as Woodland struggled back through waist-high water, joining other doomed workers in a futile but courageous attempt to plug the blowout. Woodland's and seven

other bodies were eventually recovered from 60 feet beneath the surface of the river. The men working on the first railway tunnel beneath the Hudson were making a hefty $1.50 a day at the time of the fatal incident.

As if conditions were not bad enough for the sandhogs, who already had too few safety systems to rely on, the Pennsylvania Railroad Tunnel contractor S. Pearsons & Sons decided to fire the airlock tenders in 1906 as a cost-cutting measure. These tenders were essential since they controlled the compression and decompression of the workers entering and leaving the tunnel. This downsizing triggered unwanted staff reductions when sandhogs began to die by the dozens from decompression illness. A hue and cry rose from the workers, and the New York country coroner, George F. Sharady, Jr., held a hearing on the loss of life.

Sharady's finding was that "at the very least" 50 sandhogs had died between January and May of 1906. Said Sharady, "I am amazed that there are no lock tenders in the tunnels. I have never heard of such an omission. It is a fatal oversight. The experienced lock tender is the most important man in the tunnel." The broadside against S. Pearsons & Sons was concluded with Sharady's warning, "Some dreadful calamity will result if lock tenders are not immediately restored to their duties."

Slashing the jobs of the lock tenders was saving the company $112 daily, and the Pearsons & Sons superintendent, Charles Fraser, announced that the matter was no business of the public's. It became apparent that Pearsons & Sons was working to keep deaths related to the tunnel work far out of the public eye. During Sharady's hearings, reporters with the *New York Herald* newspaper uncovered evidence that the company had hidden the true causes of their deaths and in some cases found ways of listing the place of death as being far from the tunnel in which the workers had actually died. In another example of this cover-up, the death certificates of the sandhogs John Dalton and B. Trueman, who died of caisson disease, were missing.

The sandhogs had had enough. Their attempts to organize while working on the Brooklyn caisson of the Brooklyn Bridge had withered, but they believed they had no choice but to try again in 1906 as a matter of survival. New York's Eccentric Firemen's Union announced that it would hold a sympathy strike. Electricians and explosives workers employed in the tunnel, already organized, threw their support behind the right of the sandhogs to unionize. The United Tunnel Workers was created and 500 sandhogs went on strike July 30, 1906. The leaders of other unions announced they would also walk out in sympathy with the tunnel workers, a move that threatened work on New York's other subterranean projects. As many as 5,000 workers were poised to come up from below the ground in a show of solidarity.

The strike was successful, and the airlock tenders were returned to their jobs. The United Tunnel Workers evolved into today's Local 147 in the Bronx. The union's full name is the Compressed Air and Free Air, Shaft, Tunnel, Foundation, Caisson, Subway, Cofferdam, Sewer Construction Workers of New York and New Jersey States and Vicinity. Eventually, sandhogs across America organized to protect their interests, not the least of which was their safety.

Despite the advances in safety measures, engineering practices, and knowledge about decompression illness, tunnel work—like any heavy construction trade—remains hazardous. In 1997, three deaths befell tunnel workers in Los Angeles County Metropolitan Transportation Authority's (MTA's) Red Line subway project. Brian Bailey, 36, was killed October 8, 1997, while walking at the base of an access shaft. A pair of massive grippers being swung into place to grab a debris container crushed Bailey's head.

Bailey's death ignited criticism of the contractors on the Red Line project, which had been plagued with delays and cost overruns. In two previous deaths in 1997, a worker was killed after being decapitated by a huge bucket hauling debris from the tunnel, and another worker fell 50 feet to his death. These deaths, the Los Angeles County MTA officials have pointed out, were the first to occur in the project's 10-year history. Nonetheless, the MTA also found itself criticized for the way the project had been conducted.

Sandhogs continue to work beneath ground, although their numbers are now smaller because of labor-saving machinery such as tunnel-boring machines. Despite technological advances, sandhogs will be on the job until 2020 on one of the world's most stupendous tunneling projects, still under way in New York, City Water Tunnel No. 3. Local 147's sandhogs are part of the effort to bore a water supply tunnel that will eventually span a distance of 60 miles. Its completion will allow two other water tun-

nels, completed in 1917 and 1936, tunnels that earlier generations of sandhogs helped excavate, to be inspected and repaired.

The water tunnel, which will be the third such tunnel to New York City, was begun in 1970 and will not be completed until 2020. In some places the tunnel will reach more than 800 feet beneath the surface. In a small way the project actually owes its existence to the members of Local 147, since the sandhog union mobilized to fight for the project, which was endangered by contract disputes and budget-conscious politicians.

When the Tunnel Consortium—the group of contractors that were building the tunnel—complained that the city underestimated the amount of expensive iron rings required for the tunnel's reinforcement, the dispute grew so sharp in 1975 that work on the project was halted. The contractors said the rock was not as strong as city engineers believed, necessitating more reinforcement work than the city would approve. This prompted the now-unemployed sandhogs to sue both the contractors and the city to resume the project and pay the sandhogs $10 million in damages. The work stoppage was a personal issue to the sandhogs, a total of 19 of whom had lost their lives in the tunnel by 1975. (By 2000, another five had died in the construction of the tunnel.)

Another blow to the project promised to be even more threatening. By 1975, New York City's now-famous fiscal problems had reached a crisis stage, threatening the city with insolvency. Politicians, squirming under an embarrassing glare of national attention and trying to save money, began talking of the water tunnel as an unnecessary extravagance. The New York sandhogs, much as they had a century before while laboring on the Brooklyn Bridge, turned to the press. Television crews were encouraged to film news stories inside the cavernous water tunnel in an effort to save the project and the jobs of the sandhogs.

Newspaper reporters were also given reasons for the necessity of completing the water tunnel, and two documentaries that also made the case for the water tunnel were sponsored by the sandhogs. The second documentary was also financed by another union, Local 15 of the Operating Engineers, and by the General Contractors Association. The sandhogs even lobbied Congress. By 1977, some work was resumed on the tunnel; full-scale work was again under way by 1983, thanks in no small part to the sandhogs, who fought to save not only their jobs but also a badly needed means of delivering fresh water to the world's most dynamic city.

See also BROOKLYN BRIDGE; CAISSON DISEASE; CITY WATER TUNNEL NO. 3; EADS BRIDGE; MABEY, MARSHALL; ROEBLING, WASHINGTON AUGUSTUS.

San Francisco–Oakland Bay Bridge (built 1933–1936) San Francisco Bay's other famous bridge

The bridge from San Francisco to Oakland is one of the world's great combination bridges, meaning that it is a span incorporating several types of bridging technologies. Traveling from east to west, a driver crossing the San Francisco–Oakland Bay Bridge passes over a viaduct, a series of truss spans, and then a cantilever truss bridge before entering a tunnel bored through Yerba Buena Island. The roadway then emerges to cross a pair of end-to-end suspension bridges entering San Francisco.

Although eclipsed by the fame of its sister bridge across the Golden Gate Strait, the "Bay Bridge"—as it is more commonly known—was a monumental construction task completed nearly a year before. Both bridges were crucial to the development of the region, although it could be argued that the Bay Bridge was perhaps more crucial.

The peninsular city of San Francisco curves to encircle San Francisco Bay and end at the Golden Gate Strait. The bay it shelters is a massive harbor

One of the world's great combination bridges, the San Francisco–Oakland Bay Bridge is composed of a pair of suspension bridges connected to a cantilever truss–viaduct by a tunnel at Yerba Buena Island. This view from the suspension bridge segment is looking toward Oakland across San Francisco Bay. *(Historic American Engineering Record, National Park Service)*

filled with naval installations and commercial ports. An economic godsend to the area, the bay is also a massive obstacle to vehicular traffic between San Francisco and Oakland to the east. Whereas the controversial Golden Gate Bridge connected the northern tip of San Francisco with the relatively undeveloped regions of Marin County, the Bay Bridge was designed to carry traffic between two heavily developed areas.

As San Francisco and Oakland grew, it was obvious to everyone that it was time to link the two cities. All the usual controversy erupted, including the obligatory wailing over the cost and claims that building a 4.5-mile bridge system across the bay was technically impossible. A mining engineer who considered the project viable laid the carping to rest in 1929. The engineer was Herbert Hoover, president of the United States. Hoover, along with the California governor, C. C. Young, appointed the Hoover–Young San Francisco Bay Bridge commission to study the feasibility of the project. The commission reported in 1930 that with proper engineering the project was doable.

Hoover, who had studied geology at Stanford University near the southern edge of San Francisco Bay, took an interest in the bridge and used his clout to gain the bridge rapid approval from the U.S. Army Corps of Engineers and the War Department. With the economic midnight of the Depression settled upon the land, Hoover's support was instrumental in obtaining $77 million from the newly established Reconstruction Finance Corporation.

Work began on the bridge in 1933, and across the East Bay, a viaduct curved gradually upward, resting on 14 sets of piers sunk into the shallow half of the bay. The viaduct then became a five-segment, through-truss span before it led to a cantilever truss span that reached Yerba Buena Island.

At Yerba Buena, miners were busy tunneling with the drill-blasting method to punch three bores through solid rock. Once the individual bores were completed, they were then expanded into a single gallery, in mining parlance, "broken out." The result was the largest-diameter tunnel in the world, measuring 76 feet wide and 58 feet high. The small size of the island meant the tunnel had to travel only 1,700 feet to reach the other side. The cumulative length of the East Bay bridge system is 10,176 feet, excluding the tunnel.

The grandest construction spectacle took place on the other side of the island across the West Bay with the construction of tandem suspension bridges. Because the West Bay has the deeper, navigable channel, a suspension bridge was required to provide the widest possible waterway for the passage of commercial and military shipping. Placed between the two suspension bridges would be a common anchorage for their support cables. Often described as having the measurements of half a city block, the anchorage measures 92 feet in width and 197 feet in length.

To seat the anchorage to bedrock, a caisson containing a system of large steel tubes was set on the bottom 70 feet beneath the water. Clamshell shovels were lowered into the tubes to remove the muck to allow the caisson to settle another 147 feet to bedrock. As it did so, sections were added to the tubes so the openings would not be submerged. Once it was sunk to its final depth, massive concrete pours filled the tubes, and the above-water portion of the caisson was built of concrete. Work continued until the shared caisson rose 598 feet above the bedrock. The anchorage's height from the average low-tide level of the bay was 281 feet. The anchorage required more concrete than was used in the construction of New York City's Empire State Building.

Smaller but still-massive anchorages were built at Yerba Buena Island and on the San Francisco shore, and workers also sank four massive piers into the bay to support an equal number of towers that would suspend the cables for the bridges. Once the anchorages and the towers were in place, the spinning of the two cables required by each bridge began. Each of these cables contained 17,464 galvanized wires. Bundled together, they would be compressed into a cable 28 3/4 inches in diameter. The complementary strength of the 0.195-inch diameter wires allows each cable to sustain 37 million pounds of loading.

The stiffening truss sections that would contain the two-tier roadway of the bridge were barged to the site and raised into place, where they were connected. The road deck's steel skeleton was put into place and paved with concrete.

Another major difference between the Golden Gate Bridge and Bay Bridge projects was the selection of California's state highway engineer, Charles H. Purcell (1883–1951), to oversee the project. Although consulting engineers assisted with the complex design and planning of the bridge, the project did not have a godfather like Robert Strauss,

chief engineer of the Golden Gate Bridge. The egoistic Strauss, who was ruthless in his efforts to ensure that no one but he received credit for designing a bridge he did not design, was a far cry from Purcell.

Purcell, who had originally been appointed as an adviser to the Hoover-Young Commission, was humble and quietly professional. Since 1927, Purcell had been California's state highway engineer, and most of those he selected as his lieutenants for the Bay Bridge effort were also state employees, who were charged with the task of designing the best bridge for the best price.

The toll bridge opened November 12, 1936, with tremendous ballyhoo. Designed with two decks, the bridge carried five lanes of passenger car traffic on the top level; a Key System Lines passenger train and truck traffic crossed on the lower level. Although much of the festival atmosphere was stage-managed, both rail and automobile commuters genuinely welcomed the bridge. However, this superb engineering achievement would not stand untouched or unchanged.

By the late 1950s, rail traffic on the Key System Lines had declined, and this circumstance catalyzed the first major changes to the bridge. The economically unviable rail system, combined with the exponential growth in personal automobiles, prompted officials in 1959 to strip the lower deck of its rails. The bridge was then reserved purely for vehicular traffic.

Precisely three decades later, the next changes made to the bridge were deadly and earthshaking. During the Loma Prieta earthquake of October 17, 1989, a section of the upper level deck was knocked free of its supports. With one end still attached to the deck, the other end fell like a hinged door, killing one person on the roadway below. The damage also blocked the bridge for weeks. The state wasted no time in commissioning a study of the Bay Bridge's ability to withstand another quake.

The study concluded that the suspension spans could be modified with a $540 million seismic retrofit to protect them from future quake damage. Unfortunately for the east span, California Department of Transportation (Caltrans) officials announced it could not be made resistant to temblor damage for less than the cost of a new bridge and recommended complete replacement. To pay for this work, motorists were charged a one-dollar surcharge to use the bridge in addition to the normal toll.

The decision was made to do interim seismic retrofit work on the east span, a project that was completed in July 2000. Safety retrofit work was completed on the towers of the suspension spans in 2000, but the work on the suspension bridges would not be completed until 2003.

As helpful and scenic as bridges are, they are magnets of controversy, and the replacement schemes subsequently offered by the state ignited new controversy. A new viaduct system was proposed to replace the original one that would connect to a new self-anchored, single-tower suspension bridge that will replace the cantilever truss span.

The self-anchored design eliminates the need for anchorages since the suspension cables are tied to the ends of the span's road deck. Suspenders then accept the loading of the span and transmit it to the cables. The design will also be strong enough to support a light commuter rail line should officials decide to reinstate one in the future. The bridge will also possess a bicycle and pedestrian lane. Curiously, as of 2001, there were no plans to install pedestrian lanes or a rail line on the suspension span; therefore, hikers and bikers will end their treks at Yerba Buena Island.

Although voters approved construction of a new east bridge in 1998, controversies ignited over whether the new span will be sufficiently earthquake-resistant, and environmental activists feared that the bicycle and pedestrian lane would be eliminated to make room for another vehicular lane. Construction of a new bridge may begin in 2002 and completion is scheduled for 2006. The cost of the viaduct and self-anchored span could reach $2.43 billion.

See also GOLDEN GATE BRIDGE; SEISMIC RETROFIT; SELF-ANCHORED SUSPENSION BRIDGE.

Schoharie Creek Bridge Collapse   *See* scour.

Schuylkill Suspension Bridge (built 1816) *America's first wire suspension span*   Although they were an unlikely pair of bridge builders, Josiah White and Erskine Hazard created an innovation in suspension bridge building that survives today, although their own bridge collapsed under the weight of snow and ice.

White and Hazard were owners of an iron wire factory near Pennsylvania's Schuylkill Falls who purchased the rights to bridge the Schuylkill River in the aftermath of the collapse of a chain suspension bridge. That ill-fated bridge was designed by the pioneering American suspension bridge builder James Finley.

Finley's bridges, although on the right track, were less durable than they should have been, and, as were many of the later 18th- and early 19th-century suspension bridges in Europe and America, they were often victimized by wind and excessive loading. The Schuylkill Falls Bridge was actually built by another man, who had purchased a license to construct Finley's chain bridge design. Built in 1809, it collapsed two years later under the weight of cattle.

White and Hazard erected a bridge across the 400-foot Schuylkill River only to see it collapse in 1816. The iron wire makers decided to use their own product for the next bridge, which consisted of three iron wires on each side of the span. Stirrups or hangers ran from the wires to bear the loading of an 18-inch-wide timber floor.

The bridge was erected quickly. One anchorage for the wires was a second-story window of the Hazard and White wire factory, the other a tree on the far shore stabilized by guy wires staked into the ground. This contraption was only a footbridge, and its builders sensibly limited the number of persons who could cross it to six, although larger groups often violated this rule. It measured 408 feet in length and stood 16 feet above the Schuylkill.

The bridge's main purpose was to help the iron wire factory's employees get to work, and those using the bridge were charged a toll of a penny per person that remained in effect until the bridge was paid off. The span may have been paid off faster than any other toll bridge in American history since it cost only $125. It may have been paid off just in time: The bridge collapsed under snow and ice loading during its first year of use.

See also SUSPENSION BRIDGE.

scour   Though ordinarily peaceful-looking on the surface, the currents and tides beneath the water are an awesome force, capable of moving tons of sand and rock in minutes, if not seconds. The force of water can deepen a river and deposit the silt elsewhere, and tides and currents, such as those in the Golden Gate Strait, can roll submerged boulders with ease.

Watching sonar readings from a boat moored in the middle of the Mississippi River can afford an observer the chance to see dozens of feet of riverbed disappear in seconds as a result of scouring action, only to be replaced minutes later by the reverse process, deposition. For those who build bridges, this hydrodynamic force (the wet version of aerodynamic wind erosion) has always posed a threat to the all-important foundations, or piers, beneath the surface of the water.

The problem of scour is significant and must be anticipated in the construction of any bridge. Foundations must be seated directly against bedrock. If they are not, moving water with its nearly unlimited force sweep sand and rock away from beneath a poorly placed foundation until it collapses. The surviving bridges of the Roman Empire invariably are the ones resting upon solid rock, whereas those perched atop shifting sand and gravel are no longer here.

To counteract these problems, engineers design foundations of sufficient heft and strength to make them capable of resisting the hydrodynamic effects of water rushing by at millions of gallons a minute. Each of the two submerged piers of Japan's Akashi Kaikyo Bridge weighs approximately 950,000 tons. After the poured concrete foundations were seated on bedrock, engineers ordered thousands of tons of rock dumped around the perimeters of the caissons to provide additional protection from the effects of scour. This type of rock fill is known as riprap.

Flooding and the resultant increase in water flow in rivers often create scour holes in riverbeds that may or may not be refilled naturally with sediment or gravel. This action presents a danger to medium and small bridges with shallow foundations.

Flooding and the scouring that accompanied it resulted in one of the worst bridge disasters in American history on April 5, 1987, with the collapse of the Schoharie Creek Bridge at Interstate 90 near Amsterdam, New York. When the weakened foundations of the concrete beam bridge gave way, they dropped a span into the creek. Not knowing the bridge was out, 10 people died in five cars that plunged over the edge of the collapsed bridge.

The tragedy occurred at a time when the American public began to understand that their bridges were badly in need of repair. The collapse of the Schoharie Creek Bridge prompted the federal government to begin the National Bridge Scour Program. As part of this effort, the Federal Highway Administration and the U.S. Geological Survey gather data that help engineers understand and detect the problem with greater accuracy. As part of the program, some bridges have been equipped with remote sensors, including sonar, able to detect that protective rock and gravel have been removed by scour.

Engineers have learned the hard way that building a solid foundation does not allow them to ignore the bridge and its piers. Scour is a nefarious process that continues to cause bridges to collapse in the United States. Municipalities often hire engineering firms that specialize in bridge maintenance to inspect foundations to determine whether scour is at work and recommend solutions.

According to one study, the United States still has a long way to go to protect its bridges from scour damage, particularly during flooding. During massive flooding that struck the Midwest in 1993, at least 2,400 bridges suffered damage to a total cost of $178 million. During flooding in Georgia in 1994, another 1,000 bridges suffered $130 million in damage.

Seikan Tunnel (built 1971–1988) *World's longest and deepest railway tunnel* While the British and the French were still talking about building a tunnel under the English Channel during the 1970s, the Japanese were actually accepting a similar challenge by blasting and digging their way beneath the Tsugaru Strait in northern Japan. Their effort to connect Japan's largest island, Honshu, with its sister island, Hokkaido, to the north was an official obsession that took 24 years to accomplish.

The Seikan project, a dream of Japan's national railway planners since the 1930s, was to prove as difficult, dangerous, and expensive as previous large-scale tunneling projects. Completed in 1988 after full construction began in 1971, the tunnel runs deeper than any other beneath the surface of the water (780 feet) and is the longest railway tunnel in the world, surpassing in total length the Eurotunnel between Britain and France.

The Seikan Tunnel runs for a total of 32.3 miles, whereas the Anglo-French Eurotunnel has a total length of 31 miles. However, only 14 miles of the Seikan Tunnel is beneath the treacherous Tsugaru Strait, a stretch of water separating the Japan Sea to the west and the Pacific Ocean to the east. The Channel Tunnel runs below water for nearly 24 miles, or 10 miles farther than the Seikan, making the Anglo-French tunneling effort the longest subaqueous tunnel.

The outrageously bad weather conditions that plague the Tsugaru Strait provided one of the primary motivations for construction of the tunnel. Ferry service was the traditional means of moving people between Honshu and Hokkaido, but it was disrupted up to 80 days yearly by typhoons and generally despicable weather. Public feeling about the necessity of another way to cross the strait was solidified by a September 26, 1954, disaster in which nearly 1,400 ferry passengers and crew died. It was on that day that five ferries were lost in a typhoon packing 100-mph winds. Of those killed, more than 1,100 died aboard a single ferry—the *Toya-maru*—that capsized in rough seas. The disaster spurred support for linking Hokkaido to Honshu with a tunnel, although projects that would link Japan's four home islands by bridge or tunnel had long been a national goal. The Seikan Tunnel would help fulfill that dream.

Unfortunately, the government's single-minded devotion to this concept was unswerving, even as air travel between Hokkaido and the other three home islands became commonplace. Air travel from Honshu to Hokkaido would eventually outstrip travel by ferry and by the completed Seikan Tunnel combined. Nonetheless, the project began as a great source of national pride for Japan, which was beginning to flex its recovering economic muscle after its defeat in World War II. This was only natural because the project was perhaps Japan's single most expensive and challenging infrastructure project since the rebuilding of its smashed cities and ports after the war.

Although planning for the project occurred periodically from around 1936, actual tunneling did not begin until March 1964, when a vertical access shaft was drilled straight down on the Hokkaido side at Yoshioka to enable engineers to understand the nature of the rock they would be penetrating. Engineers sank another vertical shaft on the Honshu side at Tappi two years later. Whereas mining engineers pray for rock that is soft enough to dig yet tough enough to support itself with minimal bracing, the Japanese found themselves with the worst of both worlds geologically. The young volcanic rock beneath the Tsugaru Strait was devilish to tunnel through yet permeable to water as a result of porosity and fractures. It was also prone to crumbling once penetrated by drilling and explosives. This meant that tunnel construction could not depend on the rock to support itself and would require additional time and money to keep the tunnel structurally sound and resistant to water infiltration.

Another problem faced by those building the Seikan Tunnel was that the nature of the unstable rock precluded the use of a massive tunnel-boring machine. Such a machine, much like a giant mechanical earthworm, could grind and convey the debris to the rear for removal in a single operation. Instead, Japanese workers were forced to use the more traditional method of cramming drilled shot holes full of explosives to blast rock free from the tunnel's face. A narrow-gauge temporary railway train carried workers and tools in and debris, or spoil, out during this process.

Blasting and debris removal were followed by drilling of holes and injection of grout, a cementitious compound that strengthened the walls of the tunnel while sealing cracks and fissures against water infiltration. The grout, pumped to a thickness of three times the radius of the tunnel, was essential to control the intrusion of seawater through fractured rock. This procedure was followed by the application of shotcrete, a layer of concrete applied under pressure against the interior of the tunnel for additional reinforcement.

Despite these efforts, massive flooding occurred on four occasions; the worst incidents occurred in May 1976 and again the following year. Unforeseen geologic weaknesses in the rock were to blame.

During the 1976 flooding, two miles of the service tunnel—one of three bores that constitute the Seikan Tunnel complex—and a full mile of the main tunnel had to be evacuated. Water infiltration into the Seikan Tunnel remains a problem today, although engineers placed the tunnel 325 feet below the seabed, which, in turn, lies 455 below the surface of Tsugaru Strait. As a result, the lowest part of the Seikan Tunnel is roughly 780 feet below the surface. In the two most severe flooding incidents it was estimated that water rushed in at the rate of 80 tons per minute. The water was pumped out and the service tunnel was rerouted to avoid the weak spot in the rock. In addition, more waterproofing and reinforcing measures were taken to keep the Seikan Tunnel dry. To minimize the inflow of water, the completed tunnel must be drained by four pumping stations. If the pumping stations ceased operating, it is estimated that the tunnel would fill with water within three days.

Although most think of a tunnel as a single excavation, the Seikan project included a pilot tunnel, a service tunnel, and the main tunnel. The pilot tunnel, which was drilled initially, now serves as a drainage shaft below the main tunnel. The service tunnel alongside the main tunnel serves as an escape route in case of an emergency. The tunnel is equipped with a pair of huge stations carved into rock at its midway point. The main tunnel's reinforcement after the grouting process and shotcrete application was the emplacement of curved H-shaped steel beams much like the frames or ribs used to give a submarine its internal structural strength. The tunnel was then lined with sections of reinforced concrete. Depending on the strength of the rock through which the tunnel was bored, the thickness of the final concrete support lining varies between two and three feet. Workers advancing from opposite ends of the main tunnel's path met when they holed through midway beneath the Tsugaru Strait in March 1985. The rail lines were completed in September 1986, and rail service began in 1988.

To ensure safety against fire (as great a concern in most tunnels as collapse or flooding) the Seikan Tunnel is equipped with electronic monitoring devices that detect heat in the infrared wavelength so an alarm can be sounded if a fire occurs. Escape tunnels allow passengers to go to the parallel service tunnel, and a sprinkler system is designed to spray water on any portion of the tunnel engulfed in a blaze. To cool and ventilate the tunnel and to dissipate the considerable heat of the electrical trains traveling along their steel rails, air is moved through the tunnel from ventilation shafts at one end to be expelled out the other.

The Seikan Tunnel can lay proud claim to several records, although some are of dubious distinction. The tunnel is for now the deepest and longest railway tunnel in the world; it is also among the most expensive. Its original price tag was estimated at $783 million in 1971, but costs ballooned to $6.5 billion by the time it was placed in operation in 1988—an increase of 800 percent. (However, this is not uncommon with large tunnel projects since each is essentially a unique engineering experiment subject to extreme conditions and unpleasant surprises.) The unstable rock, flooding, and numerous unforeseen engineering challenges led to the increase in costs.

The main tunnel is designed also to accept a pair of high-speed bullet trains that run on normal-gauge tracks in lieu of the narrow gauge trains in use at present. However, cost has been used as the reason for not yet employing the technologically advanced bullet trains on what is known as the Tsugaru Straits Line.

Unfortunately for Japan Railway, as the tunnel was being blasted to completion, commuter airline service became attractive to those who once took the relatively slow ferries. Ferry ridership declined by half in the 10 years after 1975, with the result that the extraordinary Seikan Tunnel would be hard-pressed to compete with a 90-minute flight between Honshu and Hokkaido by the time of its completion. However, Japan Railway officials plan to institute bullet train service in the future to compete more effectively with airline travel.

Nonetheless, the tunnel project was a prodigious feat for Japan National Railway Construction Corporation, which performed the work. During the construction of the tunnel, 34 workers died detonating, excavating, and reinforcing the shafts. Workers blasted their way from Hokkaido to Honshu using 2,860 tons of explosives while excavating 8.2 million cubic yards of rock and soil. The miners reinforced the tunnel walls with 168,000 tons of steel (enough for 42 high-rise buildings) and 2.3 million cubic yards of concrete.

Ironically, despite the existence of the remarkable Seikan Tunnel rail line, ferries still depart Honshu ports at three locations for a ride that can take anywhere from 2 to 4 hours. The submerged rail trip between the two islands takes 2 1/2 hours. Although the Seikan Tunnel may not yet be the primary means of moving people between Honshu and Hokkaido, it serves the secondary purpose of being a first-rate tourist attraction.

See also EUROTUNNEL; GROUTING; SUBAQUEOUS TUNNEL.

seismic retrofit *Preparing existing bridges for earthquakes* Although earthquakes are nothing new, the proliferation of bridge construction during the 20th century provided potential targets for earthquake damage, and nature has willingly demonstrated how destructive seismic activity can be to all types of spans.

Each time a major earthquake has caused serious damage to buildings and bridges, engineers and scientists have studied the effects and have made recommendations on ways to build future structures that resist damage better. Since California's 1971 San Fernando earthquake, scientists and structural engineers have recognized that existing structures remain at risk and have begun devising retrofit

A shattered supporting column of the Nimitz Freeway in Oakland dramatically demonstrates the destructive effects of shear forces caused by the 1989 Loma Prieta earthquake. Seismic retrofit measures that provide both flexibility and strength to bridges are a priority for important spans in quake-prone regions around the world. *(Photograph Courtesy of Craig Byron)*

improvements to bridges to prevent damage or collapse.

In seismically active areas 20th-century engineers have long been concerned about the issue of seismic activity and its effects on bridges; three major earthquakes since 1989 in California and Japan hammered home the urgency of the threat posed to bridges as never before. The Loma Prieta earthquake in San Francisco on October 17, 1989; the Northridge earthquake on January 17, 1994; and the Hanshin/Awaji earthquake that devastated Kobe, Japan, on January 17, 1995, stunned engineers and seismologists with the damage they caused to bridges.

The Loma Prieta quake killed 62 people and damaged numerous bridges, including the Oakland–San Francisco Bay Bridge and the elevated Nimitz Freeway, a double-decker span in Oakland. A section of the upper deck of the two-tiered concrete roadway of the San Francisco–Oakland Bay Bridge dropped onto the road deck below, closing the bridge, and a 6,500-foot section of the Nimitz Freeway collapsed.

However, the Hanshin/Awaji earthquake nearly six years later was truly devastating, killing 6,308 people and damaging 60 percent of the area's bridges, of which 318 collapsed completely. A massive 18-span section of the elevated Hanshin Expressway was toppled by the quake when lateral movement fractured the massive steel-reinforced concrete piers supporting it. Engineers around the world were startled because it had been widely believed that Japan had built earthquake-resistant bridges and elevated freeways.

The motion transmitted to the Earth's surface by quakes can move bridges back and forth violently. Some can avoid damage if all their components are strong enough to resist these lateral forces without yielding. Other structures can dissipate the energy through flexible connections such as elastomeric bearings that provide an elastic interface between a shifting column and a road deck. Another means of protecting large bridges is to install tuned mass dampers in their suspension towers. These huge weights are calibrated to move in a pendulum fashion to negate the destructive motion.

Other Achilles heels for bridges include their piers or columns, which may lack the brute strength to withstand shearing due to lateral quake movements. To remedy this, Japanese engineers in Kobe are installing steel jackets around the piers to strengthen them. Research has also been conducted on the use of fiber reinforced polymer composite wrappings in the same role.

Although the suspension span portion of the Oakland–San Francisco Bay Bridge was left relatively unscathed during the Loma Prieta earthquake, the collapse of the road deck section of a truss span segment closed the bridge to traffic for a month. This made it clear that an entire bridge can be made unusable if it is only partially damaged.

Bridges containing supporting piers, road decks, towers, and suspension cable anchorages may have some components damaged during a quake while others remain intact. Unfortunately, the process of adding new bearings, mass dampers, or reinforcement to supporting piers not only is terrifically expensive but in itself presents complex problems. Engineers are now trying to determine whether strengthening one component of a bridge can precipitate damage to another during a quake. In 1995, the Federal Highway Administration issued the Seismic Retrofitting Manual for Highway Bridges warning of this issue.

Because of the high costs of retrofitting bridges to endure quake damage, a careful assessment must be made as to whether a bridge is essential in the aftermath of a quake. This can allow for the seismic retrofitting of the most important bridges, or a sort of engineering triage. Some spans, if they are not essential, may be sacrificed to earthquake damage because of lack of money or time to retrofit all bridges.

California, as a result of the well-known seismic activity in its confines, has conducted an aggressive program of research and retrofitting to enable its bridges to survive the shocks and aftershocks of an earthquake. What is less well known is that much of the United States is vulnerable to severe earthquake damage, including New York City, a metropolis that is connected by what is perhaps the world's most famous collection of bridges.

Worrisome to geologists and engineers is one of the greatest series of earthquakes ever witnessed by humans, which occurred in the winter of 1811–1812 on the western edge of what was then the American frontier. Contained in the Mid-America region, essentially that portion of the United States east of the Rocky Mountains, three major earthquakes accompanied by hundreds of aftershocks took place in the New Madrid Seismic Zone between Saint Louis and Memphis.

These temblors are known as the New Madrid earthquakes, or the Mid-America earthquakes, and their occurrence is surprising to many who think of the region as immune to seismic activity. The states that might face another massive earthquake in this zone include Alabama, Arkansas, Georgia, Illinois, Indiana, Kentucky, Mississippi, Missouri, North Carolina, Oklahoma, South Carolina, Tennessee, and Virginia. Scientists say another massive quake will eventually occur, and since few structures are prepared for this type of seismic event, the results could be grim.

Some of the 1811–1812 quakes reached a magnitude of 8.7 on the Richter scale, making them far more powerful than the 8.3 magnitude 1906 earthquake that devastated San Francisco. The 1811–12 damage included the destruction of 150,000 acres of forest and an alteration in the course of the Mississippi River. Since scientists say that another massive Mid-America earthquake is inevitable, raising the spectre of possible economic devastation and a massive loss of life. The loss of life during the 1811–1812 quakes was low simply because there were few people in the region at the time.

Among the research centers dedicated to the study of earthquakes and their effects on structures are the Pacific Earthquake Engineering Research Center (PEER) at the University of California, Berkeley; the Mid-America Earthquake (MAE) Center at the University of Illinois, Urbana-Champaign; and the Multidisciplinary Center for Earthquake Engineering Research (MCEER) at the University of Buffalo. The web sites for these centers are, respectively, http://eerc.berkeley.edu/, http://mae.ce.uiuc.\edu/, and http://mceer.buffalo.edu.

The United States Geologic Survey's Earthquake Hazards Program also has a web site, whose address is http://earthquake.usgs.gov/. A massive listing of earthquake-related research sites is at http://pasadena.wr.usgs.gov/info/seismolinks.html#locale. The U.S. Department of Transportation's full version of the Seismic Retrofitting Manual for Highway Bridges can be viewed at http://tfhrc.gov//seismic/document.htm.

self-anchored suspension bridge At a distance a self-anchored suspension bridge may appear to be a cable-stayed bridge, but the similarities are only visual. Although a fairly rare type of span, the self-anchored suspension bridge obviates the need for costly anchorages by securing its suspension cables to the ends of its own road deck.

Although the absence of blockhouse-style anchorages might lead the casual observer to assume that the span is transmitting its load from the pinnacles of its towers into the foundations through compression, the bridge is functioning as a gravity-anchored suspension span. The cables, secured by the weight of the road deck, still depend on vertical suspenders between the roadway and the cables to transmit the loading of the span. The bridge deck is effectively holding itself aloft with its own dead load.

The elegance of this design is derived from using the necessary and inescapable mass of the road deck in lieu of weighty and expensive anchorages that ordinarily serve no other purpose than as huge tie-down stakes for catenary suspension cables. (Some anchorages, however, can do double duty as piers for approach spans of backspans.)

Of the few self-anchored bridges, a notable example is the Konohana Bridge in Osaka, Japan. Containing a pair of diamond-shaped towers that frame the roadway, the catenary suspension cable travels down the centerline of the bridge and is anchored at each end of the deck. Built in 1990, the main span of the bridge measures 984 feet.

A self-anchored suspension bridge is to be built as a replacement to the east span of the San Francisco–Oakland Bay Bridge. The design was selected instead of a cable-stayed bridge. One of the reasons for the choice of the self-anchored design is the elimination of expensive and obtrusive anchorages on Yerba Buena Island, where the bridge would make its landfall.

See also CABLE-STAYED BRIDGE; SAN FRANCISCO–OAKLAND BAY BRIDGE; SUSPENSION BRIDGE.

self-rescuer A portable emergency respirator that filters poison gases or provides breathable air to tunnel workers is known as a self-rescuer. In the days before there were portable and easily used emergency breathing devices, hundreds of men died of asphyxiation or poisoning by gases produced by underground fires or explosions.

If miners were to have any chance of escape, they usually had to be rescued by special teams who were equipped with respirator units, roughly similar to compressed air self-contained underwater breathing apparatus (Scuba) gear, that began to appear in the early 1900s. During the early part of the 20th century such teams were usually composed of U.S.

Bureau of Mines personnel, who had to rush to the site with their equipment. *Rescue* was a term loosely applied to such operations since miners trapped in environments in which the air was either toxic or devoid of oxygen seldom survived.

One of the major hazards in coal mines is the presence of carbon monoxide generated by an explosion or fire. Carbon monoxide displaces oxygen in hemoglobin, leading to rapid asphyxiation among those who survive the initial disaster.

As a result, research into respirators for the individual miner was conducted throughout the first half of the 20th century. Although early self-rescuers did not provide air or oxygen, they did filter out carbon monoxide.

During the 1990s a combination of technologies allowed the introduction of new high-tech "self-contained self-rescuer" devices. The Mine Health Safety Administration requires that such devices be worn by miners or stored nearby, usually in coal mines. These breathing devices, in a container the size of a lunchbox, provide goggles and a nose clip along with a respirator mouthpiece.

One such self-rescuer when opened and donned generates an initial burst of oxygen with a chlorate candle that releases oxygen in a reaction triggered by heat. Another chemical component, potassium superoxide, reacts to the miner's exhaled carbon dioxide and moisture to create a steady supply of oxygen. Manufactured by the Mine Safety Appliance Company, this new self-rescuer operates independently of the surrounding atmosphere and is capable of providing an hour's worth of oxygen.

See also HELMET MEN.

Seven Mile Bridge (built 1909–1912) *Bridge system that spanned the Florida Keys* The story of the Seven Mile Bridge is a multimillion-dollar construction epic filled with engineering success, disastrous business failure, and massive human tragedy. The construction of four different bridges that constitute the historic span in the Florida Keys was part of a monumental effort to expand Florida East Coast Railroad from Homestead, Florida, to Key West.

The personality behind the privately funded effort to build what was called the Key West Extension was none other than one of the 19th century's most famous robber barons, Henry Morrison Flagler. The millionaire businessman was a cofounder of Standard Oil of New Jersey with John D. Rockefeller (perhaps the most famous robber baron).

At 53, Flagler fell in love with Florida while spending the winter of 1883–1884 in Saint Augustine. He was also in love with the financial opportunities the state seemed to offer. With his businessman's eye he saw south Florida's potential as a tourist Mecca and began pumping millions into hotels and railroad construction. A modest rail line built by Flagler between Jacksonville and Saint Augustine became the Florida East Coast Railway, which would later be extended across the Florida Keys through his Overseas Railroad. His purchase of hotels and the development of rail service in south Florida generated Miami's first real growth spurt, encouraged expansion of the citrus industry, and caused a general development boom.

And Flagler's railway never stopped advancing. On reaching the town of Homestead near Miami, Flagler decided he would lay railroad track across the water and connect the Florida Keys, a fragmented group of islands that curved southwest from Florida's peninsula to the island of Key West. Flagler's dream meant crossing 128 miles of swamp and open sea, a challenge that made other tycoons cringe. The extension would begin at the Everglades swamp and then charge from island to island in the Keys. If this sounds a little odd now, it was considered downright bizarre at the time. Detractors of the scheme referred to Flagler's railway extension as "Flagler's Folly." The 75-year-old Flagler was undeterred by naysayers and in 1904 ordered the Florida East Coast Railway to Key West. Construction began the following year.

But levelheaded engineers and construction contractors were leery of the scheme. When Flagler advertised nationwide for contractors to bid on the job of building the railway, he received a solitary response from a contractor who wanted a cost-plus contract. An outraged Flagler instead assembled his own team of engineers and construction experts to build what was officially known as the Key West Extension. To the public it would be the Overseas Railroad.

What Flagler sought to accomplish was nothing less than monumental. Even after defeating the wetlands of the Everglades, he still had to project a rail line across the Florida Keys, which was more like a Third World region than the United States. Fresh water was nonexistent on some keys, and there were no handy suppliers of building materials or provisions. Even simple items such as rock and gravel, usually available at a nearby pit or quarry,

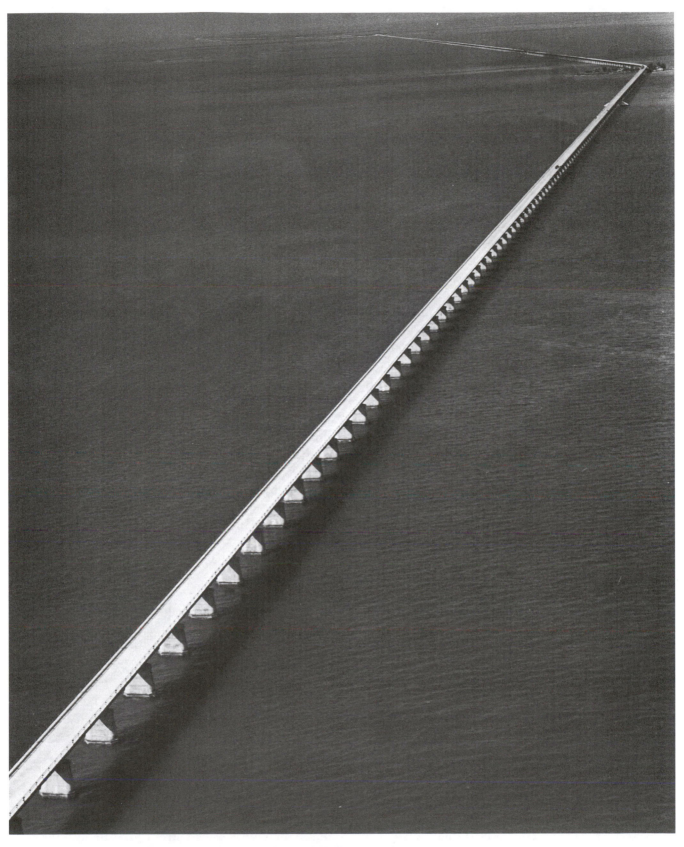

This segment of the Seven Mile Bridge reveals the challenge of linking south Florida with the Florida Keys archipelago by railroad. The railway for which it was built was never successful and eventually took a deathblow from America's worst recorded hurricane. The bridge system was converted into a vehicular highway during the Great Depression. *(Florida State Archives)*

would have to be carried in from hundreds of miles away.

The eight bridge-building crews that were organized in self-supporting units to work on sections of the extension were based on watercraft and resembled a small navy. Their heavy equipment and building materials would have to be carried on a fleet of barges, and for the first two years, even the workers lived on special houseboats, a circumstance that would have tragic consequences during the hurricane season. In a place where no other railroad or even the federal government wanted to build, Flagler poured millions.

The threat of hurricanes hung over the project for nearly six months a year. Everyone, including Flagler, knew how deadly these were and recognized that the Florida Keys were often the unhappy gateway for the powerful storms that targeted the Gulf of Mexico. In 1900, only five years before Flagler began his monumental effort to connect Homestead with Key West, a hurricane and its accompanying storm surge killed as many as 8,000 in Galveston, Texas. Efforts were made to communicate storm warnings to the work crews, but a false alarm in the Keys in 1905 created a dangerous sense of false security. When hurricane warnings were flashed to bridge-building gangs the following year, workers and management were foolishly lax in their precautions.

On October 18, 1906, the hurricane did strike the Florida Keys, where hundreds of Flagler's workers were still on barges or resting in their houseboats. High seas and powerful winds drowned 145 of the men. Some were blown into the sea, and others were set adrift on barges and small boats. Flagler's railroad responded by establishing work camps on various islands along the Keys to reduce casualties in a future hurricane. The 1906 hurricane was only a warning to Flagler and his men since the project would endure five more hurricanes before it was completed in 1912.

The Overseas Railroad continued to grow but not without difficulty. The sparsely populated Keys provided an insufficient labor pool, and Flagler's railway was forced to obtain many workers through employment agents in Philadelphia and New York. Taken to Florida at the railway's expense, the workers were to repay the company through deductions from their paychecks. Many grew tired of working in the remote Keys or, sensing opportunities elsewhere, left before repaying the company. Company

officials maintained that they simply absorbed the loss and allowed the workers to go their way.

However, allegations that Florida East Coast Railway was forcing workers to remain in the Keys against their will emerged. The construction supervisor, Joseph C. Meredith, and the chief engineer, William Krome, were indicted along with a New York employment agent on federal charges of peonage—forcing unwilling workers to remain on their jobs to repay their company debts. All were found not guilty. Despite the tropical heat, the danger of hurricanes, and the arduous work, Flagler's workers generally operated under good conditions, at least in the terms of the early 1900s. The railway even provided free medical care for a work force that ranged from 500 to 4,000 men during various phases of the project.

By January 1909, it was time to tackle the seven-mile gap between Knight Key and Duck Key, which would involve four main spans running from the northeast to the southwest. These included the Knight Key Bridge (6,803 feet), Pigeon Key Bridge (5,935 feet), Moser Channel Bridge (13,947 feet), and Pacet Channel Viaduct (9,035 feet). Some preliminary foundation work had been done on several of the segments, but in 1909 the work began in earnest. Collectively, these spans would be known as the Seven Mile Bridge.

The Knight Key Bridge began in January 1909, and workers started the repetitive process of assembling reusable wooden cofferdams for placement on the shallow seabed. The water depth ranged from only 12 feet to 30 feet across the Keys, and with the cofferdams in place, pumps removed sand to expose what local treasure hunters call hardpan, or what geologists call bedrock. The shallowness of this limestone relieved the gangs from having to undertake costly submerged excavations. Holes punched into the limestone inside the cofferdams then allowed pilings to be pounded into the rock to serve as anchors for the concrete piers.

German Alsen cement-based concrete was then poured to a depth of five feet over these pilings, using a tremie, or pipe, to pump the concrete under water. Wooden forms in the shape of the piers were erected atop these bases and filled with concrete mixed on a work barge. The piers were reinforced internally with 3/4-inch steel rod. To allow the concrete of the piers to cure in an even and complete manner, it was kept wet for 10 days after it was poured.

Using this method of foundation and pier work, construction gangs moved methodically to the southwest across the turquoise waters of the Keys. Other gangs were simultaneously working on various other segments of the Seven Mile Bridge to close the gap. Steel deck plate girders measuring 80 feet in length and weighing 41 1/2 tons each were bolted to the tops of the completed piers and later fitted with track.

This phase of the work on the Knight Key Bridge was dealt a severe blow when a hurricane struck October 11, 1909. Not only did it rip five improperly connected girders from the piers and shift others 18 inches from their alignment, it tore from their moorings four barges that contained girders and other bridge components. A barge carrying swing bridge material eventually sank, causing a loss of components that were not easily replaced. Florida East Coast Railway laid the blame for the loss of the materials and the toppled girders on the contractor, who first denied the charge and then took his men and equipment off the job. Flagler's own personnel then took on the task of installing the girders.

The storm's arrival had posed additional delays to completing the effort since many of the work barges had been sunk intentionally in shallow water when word of the storm's approach was received. In addition, a derrick that had been used to lower the girders into place had been blown into the water. When the girders blown from the piers by the storm's 105-mph winds were finally recovered, they were found to be useless, mangled by the force of the storm.

As efforts continued on the Knight Key Bridge, work was under way on the bridge segment at Pigeon Key, a five-acre island that also contained a work camp. The construction technique for the Pigeon Key Bridge was identical to that of the Knight Key Bridge except that a portion of the bridge ran across the tiny key. Foundations were sunk and piers built for 5,935 feet of bridging. This work culminated at Moser Channel Bridge, where the swing span was installed over the shipping lane known as Moser Channel.

The Moser Channel Bridge is interesting because it is the longest of the bridge segments (13,947 feet) and contained the swing bridge section. The swing bridge, mounted on a huge concrete pivot, was a through-truss span that could rotate perpendicularly to the bridge to allow the two-way passage of mar-

itime traffic on each side of its pivot. Although the Florida East Coast Railway had intended to place a swing bridge at Bahia Honda Key, the U.S. government ordered the swing bridge installed farther north at Moser Channel. Measuring 256 feet in width, the open swing bridge allowed the passage of ships between the Atlantic and the Gulf of Mexico. To protect the swing bridge's foundation pivot from a collision, huge fenders shaped like the bows of low-riding ships faced both the Gulf of Mexico and the Atlantic. By December 1911, the Moser Channel Bridge was completed and track was installed atop its rail deck.

The southernmost leg of the Seven Mile Bridge is the 9,035-foot Pacet Channel Viaduct, which extends from Little Duck Key to the northeast for nearly two miles to link with the Moser Channel Bridge. Pacet Channel Viaduct differs from the three northernmost segments of the Seven Mile Bridge because of its arch construction. Although the foundation and pier work were identical to those of the other bridges, it was decided to use 210 reinforced concrete arches instead of the beam-type construction, making the bridge a true viaduct design. Once the piers were completed in 1911, arch-shaped timber forms were prepared to serve as falseworks for construction of the arches. Secured to the piers, the curved falsework would provide a foundation for the reinforced concrete work. Once the concrete had cured, the form would be floated to the next arch under construction.

The 3/4-inch reinforcing steel set into the arches was spliced together in such a way that the arches were all connected by steel. However, because of concerns that the steel might corrode and weaken inside the concrete, it was decided to build the arches so the concrete itself would sustain the majority of the loading.

On January 21, 1912, at 6 A.M. the last piece of the Seven Mile Bridge, an 80-foot girder, was lowered atop its supporting piers by a floating derrick at the Knight Key segment. Just over 24 hours later, a final railroad spike secured the last length of track to connect the Florida mainland with Key West. An ailing 82-year-old Flagler was carried across his creation in a special train. It was a scene reminiscent of the time that the fatally ill British engineer Isambard Kingdom Brunel was carried in a special rail car to see his Royal Albert Bridge at Saltash, England, 54, years earlier. Flagler would die 16 months later, on May 20, 1913. His famed Key West Extension did turn out to be a folly and would wither financially until given a final and terrible blow by nature 22 years later.

The Key West Extension of Flagler's railroad had cost at least $20 million. And although it connected Florida's mainland with the beautiful Florida Keys, it never did the most fundamental thing that a robber baron's project was supposed to do—make money. Trade with Cuba never reached the levels necessary to make the Florida Keys segment profitable, and the transport of goods and passengers to and from the Keys was not enough to put the line in the black. By 1935, the Florida East Coast Railroad was bankrupt and in receivership.

The Great Depression, which had choked the economic life out of the United States since 1929, had hung over the land like a suffocating shroud for six years, adding in no small measure to the woes of the Florida East Coast Railway and its Key West Extension. But the railway had spurred growth and tourism in the Keys nonetheless, and now a roadway parallel to the Seven Mile Bridge was needed to allow automobile traffic across the islands.

As fate would have it, the Works Projects Administration (WPA) sought to employ hundreds of impoverished World War I veterans for the work, including some who had participated in the controversial 1932 Bonus March on Washington, D.C., to obtain early payment of a promised war bonus. These veterans became a symbol of national shame when they were brutally dispersed by an arrogant, medal-draped General Douglas MacArthur, who was then chief of staff of the U.S. Army. By 1935, some of these jobless veterans were at work building the Overseas Highway in the Florida Keys as the government sought to make amends.

Based in work camps on Lower Matecumbe Key and at Windley Key, the men were constructing roadways and bridges that were to run parallel to Flagler's unprofitable Key West Extension. Although hurricanes usually form off Africa's West Coast and build on the way to North America, a small but potent hurricane formed in the nearby Bahamas and raced toward the Florida Keys, allowing almost no time for a warning. It would turn out to be the most powerful hurricane ever recorded in the United States.

As the outer edges of the hurricane battered the Keys, residents of Matecumbe knew the storm they faced was a killer. An emergency call went out for an evacuation train, but bureaucratic bumbling delayed its departure for several crucial hours. When it finally departed, it crept slowly through screaming winds and rain so heavy that the conductor was forced to back up after unwittingly chugging past the train station on Lower Matecumbe Key. Hundreds of people struggled to board the train, but time had run out. The winds had accelerated to 200 mph as the barometric pressure dropped to the lowest level ever recorded in the United States.

As the train prepared to head north to Florida's mainland, a nightmarish disaster occurred when an 18-foot storm surge washed across the key. This wind-driven wave bowled over the rescue train and dislodged dozens of miles of track on the sandy keys. Incredibly, Flagler's Seven Mile Bridge had withstood this onslaught, but the human cost was immense.

More than 200 of the 683 veterans employed by the WPA were killed, along with 167 local citizens (including 50 members of one extended family). Some estimates place the death toll along the Keys at 1,000, but historians say the real number may never be known. For those who survived the storm surge, their reward was to watch in horror as flying debris decapitated and dismembered the living and the dead. Sand carried by the high-velocity wind scoured the flesh from those unable to find cover, making it difficult later to identify bodies whose faces had been literally sandblasted to the bone. Desperate and inadequate rescue efforts were hampered by the destruction left by the storm. The writer Ernest Hemingway, who had a home in Key West, arrived to help with the rescue efforts and later wrote a scathing article that plaintively asked who was responsible for the deaths of the veterans.

The destruction of the Overseas Railroad's land-based tracks meant the end of Flagler's dream. The rail line's right-of-way, bridges, and work camps were signed over to the Monroe County Toll Bridge Commission for $3.6 million, which would, with federal funding, convert Flagler's railway in the Keys into the Overseas Highway. More workers were brought in, and the Seven Mile Bridge segments were widened for two-lane automobile traffic. On March 29, 1938, the highway was again opened, this time to automobiles.

By 1982, the aging Seven Mile Bridge was replaced with a new concrete bridge running alongside the old one. The 70-year-old swing bridge over Moser Channel was removed because the new bridge had a segmental concrete span overpass that rose 65 feet above the water to allow ships to pass beneath it.

In the aftermath of the hurricane, a monument was erected at the Matecumbe Key town of Islamorada in memory of the veterans and the local citizens who died in the storm. The cremated remains of 300 recovered victims are interred inside the monument. The obelisk bears a bas-relief carved in limestone depicting palm trees bent by hurricane winds as waves roll over the island. It could serve just as well as a memorial to Henry Flagler's Overseas Railroad.

shaft *See* adit.

shaker *See* Henry, John.

shotcrete *Spray-on concrete for rapid reinforcement of tunnels* The history of engineering advances is often a zigzag path in which a seemingly unrelated development is applied to a different field of endeavor with spectacular results. This was the case in the evolution of a taxidermist's invention that led to the method of spraying concrete to create reinforcing tunnel walls in a short period.

Shotcrete simply stands for "shot concrete," or concrete applied through a nozzle under air pressure against embankments or tunnel walls. Shotcrete's purpose in tunnels is to reinforce weak rock, or soft "squeezing ground," which tends to compress—if not collapse completely—as the tunnel progresses. Although the basic technique of spraying concrete has been around since 1907, it was a half-century before the conservative community of mining engineers began to take it seriously as a fast, relatively inexpensive, and effective method of reinforcement.

Its unlikely origins spring from the efforts of a brilliant American taxidermist, Carl Ethan Akeley, who refined the craft of stretching the hides of dead animals over forms to a high art. While employed at Chicago's Field Museum of Natural History in 1907, he developed a method of shooting cement into animal-shaped frameworks. This invention, later refined into the cement gun for the application of Gunite, allowed Akeley to create smooth, lifelike bodies ready to be fitted with the hides of actual animals. The results were stunning, lifelike displays of wild creatures.

Construction engineers noticed this technique, and in 1909 the Cement Gun Company of Allentown, Pennsylvania, refined the technique further and designed a system to stabilize unstable sloping ground by spraying concrete on its surface. By the 1920s, it was recognized that pumping concrete against the walls of coal mines was an excellent way of fireproofing tunnels.

Tunneling, unless it is through extremely hard, stable, and cohesive rock, requires the installation of expensive and time-consuming reinforcement in the form of timbers, circular metal frames, or other methods of ensuring rock or earth does not deform or collapse. In 1954, the Austrian mining engineer Anton Brunner looked at the incredible compressive strength of concrete—its ability to endure massive weight—and decided that applying concrete to the walls of freshly dug tunnels by using Akeley's invention would produce immediate and potent reinforcement. Brunner was not taken seriously and had a real struggle convincing anyone that the technique would be faster, cheaper, and safer than conventional reinforcement.

Brunner was convinced that concrete sprayed to the sides of a tunnel would have adequate strength to ensure the structural integrity of the excavation. He wanted to apply the technique to a tunnel he was working on, but the conservative and reluctant engineers above him were hard to convince. After signing a statement taking full responsibility for the possible failure of his proposed technique, Brunner used shotcrete to reinforce a hydroelectric plant's water diversion tunnel successfully in the mid-1950s. He soon patented a process that evolved into the new Austrian tunneling method, a technique that provides for rapid reinforcement of newly dug tunnels using shotcrete.

Shotcrete is often applied initially and then backed by precast segmental liners that are assembled inside the tunnel to provide additional strength. The shotcrete method provides a rapid way of forestalling the early movement of rock or soil that can deform or collapse a tunnel.

Shotcrete has become a crucial tool for the driving of tunnels through what is termed soft ground—ground mainly devoid of self-supporting rock. It is strengthened further by spraying the concrete against a framework of steel reinforcing bar that is set against the wall. A relatively new strengthening component in shotcrete is the addition of steel or plastic fibers. These provide a greater degree of tensional strength to the concrete to help it resist any side-to-side movement, complementing its ability to withstand compression from the top down.

Although shotcrete is not called for in every tunnel, sprayed concrete has been used with greater frequency and was first applied on a large-scale tunneling project in the United States during the construction of the Washington Metropolitan Area Subway System during the 1970s. The construction of Japan's undersea Seikan Tunnel also used the rapid strength of shotcrete to reinforce the unstable and flooding-prone rock walls. The tunnel dug through unstable rock is 780 feet beneath the surface of the Tsugaru Strait.

See also AKELEY, CARL ETHAN; NEW AUSTRIAN TUNNELING METHOD.

silicosis *See* Hawk's Nest Tunnel.

Silver Bridge collapse *See* National Bridge Inspection Standards.

Simplon Tunnel I and II (built 1898–1905) *Subterranean path of the famed Orient Express* One wonders whether Europe's elite who traveled through the Simplon Tunnel on the elegant Orient Express ever pondered the struggle to build what was for 80 years the longest tunnel? Or did these passengers lounging in padded Spanish leather chairs sipping apéritifs take for granted a shortcut purchased with 39 lives? If they were like the rest of us, they probably noticed only that the alpine scenery quit going by as the train disappeared inside a mountain for more than 12 miles.

Although not as unsafe for the workers as the cursed Saint Gotthard project, which ended in 1882 after a decade of death and debacle, the Simplon Tunnel would present engineering challenges dwarfing anything encountered previously. And because the project took advantage of late 19th- and early 20th-century tunneling technology and the sometimes-tragic lessons of the massive Swiss tunnels it would eclipse, it was completed in a relatively brief seven-year period.

Simplon Tunnel was part of a progression of stunning Swiss tunnel projects, each proving to an increasing degree that long, hard rock tunnels were possible. Preceding the Simplon Tunnel were the Mont Cenis, Arlberg, and Saint Gotthard Tunnels. And although the Simplon Tunnel was eventually completed, the bad rock conditions and flooding made it a monumental challenge to the engineers and workers. It was also the first time double, parallel bores were excavated; these passages are today known as Simplon Tunnel I and Simplon Tunnel II.

Businessmen in the Swiss canton of Simplon had been dreaming since 1860 of a rail tunnel beneath Monte Leone as a passageway between Italy and Switzerland. Although a rail line of switchbacks could be negotiated up the steep mountain, the severe grade reduced the speed of a train and the load it could carry. After a group of smaller railroads merged into the Jura-Simplon Railway, an organization large enough to mount such a project, a contract was signed in 1893 with a consortium of construction firms. It would be another five years before all the details were worked out, including the roles to be played by the Swiss and Italian governments.

One of the construction firms in the consortium was Brandt, Brandau & Co., whose founder, Alfred Brandt, had invented a hydraulic rock drill that used pressurized water. Brandt's invention evolved from his experience as an engineer on the deadly Saint Gotthard Tunnel, completed in 1882, and would be used on the Simplon project.

What plagued so many tunneling projects during the 19th century—unreasonable construction contracts and poor geological data about the condition of the rock—would also nearly doom the Simplon effort. The contract called for the construction of a pair of parallel tunnels, the first to be completed in five years and the second in four years. The first tunnel completed would be known appropriately as Tunnel I. The second, completed as a smaller pilot heading, would be Tunnel II.

What amounted to a completion bond of 500,000 Swiss francs would be withheld by the owners until completion of the first tunnel, and a total of 1.5 million Swiss francs would be withheld until two years after completion of the second tunnel. If the contractor exceeded the completion dates, it would incur a 5,000-franc-per-day penalty, or an equal amount would be given as a bonus for each day the project was completed ahead of schedule. The worst element of the contract was an ironclad stipulation that the contractor's fee would cover all unforeseen geological obstacles of any kind. There would be plenty.

If the contract was self-destructive to the engineers, the wretchedly poor geological assessment of the rock was outrageously bad, limited as it was by the knowledge and technology of the late 19th century. Conditions were supposed to be conducive to mining beneath Monte Leone. Instead, the miners would find some of the worst conditions ever experienced.

The worker safety and health scandals of the Saint Gotthard Tunnel, in which 310 died, prompted a few beneficial clauses in the contract, including the requirement that just over 1 million cubic feet of air be pumped into the two headings of the tunnel every minute. The contractor was also required to remove all garbage and sewage from the tunnel while providing fresh drinking water and electric lighting. This last requirement meant the Simplon Tunnel project would be the first illuminated by something other than torches or natural gas. Last, a fund was established to pay benefits to families of workers killed during the project.

The contract obtained Swiss government approval, but elected officials, who still had a bad taste in their mouths over the flagrantly unhealthy working conditions at Saint Gotthard 16 years earlier, also required that no man be hired on the project who was not physically fit. Furthermore, it required that living conditions for the workers be hygienic.

By 1896, a tripartite agreement was concluded in which the Jura–Simplon Railway would contribute 50 million francs to the project and the governments of Switzerland and Italy, 16.6 million francs and 4 million francs, respectively. By the time all the contracts were finalized and the intergovernmental agreements were made, the year was 1898. By the time the clock began running on the five-year contract on August 13, 1898, Brandt and his fellow engineers had been working on the south portal for 12 days.

Besides developing living facilities at Brig, Switzerland, in the north and at Iselle, Italy, in the south, it was necessary to develop water resources needed to operate air compression facilities for ventilation and to provide pressurized water for the operation of Brandt's hydraulic drills.

To do this at the north portal, water would be diverted from the Rhône River by a two-mile canal. This project required blasting and tunneling since the water also had to travel through a 730-foot tunnel. For the south portal, waterpower would be obtained by construction of a reservoir trapping the waters of the Diveria River. The turbines driven by the water would turn pumps to produce water pressure at 1,700 pounds per square inch. Blowers powered by the water turbines would pump fresh air into and depleted air out of the tunnels while pipes carried pressurized water to the rock drills.

In an unusual but positive arrangement for the time, the workers were provided shower facilities at both portals, where their work clothes would be hung on hooks and then raised to the rafters above the dressing room. The clothes would be washed and mended at contractor expense.

Although electric trains would eventually be employed in the Simplon Tunnel, smaller compressed air trains would be used during its construction to prevent fouling of the air with the burning coal of steam engines. These were used to carry materials and workers to the tunnel faces.

The design of the tunnel was not only ambitious but also unique. Its 12-mile length made it the world's longest tunnel; its other unusual feature would be a pair of separate, parallel bores connected by cross passages. Until Simplon, railroad tunnels usually consisted of a single bore large enough to accommodate a pair of railroad tracks.

In the process of excavating Tunnel I miners drilled shot holes in the face of the rock; then explosives handlers blew out the rock for a pilot shaft. At the rear of this military-style assault were men with hand drills to widen the opening. With the core of the tunnel blasted out, the tension holding the peripheral rock was gone, making what remained relatively easy to chip away.

Alfred Brandt, who led the engineering effort on the tunnel, was in command of the northern heading at Brig. Consumed by the details involved in the logistics and execution of the tunneling, Brandt often worked without sleep for days. On November 29, 1899, he died of a stroke scarcely a year after work began and was immediately replaced by the Swiss engineer Edward Locher-Freuler. In command of the tunneling force on the south portal was Brandt's partner, Karl Brandau.

The necessary average rate of advance of 19.18 feet per day was not being met. Whereas the average rate of advance in the north end was 19.22 feet per day, the southern advances were crawling at 15 feet per day. The rock there was harder than expected, jeopardizing the project's timetable.

Although strong, impermeable self-supporting rock was supposed to greet the miners during their advance, what they began to find on the northern side was rock perforated by springs whose waters were as hot as 131.7 degrees Fahrenheit. This raised the mercury inside the tunnel to 82 degrees Fahrenheit, a condition counteracted by introducing a mist of refrigerated water into the air lines. Also assisting ventilation and cooling efforts was the twin-tunnel concept, which allowed air to be blown down one parallel bore and then drawn out through the other.

The troublesome southern tunnels, besides advancing a third slower than they were supposed to, encountered a series of additional problems. As they progressed, the ground beneath one segment of the tunnels buckled, requiring massively heavy stonework to tame the movement.

Worst yet, on September 30, 1901, south tunnel miners struck a subterranean spring that poured nearly 120,000 gallons an hour into the tunnel. Attempts to divert this water by mining another heading struck an even more powerful jet of water, which filled the tunnel at a rate of 158,000 gallons an hour. When it appeared impossible that conditions could be worse, miners in the smaller Tunnel II punctured rock, which poured a whopping 367,000 gallons of water per hour.

After spending 60 days overcoming this flooding, the southern tunnels advanced into a 137-foot stretch of squeezing ground whose decomposed rock oozed into the excavation with spectacular pressure. Miners used massive timbers, some measuring 16 inches square, only to watch them shatter under the force of the squeezing ground. Instead of proceeding at 15 or 20 feet per day, the engineers and miners had to spend 1 million francs and nearly half a year to cross this stretch of ground less than half the length of a football field.

To pass safely through this zone, a set of rectangular steel frames braced by powerful timbers was fabricated. The first of these modular frames was secured to solid rock and then put together frame by frame until the entire stretch of crushing ground had been negotiated. The miners managed to construct what was in effect a rectangular underground bridge of steel and timber. The structure contained 74 frames, each consisting of 14-inch girders and 20-inch square pine timbers.

This solution was in itself a problem since the opening through this reinforced steel cage was far smaller than the full dimensions of the tunnel. Falling behind, engineers decided to close the gap between the north and south headings before expanding a larger opening through the squeezing ground. A relatively soft layer of gypsum beyond the squeezing ground allowed the miners to advance at an average rate of 26.4 feet per day.

The heading of the north tunnel then had its own problems as it struck hot springs and geothermal conditions that made the rock too hot to touch. The atmospheric temperature rose above 89 degrees Fahrenheit while water gushed in at a scalding 131 degrees Fahrenheit. It was 1902, and the contractor consortium had lost the race as specified by the contract when it was 876 feet shy of its scheduled progress on the main tunnel and 3,448 feet short of the advance needed for Tunnel II.

The consortium and its workers could only move forward and made excellent progress through Monte Leone's depths even as the geothermal heat inside the tunnel began to rise alarmingly. At 4.53 miles inside the northern heading the rock temperature rose to the highest level ever recorded, a painful 132.8 degrees Fahrenheit.

More hot springs were struck, but by now rock bursts and spitting rock were becoming commonplace. These explosive releases of energy, generated when rigid rock finally snaps under the settling force of the surrounding ground, were potentially deadly and required that the miners place timber lagging against the walls for protection.

Despite the breathless race waged by the miners, masons, blasters, and engineers, the August 1903 deadline by which a full tunnel was to be completed came and went. And it was far from completed. Fortunately, the financial bacon of the contractors was pulled off the hot rock walls of the Simplon Tunnel by Switzerland's nationalization of its railroads.

The new governmental owners reviewed the situation and decreed that the completion date would be extended by another two years and added another 8.5 million francs to the project. It also slashed the late penalty from 5,000 francs a day to 4,000. Retrieved from the brink of ruin, the contractors plunged on.

At 5.9 miles into the rock, the faster-moving north heading arrived at the midpoint of the mountain and began an up-and-down course, tapping a series of hot springs on the way, as the southern heading moved northward. As miners in the northern heading sealed off the leaks and introduced chilled water spray into the air to keep the environment bearable, a landslide blocked the water from the Rhône, halting the power source for the northern effort. Since it was impossible to pump out the inflow of water or cool the working area, work was halted in the first half of 1904 and the downward leg of the northern heading was allowed to fill with water.

The miners inside the southern heading knew it was up to them alone to complete the breakthrough, and they made superb progress drilling and blasting their way through rock that for a while was devoid of the maddening springs. With only 882 feet left between the southern heading and the now-abandoned northern heading, the miners were once again toyed with by Monte Leone, when they encountered more hot springs. Naturally cold water was pumped against the scorching rock face to cool the working area, but after four months only a paltry 328 feet had been driven.

By February 1905, the southern heading now moved below the terminus of the down-angle northern heading and miners tunneled upward to connect the two. With a remotely triggered dynamite blast, the floor of the northern heading was opened to drain its heated water and potentially fatal carbon dioxide gases into the southern heading. Because of the threat of a buildup of carbon dioxide, the tunnel was cleared of people and remained so after the breakthrough detonation of February 24, 1905.

Sadly, two men who did not work as miners decided to see the breakthrough for themselves when no one was looking and were overcome by carbon dioxide gas. Although they were found and taken from the tunnel, they died a few days later. The tunnel was then officially closed to everyone to allow the still-hot rock face to cool.

Despite this success, 2,952 feet of Tunnel II's pilot heading still needed to be driven and would not be completed until December 1905. By another act of providence, the revised April 30, 1905, deadline had been extended again by the Swiss Federal Railways, halting a 4,000-franc-per-day penalty charge.

As masons lined the remnants of Tunnels I and II, miners returned to what was termed the 137-foot pressure zone, where the rectangular steel cage held the pulpy, crushing rock at bay. To widen this tunnel to its proper dimensions, miners did some of the most nerve-wracking and grueling work in the Simplon Tunnel project's seven-year history.

To widen and line this section under terrific squeezing pressure from the plasticlike rock, miners had dug 15 shafts as deep as 10 feet below the temporary reinforcing cage and began setting stone. This was done for the length of the reinforced section to create a massive foundation. The walls of the tunnel were then excavated and reinforced by thick stonework outside the cage. The arch of the tunnel was dug by using an overhead pilot tunnel and similarly lined with stone. All was held in place by temporary steel arches and other supports until the job was finished. Once the passageway was prepared, dynamite was used to clear the tunnel of the temporary supports.

In a matter of weeks the track was laid and the electrical cabling installed to power the trains. On January 25, 1906, the first train passed through the Simplon Tunnel heading south from Göschenen to Iselle. By 1919, the Orient Express, which ran from Paris, began making its way through the tunnel to Italian destinations and eventually to Istanbul. This would continue until 1977, when declining ridership forced an end to the regal service. It would be reinitiated in 1982 by an American, James Sherwood, who had once unsuccessfully proposed an English Channel tunnel. His revived train service bears the name Venice Simplon–Orient Express.

The butcher's bill on the Simplon Tunnel was not as great as it had been on other tunneling projects, but with 39 dead including the chief engineer, Alfred Brandt, and a pair of foolhardy gawkers, the tunnel had taken its tool. Of the thousands injured in some way during the project, at least 133 suffered injuries that left them disabled for life. There is no way of estimating how many departed the tunnels deafened or with dust-induced lung diseases. Tunnel II, which was completed as only a smaller pilot tunnel and lined during the original project, was fully excavated by 1922, a process that had been delayed by World War I.

Had anyone known of the zone where the ground threatened to embrace the miners to death or of the number of subterranean springs that would be tapped, it is highly unlikely that tunnel route would have been attempted. Its final length was measured at 12.3 miles, three miles longer than the next largest at the time, the Saint Gotthard Tunnel. Until Japan's 32.3-mile Seikan Tunnel between Honshu and Hokkaido was completed in 1988, the Simplon Tunnel held the record as the world's longest.

See also BRANDT, ALFRED; EUROTUNNEL; GEOTHERMAL GRADIENT; MONT CENIS TUNNEL; ROCK DRILL; SAINT GOTTHARD TUNNEL; SEIKAN TUNNEL.

Singstad, Ole (1882–1969) *Norwegian-American tunnel engineer* Ole Singstad made his mark on his adopted land by helping to design some of America's most famous vehicular tunnels, including New

York's Holland Tunnel, Queens-Midtown Tunnel, and the Brooklyn-Battery Tunnel.

Unlike bridge engineers whose accomplishments are usually soaring structures that awe in plain view, Singstad's work was usually marked only by architecturally subtle entry portals and ventilation structures that looked little different from buildings. His creations were burrowed and bored through mud and rock, ventilated with fresh air, and sealed against high-pressure water that permeated the subterranean world.

Singstad was a tough, capable professional whose honesty and integrity were beyond reproach. Hardworking and ambitious, Singstad continually sought out challenges in his professional life and eventually became one of America's preeminent authorities on vehicular tunnels, an engineering field born in New York City, where Singstad began his career.

His honesty and integrity would cost him dearly late in his career when Singstad refused to bow down to New York's Robert M. Moses, the powerful parks commissioner and Triborough Authority official who unsuccessfully sought control of Singstad's final two tunnel projects in New York City. Instead, Singstad fought for the projects, particularly the Brooklyn-Battery Tunnel, whose design he supervised. Singstad would eventually be the subject of a Moses-orchestrated smear campaign and would find his contribution on the Brooklyn-Battery Tunnel project ignored by Moses in much the same way that Charles Alton Ellis was snubbed for his role in designing the Golden Gate Bridge. On the day in 1905 when he arrived in New York City as an immigrant, Singstad could not have imagined a confrontation with someone like Moses three decades in the future. A recent graduate of Trondheim's Technical College in Norway, Singstad had immigrated to America's biggest city to practice engineering in a place that might offer more challenges than his native land. His first job was as a bridge engineer with the Central Railroad of New Jersey, and from there he took a similar position with another railroad that put him in Virginia.

As the job in Virginia was nearing its predictable end as much of the railway was completed, Singstad returned to New York, where he happened into a job as a tunnel designer. The job he lucked into involved the final stages of the subaqueous rail tunnels beneath the Hudson for the Hudson-Manhattan Railway that were completed in 1908.

By 1902, the engineer Charles Jacobs had been picked to lead a final and successful push to complete the tunnels, and it was his staff that Singstad joined. The Norwegian, who by this time had decided to seek American citizenship, gained valuable lessons in tunnel design on a project that had every financial and engineering problem a subaqueous tunnel could have. From Jacobs's staff, Singstad went to work for New York City designing the growing web of subterranean rail lines that were becoming the city's famed subway system, learning of all the most recent developments of tunneling technology.

For the hardworking and ambitious Singstad the prospect of being promoted to a higher level of responsibility and pay was dependent upon the seniority system. "It seemed the only way I could get ahead would be for some of my good friends and coworkers in Civil Service to die and make way for me," said Singstad of the situation. He left tunnel engineering for a year but returned to the field as an engineer on a vehicular tunnel being planned in Philadelphia.

Clifford Holland then recruited Singstad in 1919 to help design a revolutionary vehicular tunnel between New Jersey and the downtown area of New York. Hesitant at first because he would once again be working in what he considered to be the tedious role of a civil servant, Singstad found the job too tempting and accepted Holland's offer. The tunnel, which would later be named after Holland, who died during its construction, would be made up of two tubes, one measuring 8,558 feet and the second 8,371 feet.

The tunnel posed a series of challenges to Singstad and his fellow engineers. It would be the longest subaqueous tunnel ever constructed, and the depth of the twin bores meant the men would be working in a compressed air environment of intense pressure to hold back water leakage. It would also be the world's first tunnel to accommodate automobile traffic, and the internal combustion engine's hunger for oxygen and resultant discharge of exhaust gases created a deadly dilemma. Holland handed Singstad the problem of solving the technical challenge of ventilating the tunnel.

For the little technical guidance available on auto exhaust gases, Singstad coordinated research with the U.S. Bureau of Mines to determine the effect of these gases on humans. Data were also accumulated on precisely how much ventilation was needed to maintain a healthy supply of air.

As work on the tunnel progressed, Holland died suddenly in 1924, and Milton Freeman was appointed chief engineer on the tunnel project. Freeman then died three months later, giving the job of completing the tunnel to Singstad.

As chief engineer of the New York Tunnel Authority, Singstad was still free to be a consulting engineer on other tunnel projects, provided that this extracurricular work did not interfere with his full-time duties. Before completion of the Holland Tunnel, Singstad served as a consulting engineer on many of America's most important tunnels. His ventilation design for the Holland Tunnel became a model for future subaqueous tunnels, including the George A. Posey Tunnel between Oakland and Alameda, California, completed in 1928.

Singstad's next major subaqueous tunnel project for New York was the Queens-Midtown Tunnel. Built between 1936 and 1940 the two tubes of the tunnel measured 6,272 and 6,414 feet. Singstad designed the ventilation system to move a total of 3 million cubic feet of air through both tubes of the tunnel each minute.

Singstad's design of the Brooklyn-Battery Tunnel would be both his crowning subterranean achievement and his greatest disappointment in New York. Opposed by Robert Moses, head of the Triborough Bridge Authority, the tunnel became a test of will between Moses, who preferred a bridge, and those who desired a tunnel since it would not mar the view of Manhattan's skyline. Singstad stood his ground against Moses, publishing an unflattering, though accurate drawing of Moses's proposed bridge. Moses, for his part, characterized the bridge as a beautiful span that would hardly be noticed.

Moses's attempts to kill the tunnel project were thwarted, and construction of the Brooklyn-Battery Tunnel began in 1940, only to be interrupted by World War II. By 1945, Moses had acquired control of the Tunnel Authority, something he had long coveted, and Singstad was fired.

Work was restarted on the partially finished tunnel in the year of Singstad's firing. And although Singstad provided engineering guidance on the project he designed, his name would not be associated with it officially.

Moses managed to bully his way into power by maneuvering his supporters into the Tunnel Authority's trio of board positions by alleging the authority had mishandled its finances. He also conducted a vicious campaign to sully the reputation of Singstad. The basis of Moses's attack had been that Singstad's brother-in-law purchased property where another engineer proposed placing the Manhattan toll plaza.

Singstad, who argued against the location near his in-law's purchase, had actually been outraged to learn of his relative's dealings. While still opposing the location that would benefit his avaricious in-law, Singstad forced him to slash the property's price from $365,000 to $165,000, according to Robert Caro in his Moses biography *The Power Broker*.

Investigated both by the Port Authority of New York and New Jersey and by the Tunnel Authority, Singstad was found innocent of insider dealing. Singstad did receive a slap on the wrist for failing to disclose his in-law's activities, although Singstad knew full well that Moses was his implacable enemy and that revealing the matter would open him to an attack by the ruthless Moses. The honest and capable engineer knew full well that Moses had him in his political crosshairs and that Moses was only waiting for the moment to pull the trigger.

Moses also alleged that the Tunnel Authority was on the financial rocks during World War II thanks to bad management. The truth was that wartime restrictions on driving coupled with the disappearance of a large segment of the population into military service threatened the ability of toll tunnels such as the Queens-Midtown Tunnel to repay their bonded indebtedness to the federal government. Singstad, far from being a self-dealing public official, had skillfully brought the Queens-Midtown Tunnel in $4 million under budget, providing a nest egg of cash that could pay the interest on the government-backed bonds during the lean wartime years.

Singstad, who had never cared for civil service anyway, moved on in his career to become one of the world's premier tunnel engineers designing subterranean projects around the world. In 1944, before being cashiered by Moses, Singstad received the Abraham Lincoln Award for Distinguished Services for his efforts in the City of New York. He was also granted membership in the Royal Norwegian Society of Sciences and membership in the Order of the Crown of Belgium. Singstad was also awarded the James Laurie Prize by the American Society of Civil Engineers.

Despite being tarred by the politically vicious Robert Moses, Singstad is still justifiably revered by the Norwegian-American community as an honorable immigrant who found success in America.

See also BROOKLYN-BATTERY TUNNEL; HOLLAND, CLIFFORD; HOLLAND TUNNEL; LINCOLN TUNNEL; MOSES, ROBERT M.; QUEENS-MIDTOWN TUNNEL; TUNNEL VENTILATION.

skywalking *See* Mohawk ironworkers.

slurry shield machine When using a tunnel-boring machine to burrow through soft ground composed of sand and gravel rather than cohesive clay, miners face the prospect of the progressive failure of the tunneling face. This collapse complicates digging by the rotary cutting head of the machine and makes removing the spoil difficult.

The development of earth pressure balance machines made it possible for engineers to pump water under pressure between the working face of a tunnel and the front of a boring machine, but the liquid lacked the viscosity to suspend or hold gravel or other particles.

In 1964, the British tunneling engineer John Vernon Bartlett took a lesson from oil well drillers who used bentonite—otherwise known as drilling mud—to provide enough structural resistance to prevent collapse of a drill hole after the withdrawal of the drill pipe. Bentonite is a fine aluminum silicate clay formed from volcanic ash. When agitated or stirred, it is in a liquid state, whereas when motion ceases, bentonite immediately assumes a semisolid gel state. This characteristic is known as thixotropy.

Mixed with water and agitated by the cutting head, the bentonite becomes a viscous slurry maintained under a specific pressure that is capable of suspending and holding loose gravel in place. Seals around the circular body of the tunnel-boring machine help maintain the pressure.

Bartlett patented his concept in 1967 and the process has been in use around the world ever since. During the digging, the bentonite slurry is pumped out along with spoil from the excavation, filtered free of sand and gravel, and then returned to the tunnel face by pipe. A tunnel-boring machine equipped with Bartlett's innovation, known originally as a bentonite tunneling machine, was given its first trial use in February 1972 in London.

See also EARTH PRESSURE BALANCE MACHINE; PROGRESSIVE FAILURE; TUNNEL-BORING MACHINE.

Smithfield Street Bridge (built 1881–1883) *Pittsburgh's graceful and unusual lenticular truss bridge* Pittsburgh's Smithfield Street Bridge, also known as the Monongahela Bridge, was the third bridge built across the Monongahela River in the same location during the 19th century. The first Monongahela Bridge was a timber bridge designed by Lewis Wernwag and built in 1816. Destroyed by fire in 1845, Wernwag's bridge was replaced by a suspension span designed and built by the legendary engineer John Augustus Roebling that was completed in February 1846.

Roebling's bridge, despite its innovative design and strength, was a combination suspension and beam bridge with rigid deck sections suspended by numerous midstream piers and a cable wire cable suspension system. It served without mishap, but its design was unequal to the heavier loads placed on it during the first 25 years it was in service. The bridge, which served as one of the primary ways of moving heavy wagon loads of coal across the Monongahela, was showing signs of deterioration by the late 1860s.

Although an 1871 attempt by Pittsburgh's city leaders to build a new bridge were rebuffed by stockholders who financed the original toll bridge built by Roebling, the directors of the Monongahela Bridge Company themselves eventually relented as the condition of the bridge declined. In 1880, the Bridge Company voted to contract with a Pittsburgh engineer, Charles Davis, to design a new suspension bridge.

Work on Davis's design began in 1880 with construction of the piers. Winter halted the work temporarily. A reorganization of the Monongahela Bridge Company under the control of its majority shareholder, David Hostetter, resulted in the nullification of contracts for Davis's bridge in 1881. Hostetter desired a bridge with the capability of carrying rail traffic, something for which Davis's flexible suspension span would have been less than ideal. The Pittsburgh engineer Gustav Lindenthal was asked to submit a design, and he recommended a span that would become his first high-profile bridge-building effort.

The Austro-Hungarian immigrant, like Davis, was a mostly self-taught engineer who had distinguished himself with a number of bridge designs before being commissioned to design the Smithfield Street Bridge. Lindenthal's orders were to build a bridge 48 feet wide using the piers sunk by Davis. In addition, the bridge had to be capable of being widened to 64 feet at a later date if additional traffic required the expansion. Lindenthal decided to employ what is known as the Pauli truss, a structural innovation created by the German Frederich August von Pauli. The bridge would have a pair of 360-foot spans supported by these trusses; a third set of Pauli trusses would be added later. With the approach spans from shore the bridge would measure a total of 1,184 feet.

The Pauli truss employed lens-shaped trusses composed of an arching, hollow top chord whose lower element was a flexible chain. The original width of the bridge required only two such trusses running side by side. The trusses appeared to be gigantic spectacles staring up and down river. Lindenthal, famed for creating unique bridge designs, also strove to rid the bridge of the bulky appearance that had plagued so many spans during the 19th century. Besides using the Pauli truss instead of a conventional boxlike truss system of iron components, he opted for steel, whose greater strength would have required less structural bulk at a slightly lower cost.

Lindenthal, always mindful that the public's trust of a bridge was directly related to its sometimes misleading appearance, had both ends of the trusses terminate at massive but purely ornamental portals. Although Pauli trusses supported the bridge deck, the towers were both nods to the Gothic style of the time and added to provide the appearance of strength. Lindenthal would employ this design tactic in his construction of New York's Hell Gate Bridge more than 30 years later.

Although the design resulted in a bridge with a rather whimsical appearance, Lindenthal's creation was substantial and strong. When completed in 1883 it carried a single lane for carriages with two rail lines across the Monongahela River. When it was decided to widen the bridge in 1891 with the addition of a third set of Pauli trusses, the two lanes of rail traffic were positioned on each side of the bridge's centerline. Later the two rail lines were placed on the same side of the bridge and vehicular traffic was allowed on the other lane. The ornate portals were replaced with simpler facades in 1915 and remain in place today.

In 1934, a lightweight aluminum rail deck was put into service beneath the vehicular lanes to reduce the dead load of the bridge and increase by four times the live load it carried. This was the first large-scale use of aluminum for a bridge road deck. Subsequent deck rehabilitation work in 1967 replaced the deck's asphalt coating with a thinner polyester-sand layer, carving another 97 tons of dead weight from the superstructure. In 1985, passenger rail traffic was removed from the bridge, and between 1994 and 1995 additional renovation work on the bridge finally eliminated the rail lines and provided for a total of three vehicular lanes.

The bridge was purchased by the City of Pittsburgh in 1896 and was acquired by Pennsylvania in 1958. It stands as the oldest river bridge in Allegheny County and is considered the oldest through-truss bridge in the United States. In 1996, the bridge was returned to its original three-color configuration as set out by Lindenthal in his original design. It is considered one of Pittsburgh's most beautiful structures and has been designated a National Historic Landmark by the federal government and a National Historic Civil Engineering Landmark by the American Society of Civil Engineers.

The designs of Lindenthal's Smithfield Street Bridge and Isambard Kingdom Brunel's Royal Albert Bridge at Saltash are often compared because Brunel's bridge also uses the seldom-seen lenticular truss concept.

See also COLOSSUS; HELL GATE BRIDGE; INSPIRED CARPENTERS; LINDENTHAL, GUSTAV; ROEBLING, JOHN AUGUSTUS; ROYAL ALBERT BRIDGE AT SALTASH; TRUSS BRIDGE.

Snowy River Hydroelectric Scheme *See* rock bolts.

solid-ribbed arch A relatively rare type of design, a bridge containing a series of parallel arches supporting a bridge roadway comprises a structure referred to as a solid-ribbed arch. When viewed from below, such arches resemble the structure of a rib cage as they curve side by side to support a road deck. Bridges of this type can be constructed of concrete, steel, or iron.

Sommeiller, Germain (1815–1871) *Mont Cenis Tunnel designer and rock drill innovator* The engineer who would design and direct construction of the first major alpine tunnel, and what was the world's longest tunnel, was an orphan of humble beginnings who spent the first part of his career as a math tutor.

Born in Faucigny, Sardinia, in what is now France and orphaned at 15, Germain Sommeiller was raised by his sister, whose savings and encouragement made it possible for him to attend a local college. In 1836, Sommeiller left for Turin University in northern Italy, where he earned an engineering degree. No doors were thrown open for Sommeiller, who, unable to find employment in his field, gave math lessons to other students seeking university admission. With no other opportunities open to him, Sommeiller entered the Sardinian army as a junior officer.

Soon after, Sommeiller was offered a position with Sardinia's Corps of Civil Engineers, and at the age of 30 he began practicing his chosen profession. His abilities were at last recognized, and Sommeiller then found himself assigned to the Transport Department. In 1846, the government sent him to Belgium to study railway engineering. When Sommeiller returned in 1850 he assisted with the design of the rail line between Turin and Chambêry.

Although his career had a slow start, Sommeiller made up for lost time. In the 1850s, with the help of a fellow engineer, Severino Grattoni, he designed an engineering proposal for the world's longest tunnel, to run beneath Mount Fréjus between Modane, France, and Bardonécchia, Italy. His scheme was approved, and in 1857 the Sardinian government appointed him chief engineer for the Mont Cenis Tunnel, so named because it lay near the Mont Cenis Pass.

Sommeiller seized control of every detail of the project. When an early steam-powered drill selected for use on the tunnel needed improvement, Sommeiller stepped in and redesigned the drill not once but twice. He would eventually mount multiple drills on a carriage to speed the process of drilling shot holes. This innovation created the first mining jumbo, or moving framework, carrying multiple drills. His inventive talents were again put to the test when an inadequate compressed air system for powering the drills and providing air to the miners prompted the engineer to design a newer and better one.

He even set a standard for working conditions when he provided livable housing and hygienic conditions for thousands of workers recruited to the alpine villages near the portals of the 8.5-mile tunnel. During the tunneling project, Italy ceded Sommeiller's home region to France in 1859, prompting the engineer to choose Italian citizenship and became politically active, serving as a member of the Italian Parliament.

The demands of tunneling on both workers and engineers are great since the task, by its very nature, is an unending problem-solving process that entails fighting massive and often dangerous geologic forces. Sommeiller met those challenges in a brilliant, humane, and capable fashion, but, as it did for so many other tunnel builders of the 19th century, the strain proved too much. Sommeiller died July 11, 1871, a mere two months before the Mont Cenis Tunnel was officially opened.

Either because of the copious amounts of alcohol consumed during the celebration or because Sommeiller had somehow alienated the powers that be during his Herculean efforts to complete the tunnel, his name was not even mentioned in the speeches at the September 17, 1871, ceremony.

Sommeiller, however, was remembered in the place of his birth in what is now France. In the town of Annemasse, just below the southern tip of Lake Geneva and only eight miles northwest of his birthplace, Faucigny, a street that is home to industrial and high-tech businesses now bears his name, rue Germain Sommeiller.

See also JUMBO; MONT CENIS TUNNEL.

spalling Spalling is the process of breaking off chips from the surface of rock or concrete. In tunneling, spalling due to the pressure of the surrounding rock as it settles can be observed on the rock interior of an excavation. The violent and potentially harmful release of large sections of rock caused by such geologic forces is referred to as rock burst. When stiff rock shatters but only expels smaller or lower-velocity spall or chips, it is referred to as spitting rock.

The presence of spalling is also something looked for during the inspection of reinforced concrete bridges. The chipping off of concrete is usually indicative of corrosion that has caused the internal steel reinforcement to swell. This swelling occurs with such force that it forces off concrete. Extensive spalling of concrete in this fashion can doom a bridge by robbing it of structural integrity.

Concrete spalling is also observed in tunnels with concrete linings that have been exposed to extremely high levels of heat resulting from a fire. A November 18, 1996, fire that began on a truck being shuttled through the English Channel's Eurotunnel reached temperatures sufficient to spall large chunks of concrete from the tunnel's 15-inch lining. Heat from fuel fires resulting from auto collisions can also create a similar spalling effect on concrete bridges, necessitating expensive repairs.

See also ROCK BURST.

steel *Iron's successor* The use of iron ushered in a new era of bridge-building during the 18th century. Unfortunately, iron's impurities, including carbon, weakened it and made it brittle, preventing the abundant metal from being an ideal structural material. The superior iron derivative of steel had been produced for centuries, such as Damascus steel used in edged weapons, but the early methods of producing steel made the stronger metal prohibitively expensive.

The production of steel essentially revolved around the reduction of carbon and other impurities in pig iron that made it weak and brittle. But, until the mid-1800s there was no industrial-scale method for processing steel efficiently and cheaply. A seemingly simple process, perfected almost simultaneously during the 1850s in America and England, would provide new capabilities to bridge and tunnel engineers.

In 1851, an American, Henry Kelley, discovered a process that would oxidize carbon and other impurities from iron, making it stronger and more ductile. Just four years later Henry Bessemer in England would perfect a similar process. Bessemer's patent on the process would be disputed successfully by Kelly since the American's perfected innovation predated Bessemer's. Although Bessemer remains the far more famous of the two, Bessemer actually had to pay Kelly for use of what is generally known today as the Bessemer process.

One of the primary means of creating steel is to reduce the amount of carbon that makes up about 4 percent of pig iron, an initial form of manufactured iron that absorbs carbon during the heating process. Oxygen injected into molten iron removes the carbon impurities in the form of exhaust gas to reduce the level of carbon to less than 1.7 percent. The relatively inexpensive process converted iron into economical steel that eventually became the material of the Industrial Age.

But the superiority of steel did not mean it would catch on as quickly as one might think. Even 15 years after Kelley and Bessemer's discoveries on how to make strong and cheap steel, little of the new metal was being made. By the late 1860s iron was still the preferred material for everything from stoves to bridge components. It is not that steel was not superior; iron had a long history of development with generations of engineers comfortable with its capabilities, despite its inferiority to steel. Of all people to jump-start the world's love affair with steel as a bridge-building material, it would be an engineering maverick who had never designed a bridge in his life.

When James Buchanan Eads designed the bridge across the Mississippi that would later bear his name, he had pursued a career as a salvage diver in the Mississippi River and was not a lifelong bridge engineer wedded to traditional bridge-building materials of iron or stone. Eads simply wanted the best structural material, and as far as he was concerned, test results showed that steel was it. The Eads Bridge, built between 1867 and 1874, was for the most part to be constructed of this relatively new material. Eads's accomplishment was noted around the world and iron slowly began to disappear from bridge projects. When the iron columns of the Tay Bridge collapsed in 1879, engineers took note, and the ascendancy of iron as a structural material of choice would endure for only 20 more years.

Steel, which continued to be strengthened and refined, was soon used in arch bridges, cantilever truss spans, and the wire of suspension bridge cables perfected by John Augustus Roebling, who also designed the Brooklyn Bridge. When it was discovered that concrete could be reinforced with iron to make it an acceptable structural material for bridges in the 19th century, steel would later become the reinforcement of choice. Steel would also make its way underground to provide framing to support unsteady tunnels while also providing internal reinforcement to concrete tunnel linings.

The development of prestressed concrete by Eugène Freyssinet in the first half of the 20th century provided yet another role for highly refined steel cables, which would be placed under great tension inside a concrete beam to strengthen it, yielding an improved material for bridge construction.

See also CONCRETE; EADS BRIDGE; FREYSSINET, EUGÈNE; IRON.

steel-driver *See* Henry, John.

Steinman, David Barnard (1886–1960) *Brilliant bridge designer, self-promoter, and designer of the Mackinac Bridge* David Barnard Steinman was raised in the slums literally in the shadows of the New York approaches to the famed Brooklyn Bridge. His parents were poor immigrants, about whom little is known, as a result of Steinman's own efforts to treat his early years as if they had not occurred. What is known is that Steinman was a brilliant student driven to succeed who worked his way through college until he managed to earn a Ph.D. in engineering.

Although perhaps embarrassed by his humble beginnings, Steinman was consumed with a near-fanatical desire for education that could pave his way out of New York City's dingy slums. The metaphorical significance of his means of leaving his impoverished life behind—the design of great bridges—is immediately obvious. His prowess as a mathematician emerged early as the five-year-old Steinman amazed teachers with his ability to multiply massive numbers in his head.

The impoverished Steinman entered City College of New York, where working-class youths sought higher education, and he graduated summa cum laude in 1906. During his time at City College, Steinman wrangled permission to clamber about the Williamsburg Bridge, then under construction, evidence that bridges were very much on the teenager's mind. Setting his sights on an advanced degree in engineering, Steinman applied to Columbia University. Once accepted, he fought a long financial struggle to pay for his tuition with a combination of scholarships and part-time jobs.

Steinman taught in night school and took any engineering work that came his way while a student. To obtain his civil engineering degree Steinman authored a thesis on the design of the "Henry Hudson Bridge," a project Steinman's own engineering firm would actually accomplish during the 1930s. In 1910, Steinman became America's youngest civil engineering professor when he accepted a teaching position at the University of Idaho. In 1911, Steinman earned his Ph.D. in civil engineering from Columbia. Although he yearned to build long-span bridges, his only bridge-building accomplishment was construction of a small wooden span with a troop of Boy Scouts across an Idaho stream.

However, Steinman found success as an academic if not as a practical engineer. His Ph.D. thesis, which compared the merits of suspension and cantilever spans, was later published as a textbook, *Suspension Bridges and Cantilevers: Their Economic Proportions and Limiting Spans*. This success whetted Steinman's appetite for writing. Soon afterward he translated a pair of important German books on bridge design theory by the Czech engineering professor and bridge designer Josef Melan. The brilliant Melan had developed the deflection theory that launched an era of thinner and lighter bridges during the 20th century.

The slightly built man with the round face from the gritty confines of New York may have seemed oddly out of place in the wide open spaces of Idaho, but Steinman would not remain in his bucolic setting for long. In July 1914, Steinman obtained a position with the august bridge builder Gustav Lindenthal, who was then building the Hell Gate Bridge in New York. A notorious self-promoter, albeit one with the intellectual goods to back up his claims, Steinman managed to obtain a position as "special assistant" to Lindenthal, ranking just below the more reserved Swiss immigrant Othmar Hermann Ammann, who would later rank with Steinman among America's most esteemed bridge engineers.

Steinman calculated the stresses that were anticipated in what was the largest steel arch bridge built to that time. His chore was made even more urgent by the 1907 collapse of an unfinished cantilever span of Canada's Quebec Bridge, an event blamed on an inadequate understanding of the stresses on the structure. Later, Steinman presented a paper to the American Society of Civil Engineers detailing his calculations. Curiously, he met with mild criticism from Ammann, who was also presenting a professional paper at the same time on the overall Hell Gate project. Ammann intimated that Steinman's work had a narrow focus. This made it appear that that Steinman's calculations could not be applied in a general fashion to other projects, something Steinman denied during their discussion.

Steinman was the odd man out while working for Lindenthal since Ammann was, at the time, Lindenthal's fair-haired assistant. When Ammann returned to Switzerland in 1914 for military service, Steinman was promoted to fill Ammann's shoes, a situation that was immediately reversed when Ammann returned. When the United States entered World War I in 1917, funding for bridge projects dried up overnight. Steinman was let go, although Ammann remained in Lindenthal's service, managing a clay mine of which Lindenthal was a part owner.

For the second time in his career Steinman entered academia, as a professor of engineering at the City College of New York in 1917. With funding for large-scale bridge construction still scarce, Steinman was surprised to be asked by Holton Robinson to help prepare a design for a bridge to connect the island of Santa Catarina with the mainland of Brazil. Steinman left his teaching job and set up his own one-man engineering firm. No job was too small for Steinman. Holton and Steinman eventually formed an engineering firm bearing their names. The Brazilian span, known as the Florianópolis Bridge, was an eyebar chain suspension bridge that at the time was the largest in South America.

Steinman and Robinson obtained an important commission in the 1920s when they were asked to design the Carquinez Strait Bridge across San Francisco Bay. The cantilever truss bridge possessed a pair of spans bridging 1,100 feet. By the late 1930s Robinson & Steinman were hired to design the Saint Johns Bridge in Portland, Oregon, an impressive suspension bridge with a 1,200-foot clear span over the Willamette River. Both engineers inspected the bridge from above and below on its completion in 1931 from a biplane flown by a stunt pilot. This bit of showmanship seems to be pure Steinman.

All the time Steinman and his partner, Robinson, were designing, Steinman continued to write. He authored professional papers, provided his opinion liberally, and would eventually author a number of well-received books on bridge construction. Among these were *Bridges and Their Builders* and *The Builders of the Bridge*. The latter, published in 1945, recounted the biographies of the extraordinary immigrant engineer John Augustus Roebling and his son, Washington, and the epic story of their efforts to build the famed Brooklyn Bridge, which towered above Steinman's own childhood. Throughout his life, Steinman made it clear that Roebling was a heroic figure in the milieu of American engineering.

By 1925, the garrulous Steinman was president of the American Association of Engineers and used his position to push for higher standards for engineers in both professional and ethical matters. Steinman was also a vigorous proponent of the state certification of engineers and suggested that only licensed engineers be allowed to join professional organizations, a position that put him at odds with some of his colleagues. By 1935, licensing of engineers was required in 32 states.

Author, public speaker, and showman, no engineer was further from the reclusive, socially stunted stereotype of the engineer than Steinman, who basked in the limelight that he frequently sought, voiced strong opinions that often irked his colleagues, and reveled in controversy. This willingness to embrace controversy was amply demonstrated during the late 1930s when a slavish adherence to slender bridge design lured many bridge designers, including Steinman, into building bridges unstable in high winds.

Steinman, however, was quick to recognize this miscalculation. When it became apparent that the 750-foot suspension span portion of his Thousand Islands Bridge connecting New York with Ontario was susceptible to wind-induced movement, Steinman quickly moved to stabilize the bridge as it neared completion in 1938. His answer was to attach diagonal cable stays from the road deck below to the suspension cables. His magnificent Deer Isle Bridge in Maine had a suspended span of 1,080 feet and its light construction also invited serious movement, prompting Steinman to incorporate cable stays on that bridge also.

Steinman, struggling like his colleagues to build cheaper, lighter bridges during the Depression, was as much in the dark as his colleagues about the complex interaction of the wind and long-span suspension bridges. When the Tacoma Narrows Bridge designed by the thin bridge guru Leon Salomon Moisseiff began to undulate, Steinman provided his unsolicited advice to the engineer in charge of the completed bridge and was ignored. Ammann's own lightweight Bronx-Whitestone Bridge, completed in 1939 and bearing design features contributed by Moisseiff, was also experiencing unwanted movement.

Once again, two of America's most highly regarded bridge designers were in conflict: Ammann chose to use diagonal stays running from the top of his suspension towers to the floor to stabilize the Bronx-Whitestone Bridge, a method that Steinman questioned as less efficient and reliable. Ammann and Steinman authored articles in the professional periodical *Engineering News Record* expressing their views on the matter of taming the movement of the Tacoma Narrows Bridge, which had collapsed the month before.

As the engineering professor and author Henry Petroski points out in his superb overview of bridge-building in America, *Engineers of Dreams,* Ammann and Steinman traded shots over the Tacoma Narrows Bridge solution. Ammann labeled Steinman's research approach as overly complex "guesswork."

Despite the unusually sharp exchange in letters to the *Engineering News Record,* both the Deer Isle Bridge and the Bronx-Whitestone eventually needed additional alterations to calm their unwanted movement. Triangular fairings whose sharp edges faced out from the stiffening plate girder deflected the wind from the blunt edge of the Deer Isle Bridge's plate girder. Ammann's bridge would later acquire a substantial deck-level truss railing, and eventually be fitted with a tuned mass damper to compensate for motion.

Compared with Ammann's seriousness, there was something brash and slightly refreshing in Steinman's manner. He loved building bridges and relished the fame that that accomplishment gave him, but he also actively sought the limelight. His ego was considerable, but Steinman seemed to have something of the performer in his makeup. He was not above public readings of his poetry based on bridges that he considered poems in themselves.

In this, Steinman had something in common with the famed British engineer and bridge designer Thomas Telford as well as the American bridge engineer Joseph Strauss, who was chief engineer of the Golden Gate Bridge: All three fancied themselves poets. All three also proved that as poets they were great bridge builders. Steinman's verses were often flowery odes to bridges of his own construction. Perhaps being a visual poet in the construction of bridges did not adequately express his personality, although it should have.

Steinman penned a poem in celebration of the 1957 completion of what is arguably his greatest design, Michigan's Mackinac Bridge. Wrote Steinman in "The Bridge at Mackinac," a poem not destined to become a classic:

> There it spans the miles of water,
> Speeding millions on their way—
> Bridge of vision, hope and courage
> Portal to a brighter day.

Although no Kipling or Tennyson, Steinman seemed to have fun designing bridges and making public appearances. He became something of a self-appointed ambassador and bon vivant of bridge engineering, a world that to many was a black art practiced by mysterious men with goatees, starched wing collars, and thick European accents.

At the end of World War II, Steinman was actively promoting his dream of bridging the Verrazano Narrows with a mighty suspension bridge he dubbed Liberty Bridge. Steinman produced brochures and touted the project to the public, their officials, and anyone he could buttonhole. His dream was not to be: His old rival Ammann, whose considerable skills as a bridge designer had secured the confidence and trust of New York's public works chief Robert M. Moses, would build the Verrazano Narrows Bridge.

Steinman did have one last engineering hurrah left with his design of the massive Mackinac Bridge spanning Michigan's Straits of Mackinac where Lake Huron and Lake Michigan meet. The project was a stunning triumph of engineering in terms of not only the sheer size and scale, but also the challenge of building a bridge across a body of water whose wild weather and wave action are closer to those of the open ocean. When completed in 1957, the bridge had a central suspended length of 3,800 feet, only 458 feet shorter than Ammann's Verrazano Narrows Bridge, completed in 1964.

See also AMMANN, OTHMAR HERMANN; LINDENTHAL, GUSTAV; MACKINAC BRIDGE.

Stephenson, George (1781–1848) *Locomotive and mine safety lamp inventor and father of the bridge engineer Robert Stephenson* Born in poverty in Wylam, Northumberland, England, George Stephenson was wealthy in ambition, intelligence, and inventiveness. He followed his Scottish-born father's trade by becoming an operator of a steam engine water pump used to prevent the coal mine at Newcastle on Tyne from flooding. Stephenson attended school at night to become literate and furthered his education at a local library.

While continuing to learn, Stephenson also became a superb steam engine mechanic. He married and took on myriad odd jobs to make ends meet. He acquired better jobs as a mechanic and fathered Robert in 1803, but his good fortune was clouded by the death of his wife the following year. Stephenson was insistent that his son attend school and the two studied together.

Despite his reverence for education and his penchant for working mathematical problems in his spare time, Stephenson was no milquetoast, as he demonstrated on several occasions. A bully who was despised and feared by Stephenson's fellow workers at a coal mine accused a young Stephenson of being a poor worker. Stephenson's response was immediate and final. He gave the insulting coworker a serious thrashing that elevated Stephenson to something of a local hero. In 1814, a fire inside a coal mine sent Stephenson dashing inside to take charge of a successful effort to extinguish the underground blaze, again bolstering his reputation as a leader.

Constantly tinkering and studying, Stephenson sympathized with the miners who risked being blown to bits by explosive underground gases while using candles and coal oil lamps for illumination. Around 1815 he invented the geordie lamp (so named after the nickname given local miners), which not only prevented the open flame inside from exploding volatile gases such as methane, but also provided a warning to miners by burning differently if a flammable gas entered the lamp in small amounts.

Just two years before designing his lamp, Stephenson visited a coal mine where a steam engine had been modified to pull carts filled with coal from a coal pit. To gain traction a cogwheel was used to pull the device along wooden rails. Stephenson, fully versed in steam engine technology, designed a version of the device and boosted its power by drawing the exhaust steam through a smokestack to pull additional air into the firebox. The result was the first practical locomotive, and the age of railroads was born.

His son, Robert Stephenson, studied engineering at his side and assisted him in designing new locomotives, including the "Rocket," which won the historic speed race with other locomotives in 1829. Stephenson began designing not only locomotives but also railways as well. Much of the early use of trains was dedicated to hauling the energy source of the period, coal.

Stephenson remained close to his son and often helped the younger Stephenson solve technical problems related to various tunneling projects. When the Kilsby Tunnel under construction in 1833 flooded, Robert Stephenson became distraught since there seemed no way to eliminate the seemingly endless amount of water. George Stephenson recommended the sinking of additional shafts to reach the groundwater, which was then pumped to the surface by huge steam-powered pumps. After months of pumping, the problem was finally remedied and Robert Stephenson was able to finish the tunnel.

During their work establishing railroad lines, Robert Stephenson became a proficient bridge designer, building dozens of bridges; later he achieved lasting fame for his High Level Bridge and his Britannia Bridge with its enclosed, tubular spans.

In his later years the elder Stephenson retired to his farm, where he applied his inventiveness to solving the problem of the crooked cucumber. Displeased with the curved shape of the fruit, Stephenson developed glass cylinders that forced his garden's cucumbers to grow straight.

The unassuming Stephenson declined to accept a knighthood later in his life, a step also taken by his son.

See also SAFETY LAMP; STEPHENSON, ROBERT.

Stephenson, Robert (1803–1859) *Bridge- and tunnel-building son of the locomotive inventor George Stephenson* Although he lived in the giant shadow of his famous father, George Stephenson, Robert Stephenson proved himself a capable engineer in his own right, establishing a reputation for constructing rail lines and the bridges and tunnels they required.

Stephenson's career as a bridge designer was made possible by the accomplishments of his brilliant, mostly self-educated father. George Stephenson was a man of unparalleled industriousness who had risen from the grinding poverty of England's coal mines to master steam engine technology and build the "Blucher," the first practical locomotive.

The younger Stephenson, thanks to his father's financial success, attended school as a child and spent six months at Edinburgh University before returning home to learn at his father's elbow, an education that money could not have purchased. In 1821, the Stephensons teamed up to design the Stockton and Darlington Railway, whose steam engines the elder Stephenson had convinced the owner to use instead of horses.

In 1829, Robert Stephenson helped his father design the locomotive "Rocket" for entry in a race with other locomotives. "Rocket" attained the unheard of speed of 36 mph, easily winning the contest. Stephenson eventually left England for a short period to prove himself as a mining engineer in Colombia, but the venture proved less than profitable. He subsequently returned to England, where he embarked on his career designing railways, bridges, and tunnels.

Among his most impressive bridge designs are those of the High Level Bridge across the Tyne River at Newcastle. This massive span utilized six iron arches to carry a trio of rail lines on an upper deck and vehicular lanes on a lower roadway. Completed in 1849, the bridge was built at virtually the same time that Stephenson was overseeing the building of the Britannia Bridge across the Menai Strait between Wales and the island of Anglesey.

Although the High Level Bridge was a breathtaking structure, the Britannia Bridge revealed that Stephenson possessed some of his father's inventiveness. Unable to use an arch bridge, which would provide insufficient overhead clearance for shipping, Stephenson in 1845 hit upon the idea of creating a new structural form, the hollow-box beam or girder. Having heard of an iron ship whose structure did not fail when it was partially suspended in air during launching, Stephenson decided to pattern his girder on the hull of a ship.

The elder Stephenson, whose old friend from his coal mining days, William Fairbairn, had become an expert on iron shipbuilding, recommended Fairbairn to his son as a collaborator. Fairbairn, in turn, enlisted the London University mathematician Eaton Hodgkinson to calculate the stresses on what was originally to be a tubular girder, though an enclosed, rectangular beam was eventually determined to be the best shape. Reinforced with iron beams on the roof and floor as well as iron reinforcement on the vertical sidewalls, what have become known as tubes were developed.

To bridge the Menai Strait, Stephenson employed four of the massive tubes connected end to end and support them with a trio of 200-foot towers. The tubes were so strong that Stephenson discarded a plan to reinforce what was in effect a box-shaped, elevated iron rail tunnel with suspension chains. Excessively heavy and expensive, Stephenson's Britannia Bridge was a sensation nonetheless. Although the hollow-box girder has been hailed as a superb technical achievement, the Britannia's tubes were replicated in only a handful of subsequent bridges. The use of truss-reinforced decks and cantilever truss bridges requiring less iron made his novel concept commensurately less attractive.

Despite this, the tubular bridges did enjoy a brief period of popularity, and a massively long tubular bridge was built at Montreal by 1860. A far smaller bridge, known as the Conway Tubular Bridge, was built at the Conway Castle in Wales. Although measuring only 424 feet, the bridge remains in use. The Victoria Bridge in Canada was dismantled near the end of the 19th century, and the tubes of the Britannia Bridge were damaged by fire and removed.

At the time his tubular bridge design was considered a huge success, since, unlike many bridges of the period, it did not fall down. While building the High Level Bridge and the Britannia Bridge, Robert Stephenson endured the bitter taste of failure with the fatal collapse of a bridge of his design that entered service only eight months before it collapsed. Known as the Chester Bridge, or sometimes referred to as the Bridge over the River Dee, the new bridge was given a clean bill of health in a safety inspection on October 20, 1846, and was allowed to enter service. Soon after it went into operation, locomotive drivers complained that the bridge vibrated excessively when crossed by trains. Stephenson inspected the bridge and was unable to pinpoint any problems.

But there were problems, and on May 24, 1847, a passenger train was attempting to cross the span when the locomotive driver felt the rails' dropping from beneath his wheels. He accelerated enough to get the engine off the bridge, but five passenger cars fell into the river, killing five and injuring 18. The ensuing investigation showed that an iron girder had collapsed.

Stephenson was not at his best during the inquiry and insisted to unbelieving officials that the train had derailed, causing the collapse of the bridge. The inquiry did not buy Stephenson's self-serving claim but did hear straight-from-the-shoulder testimony from Stephenson's friend Isambard Kingdom Brunel, who expressed distrust of iron as a bridge material because of its unpredictable nature. The incident threatened to extinguish Stephenson's career, but his subsequent achievements in the Britannia and High Level Bridges salvaged his public image. Few mentions are made of Stephenson's Chester Bridge in popular references to the engineer, who despite his obvious accomplishments, was all too human when it came time to fess up to shortcomings.

Stephenson's brief experience as a mining engineer did not go to waste during his career as a railway builder. In 1833, while he was chief engineer of the London & Birmingham Railway, landowner opposition to a proposed rail line forced the relocation of the line, requiring Stephenson to tunnel more than 7,000 feet through the Kilsby Ridge. As a result of incomplete geological data, the tunnel was driven through porous ground that flooded the excavation with copious amounts of water, threatening the project. The flooding proved so difficult to control that Stephenson considered abandoning the tunnel.

After consultations with his father, who had made numerous developments regarding steam-powered pumps, Stephenson employed a battery of huge pumps that ran day and night for 19 months to control the flooding. Construction of the tunnel required nearly 1,300 men to work for two years. To waterproof the tunnel, a lining containing 30 million bricks laid to a thickness of 27 inches was necessary. By the time the tunnel was completed, its cost had risen to three times its original estimate of 100,000 British pounds. The Kilsby Tunnel was finally completed in 1837.

Stephenson's fame and reputation grew until the end of his life. He not only continued to design bridges but also arbitrated disputes between contractors and railways, having become something of a technological Solomon. His Britannia Bridge's hollow-beam concept was replicated on a grand scale with the construction of the Victoria Bridge across the Saint Lawrence River in Montreal. The Victoria Bridge rested on 60-foot piers angled upstream to stand against massive wintertime ice jams. The piers supported two dozen rectangular tubes measuring 242 feet each. The entire length of the bridge with its approaches measured nearly two miles.

Built between 1854 and 1860, it was a project Stephenson did not live to see completed. Like his friend and colleague Brunel, Stephenson died in 1859. Both men had remarkably similar backgrounds in that both were the sons of two of England's most famous engineers and lived to become the most famous civil engineers of their own day. In the end, Stephenson would not see the completion of his most impressive bridge, the Victoria; nor would Brunel see the completion of the Clifton Bridge five years after his own death.

See also BRITANNIA BRIDGE; BRUNEL, ISAMBARD KINGDOM; FAIRBAIRN, WILLIAM; HOLLOW-BOX GIRDER; STEPHENSON, GEORGE.

Stirling Bridge (circa 1297) *Braveheart's victory bridge* Some bridges are notable for the unique beauty of their design, and some are fascinating technological icons because of the engineering wizardry that enabled them to leap wide waters. And some, as plain and simple as they may be, are cast by fate to play a historical role. Such is the case of the narrow and rickety wooden bridge that allowed Scotland's "Braveheart," Sir William Wallace, to defeat English invaders in a bloody engagement known as the Battle of Stirling Bridge.

Not much is known about the bridge that crossed a narrow segment of Scotland's Forth River. What is known is that the narrowness of its road-way was equaled only by the stunted tactical acumen of the English commander who used the inadequate bridge in an attack on Wallace. If the history of the bridge's origin is a bit muddy, little more is known about Wallace, whose exploits were made famous by the filmmaker Mel Gibson in his 1995 *Braveheart*. What has been lost in the shuffle between Hollywood fiction and historical fact is that a bridge played a central role in Wallace's most spectacular victory against Scotland's English oppressors.

Scotland was having succession problems when its king, Alexander III, died in 1286 of a broken neck. This was bad news since Alexander had no direct heir to his throne and determining who would be king in Scotland could entail some very nasty business. England's King Edward I, always seeking to widen his rule, arranged the marriage of his son and Alexander's granddaughter, Princess Margaret, who was to ascend the throne. The princess inconveniently caught a fever and died while traveling from her home in Norway, throwing another monkey wrench into Scotland's political situation. Edward helped decide the matter by appointing a puppet king and essentially ruled Scotland himself, peeving plenty of nationalistic Scots.

Scotland, which had been peaceful up to his time, was soon filled with murderous English soldiers, taxmen, and of course the figurehead Scottish king. England's heavy-handed tactics finally earned it an enemy that became its worst nightmare. Legend has it that William Wallace's wife (or perhaps girlfriend) was murdered while English soldiers pursued Wallace for outlaw behavior. An enraged and tremendously strong Wallace caught up with the English patrol soon after and slaughtered them with the help of his comrades. This act of defiance around 1296 ignited Scotland's resistance to Edward, who dispatched the Scottish nobleman Robert the Bruce to quell the rebellion. Bruce refused to turn against his own countrymen, prompting his own rebellion against Edward and England.

An English army—estimated at 50,000—marched north to Scotland, where Wallace's extraordinary skill as a leader had, in the meantime, made him the commander of Scotland's rebel forces. John de Warrenne, the earl of Surrey and Sussex, led the English troops toward Stirling Castle, where Wallace had arrayed his Scots behind the Forth River. Warrenne sent priests to offer Wallace surrender terms that Wallace flung back into Warrenne's teeth with the reply "Let thy masters come and attack us; we are ready to meet them beard to beard."

Warrenne's patience was exhausted, and he decided he would smash Wallace in short order. The day was September 11, 1297, and the Battle of Stirling Bridge was about to begin. Directly before his army was a bridge across the Forth, a timber span barely wide enough for two men to cross side by side. The Forth River is not very wide at Stirling, and Warrenne was warned by a Scottish turncoat to avoid the bridge and wade across a shallow segment of the river nearby where hundreds could cross at once. Warrenne, in a hurry to kill Wallace, disregarded this sensible advice and ordered his troops to attack directly across the bridge, a move that defied common sense.

Wallace's men waited across the Forth River as the English forces strained to cross the narrow, congested bridge. When a portion of the English army was across and the bridge was loaded with more troops as thousands more milled about waiting their turn on the far shore, Wallace ordered his men to strike. The Scots fell upon what there was of Warrenne's force on the Scots' side of the river and began the sweaty work of slaughter. The English were fighting for their lives when a second Scottish force maneuvered between the bridge and the trapped English troops, cutting off any hope of retreat or reinforcement. Warrenne decided to order his personal regiment across the bridge to reinforce those across the river and compounded disaster with catastrophe: Under the weight of struggling horses and men, the bridge suddenly collapsed, throwing the troops into the chilly Forth River.

Wallace and his troops pursued the English across the river, butchering those they could catch. Of course, crossing the river was no problem for Wallace's Scots, who used the fording spot that Warrenne had refused to employ.

Although a brilliant and a complete victory for the Scots, the battle did not drive Edward and his English armies from Scotland. Wallace and his forces were dealt a decisive defeat by Edward's forces at the Battle of Falkirk on July 22, 1298, and Wallace either went into hiding or fought as a guerrilla leader in Scotland. Edward was not through with Wallace and eventually had him arrested near Glasgow on August 5, 1305. Hauled to London, Wallace was immediately found guilty and sentenced to be hanged, drawn, and quartered.

Wallace was to be repaid for his "treason" against Edward and for his victory at the Battle of Stirling Bridge. First he was hanged, but before dying he was cut down, his entrails were pulled from his abdomen, and his genitalia were sliced off and burned before his eyes. The executioner's last task was to hack into Wallace's chest cavity and remove Wallace's beating heart. To finish the job, the Scottish patriot was decapitated and then dismembered. Needless to say, the punishment of being drawn and quartered was an invention of King Edward I. Wallace's head was displayed on London Bridge over the Thames River.

Though his body parts were ordered scattered by Edward, what is believed to be Wallace's sword still resides in Scotland at the National Wallace Museum in Stirling, Scotland, and is available for public viewing. As for the bridge at Stirling, it was eventually replaced by a stone span in the 15th century that can be seen today as well. It is a solid four-arch stone bridge, a far cry from the wooden bridge it succeeded. It is known locally as the Old Stirling Bridge, though it is obviously younger than the one it replaced.

See also OLD LONDON BRIDGE.

Stone, Amasa (1818–1883) *Prolific bridge builder vilified after the Ashtabula Bridge disaster* Amasa Stone was a New England cabinetmaker married to the sister of William Howe, the inventor of the Howe truss, which revolutionized American bridge design. Howe's trusses made bridges capable of withstanding heavier loads and were a boon for the railroad industry as bridge-building exploded across America.

In 1841, Stone purchased the rights to employ Howe's truss in New England for $40,000 and embarked on an enterprise that spawned a web of bridge-building companies across America. Howe was satisfied to remain at home while the royalties from the use of his truss design flowed in. Howe's truss, realized Stone, would make it possible to build stronger, more rigid bridges.

Recruiting other men, most of whom were his relatives, Stone began building railway bridges, and by 1849 had moved his operations from New England to Cleveland, Ohio. In Cleveland, Stone and his partners constructed the rail lines and bridges for the Cleveland, Columbus, and Cincinnati Railroad.

The Howe truss, modified through the years by Stone, became the primary reinforcing truss employed on the bridges used by the Cleveland, Columbus, and Cincinnati Railroad. These wooden bridges were usually truss-reinforced roadways supported by intermediate masonry piers.

Andros Stone, Stone's younger brother, formed a partnership with Lucius Bolles Boomer to form the once-famous bridge-building firm of Stone and Boomer, based in Chicago. This firm also made frequent use of the Howe truss as timber railway bridges flew across rivers with the expansion of railroads. Stone and Boomer eventually built bridges for 24 different railroads in Illinois, Wisconsin, and Missouri.

Stone's fortunes multiplied. He not only became Cleveland's first millionaire at a time when being a millionaire really meant something, he also became president of the Lake Shore and Michigan Southern Railroad.

Flush with success, Stone decided to take a hand in designing a bridge across Ohio's Ashtabula River using a Howe truss made completely of wrought iron rather than wooden diagonals and iron tensioning rods. A railroad engineer in his employ warned Stone that the design might not be sufficiently strong but deflected the advice. The two-span Howe truss bridge was built in 1865 and served without problems for 11 years.

On December 29, 1876, a Lake Shore and Michigan Southern Railroad train with two locomotives and 11 cars was crossing the bridge during a snowstorm when the span gave way. The two locomotives made it across safely but the passenger cars equipped with wood-burning stoves crashed into the ravine below. Many were killed outright in the crash but others died from a fire sparked by the stoves that engulfed the wooden rail cars. Of the 157 aboard the train, at least 85 died. Only one person in the rail cars escaped injury.

Stone and his railroad were found responsible for the accident, and Stone became the subject of national scorn since his design was poorly thought out and was considered somewhat experimental. The wealthy bridge builder, perhaps in an attempt to make amends, provided the funding to allow the Western Reserve Academy in Hudson, Ohio, to move to Cleveland in 1881 to become the Adelbert College of Western Reserve University. The school later became the Case Western Reserve University.

Already saddened by the death of his son, Stone committed suicide with a pistol in 1883.

See also EADS BRIDGE; HOWE, WILLIAM.

Strauss, Joseph Baermann (1870–1938) *Chief engineer of the Golden Gate Bridge and inventor of the improved bascule bridge* Arguably one of the most eccentric bridge builders in American engineering history (which is full of eccentric bridge builders), Joseph Baermann Strauss is erroneously considered by many to be the sole designer of the Golden Gate Bridge. Although he was eventually made the chief engineer to oversee the design and construction of the bridge, Strauss's own unattractive design for the bridge was scrapped as unfeasible. Curiously, Strauss fought for a dozen years to obtain the project only to spend as much time publicizing his affiliation with the bridge as he did working to build it. All in all, the story of one of America's shortest bridge builder (he measured five feet, three inches) and his role in constructing what was once the nation's longest bridge is a very odd one indeed.

Although other men would play a larger role in designing the bridge and supervising its construction, Strauss seemed most intensely focused on making sure no one got credit for building the bridge but he. It was surprising that Strauss seemed so distant from the project after it started since he spent a dozen years lobbying, politicking, cajoling, and sweet-talking anyone in sight to encourage its construction.

Not only a born promoter, Strauss was also the quintessential outsider. He was a Jewish businessman sensitive to slights in a nation possessing a solid layer of anti-Semitism. He was also from the Midwest and traveled from his home in Chicago to San Francisco to promote the bridge to its West Coast citizenry. He was even an outsider in his profession of bridge-building since Strauss, claiming to be an engineer, possessed neither an engineering degree nor a membership in the American Society of Civil Engineers. However, the inventive Strauss did manage to become a member of the American Society of Mechanical Engineers.

The author John van der Zee diagnosed Strauss as someone obsessed with obtaining acceptance and validation. His hunger for this approval was so great that Strauss eventually became insistent that no one but he be credited with the design of the bridge. No bridge builder in America would have spent 12 unpaid years trying to secure a bridge job, but Strauss needed the Golden Gate Bridge for more than just employment. His position as chief engineer on the Golden Gate Bridge was proof Strauss was the equal of any bridge engineer.

Born in Cincinnati, Ohio, on January 7, 1870, Strauss grew up in the same city where the pioneering suspension bridge builder John Augustus Roebling constructed one of his masterpieces. Living in the shadow of the Cincinnati-Covington Bridge—completed three years before Strauss's birth—apparently may have had an effect on Strauss, who entered the University of Cincinnati with the dream of becoming a bridge builder. Strauss also sought the fame of Roebling, whose masterpiece was the Brooklyn Bridge. Since the school did not have an accredited engineering department, he earned a bachelor of arts degree in its liberal arts school. However, one of his areas of study was engineering, and his graduation thesis was an ambitious proposal to bridge the Bering Strait.

The son of a father who was a portrait painter and a mother who was a pianist, Strauss considered poetry his personal art form. Named the class poet in his senior year, he penned not just a poem but an epic for his fellow seniors. During his university's commencement exercises, Strauss read the epic poem along with his Bering Strait bridge thesis. He would continue to write poetry for the balance of his life. Most of it ranged from bad to really bad, indicating Strauss was about as imaginative a poet as he was an innovative bridge designer.

On his graduation in 1892, Strauss headed for employment with the New Jersey Iron and Steel Company, a bridge-building firm. While there he held a variety of positions, learning every aspect of the bridge-building business. Strauss then worked for Chicago's public works offices, and in 1902 he struck out on his own as a bridge design consultant. In this capacity he repaired and modified numerous bascule bridges, otherwise known as drawbridges, in the Windy City. These bascule bridges were ugly and usually small but in great demand in America's growing cities.

Strauss decided he could improve on the original design of these bridges and did so. He substituted cheap concrete for the expensive iron counterweight; added a trunnion, or pin; to serve as the movable span's hinge; and made the system for operating these bridges simpler and more reliable. As the automobile entered American life, these bridges became even more popular.

Around 1919, San Francisco's public works czar, Michael Maurice O'Shaughnessy, asked Strauss to describe how he would build a bridge across the city's Golden Gate Strait. Strauss became mesmerized with the challenge of doing so and immediately designed a bridge that was as hideous as it was unfeasible. He spent a dozen years lobbying San Francisco area citizens to support construction of his concept. Later, when the project became a reality with Strauss at its helm, other engineers more familiar with large-scale bridge projects helped design a far more beautiful and workable bridge. Strauss recognized his own design had to go but falsely maintained the belief that he was one of the principal designers of the final concept. His ugly and unworkable bridge was voluntarily discarded in lieu of the final design, which sprang from the minds of other engineers, most notably Charles Alton Ellis and Leon Moisseiff.

What Strauss seemed to desire was recognition, and being named chief engineer of the bridge project was the prize that the wealthy builder of short, ugly bridges desired. Strauss's connection with the bridge became really bizarre when his dream came true and he was appointed chief engineer. With his great victory in hand, Strauss disappeared without warning for two months, claiming a nervous breakdown. During his "breakdown" Strauss sailed from California to New York, where he granted newspaper interviews. From there, Strauss headed for his wedding to his second wife, a singer from Los Angeles.

On his return to San Francisco, Strauss emerged mostly to defend the highly controversial Golden Gate Bridge project from detractors who worked full time to derail the effort. Left behind to work literally night and day to come up with a workable design for the bridge was the brilliant engineer Charles Alton Ellis. Strauss had hired Ellis to help him enter the realm of long-span bridge design.

Mean-spirited and ruthlessly protective of his role as the designer of a bridge he did not design, Strauss eventually fired Ellis for allegedly taking too much time with complex calculations to guarantee the safety of the bridge. Coincidentally, Ellis's December 1931 firing occurred just months after the engineer gave a speech to the American Academy of Science in which he innocently described his ample role in designing the bridge.

In fairness to Strauss, however, his major and perhaps pivotal role in advancing the Golden Gate Bridge project was to serve as a lobbyist at large for the effort. He was a formidable opponent when he engaged the seemingly endless foes the bridge effort seemed to spawn. In the early days of touting the bridge, he spent his own money on travel and engineering designs to whip up public support for the project, a task that resulted in making him one with the bridge in the mind of the public. He admitted in later years that he lost perhaps more money than he made on the bridge project. Strauss claimed he turned down many bridge projects to focus on the torturous effort to sell the Golden Gate Bridge concept to taxpayers and politicians.

Despite his efforts on behalf of the bridge, Strauss did not have a lock on the job until he agreed to hire as consultants the world famous bridge designers Othmar Ammann and Leon Moisseiff—engineers who possessed the topflight credentials he lacked. When the bridge was finally completed, Strauss made sure that Ellis's replacement as chief assistant engineer was given a minimal amount of credit for the work done. Ellis's replacement, an engineer named Clifford Paine, and Strauss had a terrific row when Paine wanted to be listed on the bridge's completion plaque as "assistant chief engineer." Strauss refused, still hogging as much of the limelight as he could. Despite a bitter argument, Strauss would list Paine only as "principal assistant engineer."

Completion of the Golden Gate Bridge by May 1937 gave Strauss a chance to revel in his victory and perpetuate the myth that he single-handedly designed what remains one of the world's most famous bridges. Strauss the "poet" penned some of his tortured verse to celebrate the construction of the bridge in the face of opposition:

Launched 'midst a thousand hopes and fears
Damned by a thousand hostile seers,
Yet ne'er its course was stayed;
At last the mighty task is done.

Strauss, affected by a cardiac disorder as he reached his 60s, died on May 16, 1938, at the age of 68. His death occurred only 12 days before the first full year of the Golden Gate Bridge's operation.

See also BASCULE BRIDGE; ELLIS, CHARLES ALTON; GOLDEN GATE BRIDGE.

Strömsund Bridge   *See* cable-stayed bridge.

subaqueous tunnel   Although hidden springs can inundate tunneling efforts in mountains and groundwater can seep into mines below the water table, those tunneling beneath the ground and rock under bodies of water must counteract the constant force of water pressure.

This water pressure, which is dictated by its depth above the tunnel, can work to crush an unreinforced excavation or push its way through porous rock or soil. If it does not drown those inside, it at least creates a difficult and hazardous work environment.

For this reason, no confirmed effort to push a tunnel beneath a river or strait is recorded until the development of the tunneling shield by Marc Brunel that allowed the 1843 completion of a subaqueous tunnel beneath the Thames in London. His iron shield supported the excavation as miners dug at the face of the tunnel while masons in the rear reinforced the tunnel walls with a brick lining. Despite Brunel's innovation, the project was plagued with blowouts and massive intrusions of water, taking 18 years to complete.

The railroad engineer DeWitt Clinton Haskin adapted the use of a compressed air atmosphere, formerly reserved for underwater caisson work on bridge piers, to tunneling in 1874, when he attempted to excavate a rail tunnel beneath New York's Hudson River. Haskin ignored Brunel's tunneling shield, believing that air pressure alone would prevent collapses and flooding. Haskin's experiment was disastrous. A July 21, 1880, blowout that flooded the tunnel killed 20 sandhogs, although the flooding occurred near the entrance of the tunnel and not at its working face. Haskin abandoned the project two years later.

The South African James Henry Greathead pioneered the use of compressed air in conjunction with an improved tunneling shield originally designed by his mentor, the English engineer Peter W. Barlow. With air pressure controlling flooding and Greathead's own version of the circular shield providing structural rigidity to the tunnel, work progressed in a relatively safe and rapid fashion.

Greathead's innovation ushered in a golden age of subaqueous tunneling and eventually allowed the completion of the Hudson River tunnel originally started by Haskin. Unfortunately, the use of compressed air carried health risks to sandhogs, who suffered from the effects of caisson disease, or decompression sickness, for decades before science understood its causes. A number of famous subaqueous tunnels soon followed in America, among them the Holland Tunnel, Queens-Midtown Tunnel, and Brooklyn-Battery Tunnel. All of these were excavated by using compressed air and tunneling shields.

The deepest subaqueous tunnel in the world by the beginning of the 21st century was the Seikan Tunnel, which runs 780 feet beneath Japan's Tsugaru Strait. The water pressure at this depth was so great that the air pressure required to prevent flooding would have killed the workers. Instead, workers grouted the fractured rock lining the tunnel with a cement mixture to make it watertight enough to preclude flooding until a concrete lining was installed.

The world's most famous subaqueous tunnel, Eurotunnel beneath the English Channel, was also driven without the use of compressed air through a type of rock known as chalk marl. Although most of it was impermeable, porous sections of the tunnel's route required extensive grouting to prevent flooding.

A method of creating subaqueous tunnels without excavating beneath a river or seabed is the use of immersed tube tunnel technology, in which prefabricated steel or concrete tubes are sunk beneath the water into open trenches and joined with watertight connections.

See also BROOKLYN-BATTERY TUNNEL; EUROTUNNEL; GREATHEAD, JAMES HENRY; HOLLAND TUNNEL; HUDSON TUBES; IMMERSED TUBE TUNNEL; QUEENS-MIDTOWN TUNNEL; SEIKAN TUNNEL.

submerged floating tunnel *See* immersed tube tunnel.

Sunshine Skyway Bridge collision *See* bridge collision mitigation.

suspension bridge The history of the modern flexible suspension bridge is a shaky one. In the 1800s, suspension bridges collapsed with such regularity that the type became highly suspect. Famous engineers who wowed the public with their masterpieces would later learn their magnificent creations had been reduced to rubble by a gust of wind or had simply snapped and carried people away to their deaths.

When the Basse-Chaine Bridge in Angers, France, collapsed on April 16, 1850, it killed not only 226 marching soldiers but also much of Europe's enthusiasm for building suspension bridges. Suspension bridge failures in the United States in the first half of the 1800s also raised concerns.

But the engineering allure of the suspension bridge was far too strong for the concept to be abandoned. The promise of the suspension bridge was a span that could cross great distances while weighing far less than traditional stone arch bridges. Engineers were bewitched by the prospect of suspending a relatively inexpensive roadway in midair by slender chains or cables instead of costly foundations and piers.

Perhaps this idea does not sound exciting to modern folks who have watched video transmissions from Mars, but to 18th- and 19th-century people, who had to rely on unpredictable ferries, this was heady stuff. Major suspension spans not only were necessitated by the transportation requirements of the Industrial Revolution but were being made possible by engineering and materials developments of the period.

A suspension bridge is based on a concept that seems implausible to the uninitiated: a bridge held aloft not by its foundations and towers but by flexible, even wobbly, cables anchored at both ends. Not only is the concept workable, but the suspension bridge type remains one of the cheapest and strongest alternatives for spanning wide waterways. As for most efficient engineering techniques, the theory behind the suspension bridge is succinct and sensible. Whereas traditional bridges required numerous and massively expensive piers to support successive spans, most suspension bridges require not more than two such foundations, between which the center span is supported by the cables.

Though sometimes considered a recent form of bridge construction, suspension bridges have been around for at least 2,000 years, proving that "modern" designers did not recently corner the market on innovation. Some of the original suspension bridges were admittedly crude by today's standards: Instead of precisely manufactured steel cables held aloft by massive towers to suspend tons of roadway, the original suspension "cables" were vines or handmade rope, as they still are in remote areas of the Third World.

The basic elements of suspension bridge design have not changed for centuries: the piers, the support cables, stone or concrete anchorages where the ends of the cable are secured to hold the weight of the road deck, and towers for elevating the cables.

Seemingly rickety (because of the lightweight appearance of their design and tendency to sway), the early, rudimentary suspension bridges became a handy Hollywood prop for such cinematic fare as Tarzan movies. Who has not seen someone in a B movie chop through the rope of a wildly swaying suspension bridge to cascade pursuers into a gorge? If nothing else, the engineering principles portrayed are accurate. The rope or vine is carrying the weight of the span just as the steel cables do on the Brooklyn Bridge.

Construction of reasonably substantial suspension bridges dates back at least to 206 B.C. in China, but Europeans did not devise such a span until 1741, when the 70-foot Winch Bridge was built across the Tees River in England. Unlike subsequent suspension bridges with their supporting cables above the road deck, the Winch Bridge had chain cables that ran beneath the wooden roadway.

European engineers became intrigued with conquering suspension bridge design and continued to build them, although they were initially known as chain bridges since iron links were used in lieu of supporting cables of iron and later steel. Since the complexities of building a suspension bridge were beyond the ken of engineers and entrepreneurs, many spans, such as the Basse-Chain Bridge, collapsed.

But there were successful early suspension bridges. As suspension bridges swayed and sometimes fell, one of the most successful—with some weaknesses—was built. This was the Scottish engineer Thomas Telford's 1826 Menai Bridge in Wales. The slender road deck of what was then the longest single span in the world showed the true promise of the suspension bridge. It was light, inexpensive, and beautiful. But an ignorance of aerodynamics was revealed when the week-old bridge was violently shaken by a strong wind. In a second storm in 1839, winds flounced the road deck so badly it had to be replaced. After the Basse-Chaine Bridge collapse 21 years later, many Europeans shied away from the design.

In America the suspension bridge would come of age. In 1801, the American James Finley built his first chain suspension bridge over Jacobs Creek in Pennsylvania. Finley built numerous bridges, most aimed at providing rural Americans ways of traveling and moving goods. When several of his bridges collapsed—one under the weight of cattle and another under heavy snow—suspension bridge construction slowed but did not stop.

In 1816, a pair of American iron wire makers, Erskine Hazard and Josiah White, designed and built a-408-foot suspension bridge walkway across Pennsylvania's Schuylkill River. Their innovation was the use of bundled wires to support the bridge in lieu of chains, a system that was to prove far stronger than iron links. Those crossing the span were charged a penny to defray its cost. The Swiss bridge engineer Marc Séguin was thinking along the same lines when he determined that cables of bundled wires had twice the strength of iron chains. He built his first wire bridge in Geneva in 1823.

As in Europe, interest in suspension bridges had cooled in America as a result of their baffling failures. But during the early 1800s, a pair of Pennsylvanians from different backgrounds began to study the problem of suspension bridge design. The American-born Charles Ellet would design some of the boldest suspension bridges in America. His contemporary was the Prussian immigrant John Augustus Roebling, who held a diploma from one of Germany's finest engineering schools. Roebling would eventually eclipse Ellet in fame by designing a series of superb bridges, including the Brooklyn Bridge. Roebling would also be the world's first long-span suspension bridge designer who could truthfully claim that none of his bridges had been shaken to pieces.

Although Ellet was a brilliant promoter, the bridge designer David Barnard Steinman has pointed out that Ellet lacked the "sound foundation" of engineering knowledge to build truly capable bridges. Ellet's second bridge would be his last, when his 1,010-foot Wheeling Bridge over the Schuylkill River collapsed in a strong wind in 1854.

Roebling's suspension designs were so robust that the 821-foot Niagara River Bridge he completed in 1854 was sturdy enough to carry both a railway on a top deck and a roadway on a lower deck. The bridge stood for half a century before being dismantled to make way for a new one. But it was Roebling's Brooklyn Bridge design that fully legitimized the suspension concept. An accident led to his death before construction began, and his son, Washington Roebling, completed the Brooklyn Bridge in 1883.

The elder Roebling's suspension bridges were made successful in no small part by his understanding that adding weight to the road deck of a bridge reduced its susceptibility to wind action. With his intuitive understanding of the destructiveness of wind, Roebling also employed diagonal stays, or cables that ran from the tower to the bridge deck, to dampen unwanted movement. Although Roebling's mid-1800s recognition of how to make a bridge endure wind forces ensured that none of his own bridges collapsed, it would be a lesson forgotten by the 1930s as engineers created suspension bridges that were progressively thinner and lighter.

Completed in 1937, the Golden Gate Bridge was heavily influenced by the consulting engineer Leon Moisseiff, who advocated lighter and narrower suspension spans. The world's most famous bridge needed its deck reinforced by the 1950s as a result of excessive wind movement. The November 7, 1940, collapse of the overly light deck of the Tacoma Narrows Bridge (which Moisseiff designed) hammered home the necessity of applying aerodynamics to bridge design and marked a return to heavier bridge decks.

However, these mistakes resulted in a return to the basics of suspension bridge design recognized by Roebling and sparked a new generation of suspension bridges that were far more stable while also far longer. The suspension bridge remains the solution of choice when engineers need to extend roadways across wide expanses of water. An example is Japan's Akashi Kaikyo Bridge, completed in 1998 with a center span of 6,528 feet, dwarfing the 1,595-foot length of the Brooklyn Bridge's main span. The Akashi Kaikyo's center suspension span was, for the time being anyway, the longest in the world at the dawn of the 21st century.

See also AKASHI KAIKYO BRIDGE; BRIDGE AERODYNAMICS; BROOKLYN BRIDGE; MOISSEIFF, LEON SALOMON; ROEBLING, JOHN AUGUSTUS; TACOMA NARROWS BRIDGE.

T

Tacoma Narrows Bridge, "Galloping Gertie" (built 1938–1940) *Fatally thin* Bridges are built not only on bedrock but also on the success and failure of previous bridge designs. In the case of the catastrophically flawed Tacoma Narrows Bridge, the advanced design was a little too advanced for its own good and foolishly took the concept of a thinner and lighter suspension bridge that had gained ground during the first decades of the 20th century too far. Because it was the brainchild of one of the most highly respected bridge designers of the 20th century, warnings from "lesser" engineers about its flaws were ignored.

The dramatic collapse of the bridge dumbfounded America's elite clique of bridge designers, including its principal designer, Leon Salomon Moisseiff—an engineer respected to the point of reverence by America's civil engineering community. Moisseiff was a central figure in the perpetuation of a trend to build thinner and lighter suspension bridges, a move he helped initiate. Moisseiff and his engineering colleagues were so thunderstruck by what happened that it took the aerodynamicist Theodore von Kármán, an expert far removed from the civil engineering community, to provide a possible answer to the disaster, whose causes remained in dispute more than a half century later.

Although its collapse occurred a decade before the midpoint of the 20th century, the sight of its destruction is familiar to many today because it was captured on film by Frederick B. Farquharson, a University of Washington engineering professor. Farquharson's film was spliced into newsreel reports of the day, shocking a nation that had come to expect success from its bridge builders. And although comparatively smaller bridges have collapsed, the dramatic destruction of the Tacoma Narrows Bridge failure was the only wind-related collapse of a major long-span bridge in America in 100 years.

Tacoma's ill-fated bridge had ordinary beginnings. It was obvious that a bridge was needed to span the Tacoma Narrows of Puget Sound separating Tacoma from a sprawling peninsula to the west containing Bremerton and several military installations. In 1932, the civil engineer Elbert M. Chandler, who once headed the State Reclamation Service, organized the Tacoma Bridge Company in an effort to secure governmental funding for the project. As America grew, settlements were now flexing their muscles as cities, and many were seeking bridges not only to assist the flow of traffic but to channel development to certain areas. The Great Depression also spurred cities to seek infrastructure projects to provide employment for the jobless.

The Tacoma Narrows held many dangers, including its 200-foot depth and an 8.5-mph tide that could play havoc with any effort to build submerged foundations for support piers. Above the water there were other problems that could affect worker safety. The winds that blew through the Narrows were frequently strong, and 40-mph gusts not uncommon. The length of the bridge with its approaches would be 5,000 feet—nearly one mile—with the suspended center span measuring a whopping 2,800 feet. When built, the bridge would be the third-largest suspension span in the world.

Not surprisingly, Chandler found the depression-crippled coffers of Pierce County unable to fund the bridge. He then mounted another unsuccessful appeal for funds from the federal Reconstruction Finance Corporation. The federal agency turned down the proposal because it did not believe the bridge could produce adequate toll revenue with which to repay its construction costs. Undeterred, Chandler decided federal money might flow a bit more freely with the election of Herbert Hoover's successor, Franklin Delano Roosevelt. Unfortunately, Roosevelt's Public Works Administration, despite lobbying by Washington's U.S. senator, Harold Bone, stood firm against financing a bridge across the Narrows.

Persuaded by the state legislature's creation of the Washington Toll Bridge Authority, which could finance the project, the Public Works Administration in 1938 granted the state $3 million toward the construction of a bridge. When the span was initially designed by the Washington Department of Highways bridge engineer Clark H. Eldridge, it had a beefy and rigid structure with a 25-foot-thick web truss beneath its road deck. Strong and impressive, the bridge also carried an $11 million price tag. With money still a crucial factor, it was decided that Eldridge's strong and wind-resistant design was far too expensive and the plans were discarded. The new design could cost no more than $6.4 million. Leon Moisseiff, fresh from his triumph as one of the consulting engineers on the Golden Gate Bridge, completed in 1937, was asked to design a less expensive bridge across the treacherous Tacoma Narrows.

Moisseiff's concept was an amazingly slender bridge that required far less material, allowing it to fit nicely with the tight budget under which it was to be built. Gone was Eldridge's 25-foot-deep stiffening truss, and in its place, Moisseiff had only eight-foot-deep plate girders reinforcing the road deck. Moisseiff's calculations led him to believe the depth of the trusses was not significant in reducing deflection, or movements of the road deck. The traditional stiffening truss reinforcement was a latticework of steel that not only provided strength and weight but also allowed wind to pass through relatively unimpeded, whereas the comparatively narrow but solid plate girders could not. Moisseiff saw no significance in this aerodynamic difference, which may have played a major role in the destruction of the bridge.

The ratio of the width of the bridge to its length was far lower than that of any other suspension bridge built, raising more concerns about its stability. T. L. Condron, supervisory engineer of the Washington Toll Bridge Authority, expressed unease with Moisseiff's design in a letter that is almost confusing because of its apologetic diplomacy toward the famed designer. In it, Condron warned that the width to length ratio of the proposed Tacoma Narrows span was smaller than that of the Golden Gate Bridge, which had the lowest length to width ratio up to that time.

The central span of the Tacoma Narrows was 72 times longer than its 39-foot road deck width, whereas the Golden Gate Bridge was only 47 times longer than its central span width. Condron revealed that he traveled to confer with a University of California at Berkeley engineering professor about the issue and was assured the design seemed feasible. Nonetheless, the suspicious Condron asked his superiors to consider widening the roadway and adding weight (and cost) to the structure. The money crunch and Moisseiff's reputation ensured that this proposal was rejected. His letter was written on September 21, 1938, just months before construction began on the bridge.

Despite Condron's polite reservations, he nonetheless concluded the bridge was sound, swayed not only by Moisseiff's reputation and the faith of others in the designer's acumen but also by the economic pressures to keep the bridge as cheap as possible. Men were out of work, and the bridge, although nice to have, was also needed to provide employment. Additional expense added to the bridge's tab by overly cautious engineers like Condron would delay the project and scotch employment for impoverished Depression era workers.

If the design were not dicey enough, the contour of the Narrows's submerged waterway added to the problems of construction. When glaciers began advancing and retreating across the region 2 million years ago, mile-thick walls of ice gouged out Puget Sound and its inlets. The result is that the Tacoma Narrows bottom is fairly deep and that placement of the foundations would have to be relatively close to shore. Open caissons would have to be sunk in the powerful current and held in position while clamshell buckets excavated through as much of 90 feet of the floor of Puget Sound to reach an adequate layer of bedrock.

At the time, the pier work for the Tacoma Narrows Bridge was the deepest of its kind, as the east pier sank a total of 230 feet beneath the water. Work on the bridge commenced and caissons were sunk on the east and west sides of the Narrows. Acceptable bedrock for the east pier was reached after digging 90 feet beneath the channel's bed, lying 140 feet beneath the surface. The pier then rose 247 feet from its base. The west pier's excavation started at 120 feet below the water's surface and was excavated 55 feet to bedrock. It stood only 198 feet in height. The necessary concrete anchorages to hold the bridge's cables were constructed on each bank.

With the foundations completed, construction of the towers proceeded on schedule and both rose 425 feet above their bases. The two wire cables were spun and hung across the four steel saddles emplaced atop the legs of both towers. The strands comprising these cables were looped on pulleys connected to eyebars set in the heels of the concrete anchorages on each shore.

As the road deck began to take shape beneath the cables and steel suspenders, workers atop the windswept bridge began to suffer from motion sickness as the span oscillated up and down between the two towers. The bridge soon acquired the nickname "Galloping Gertie," a fine piece of workingman's slang originally used to describe a spirited and uncontrollable horse. Later, when the wildly gyrating bridge would be seen on film by millions, more than the workers would refer to the bridge by the moniker. Nonetheless, the work proceeded rapidly and the bridge was completed in 26 months. It opened to the public on July 4, 1940.

The completion of the bridge and the added weight of the concrete roadbed did nothing to alleviate the movement of the bridge's road deck. The bridge, which was the pride of the Tacoma community, became famous as something of an amusement-park ride for motorists crossing the undulating structure. Vehicles would disappear in a fold of the rolling span only to reappear periodically as they ascended another concrete wave. The behavior of the bridge caused anxiety among motorists, but officials continually reassured them the bridge was safe.

Nonetheless, the University of Washington engineering Professor Frederick Farquharson was asked to study the bridge and develop recommendations on how to resolve the stability issues. Farquharson first recommended the installation of steel cables running from the sides of the bridge to 50-yard-long blocks of concrete. The bridge still moved up and down, ripping the cables loose. Farquharson suspected the bridge's solid girders beneath its road deck were catching too much wind and needed to be ventilated. He suggested that holes be drilled through the plates to allow the air to pass, but time did not allow this since the bridge collapsed before it could be done. The behavior of the bridge was confusing since observers noted that the speed of the wind and the motion of the flouncing bridge did not correlate. At times a less severe wind caused worse oscillations in the bridge than a stronger one.

On the morning of November 7, 1940, a windstorm blew through the Tacoma Narrows, causing the bridge to roll and flex more than previously seen. This time, worried officials were not so dismissive of complaints about the bridge's instability and closed it to traffic. By 10 A.M. the winds were moving at 38 mph, and eventually they reached a high of 42 mph, well below the level of simple wind pressure the bridge's superstructure had been designed to endure. Notified of the activity, Farquharson rushed to the scene to film the bridge's behavior with a 16-millimeter movie camera borrowed from a local photography shop. All 2,800 feet of the central span was flexing like a thin, flat piece of supple plastic being twisted at both ends by huge invisible hands. A vertical suspender cable had given way, causing the road deck to become unbalanced and added to the gyrations. The flexing was so severe that the edges of the central span flounced upward 28 feet, making it possible to see the underside of the road deck.

Caught on Farquharson's footage was a small drama in which the intrepid Tacoma newspaperman Leonard Coatsworth decided to drive on to the flouncing structure accompanied by his dog, Tubby. "Just as I drove past the towers, the bridge began to sway violently from side to side," said Coatsworth after narrowly escaping death. "Before I realized it, the tilt became so violent that I lost control of the car. . . . I jammed on the brakes and got out, only to be thrown onto my face against the curb. Around me I could hear concrete cracking. I started to get my dog, Tubby, but I was thrown again before I could reach the car. The car itself began to slide from side to side on the roadway."

"On hands and knees most of the time, I crawled 500 yards or more to the towers. . . . My

Washington state's Tacoma Narrows Bridge collapses after the lightly built span fell victim to moderate winds in 1940. This shocking event reversed a trend toward thinner and lighter suspension bridges, although the principles guiding its design were considered state-of-the-art at the time. *(MSCUA, University of Washington Libraries, Farquharson # 12)*

breath was coming in gasps; my knees were raw and bleeding, my hands bruised and swollen from gripping the concrete curb. Toward the last, I risked rising to my feet and running a few yards at a time. . . . Safely back at the toll plaza, I saw the bridge in its final collapse and saw my car plunge into the Narrows." Tubby, Coatsworth's dog, was the only casualty claimed by Galloping Gertie.

As Farquharson filmed the drama, the bridge began to break apart. At 11 A.M., the steel plate girders began to peel away from the road deck, sending a 600-foot section of reinforcing plate girders and roadway into the waters below. Then the remainder of the concrete and steel road deck began

to fall. Within 10 minutes the central span had all but disappeared. The loss of the central segment robbed the suspension cables of the weight they were designed to hold. As a result, the approaches behind the towers sagged downward 60 feet, bending the tops of both towers back toward their respective shores a full 12 feet. Even the 50-yard-long blocks of concrete attached to stabilizing cables added to the bridge had been jerked into the water by the collapse.

In the engineering postmortem that followed, the federal Public Works Administration appointed a commission to study what had happened to the bridge. Included on the commission were Othmar

Ammann—who helped design the Golden Gate Bridge with Moisseiff—and Theodore von Kármán, a leading expert on aerodynamics. Von Kármán—who later cofounded the Jet Propulsion Laboratory—decided the culprit was not the lightness of the bridge but its aerodynamic qualities. He believed that the structure reacted with the wind in a damaging way and perhaps fell victim to vortices—swirls of wind that can create turbulence on the lee side of a structure. This turbulence would set up a force that could help oscillate the bridge.

Ammann was hard-pressed to lay blame on Moisseiff, with whom he had cooperated on several bridge designs, noting that the bridge had been built according to prevailing, tested, and accepted engineering concepts of the period. The Bronx-Whitestone Bridge designed by Ammann with help from Moisseiff as a consulting engineer and completed in 1939 was also prone to an unusually large amount of flexing in high winds. It was eventually strengthened with a stiffening truss atop its road deck.

The discussions of what happened to the bridge became hopelessly complicated, with experts still in disagreement about the real causes. Moisseiff, who was 68 at the time of the collapse, was asked whether he knew why the bridge failed. He answered simply that he did not. At about the same time, Moisseiff had designed a span across the 8,615-foot Mackinac Strait that was a longer version of his all-too-thin-and-narrow Tacoma Narrows Bridge. Needless to say, when a bridge was built across those straits in 1957, it was an extremely hefty design.

It was obvious that the trend of making bridges thinner and lighter—concepts spurred on by the enforced economies of the depression—had gone too far. It was also obvious that aerodynamic elements of long-span bridges would have to be planned carefully as well.

Precisely 13 months after the collapse of the beautiful but anorexic bridge, the United States was at war and there was not time, money, or material available for rebuilding. A replacement bridge would not be constructed for a decade.

An odd footnote to the destruction of the Tacoma Narrows Bridge emerged when the state of Washington decided to ask the company that insured the structure to pay off on the policy taken out on the bridge. This was news to the insurance company, which was astonished to learn it had "insured" the bridge. The company soon learned that its agent had made off with the premium without reporting the transaction. The insurance company had to pay Washington state $4 million, and the agent received a jail sentence.

See also BRIDGE AERODYNAMICS; MOISSEIFF, LEON SALOMON.

Tatara Bridge (built 1994–1999) *The world's longest cable-stayed span* Although Japan has nearly bankrupted its public works budget with grandiose projects to connect its main island with its surrounding home islands, the result has been the erection of some of the most stunning bridges in the world along with a number of engineering firsts. One of the most recent is the 1999 completion of the $605 million Tatara Bridge between Ikuchi and Omishima islands.

In the early 1970s, it was intended that the bridge be a suspension span, but by 1989, when efforts to plan and build the bridge once again became active, it was decided that a cable-stayed span would cross the Seto Inland Sea. Concerns that the massive anchorages of a suspension bridge would mar a national park on one shore prompted officials to select a cable-stayed design. Although a cable-stayed bridge would not be less expensive than a suspension span, it was decided that it could be built more quickly while eliminating the eyesore anchorages.

Japanese engineers began conducting numerous studies including wind tunnel testing to determine the best design and materials. Because Japan is prone to typhoons with winds that can exceed 100 mph, the shape of the towers was carefully considered. Eventually, it was decided that the steel towers would have a modified diamond shape looking much like an upside-down tuning fork. Instead of being fused together, the apex of the towers would contain a space to help break up vortices created by strong winds. The towers would rise 721 feet above the water to accept a total of 168 cable stays. The towers were built in a modular fashion, each composed of 23 segments weighing 140 tons each. The segments were shipped by barge to the site and then lifted into place.

The zinc-coated steel stays run from the towers to the outside edge of the deck, where they are bolted to internal cable anchorages. The 65-foot-long deck sections were fabricated away from the construction site and transported by barge. Traveling cranes on the deck lifted the sections, and the center span was built in a cantilever fashion: As the deck sections moved toward one another, additional stays were attached and the deck progressed toward the center from both towers. The steel-box girder deck is remarkably thin, measuring only 8.8 feet in depth. The width of the deck is just over 100 feet, giving the bridge a slender and graceful profile.

During the construction of the Tatara Bridge, engineers worldwide became aware of an aerodynamic problem involving stays on some bridges that are covered with a polyethylene sheath. When rain clings to the sheath, it can change the aerodynamics of the cables, causing excessive vibration. Since such sheathing would protect the Tatara Bridge's stays, engineers specified that dimples be incorporated onto the surface of the sheath to counteract this problem.

Since the Tatara Bridge is located in a region susceptible to typhoons and considerable seismic activity, the bridge has been engineered to endure both: winds of up to 140 mph and a magnitude-8.5 earthquake, the upper limits of the natural forces the bridge is expected to encounter.

See also BRIDGE AERODYNAMICS; CABLE-STAYED BRIDGE.

Tavanasa Bridge (1905) At 167 feet, the span of the reinforced concrete arch bridge crossing the Rhine at Tavanasa, Switzerland, is far from the longest bridge in the world, but its construction in 1905 marked the beginning of the influential designer Robert Maillart's abandonment of 19th-century bridge styles.

Although most bridges had side walls concealing the arch supporting the bridge, Maillart noted in his Züoz Bridge built in Switzerland in 1901 that heating and cooling of the reinforced concrete had caused vertical cracks in its concrete side wall known as a spandrel. Although this cracking had no real impact on the structure, Maillart decided the walls could be eliminated as unnecessary to prevent the cosmetic problems.

The Tavanasa Bridge would be a reinforced concrete arch bridge containing three hinges to absorb expansion and contraction. These hinges were placed at its abutments and at the center of the bridge. The spandrels, which had contributed weight to the structure, increasing its dead load, would not be built. This would give Maillart's Tavanasa Bridge his characteristic triangular openings at both ends of

the supporting arch. The bridge's silhouette is very similar to that of Maillart's famed 1930 Salginatobel Bridge just west of the one at Tavanasa.

The arch itself was a design Maillart developed for the Züoz Bridge that contained Maillart's invention of the first hollow-box supporting arch. The horizontal deck of the Tavanasa Bridge appears to connect to a stone masonry bridge abutment, whereas it is actually supported by a relatively thin concrete cross wall beside the abutment.

Unfortunately, the Tavanasa Bridge crosses the Rhine River inside a steep alpine valley that occasionally experienced avalanches. In 1927, a heavy loading of snow broke free and careened down the valley, smashing the bridge. Analysis later showed that the bridge itself had been well engineered and was strong enough for ordinary use. However, neither it nor any other bridge could have stood against a wall of snow weighing thousands of tons. A new bridge was built at the location, but Maillart was not the contractor.

See also HOLLOW-BOX ARCH; MAILLART, ROBERT; PRESTRESSED CONCRETE; SALGINATOBEL BRIDGE.

Tay Bridge disaster (built 1871–1878) *Britain's worst railway accident* When it came to bridges, engineer Sir Thomas Bouch was not a John Augustus Roebling or an Isambard Kingdom Brunel —other 19th-century engineers who meticulously planned and built their bridges. Bouch blindly trusted assistants and material suppliers and in the case of his Tay Bridge seemed unconcerned with maintenance after its completion.

The result was that the bridge failed on a stormy night, killing all 75 aboard a crossing train. The bridge collapse, as do most highly publicized and shocking disasters, spawned a controversy that continues to be debated.

The ill-fated Tay Bridge became a reality because of cutthroat business competition between Caledonian Railway and Bouch's employer, North British Railway. Bouch, an unimaginative designer of rail bridges, had proposed bridging the River Tay for years, although his employer was hard to convince.

North British Railway eventually took an active role in obtaining government permission to bridge the Tay and was granted a charter to do so in 1870. Because this bridge had been Bouch's pet project for so many years, no thought was given to soliciting bids from other engineers.

This period drawing shows the aftermath of the 1878 collapse of Scotland's virtually new Tay Bridge, whose High Girders gave way beneath the weight of a passenger train that may have been traveling far too fast. Helmet divers prepare to descend into the cold water to search for the bodies of the 75 who died. *(Author's Collection)*

Work began on July 22, 1871, and progressed smoothly when, out of the blue, the contractor Charles de Bergue bafflingly committed suicide. The bridge-building firm of Hopkins, Gilkes & Co. Ltd.—which had built many railroad bridges for Bouch in the past—assumed the contract and the project was on again.

The Tay Bridge was to be the longest such bridge ever, crossing two miles of River Tay estuary. Despite the massive length, the design was fairly straightforward. A total of 85 piers would provide support for 85 massive cast iron spans, the center 13 elevated to allow for river traffic. These became known as the High Girders. Of these High Girders, 11 would measure 245 feet and two would be 227 feet in length.

Although all the foundations for the piers were to be sunk to bedrock by using compressed air caissons, this would not be done because the bedrock was as deep as 157 feet in some sections of the river. Caissons would be used where the bedrock was shallow, and pilings would be driven to provide a platform for the piers where it was not.

As a result, the supporting columns for the spans would have to be reduced in weight by becoming skeletal iron supports in lieu of solid masonry. These iron columns would be set on masonry bases and cross-braced by a latticework of iron plates. These iron columns would then be riveted to the iron girders crossing the river. The High Girders were enclosed in a though-truss section of iron for additional reinforcement.

The building of the bridge had not been exceedingly costly in terms of the lives of its workers for the time: Of the nearly 700 men who labored on the bridge for eight years, only 20 died.

The worst single incident occurred in the pre-dawn hours of August 26, 1873, when a blowout in a caisson killed six workers. Of the 10 working that night, six working at various levels of the caisson died. Although the bodies of three men were quickly located in the upper reaches of the caisson, helmet divers had the eerie job of entering the flooded caisson's lower levels to retrieve the remaining three victims.

Year after year the bridge grew until virtually completed in September 1877 at a final cost of 300,000 pounds. It was then that Bouch decided it was time to test the bridge by driving a fully loaded passenger train across it.

Bouch himself rode in the engine with the conductor, Dugald Drummond, who accelerated the engine to 25 mph as it chugged across the bridge. Halfway across, a terrible noise shook the passenger cars, causing all aboard a moment of fear. Someone looked out a window and saw that a piece of scrap lumber had been left in the bridge's superstructure and was striking the cars as they passed. The train crossed the Tay and returned without additional problems.

After a few more months of finishing touches, North British Railway had an inspection in 1878 to certify the bridge as ready for use. The inspector was the retired major general Charles Scrope Hutchinson, a veteran of the British Army's Royal Engineers. His inspection lasted three days and was as thorough as a one-man examination could be. Hutchinson recommended that no train cross the bridge traveling more than 25 mph, the same speed Drummond had decided to travel the previous year. The bridge was then open for business.

Although it was well known that no train was allowed to exceed 25 mph on the span, some passengers begin to suspect the train was going much faster. One passenger, William Robertson, who was an engineer, began to time the train's two-mile journey across the bridge and clocked it at 42 mph. Another passenger, Alexander Hutchinson, also found the train traveling far too fast and observed that the air pushed by the train through the central span's iron latticework forced the girders to move up and down.

On Sunday, December 28, 1879, a horrific storm began to build and shove its way down the River Tay. By 5 P.M., the wind velocity was estimated at 78 mph, placing the storm in the category of a hurricane.

It was into this storm that North British Railway engine No. 224 pulled six cars. Aboard the passenger cars were 70 travelers, including women and a number of children. The remaining five aboard the train were North British Railway employees, including the engine crew. The train made its last stop on the way to Dundee and gained speed as its steam engine huffed and puffed across the bridge. It was then that witnesses saw a shower of sparks through the iron latticework of the high girders as the train and 3,000 feet of bridge plummeted into the storm-tossed water below. The Tay Bridge, nearly eight years in the making, had stood only 19 months under the weight of full-time rail service. The bridge and those depending on her died together at 7:20 P.M.

Helmet divers descended to the wreckage and groped along in darkness, trying to gain information about what had happened and to locate the dead. At first the North British Railway offered five pounds for each body located, but this amount was eventually reduced to two pounds. Eventually, 46 of those aboard were found and 29 of the victims were consigned forever to the waters of the Tay.

Only six months before the disaster, Queen Victoria had knighted Bouch for his triumph in designing the Tay Bridge. Now he was the focus of an official inquiry whose findings would shock Great Britain. Foundry workers revealed that many of the girders and columns contained casting imperfections that were camouflaged with a melted paste of beeswax and iron powder. Testimony also revealed that no one involved with the bridge project was an authority on structural iron.

Although building materials are now tested and analyzed as part of quality control, the iron for the Tay Bridge was accepted as good as a matter of faith. It was also revealed that the decision on the thickness of the iron used for the supporting columns was left to the foreman of the foundry, who had no role in designing the bridge. Predictably, all North British Railway locomotive drivers claimed they never exceeded the 25-mph speed limit on the bridge. Contradicting this testimony were passengers who streamed in to tell of clocking the train's excessive speed on numerous occasions.

The public was in for another shock when they learned that Henry Able Noble—a man trained as a brickmason—was in charge of maintaining an iron bridge with an inadequate crew of workers. Although ignorant of the properties of iron, Noble noticed thin slits as tall as a man in the iron columns and reported them to Bouch, who ordered the fractured columns encased in iron bands. When cracked brickwork was also found in the bases holding the iron columns, Bouch prescribed the same technique: constricting bands of iron.

Noble testified that during the 19 months he oversaw maintenance of the bridge, not a single contractor, railway official, or civil engineer had taken a look at the span. Amazingly, Noble said he was never given any instructions about the type of failure to look for nor specifically told to make any repairs. He also revealed he used his own money sometimes to patch up the world's longest railway bridge.

When Bouch testified to the Court of Inquiry, he had to admit the bridge had been designed to withstand wind pressure of only 10 pounds per square foot. This dismayed the court when other testimony revealed that in America, large bridges were designed to withstand no less than 50 pounds of air pressure per square foot, and those in France were designed to hold fast against 55 pounds of air pressure.

When the court issued its report, it flatly stated, "Sir Thomas Bouch is, in our opinion, entirely to blame." As for the shoddy iron allowed in the bridge, the inquiry board stated, "It is difficult to understand how the numerous defects should have been allowed to pass if there had been proper and competent persons to superintend the work." It also blasted the railroad for failure to control the speed of its trains.

The wreck commissioner and inquiry board member Henry Cadogan Rothery, who pushed for the toughest possible wording in the report, concluded that the bridge had previously been weakened by storms and by the strain caused by trains moving at an unsafe speed. The last two straws on the back of the bridge were the crossing of the train and the powerful winds of December 28, 1879. Despite these findings, engineers are still debating the cause of the disaster more than a century later.

North British Railway, having determined the Tay Bridge was an economic boon to business, built another, stronger bridge between 1882 and 1887. The new bridge was designed by John Fowler and Benjamin Baker and eerily stands today alongside the still-visible stumps of the 85 bases that once supported the suspect iron columns of Bouch's ill-fated bridge. Much of the iron salvaged from Bouch's bridge was inspected and used in the second bridge project. Bouch, who was ill at the time of the disaster, died soon after the inquiry.

Bouch's final indignity was to be cashiered as designer of the Forth Bridge, and that project given to Baker and Fowler. Drawing lessons from Bouch's failure, Baker and Fowler designed what is considered one of the most overbuilt rail bridges. The modern-looking bridge is made out of massive lengths of tubular steel—not iron—and continues to carry rail traffic over the Firth of Forth.

See also FIRTH OF FORTH BRIDGE; IRON.

Telford, Thomas (1757–1834) *England's self-taught bridge creator* All the odds were stacked against Thomas Telford's being anything other than a working-class Scottish lad raised by a widowed mother. Even before his entering the world, Telford's prospects were dimmed by the death of his sheepherder father. Thanks in no small part to the support of an extended family and his own innate brilliance and dedication to work, Telford made himself one of the most important civil engineers of the 19th century.

He built what at the time was the world's longest suspension span, the Menai Bridge, and was the first mainstream bridge builder in England to make use of iron as a structural material, a move that encouraged others to do the same.

The stratified society that still pervaded England gave opportunities to Telford and other gifted engineers of humble beginnings who demonstrated the ability to advance mechanization in Great Britain. Telford was equal to the task. Good-natured and so devoted to his work that he never took the time to marry, Telford deftly mastered the science of civil engineering without a day of college and promoted improved construction techniques for bridges, canals, and harbors.

Thanks to Scotland's egalitarian educational system, Telford received his basic education before his hovering relatives decided the youth should be apprenticed to a stonemason, and by his early 20s he had mastered that demanding trade. Released from his apprenticeship, Telford would find work in Edinburgh's construction industry until 1782, when he moved to London. By 1785, he was appointed superintendent of a Royal Navy harbor construction project, and from there he wangled a job as surveyor of Shropshire County, England. Shortly afterward, Telford established himself as a designer of canals, which predated railroads as the primary inland means of transporting goods during the 19th century.

But Telford's most famous engineering legacy would be his bridges, of which he constructed scores throughout Great Britain. Like all great bridge engineers, Telford came to loathe overly heavy bridges and began to see that superior designs were in fact lighter and more attractive if not actually artistic in appearance. Telford, unafraid to use new materials and concepts based on experimentation rather than high-flown theory, decided to make widespread use of iron as a bridge-building material when his contemporaries hesitated to do so.

However, Telford would not design the first bridge using iron. The world's first iron arch bridge was actually built in Coalbrookdale, England, as a means of transporting materials to and from an iron factory. Known as the Iron Bridge at Coalbrookdale, it was built in 1779 by iron makers, Abraham Darby III and John Wilkinson, as something of a lark. The architect Thomas Pritchard helped them design the bridge, for which they fabricated its iron structure. This daring bridge with its enchanting appearance—which was little more than a marketing gimmick to prove the worthiness of iron—still stands.

The experimental bridge gave the world a less expensive and far lighter span than a similar bridge of stone, but everyone else preferred to stay with stone—everyone, that is, but Telford.

Telford studied the Iron Bridge at Coalbrookdale, decided he could build a better one, and set about designing the Buildwas Bridge over England's Severn River, using iron made by Darby and Wilkinson. The bridge was completed in 1796, and by 1800, Telford had designed four more iron bridges, as he became more comfortable with iron's structural attributes. The strength of iron allowed Telford to flatten the arches of the bridges, making them more appealing visually. Telford's successful iron bridges convinced others of iron's utility, and iron arch bridges began popping up in France and England.

By the early 1800s, Telford began to study the design of the suspension bridge—a bridge type in existence for centuries in one form or another, but notoriously unstable and prone to collapse as a result of high winds or heavy loads. Telford deter-

mined that flat iron links, developed by a Royal Navy captain, Samuel Brown, would support the weight of a suspension bridge. Telford designed a suspension bridge using these links and sought permission to build it over the Menai Strait separating Anglesey Island and Wales. The radical design provided a desperately needed way to link the Anglesey Island port of Holyhead with the Welsh mainland, and its construction was approved. Built between 1818 and 1826 at Bangor, Wales, the bridge had a roadway that rose 98 feet above the strait and a center span 580 feet in length—the longest suspension bridge built up to that time.

Although the bridge was repaired twice within a decade of its completion and was mostly rebuilt in 1939, it continues to carry automobile traffic over the Menai Strait. It is considered among the world's most beautiful and historic bridges, though its suspension system and roadway are not original.

Telford was a prolific engineer who was eventually accorded near-hero status in England. In 1800, he proposed construction of a 600-foot iron arch bridge to replace the decrepit London Bridge completed nearly 600 years before. But his design was too radical for London officials, who opted for a more classical stone masonry bridge design by John Rennie, Sr., another self-taught Scottish engineer of humble beginnings.

Despite his lack of a formal technical education, Telford was so well respected by his fellow engineers that he was elected the first president of England's Institution of Civil Engineers in 1820. He worked to make it a viable organization for the promotion of professionalism among engineers and obtained a Royal Charter from the British crown to legitimize its efforts. He died on September 2, 1834, at 77 and was interred at London's Westminster Abbey.

See also IRON BRIDGE; LONDON BRIDGE; MENAI BRIDGE; PAINE, THOMAS; RENNIE, JOHN, SR.; SUSPENSION BRIDGE.

Thames Tunnel (built 1825–1843) *The world's first subaqueous tunnel* A tunnel beneath the Thames River had long been a dream among London developers who wanted a way of crossing the river without resorting to ferries, but soft ground, subaqueous tunneling had never been successfully attempted anywhere because of the problems of cave-ins and flooding.

A 1798 plan to bore through the clay and sand beneath the Thames was put forward by a canal engineer, Ralph Dodd, who proposed the digging of a 2,700-foot tunnel linking Gravesend and Tilbury, but the plan came to naught. Robert Vazie believed it possible to tunnel between Rotherhithe and Limehouse. Vazie's 1802 proposal was received so favorably that investors eagerly contributed to the scheme, and the Thames Archway Company was formed in 1805 to execute his plan.

Recognizing the terrible risk of flooding should the river break into the excavation, Vazie wanted to resort to a tried-and-true method of digging a smaller tunnel or a drift beneath the main tunnel to serve as a drain for unwanted water. This would have allowed floodwater to be pumped out as the main tunnel was dug. Vazie's sound suggestion was rejected because it was too expensive and Vazie's effort did not get to first base.

The interior of one of the two galleries of the Thames Tunnel as it appeared during controversial strengthening and reconstruction work in 1996. The work was completed in 1997, readying the tunnel for what London transit officials hope will be another 150 years of service. *(Courtesy Taywood Engineering)*

An 11-foot vertical shaft constructed on the Rotherhithe shore to allow work on the horizontal tunnel was flooded so frequently and completely that Vazie reached a depth of only 42 feet before the money ran out. A second attempt to complete the shaft took the miners down 76 feet before water again filled the shaft.

A frustrated Thames Archway Company then promoted Vazie's assistant, Richard Trevithick, the inventor of the high-pressure steam engine and the son of a mining engineer. Trevithick turned to the men from his own region of Britain, hiring Cornish miners, who were considered among England's most skillful.

With his tough miners, Trevithick went back into the shaft and began to dig beneath the Thames, fighting flooding all the way. The drift they were excavating was only five feet in height and a narrow three feet in width, forcing the miners to work in terribly cramped conditions. Miraculously, Trevithick and his tough-as-nails miners neared the opposite bank of the Thames by January 1808. In fact, because the level of the Thames is affected strongly by tidal action, the tunnel had made it beneath the low-tide mark of the shore.

But the Thames was not going to allow itself to be defeated so easily. An extraordinarily heavy tide raced in, placing too much water pressure above the tunnel, and a massive breakthrough flooded the shaft. The flooding overwhelmed the pumps, forcing Trevithick and three of his intrepid miners to flee as water rose in the narrow tunnel. All escaped with their lives but only barely.

As the water was rising inside Trevithick's tunnel, the money and patience of the investors were running low. The Thames Archway Company unsuccessfully sought some way to make Trevithick the scapegoat, but he had done everything possible with the available technology to complete the tunnel.

As engineers began to cluck that soft ground, subaqueous tunneling was impossible, Trevithick decided to build cofferdams to hold back the Thames and allow the installation of prefabricated tunnel sections. Thames Archway Company said no to the harebrained scheme with no way of knowing that Trevithick had recommended a completely sound proposal.

The technical innovation that Trevithick needed was only a few years from development. In 1818, the French immigrant Marc Isambard Brunel took a lesson from a wood-boring worm called the *Teredo*

to design a tunneling shield. This was nothing more than a three-story series of iron compartments in which miners would dig 4 1/2 inches at a time before using screw jacks to shove their iron box forward. Workers behind the miners inside the shield would reinforce the soft tunnel walls with brick.

Word of the concept reached I. W. Tate in 1823. Tate, a proponent of the tunnel attempted by Vazies and Trevithick, pressed Brunel to apply his invention to a new Thames tunneling effort. Faster than a subaqueous tunnel can fill with water, a new consortium, known as the Thames Tunnel Company, was formed, and by 1825 Brunel was in the tunneling business. Brunel would sink his vertical access shaft on the south side of the Thames at Rotherhithe and tunnel north to London's Wapping section.

Brunel demonstrated his brilliance in tackling a complex problem by eliminating the hazards that beset his tunneling predecessors. First he selected his tunneling route and depth on the basis of borings so he would tunnel through relatively watertight clay rather than porous quicksand.

Brunel also designed a reinforced access shaft liner composed of a massive brick tube reinforced with vertical iron rods that connected to cast iron "curbs" at the top and bottom edges of the liner. This innovation created a virtually indestructible tube that descended into the ground as the vertical shaft was excavated. As it sank, more brick was added to the cylinder. In this way Brunel was able to prevent the breakthroughs and flooding that ruined Vazie's shaft.

Laying the first bricks of the shaft liner were first Brunel and then his son, Isambard Kingdom Brunel. The 19-year-old Isambard would be appointed engineer in charge. With bricklayers placing a thousand bricks a day apiece, the shaft was soon finished and workers began digging inside the completed brick tube, 50 feet in diameter. As dirt was hauled away, the brick cylinder descended into the ground and reached the required 42 feet of depth in only 21 days. The project fascinated the public, including such dignitaries as the duke of Wellington. All were surprised to see the 1,000-ton masonry liner slowly sink into the earth. When the shaft liner was in place, the effort to drive a tunnel horizontally beneath the Thames began.

Brunel's tunneling shield was installed at the base of the shaft, and by dim candlelight miners began to dig at the clay protected by an iron awning overhead. After digging away 4 1/2 inches of clay, the shield was moved forward this distance by jacks.

Masons then laid a heavy brick lining. Since the shield was moved only 4 1/2 inches at a time, this was the only portion of the tunnel that not reinforced by the shield or brickwork, vastly reducing the risk of flooding or cave-in.

A total of 12 shields were employed side by side in the tunnel. Each shield stood 21 feet, four inches, tall and three feet wide and accommodated three levels of miners working in their own iron compartments. A total of 36 miners worked to bore through the twin arches of the tunnel simultaneously. The tunnel's internal shape would be of a pair of double arches with a supporting brick wall between them. The shields were installed and moving forward by November 28, 1825.

Although Marc Brunel took all the precautions he could, the geologists who analyzed core samples of the riverbed and embankments could have been reading tea leaves for all the good they did the project. These 19th-century consultants predicted that a solid stratum of impermeable clay would make for relatively dry digging. To stay within this layer of clay, the tunnel would have to be no more than 14 feet beneath the riverbed. Brunel started the tunnel by driving it just below the surface of the mud but discovered there was no solid layer of clay. As Vazie and Trevithick had, Brunel ran into weak and porous pockets of sand and gravel that were subject to breakthroughs and water inflow.

Adding to Brunel's woes were the directors of the Thames Tunnel Company, who chafed at the slow progress of the tunneling and doubled the distance the frames would be jacked forward to nine inches at a time. This horrified Brunel because it doubled the amount of unreinforced riverbed muck that was to be dug out through the open fronts of the shields. When the shields were advanced, that unprotected distance would remain at the rear until bricked up. It did not require an engineer to understand that such a move would drastically increase the risk to those working in the tunnel. Brunel's recommendation to build a drainage tunnel was again called too expensive.

Despite the basic decency of Marc Brunel as well as his son, Isambard (who worked alongside the laborers inside the tunnel for up to 36 hours at a time), conditions beneath the Thames were frightening. Gas lamps provided weak light while men used hand tools to claw muck from the working face of the tunnel. The men digging in the lowest frames of the shields usually did their work knee deep in water

seeping through the earth, an unpleasant task since the Thames at this time was nothing more than a large sewage ditch.

Respiratory illnesses plagued those working in the dank conditions as they watched for the constant threat of massive flooding. This fear plagued Marc Brunel, who envisioned dozens of men's being drowned en masse if the river poured through the weak ground.

One of the most horrific things that befell those laboring beneath the Thames River was a strange, often fatal illness that was frequently accompanied by blindness. One of Isambard Brunel's assistant engineers on the project was Richard Beamish, who was blinded in one eye by the strange malady. Isambard Brunel was laid low by the fever in 1826 but recovered. Another of Isambard Brunel's assistants died of the fever in the first month of 1827. The assistant, Riley, lasted only two months working beneath the Thames. The cause may have been toxic hydrogen sulfide gas given off by decaying biological matter in the river.

Marc Brunel worried incessantly about a catastrophe that would kill a sizable number of sightseers who regularly visited the tunnel to observe the Dantesque sight of miners' clawing their way through the mud in the flicker of candles and oil lanterns.

On the night of May 18, Brunel's nightmare became a reality just as his son, Isambard Brunel, left the tunnel for home. On that night, a heavy tide crept in, increasing the depth of the water and therefore the water pressure over the tunnel. Beamish, who was taking Isambard Brunel's place, wore a raincoat, expecting a wet night. His precaution was little help when a powerful jet of water shot through the tunnel face, blasting a miner backward out of the shield. Beamish, standing on the scaffolding at the rear, caught the man before he was thrown to his death. All inside the tunnel scrambled toward the access shaft as the water roared in.

Water smashed through the tunnel as workers clambered desperately up the stairs of the access shaft. Above, Isambard Brunel was helping the last of the men up when he heard the voice of an elderly water pump repairman cry for help. Brunel grabbed a rope and dove into the dark water and surfaced with the drowning man, whose name was Tillett.

Isambard Brunel then commandeered a diving bell and with an assistant sat in the open-bottomed device as it was dragged along the floor of the river. What they discovered was that the riverbed had been dredged for gravel, making the muck above the tunnel dangerously thin, allowing the river to penetrate the excavation. During the inspection the assistant in the open-bottom diving bell lost his footing near the edge of the gaping black hole and was nearly sucked out of sight. Brunel extended his leg to the man, who was hauled back into the safety of the diving bell.

Isambard Brunel then ordered bags of clay dumped into the depression to make the ground relatively watertight. By June 11, 1825, nearly 20,000 cubic feet of clay had plugged the hole. As two directors of the Thames Tunnel Company and three other men also inspected the eerie, partially flooded tunnel, one of the directors stood up, struck his head, and fell, capsizing the boat. While the two directors managed to cling to the wall and two workers swam to safety, one worker drowned, the first to die by accident in the Thames Tunnel.

By November 1827, the hole in the riverbed plugged and the miners once again at work, Marc Brunel decided to hold a banquet in the completed section of the tunnel, which was separated into two galleries by a supporting wall punctuated with arches that served as crossover passageways. On one side the laborers would be feasted, and on the other the Tunnel Company directors, investors, and other dignitaries would be wined and dined.

Just two months later, on January 12, 1928, Isambard Brunel and two miners were working in the frames of the shields when a powerful column of water spurted from the ground and washed all three down the tunnel. The two miners were killed outright and Brunel's leg became trapped until he could wrench it free. Brunel's unconscious body was flushed to the surface of a secondary access shaft like that of a buoyant rag doll. In addition to the two men who died near Brunel, four other workers drowned in the catastrophe.

Injured and ill, Isambard would not return to work on the tunnel. An economic slump injured the Tunnel Company's ability to raise more funds to repair the breakthrough. Seven years passed before the company was granted 246,000 British pounds to continue the work. Marc Brunel redesigned his revolutionary tunneling shield to make it more effective and safe. Fueled by government money, the Sisyphean task was renewed in 1835, and by 1840 the twin-arched tunnel was approaching the northern shore. A second brick access shaft was sunk 70 feet below the surface of the opposing shore to accept the approaching tunnel on the north bank. In what was considered an engineering miracle at the time, the Thames Tunnel finally reached across the Thames River in March 1843. Marc Brunel was knighted for his efforts, and the Tunnel Company opened it up to pedestrian traffic, charging a penny per person to stroll through the 1,197-foot tunnel.

Despite the flooding and the deaths of at least 10 men during the construction of the Thames Tunnel, it managed to provide the public with virtually problem-free service during its first 150 years of operation. In its early days, the tunnel was an unmitigated financial disaster. Although it was filled with shops and pedestrians soon after it opened, making it perhaps the first indoor (and underwater) shopping mall, interest in it declined. By the 1860s, it had become a haven for prostitutes and thieves and was considered unsafe by London's decent folk. Facing bankruptcy in 1869, the Thames Tunnel Company sold its tunnel to the East London Railway Company at a tremendous loss.

The tunnel was eventually integrated into the London Underground subway system since each of the tunnel's two galleries was capable of accommodating a rail line. Because of the shallowness of the earth between the tunnel and the river above, those inside could hear propeller and engine noises of passing steamships.

By the 1990s, experts determined that time had taken its toll on the tunnel, and the possibility of a breakthrough was real enough to encourage transit officials in London to formulate a renovation plan. Because it was still considered an important part of London's underground passenger rail system, it was decided to remove the original tile work and reinforce the tunnel with two layers of shotcrete, or sprayed concrete, the initial layer mixed with steel fibers and the final layer mixed with plastic fibers.

On March 24, 1995, a day before the massive renovation and strengthening work was to begin on the tunnel, Britain's heritage secretary listed the tunnel as a historic site. What followed were 13 months of debate and controversy between those who wanted to preserve the original's tunnel's appearance and transportation officials who wanted to make it safer.

The historical preservation requirements and the delays they caused boosted the cost of the renovation from 6.3 million pounds to 23.2 million pounds by the time the work was completed in 1997. To celebrate the renovations, a formal dinner was hosted in the tunnel on March 25, 1997—nearly 170 years after Marc Brunel held a banquet inside the tunnel for dignitaries and miners.

See also BRUNEL, ISAMBARD KINGDOM; BRUNEL, MARC ISAMBARD; SUBAQUEOUS TUNNEL; TUNNELING SHIELD.

Thomé, de Gamond, Aimé (1807–1876) *The English Channel tunnel's greatest promoter*

History may not be quite sure what to do with Thomé de Gamond, a French businessman, farmer, engineer, and dreamer. Apparently born to a family of some means in Poitiers, France, Thomé had the misfortune of not only being orphaned at 16 but hounded from France with his uncle, with whom he lived. The uncle's opposition to the restoration of the French monarchy necessitated their flight.

Thomé was educated in Holland and Belgium and became enthralled not only with engineering but also with the most exciting construction challenge of the day: the design of an English Channel crossing. After his exile, Thomé returned to France, where he was certified as a hydrographic and mining engineer, as a lawyer, and as a physician. In his spare time, Thomé was an engineering officer in the French army reserves.

By 1837, Thomé had designed seven different bridges and tunnels he hoped might connect England with France. When his ideas were ignored or failed to attract support he simply designed new bridges and tunnels.

Some of his subsequent ideas revealed his naïveté regarding engineering. An 1840 proposal would have thrown a dam across the entire width of the English Channel's Dover Strait, leaving a trio of openings for ship traffic. Tidal action would have wreaked havoc with this design, and the inevitable collisions between ships and this massive wall in Channel fog would have doomed an untold number of sailors.

Undeterred, Thomé proposed a tunnel design in 1856 that was to have a remarkable resemblance to another promoted 130 years later. This tunnel would emerge through an artificial island and receive railroad traffic through a spiral ramp. The 1986 EuroRoute design for a vehicular tunnel across the English Channel also called for an artificial island and spiral ramps leading automobile traffic beneath the water.

At the age of 48, Thomé risked his life to dive 100 feet to the seabed of the English Channel, where he groped about in utter darkness to determine the geologic features a tunnel might encounter. Thomé claimed he plunged into the chilly waters of the Channel stark naked, protected only by buttersoaked lint plugs stuffed into his nostrils and ears. Thomé wrote that he filled his mouth with olive oil to prevent the water from entering his lungs. (How he kept the olive oil out of his lungs is not recorded.)

Following the technique of sponge and pearl divers, Thomé would then hold on to bags filled with stones for ballast. After colliding with the seabed, the intrepid Thomé claimed he would investigate conditions as long as his breath prevailed. He would then release the stones and rocket to the surface with the aid of inflated pig bladders. The sight of a naked, shivering man erupting to the surface while clinging to a dozen swollen pig bladders must have been memorable for ferry passengers.

During one dive into the murky and chilly depths of the Channel, Thomé was bitten on the chin by a sea creature that may have been a small fish or an eel. In retrospect, it is miraculous that Thomé did not suffer worse consequences as a result of his crude diving technique.

As the years progressed, Thomé's plans grew more and more grandiose. One scheme was a rail tunnel that would be entered by trains spiraling down a 1,000-foot-diameter shaft. If impractical and unrealistically expensive, Thomé's designs would have nonetheless been visual masterpieces. After nearly 40 years of promoting Channel-crossing schemes, one of Thomé's plans supposedly gained approval from England's Queen Victoria, who suffered seasickness during Channel crossings.

He also managed to enlist the support of two Englishmen, the mining engineer William Low and the railway engineer Sir James Brunlees. All three had independently concocted Channel tunnel plans, and they pooled their ideas to produce a single proposal.

Sadly, the greatest Channel tunnel dreamer in history never saw his tunnel built. Thomé impoverished himself with his schemes, having spent 175,000 francs in his efforts to build his beloved tunnel. At the time of his death in 1876, it seemed that Thomé's writings and drawings would be the only legacy of his schemes. However, his affiliation with Low would later bear fruit when Low was able to inspire the English railroad magnate Sir Edward Watkin to begin a tunneling effort in 1880. The effort seemed destined for success until halted by negative public opinion in England.

Thomé's efforts to promote a Channel tunnel gained him a bit of fame that has survived the passage of more than 100 years. A restaurant near Calais overlooking the English Channel has been named after him. The Thomé de Gamond Restaurant provides a view of England's famed Cliffs of Dover now linked to France by the Eurotunnel, which was completed in 1993.

See also EUROTUNNEL; MATHIEU, JACQUES-JOSEPH; WATKIN, EDWARD.

Throgs Neck Bridge (built 1957–1961) The Throgs Neck Bridge between the Bronx and Queens over the East River was born of the negative experiences of the collapse of the too-thin Tacoma Narrows Bridge in 1940 and the wobbly but intact Bronx-Whitestone Bridge completed in the same year.

It would be the next to last of the great crossings designed by the brilliant civil engineer Othmar Hermann Ammann. The robustness and stability of the Throgs Neck Bridge would constitute a display of not only style and taste but also strength. It was the first great long-span suspension bridge designed by the prolific Ammann since the erection of his Bronx-Whitestone Bridge, which eventually required stiffening to prevent its wind-induced movement from unnerving motorists. The thin-bridge guru Leon Salomon Moisseiff, who helped Ammann design a slender and lightweight profile for the Bronx-Whitestone Bridge, had designed the ill-fated Tacoma Narrows Bridge. Although the Tacoma Narrows span was destroyed by a relatively mild gust of wind, the Bronx-Whitestone Bridge exhibited only disconcerting movement in strong winds, dampening trust in the bridge even if it did not threaten safety.

When the New York City building czar Robert M. Moses pushed through an expressway requiring the Throgs Neck Bridge, the job of designing the bridge was given to Ammann's engineering firm, Ammann and Whitney, with Ammann as the chief engineer. The year was 1955, and the era of dangerously slender and light bridges had crashed 15 years earlier with the destruction of the Tacoma Narrows Bridge. Ammann, mindful of the problems that slender and light bridges posed, was to make the Throgs Neck Bridge the essence of stability.

Moses's desire to build the Throgs Neck Bridge stretched back years. The bridge would sit two miles east of the Bronx-Whitestone Bridge, hurriedly completed to easier access to the 1939–1940 World's Fair. Similarly, the Throgs Neck Bridge would provide additional access to the 1964 World's Fair—also at Flushing Meadow—of which Moses was president.

In 1955, Moses was chairman of the Joint Study of Arterial Facilities, who determined a new crossing was needed between the Bronx and Queens. Characteristically, Moses shoved the project past critics, many of whom were residents living at either end of the proposed bridge at Bayside, Queens, and Fort Schuyler, in the Bronx. Public opposition to the project was understandable since 421 homes had to be moved to a new location. Ammann, who had soldiered for Moses as the chief engineer of the Triborough Bridge and Tunnel Authority during the 1930s, was tapped to design the bridge. Construction began October 22, 1957.

Ammann's design was to be similar in appearance to the Bronx-Whitestone span but with major structural differences. Instead of a flexible plate girder measuring a narrow 11 feet in width beneath the road deck, the Throgs Neck Bridge would possess a 28-foot-deep stiffening truss that added weight and rigidity to the span. (A truss railing had to be retrofitted to the Bronx-Whitestone Bridge during the 1950s to tame its excessive movement.) Like the San Francisco–Oakland Bay Bridge, which possessed an offshore concrete anchorage, the Throgs Neck span would require one such anchorage just offshore from Queens and another at the shoreline of Fort Schuyler in the Bronx.

The wide expanse of water that had to be crossed by the bridge necessitated a bridge system measuring a total of 13,400 feet. A viaduct carried atop a series of concrete piers spanned more than 11,000 feet of this distance. Whereas the Bronx-Whitestone Bridge's central span measured an impressive 2,300 feet, the narrowness of the navigable channel crossed by the Throgs Neck Bridge meant the central suspended span of the newer bridge need be only 1,800 feet. A pair of 550-foot side spans would extend between the two suspension towers and the anchorages.

Each anchorage contains 170,000 cubic yards of concrete, weighing 344,000 tons, excluding the weight of the encased reinforcing steel. Both anchorages are seated on foundations excavated to bedrock and measure 140 feet wide and 200 feet long. They rise 150 feet above the water. The two

This overhead view of the Throgs Neck Bridge looking toward the Bronx from Queens reveals a massive suspension span designed by Othmar H. Ammann with its long viaduct approaches. Completed in 1961, the bridge has clean lines that bear a resemblance to the lighter and wobblier Bronx-Whitestone Bridge built by Ammann 22 years earlier. The Throgs Neck Bridge is a far heavier and more stable span. *(Historic American Engineering Record. National Park Service. Photograph by Jet Lowe)*

piers supporting the towers contain 60,000 cubic yards each.

Although not nearly as daring as the Bronx-Whitestone Bridge, the shorter length of the suspended central span combined with its wider profile and greater rigidity made for an exceedingly stable bridge. The towers of the Throgs Neck span rise to 360 feet, making them roughly 17 feet shorter than those of the Bronx-Whitestone. The towers possess the signature appearance of Ammann's design touch in that they are not supported by cruciform supports between their legs. At the top of the Throgs Neck towers, ruler-straight struts

bearing flattened arches on their underbellies connect the legs of each.

Saddles sit atop each tower leg bearing the vertical loading of the curved steel suspension cables that transmit the weight of the suspended span to the concrete anchorages. The two suspension cables measure 23 inches in diameter each and are spun out of 13,300 miles of high-strength steel wire. The combined weight of the cables and suspenders is 4,000 tons. The road deck of the bridge measures 93 feet and carries a total of six lanes of traffic.

The bridge was opened to the public January 11, 1961, and carried 63,000 cars daily between the Bronx and Queens. Although the plan was for the bridge to relieve traffic on the Bronx-Whitestone Bridge, the additional crossing actually encouraged a greater flow of traffic. By the late 1990s more than 100,000 cars a day were crossing the span.

See also AMMANN, OTHMAR HERMANN; BRONX-WHITESTONE BRIDGE.

tied arch bridge   *See* arch bridge.

timber bridges   Timber or wood bridges, among the oldest types of spans built, continue to be constructed around the world, including in the United States, and new technology may be opening a door for a new era of timber bridge construction.

Considering the handiness of timber and its suitability as a strong structural material, not surprisingly timber bridges were a predominant feature of areas where sufficient forest resources were available. Although stone was the preferred material for long-span bridges over deep and wide waterways before the use of iron, steel, and concrete, timber bridges offered many advantages to bridge builders.

Sizable trees of the appropriate grade and species growing near the site of a bridge project provided easy access to building materials that merely needed to be logged and shaped. Stone bridges required the quarrying and painstaking finishing of the stone by stonecutters. Then, the heavy stone often had to be transported long distances to the work site and assembled by skilled masons. In addition, wood's considerable compressive strength made it possible for timbers to serve as both pilings driven into soft ground as foundations and piers to support the bridge deck, which could also be made of wood.

Even when timber was not readily available, particularly in poorly forested nations such as 19th-century Great Britain, wood was still considered such a superb bridge-building material that it was imported from distant shores. The famed English bridge and tunnel engineer Isambard Kingdom Brunel, who needed to construct more than 1,000 miles of rail lines rapidly for the Great Western Railway, imported pine from the Baltic region of Europe. Strong and resistant to rot, the pine could be shaped into modular components to allow the rapid construction of desperately needed railroad bridges.

Since it is a biological material, wood is prone to rot and degradation caused by bacteria. And, because it will burn, an imposing and capable wooden bridge can be destroyed in minutes by fire. For these reasons, wooden bridges were considered the least substantial of long-term structures. Another drawback is that timber spans also required extensive labor and expense for maintenance and repair, a criticism often leveled at the numerous timber bridges erected by Brunel. However, the speed with which a timber bridge can be constructed made the wooden spans attractive, particularly to railroad interests during the Industrial Revolution.

Although timber is liable to rotting, fire often posed the greatest threat to bridges. One of the most impressive timber bridges ever built fell victim to fire near Philadelphia in the early 19th century. Considered one of the world's greatest engineering achievements was the Upper Ferry Bridge, also known as the "Colossus." Designed and built by the German American Lewis Wernwag in 1812, the bridge arched 340 feet across Pennsylvania's Schuylkill River. At the time, it was considered the second longest span in the world. The bridge served admirably until destroyed by fire in 1838. The brilliantly designed structure was not rebuilt by its owners and was replaced in 1842 by a wire cable suspension bridge designed by Charles Ellet, Jr.

Thanks to the abundance of timber in much of the eastern United States, timber bridges proliferated. Although some might assume that covered bridges of timber were the original type in America, they actually did not appear until 1805. Although picturesque, covered bridges were given roofs and sidewalls not to enhance aesthetic appeal but to protect the bridge from precipitation and the deterioration that accompanies rot. It was estimated that if covered, a timber bridge had its lifespan extended from 10 years to 50 years.

The first covered timber bridge is believed to be one designed by Timothy Palmer and built over Pennsylvania's Schuylkill River in 1805. Known as the Permanent Bridge, it was originally built without a roof. It was decided after its completion to add one on the basis of the logical assumption that a wooden bridge protected from rain would last longer. Palmer concurred with this addition to his bridge, believing it would expand its useful life fourfold. The idea caught on, and covered bridges appeared from New England to California during the remainder of the century.

It is estimated that as many as 500 covered bridges remained in use on America's roadways by the 1930s, but the number of timber bridges constructed for both road and rail use declined during the last half of the 1800s. Bridges that were built of timber during the remaining two thirds of the 20th century used lumber treated to forestall rot and eliminate the need for expensive roof and wall construction.

The use of timber in major bridges on primary roads declined during the 20th century, as high-strength concrete and cheaper steel became available. Nonetheless, many urban sections of the United States continued using timber bridges treated to resist rot since they were relatively inexpensive and easy to build. In Houston, Texas, one of America's fastest-growing cities, timber bridges carried a considerable volume of traffic in many suburban areas until the 1960s.

Do not assume that timber bridges have disappeared from the American scene. More than 14 percent of the bridges in America contain timber components, according to the National Bridge Inventory. Of these, 7 percent are solely of timber components and another 7.3 percent have timber road decks supported by a steel superstructure. This amounts to roughly 80,000 bridges nationwide.

Timber bridges remain in place in mostly rural areas on secondary roads, or roads that experience a low volume of heavyweight traffic. Timber bridges are often constructed across creeks and sloughs on private property since the volume and weight of traffic they must carry are minimal and the cost of construction is reasonable. There are no official figures on how many of these private bridges might exist, but they could easily number in the thousands.

However, technology developed since the 1930s may lead to a new era of timber bridge building. These developments involve the pressure lamination of wood, which has created composite materials with greater strength than ordinary timber. In recent decades these wood products have been applied to bridges with considerable success. Glue-laminated, or Glulam, components are made of wood sheets bonded together under pressure by using waterproof adhesives to form bridge decks. The Keystone Wye Bridge in South Dakota, constructed in 1968, employs glue laminate wood beams and it spans a distance of 160 feet.

Another technique is to place layers of wood together and compress them with steel bars in a technique known as stress-laminating. Developed during the 1970s, this technique is similar in concept to prestressing concrete and allows the use of weaker woods in bridge construction. Another technique combines glue lamination with stress lamination to create high-strength wood bridge components.

Lamination of wood not only makes it stronger but also creates a building material of uniform strength and predictable characteristics, essential properties in engineering a structure as complex as a bridge. Because of these and other developments, the Timber Bridge Initiative Act was passed by the U.S. Congress in 1989 to encourage the construction of timber bridges by the United States Forest Service. The Federal Highway Administration (FHWA) established a similar program to encourage construction of high-tech timber bridges across the nation. Between 1993 and 1997 the FHWA funded timber bridge projects to the amount of $37 million.

It appears the timber bridge may be making its way back from near-extinction, as FHWA experts believe that new types of wood bridges can carry the same loading as bridges of steel and concrete while using one of the oldest bridge materials.

tremie Thanks to the ability of concrete to cure or harden underwater, it is an engineering godsend for the construction of bridge piers or any other foundation work that must be done while in submerged conditions. Because concrete is more dense than water, it can be carefully poured so that it displaces the water above without disintegrating into its components. To ensure that concrete flows to the bottom of a submerged form, the concrete is poured through a tremie, a pipe with removable sections.

As the level of the poured concrete rises, the tremie is raised while sections of the pipe are removed until the concrete reaches the desired level inside a cofferdam or caisson. Tremie piping can be rigid or flexible. Concrete can be poured from a bucket or a hopper into a funnel at the top of a tremie or can be pumped under pressure. In the construction trades, concrete poured by using a tremie is sometimes referred to as tremie concrete. By using a tremie, the concrete can be poured underwater without allowing its aggregates of rock and sand to separate from the cement, seriously weakening or destroying the mixture.

The miraculous properties of concrete, and the use of tremies to place the material where it is needed underwater, have actually allowed teams of divers to pump concrete into damaged submerged reefs in an effort to rejuvenate them by encouraging new coral growth. One such project was undertaken near Miami in the mid-1990s when a ship gouged out a section of reef.

Massive tremie operations in which fleets of concrete trucks arrive to deliver their loads to concrete pumping trucks are necessary for the massive foundations of modern bridges. The concrete is transferred to a pumping truck, which then moves the material under pressure through the tremie. During the Dame Point Bridge project in Florida, concrete was pumped underwater to create a seal between the bedrock of the Saint John's River and the foundation for the bridge piers.

The use of tremie piping in modern times began soon after the 1824 rediscovery of concrete, a construction secret that had been used centuries before by Roman Empire engineers in the construction of bridges and buildings.

See also CONCRETE.

trenchless technology   *See* microtunneling.

Triborough Bridge and Tunnel Authority   *See* Moses, Robert M.

truss bridge   A truss is an ancient design innovation based on the proven concept that a triangular framework supports itself and is resistant to collapse. When applied to bridges, this rather simple concept has made some of the world's most important structures possible and continues to play a major role in today's bridge-building.

A truss is essentially a repetitive triangular-shaped supporting structure that forms bracing. Though it was a concept originally developed for the construction of wooden buildings, it has been applied to bridges for more than 200 years in one form or another. Whether the truss material is timber, iron, or steel, it provides a greater amount of strength for less weight than most other stiffening and reinforcing methods.

Myriad truss designs abound, ranging from the simple to the dizzyingly complex. Truss bridges proliferated in the United States during the growth of railroads in the 1800s, when many were constructed of timber. As bridges grew longer and more daring, truss-reinforced road decks became essential to provide lightweight strengthening to thinner and lighter bridges. In this method, a deck truss is placed along the underside of a bridge's roadway to add stiffening.

Many bridges, including the Golden Gate Bridge, use a truss-stiffened deck, which remains one of the most common types of road deck reinforcement. The latticework of reinforcement provides additional weight and tremendous stiffening to bridges prone to the effects of high wind. A side benefit of truss reinforcement is its open design, which also provides less resistance to wind.

The use of a narrow monolithic plate girder beneath the roadway instead of a ventilated deck truss on the underside of the Tacoma Narrows Bridge is partially blamed for the span's November 7, 1940, collapse five months after its completion. Some experts believe the bridge—which was also far too light and flexible—would have been better served had its structure provided less wind resistance. The bridge that replaced it 10 years later had a substantial deck truss whose open design allowed wind to pass through.

Other bridge designers have opted to create truss walls rising from each side of a road or railway deck in a configuration known as a pony truss. A variation of this design placed additional truss reinforcement across the top of side trusses. This kind of arrangement allows traffic to pass through a cagelike tunnel of truss reinforcement that makes the bridge extremely stable and strong. This design is known as a through truss for obvious reasons.

There are few types of bridges to which truss reinforcement could not be applied. Gustav Lindenthal's magnificent Hell Gate Bridge between Queens and the Bronx is an arch bridge with strong truss reinforcement of its supporting overhead arch. New York City retrofitted a road-deck-level truss reinforcement that looked much like massive handrails on the Bronx-Whitestone Bridge after the collapse of the Tacoma Narrows suspension bridge. The ill-fated bridge shared a lack of truss reinforcement with the Bronx-Whitestone Bridge. Although the Bronx-Whitestone Bridge's chief engineer, Othmar Ammann, believed his bridge was rigid enough, New York's Triborough Bridge and Tunnel Authority requested that Ammann strengthen it nonetheless.

He did so with one of the most effective means at hand, a truss.

See also AMMANN, OTHMAR HERMANN; HOWE, WILLIAM; MOISSEIFF, LEON SALOMON.

tungsten carbide *Metallurgical breakthrough for tunneling* Pushing a tunnel through rock involves either blasting or tunnel-boring machines, and both methods require materials hard enough to penetrate rock. For blasting, drill bits must be hard enough to penetrate the rock efficiently to make room for explosives that perform the real job of excavation. Tunnel-boring machines need cutters that can grind away rock at a high rate of speed. The answer to these requirements was the creation of tungsten carbide.

Steel drills and cutters, though hard, eventually fell victim to the hardness of the rock they were working against. This problem led to the downfall of the Brunton tunneling machine developed during the 1870s. Steel drill tools had to be withdrawn from service, transported from the tunnel, and resharpened. This process was monumentally expensive and time-consuming in a business in which progress was measured by the cumulative depth of holes drilled and then blasted per day. A curious metal derived from mineral deposits was to provide relief.

Discovered in 1779 by an English scientist, Peter Woulfe, and isolated four years later by two Spanish researchers, tungsten actually gained its name from the Swedish language, in which it is known as *tung sten,* "hard stone." With the highest melting point of any metal (6,170 degrees Fahrenheit), it was all but impossible to work with until a French chemist, Frédéric Henrí Moissan, developed the electric arc furnace, capable of generating 6,332 degrees Fahrenheit.

Moissan, who was to receive a Nobel Prize in chemistry in 1906 in part for his creation and use of the electric furnace, was able to combine many metals with carbon, including tungsten. The result of his carbon and tungsten mixture was an extremely hard metal, tungsten carbide. Amazingly hard but far too brittle for practical use, the metal had to be made more cohesive. The chemist P. Williams, working in Moissan's laboratory in 1898, carried the process a step further by creating tungsten carbide powder from tungstic oxide, carbon, and iron. Using a trio of chemical steps, the tungsten carbide powder was separated from the mix-

ture by Williams, who studied its properties and found it to be only slightly less hard than diamond.

An early application of tungsten occurred in 1909 when an engineer, William D. Coolidge, developed a process for producing ductile tungsten carbide wire, creating a truly practical incandescent lamp filament.

Sintering tungsten carbide (reducing it to a microscopically fine powder) and mixing it with other powdered metals such as cobalt, nickel, or iron result in a superhard metal that is no longer brittle. This 1914 discovery by a German researcher, Karl Schröter, in Berlin led to the earliest use of sintered tungsten carbide as a tool metal. By 1928, a German firm was exporting tungsten carbide tools.

Tungsten carbide's hardness made it the logical choice for drilling tools, but its hardness and resistance to temperature made cutting tips of the material difficult to weld to rock drill steels, as the rods are known. During the 1940s, the Swedish rock drill machine manufacturer Atlas Diesel and the incandescent light bulb manufacturer Luma took on this challenge and found ways of successfully joining the cutting tips to drill tools. This work was further perfected when Atlas cooperated with another Swedish firm, Sandvik. By 1947, these tools were being marketed globally.

Hard enough to drill through rock, these new drills were most valuable for the longevity of the tungsten carbide drill tip, since they lasted longer, saving the time necessary to replace them. Tungsten carbide–tipped drill rods have become so refined that they last 30 times longer than all-steel drill rods.

Tungsten carbide tips or inserts have become the tool of choice for cutting rock, and barring a stunning new development in materials technology, they will retain that role. With the introduction of efficient tunnel-boring machines in the 1950s, tungsten carbide inserts found a new use. The boring machines, which contain an array of rock cutters on their circular faces, must also be capable of tunneling rapidly and continuously to hold down costs. Tungsten carbide provides a material that reduces the necessity of shutting down the boring machine to replace a dulled cutter.

See also BRUNTON TUNNELING MACHINE; ROCK DRILL; TUNNEL-BORING MACHINE.

tunnel-boring machine *Mechanical mole* The dream of mining engineers for centuries has been development of machines powerful enough to dig large-bore tunnels rapidly and safely through solid rock or soft earth. Although tunneling shields protected miners from cave-ins as they shoveled out soft earth and crude rock-tunneling machines were developed, these 19th-century innovations left much to be desired.

Although a rudimentary soft earth tunnel-boring machine made its appearance soon after 1900, the excavation demands of hard rock tunneling were more severe, and viable full face, hard rock tunnel-boring machines did not appear until 1954 in the United States.

An early soft earth boring machine was the Thomson excavator, which used a dredging ladder that operated as a giant chainsaw equipped with buckets to claw and remove earth. The dredging ladder was operated by an electric motor and fit inside a tunneling shield to protect the miners from cave-ins. Developed in 1898 by Thomas Thomson, the machine saw limited use on one Central London Railway tunnel.

A better version of the tunnel-boring machine was developed by a British railway engineer, John Price (1846–1913), who originally proposed his rotary tunneling machine in 1896. This was the first true soft earth tunnel-boring machine in concept in that it was a self-contained tunneling shield equipped with a revolving cutting head. At the rear of the machine were 10 hydraulic jacks designed to shove the machine forward as it advanced. By 1906, Price's machine had been refined and was used in the excavation of London's Rotherhithe Tunnel.

Its steel shield protected workers from a soft earth collapse and allowed them to reinforce the tunnel behind the machine. The circular steel cutting head rotated and allowed earth to be channeled out the rear of the machine. The boring machine's design also allowed carts to move close behind, making the mucking operation more efficient. Differentiating the thrust of the hydraulic jacks allowed the machine to be steered.

The real triumph was that Price's device allowed miners to drive up to 180 feet each week in the soft clay. This innovation was immediately noticed by other tunnel builders, who also realized the machine required only 10 workers compared to the 14 needed on a conventional tunneling shield. Known as a rotary tunneling machine, Price's invention proved itself in soft ground and was manufactured in large numbers for use around the world.

Modern soft earth tunnel-boring machines are good choices for long-distance, large-diameter tunnels, although they may not be economical for smaller tunnels. Modern models designed for wet conditions can be controlled by computers.

James S. Robbins and Associates of Seattle designed the first successful full-face hard rock tunneling machines around 1954 for the Oahe Dam project in South Dakota (1948–1962). The machines were needed to dig six diversion tunnels to allow the Missouri River to flow past the dam during its construction and five additional tunnels to route water into the dam's hydroelectric turbines. The 19-foot-diameter diversion tunnels and the 24-foot power tunnels were such an undertaking that the U.S. Army Corps of Engineers overseeing the project and its contractors needed something better than drill-blasting. The Robbins firm was asked to develop machines that could bore through the region's shale and move the spoil to the rear for mucking out.

The first tunnel-boring machine for rock was the Model 910, which possessed all the basic features of today's hard rock boring machines. It was a cylindrical shape armed with rotating cutters tipped with tungsten carbide cutting heads. The cutting head ground out the shale to the precise dimensions of the tunnel and was powered by a pair of 200-horsepower engines. A 25-horsepower motor provided hydraulic pressure that moved the machine forward. The advances were stunning: a record 140 feet recorded in a single 24-hour period.

A second tunnel-boring machine developed for the Oahe Dam project was larger and even more powerful, but a design flaw in the cutting head allowed a cavity to develop ahead of the machine, resulting in rock falls. Its record rate was only 93 feet in a 24-hour period.

Robbins and other machine makers learned their lesson and began producing improved hard rock tunnel-boring machines. Like soft earth machines, rock-tunneling machines can be built to endure wet ground. Unfortunately, each tunnel-boring machine is capable of boring a tunnel of only one diameter, meaning that some machines are only used once. However, as more tunnel-boring machines have been produced and then made surplus, machines are now being reconditioned and leased to projects for which they are suited.

Tunnel-boring machines have grown in size and complexity during the past 100 years. Modern machines are equipped with control rooms where operators manage the boring process while computerized systems keep the boring machine on course. Tunnel-boring machines have become so large that they must usually be transported to a work site in pieces and then assembled underground. Their removal is a reversal of this process.

As tunnel-boring machines grew larger, they have also grown smaller. Miniature versions of tunnel-boring machines were developed during the mid-1970s for the purpose of digging smaller tunnels for sewer and electrical lines. This form of tunneling, known as microtunneling, has become a major industry around the world. Miniaturized tunnel-boring machines can be operated from aboveground by remote control. This method eliminates cut-and-cover techniques that disrupt traffic and sometimes sever underground utility or telecommunication lines.

See also EARTH PRESSURE BALANCE MACHINE; SLURRY SHIELD MACHINE.

tunneling shield A device patented in 1818 by a French expatriate inventor, Marc Brunel, provided a structural shield around miners as they excavated their way through soft earth. His shield was fabricated of iron and resembled a cell of rigid workspaces in which men excavated the ground ahead. The shield was essential for subaqueous tunneling, an endeavor that had been virtually impossible as a result of the dangers of flooding and collapse. Using two versions of his shield, Brunel excavated the world's first subaqueous tunnel beneath London's Thames River between 1825 and 1843.

An English engineer, Peter W. Barlow, designed a far narrower and lighter circular shield during the 1860s. The shield gave miners additional protection against cave-ins and also protected workers who were assembling cast iron rings with which to support the tunnel. Like Brunel's shield, Barlow's original design was moved forward by screw jacks.

Eventually, compressed air shield tunneling, which further prevented uncontrollable infiltration of water into subaqueous tunnels, came into use. James Henry Greathead, a protégé of Barlow's, is credited with creating the modern tunnel shield and coupling its use with compressed air. The Greathead shield was used with virtually no changes through much of the 20th century.

See also BRUNEL, MARC; GREATHEAD, JAMES HENRY; SUBAQUEOUS TUNNEL; THAMES TUNNEL.

tunnel-lining The excavation of a tunnel is far from the last step in its construction: Making the tunnel resistant to water infiltration and reinforcing the soft ground or rock through which it travels are essential.

Tunnel-lining can be accomplished in a variety of ways, including installation of segmental sections of cast iron, steel (rarely), or concrete to make a series of interconnected rings. Shotcrete, or sprayed concrete, is another method of tunnel-lining that can be applied very quickly after excavation.

Under certain conditions, several types of lining might be combined, particularly if a segment of tunnel is exceedingly weak or plastic in behavior. An iron lining may be employed and backed by rings composed of steel-reinforced concrete segments.

The lining serves several functions besides giving the tunnel a smooth, finished look. Tunnel-linings, usually of concrete, are used in water tunnels to protect the actual excavation from erosion.

Tunnel-linings can be made extremely watertight by the application of cementitious grout that is pumped under pressure between the lining and the tunnel wall. This fills voids to keep the pressure on the lining uniform and to reduce water seepage.

The use of steel-reinforced concrete segments that form a ring has increased since the 1950s. Such linings offer great strength as well as a smooth, finished look. During the 19th century, tunnel-linings were traditionally of mortar and brick. Sometimes curved iron plates were used as a rapid means of lining a tunnel.

Iron rings came into use during the 19th century and were an integral part of the Greathead shield method of tunnel construction, in which lining segments were erected behind the shield with each incremental advance.

Before the introduction of segmental concrete linings that are bolted together, concrete was poured into forms. This was the way the 36-inch concrete linings of the four Hoover Dam diversion tunnels were built during the 1930s. Injection grouting pumped between the lining and the tunnel wall followed this concrete work.

See also EAST RIVER GAS COMPANY TUNNEL; EUROTUNNEL; HOOVER DAM DIVERSION TUNNELS.

tunnel warfare *Underground assaults on enemy fortifications* It was only natural that military forces would adapt tunneling technology to warfare, particularly in protracted conflicts with an enemy

Sandhogs are encircled by the iron lining rings in a segment of the Lincoln Tunnel beneath New York's Hudson River during the late 1930s. The iron segments are bolted together until they form a supporting lining of immense strength, one of the more expensive methods of protecting a tunnel against collapse and leakage. Grout pumped behind the iron rings ensures a solid, watertight fit. *(National Archives and Records Administration)*

based in a fixed fortification. Since frontal attacks against the walls of fortified cities were often costly if not impossible, it occurred to commanders that attacking in a subterranean fashion was the next logical step.

Early military tunnels were driven beneath enemy fortifications and reinforced by timber bracing. When the tunnel made its way beneath a defensive line or inside a fort, the timbers were intentionally set ablaze to cause the mine to collapse. Fire either brought a section of the fort's wall down or created a breach in the ground through which attacking soldiers emerged. This may be one of the first examples of crude "stealth" warfare. Its use is recorded by the Assyrians in 880 B.C. and was used

by Caesar during his siege of the French city of Marseilles in 49 B.C.

Tunneling for the purpose of collapsing enemy structures was such a common practice through the centuries that the military technique added the word *undermine* to the language, although the term is now used mostly in the context of eroding another person's or organization's political or economic influence.

The earliest known use of explosives for the destruction of fortifications in conjunction with tunneling occurred in 1403, when Florentine troops detonated a black powder explosive beneath the walls of Pisa during a war between the two city-states. Leonardo da Vinci, one of the world's most famous combat engineers, helped to perfect this technique in the 1500s.

In World War I, both German and Allied forces conducted tunnel warfare operations against the other's trenches. The British, employing engineering troops composed of skilled coal miners, dug massive tunnels beneath the elaborate trench systems of the German Imperial Army on the Western Front. The British hoped that by detonating charges beneath German trenches they could generate a breakthrough. On July 16, 1916, British military miners detonated 200,000 pounds of modern high explosives beneath a line of German fortifications before an attack.

The attack, although not successful, did create a massive eruption of earth and added to the anxieties of the surviving German troops, who had to wonder what was going on beneath their feet.

Tunnels have also been considered as a conduit for attacking forces. Since the end of the Korean War, several massive tunneling operations leading from North Korea to South Korea beneath the Demilitarized Zone patrolled by both nations have been uncovered. On April 22, 1997, Peruvian antiterrorist forces who sprang from a tunnel driven beneath the Japanese ambassador's residence freed 71 hostages. The 14 Tupac Amaru Revolutionary Movement terrorists were completely surprised and all were killed.

Tunnels graduated from earthen bomb canisters in war to means of gathering military intelligence. In 1952, during the chilliest portion of the Cold War, the CIA dug a tunnel from West Berlin to East Berlin, where it tapped into subterranean East German and Soviet telephone cables. Unknown to the Americans, the Soviet Union was aware of the tunnel all along because it was informed of the plan by nothing other than a "mole," or a double agent who "tunneled" into his own country's intelligence service to aid the enemy. The mole was the British agent George Blake, who was converted to Soviet sympathies during his Korean War captivity. The Soviets, eager to protect their mole, decided that closing the tunnel would reveal Blake as the source of the tip. The Soviets later claimed they knowingly allowed it to operate for five years before closing it in 1957.

As does submarine warfare, subterranean warfare has its own assortment of countermeasures. Troops concerned they might be the targets of tunnel warfare would place their ears to the earth to listen for sounds of digging. The U.S. Marines in the Vietnam War at the besieged base of Khe Sanh in 1968 used stethoscopes to listen for North Vietnamese Army miners. During protracted siege warfare, the defenders of a fortification might dig their own tunnel toward a suspected enemy tunnel to intercept it or blow it up. These countermines also help defenders hear what was going on beneath the earth around their positions.

During the Vietnam War, American and South Vietnamese troops contended with a vast system of tunnels dug by the Vietcong. This tunnel system held supplies, field hospitals, and kitchens. Americans known as tunnel rats entered these booby-trapped and pitch black tunnel systems to locate supplies and to plant explosives for their demolition.

Tunnels in war can also serve as subterranean forts. Between the two world wars, France designed and partially built a massive line of strong points between itself and Germany. The system of forts, known as the Maginot Line, had complexes of tunnels and fortified fighting positions for housing troops, munitions, and supplies. During World War II, Switzerland also prepared massive underground complexes in the Alps, known as the National Redoubt, where it intended to conceal troops for a protracted fight against Nazi Germany in case of invasion.

See also MAGINOT LINE; NATIONAL REDOUBT: SWITZERLAND.

turner   *See* HENRY, JOHN.

Tweed, William Marcy ("Boss Tweed") (1823– 1878) *Corrupt Democratic boss of New York City* William Marcy Tweed, commonly known as Boss Tweed, rose to power in New York City politics during the late 1850s and by 1861 had what would be a decade-long stranglehold over public policy and spending. During his reign over the Democratic Party organization known as Tammany Hall, Tweed manipulated contracts and siphoned off $40 million to $200 million in New York City funds. His own personal take totaled at least $2.5 million.

Who says "public service" does not pay?

His rise to nearly undisputed political power in New York occurred at a time of two important transportation projects: the Brooklyn Bridge and Alfred Ely Beach's privately financed pneumatic subway tunnel.

The Brooklyn Bridge, Tweed would assist in his own greasy way. Unfortunately for Beach's marvelous private subway effort to save New York from choking on its surface traffic, Tweed would deem it unprofitable for him personally and would use his political clout to kill it.

Helping to breathe life into one project while strangling another was no problem for the former bookkeeper and volunteer fire company foreman. Tweed had so completely seized control of New York's political innards that nothing happened in the city unless it had his blessing, which was always available for a price.

Tweed's name is as synonymous with political corruption as the name of Benedict Arnold is with treason. So great was Tweed's power that even honest politicians and businessmen were forced to pay homage and no small amount of cash to Tweed in order to conduct political and fiscal business in the city.

The Brooklyn Bridge's construction was allowed to proceed because Tweed and his cronies were given what amounted to a controlling interest in the project with a gift of stock. Fortunately for the Brooklyn Bridge, Tweed's political wings were clipped early in its construction, but not until he had spent two years (1869–1871) as one of six members of the New York Bridge Company overseeing the work.

Tweed's plan, which nearly resulted in his control of the Brooklyn Bridge, revolved around a construction company owner, William Kingsley, and a New York state senator, Henry Cruse Murphy. Kingsley, it is said, convinced Murphy that a Brooklyn Bridge was possible and necessary during a marathon discussion in the home of Murphy in 1866. Both became devoted to seeing the bridge completed and both undoubtedly realized that Tweed's approval had to be obtained.

Here the story gets a bit murky. Tweed later claimed that he and his cronies received as much as $65,000 from Murphy and Kingsley as well as stock that would give them control of the bridge. Both Murphy and Kingsley were highly respected men; their transgression at the time of the alleged payoff reflected the influence-peddling that constituted the real-world cost of doing business in New York.

Tweed provided this testimony after fleeing to Spain in 1875 and being returned a year later to face trial for his many misdeeds. Tweed claimed the deal was struck when the Bridge Company needed funding from New York. Murphy agreed to pay $55,000 or $65,000 to Tweed (Tweed could not remember the exact amount) to purchase the desperately needed votes of New York City aldermen. Tweed was to be the bagman for the caper, while undoubtedly taking his own cut after paying off the requisite politicians.

Tweed also claimed that Kingsley made the cash delivery in a carpetbag and revealed that Kingsley gave him enough shares to ensure that Tweed and his cronies controlling interest in the bridge; Kingsley paid 80 percent of the cost of the stock.

Kingsley and Murphy vociferously denied Tweed's allegations, although the historian David McCullough in *The Great Bridge* offers that the dignified and capable Murphy may have realized that clasping the cloven hoof of Tweed was the only way to get the bridge built. Kingsley, knowing New York politics intimately, was probably not averse to a kickback here and there to get a project moving. Although questionable disbursements were made for materials and services during Tweed's tenure on the Bridge Company board, his mounting legal woes by 1873 ended most of his involvement with the bridge.

Tweed, who fled to Spain in 1875 to avoid prosecution for corruption, was recognized a year later and arrested by a passer-by who identified the fugitive from a caricature of Tweed in one of Thomas Nast's brilliant political cartoons in *Harper's Weekly*. Tweed had always despised the caricatures and undoubtedly hated them even more after his arrest.

Although Tweed was unable to profit as fully as he wanted from construction of the Brooklyn Bridge—completed eight years after his conviction—he did manage to choke the life out of Alfred Ely Beach's pneumatic subway.

Beach, the publisher of *Scientific American*, was also a patent attorney and the inventor of several mechanical creations. Honest, intellectual, and disgusted with crooked politics, Beach knew Boss Tweed would never allow him to build an underground subway. Beach craftily (and falsely) sought legislative permission to build a pair of 4 1/2-foot tubes for the purpose of pneumatic "mail" delivery beneath the streets of New York. Beach then sought a last-minute amendment to allow him to build a single larger tube, slipping the scheme past Tweed and his "Black Cavalry" of legislators, who voted any way Tweed dictated.

Beach knew he would have to build his pneumatic subway beneath New York with great secrecy.

In this way, the gutsy but bookish Beach intended to avoid greasing Tweed's palm, which was already morally calloused from taking payoffs. In 1869, Beach rented an empty storefront across from New York's City Hall and began surreptitiously digging the tunnel that would house his air-powered subway car. In February 1870, Beach unveiled his 312-foot subway line to a grateful public accustomed to risking their lives to dodge horse and wagon traffic on the streets above.

Beach's subway, though successfully completed and fully operable, would be suffocated politically by Tweed, who saw personal gain in the construction of elevated railways. Beach's delightful plan would have provided an air-powered car for a quiet ride beneath the cacophonic and garbage-strewn streets of 19th-century New York.

Tweed, displeased with Beach's little subway, put the kiss of death on the enterprise by pulling strings to block a legislative request by Beach to expand the line. For two years Beach's legislation requests were thwarted. By 1873, when revelations about Tweed's misdeeds were wrenching power out of Tweed's hands, Beach was able to obtain legislative permission to expand his subway. But by then it was too late: A financial crisis evaporated the financial support Beach hoped to garner from investors. Tweed won, and New York's bedraggled commuters lost.

Tweed died in prison on April 12, 1878, at the age of 55. Beach's subway became a forgotten oddity, one of the last victims of Boss Tweed's corruption. The Brooklyn Bridge became one of America's most successful infrastructure projects after almost becoming the property of one of the nation's most crooked political manipulators.

See also BEACH, ALFRED ELY; BROOKLYN BRIDGE.

ventilation *Giving tunnels a breath of fresh air* So crucial is proper ventilation to the safety of workers that the Mine Safety Health Administration (MSHA) regulations regarding coal mines state, "No aspect of safety in underground coal mining is more fundamental than proper ventilation." Ventilation systems in mines, transportation tunnels, and all underground excavation operations are expressly required and strictly regulated by MSHA and by the Occupational Safety and Health Administration (OSHA).

Not only does proper ventilation play a role in providing fresh air to those who work underground, but it also reduces the presence of flammable gases such as methane or explosive particles such as coal dust. Ventilation is also a primary means of reducing unbearable temperatures caused by geothermal heating that occurs as workers dig deeper into the ground.

Amid all the other hazards of mining and tunneling is the life-and-death issue of air quality. Not only can tunnels be excruciatingly hot and dangerously dusty, but the exertions of workers or the operation of machinery powered by internal combustion engines can deplete the air of oxygen. If this were not bad enough, gases lurking for millions of years inside the buried rocks of tunnels are ready to poison those who set them free.

These gases can range from gas carbon dioxide to flammable methane. The odorous and caustic gas hydrogen sulfide is also lurking below ground. Although some gases may not be poisonous enough to kill with their toxicity, their potential to dis-place oxygen from the surrounding air makes them deadly nonetheless. One of the deadliest is carbon monoxide, which can kill within minutes in small concentrations.

In recent years the growing awareness of the deadly and debilitating disease silicosis caused by silica dust has prompted more concern over proper ventilation for rock drillers and others who generate large amounts of dust underground.

The operations of coal mines can produce massive quantities of coal dust that if mixed with the proper concentration of air can be devastatingly explosive. Ventilation is a fundamental way of preventing coal dust fires and explosions, which have killed hundreds of coal miners in some of America's worst workplace disasters.

Balanced against these explosive gas and dangerous dust hazards in tunnels is a most prosaic gas known as dead air, or air whose oxygen has been depleted by the process of natural oxidation within rock or by the respiratory action of workers. Ventilation is essential in keeping oxygenated air circulating in every section of an excavation.

The most efficient and safe means of eliminating the hazards of toxic or flammable gases and dust is the utilization of a ventilation system capable of pumping sufficient air to the working face of a tunnel while removing vitiated or exhausted air and suspended dust to the outside. This can be done by fans circulating air into the tunnel through ducting or shafts leading to the working area. In those rare instances in which twin tunnels are driven side by side, the bores can serve as a two-way ventilation

337

duct, one carrying air to the heading of the tunnel and another drawing it away.

Although internal combustion engines are used during tunneling and mining, the use of a tunnel for vehicular traffic presents a situation in which massive amounts of carbon monoxide and other harmful gases are expelled while great amounts of oxygen are consumed. The first long subaqueous vehicular tunnel in the world was New York's Holland Tunnel, which presented designers with air quality issues that had not been dealt with previously.

The answer was to construct ventilation buildings at each end of the tunnel and to force air through a duct built into the crown of the tunnel with 84 huge fans. Half the fans blew the air through the duct from one shore while the other half pulled the air to the far shore.

Adequate ventilation is crucial to safety in case of fire because more deadly gases are expelled into an enclosed tunnel along with a massive volume of heat. Strong ventilation was essential in preventing deaths in the May 13, 1949, fire in which a truckload of carbon disulfide burned in the Holland Tunnel, causing $43 million in damage. The November 18, 1996, Channel Tunnel fire turned into a massive underground conflagration, although a superb ventilation system prevented any deaths.

See also AFTERDAMP; CARBON MONOXIDE ASPHYXIATION; FIREDAMP; GEOTHERMAL GRADIENT.

Verrazano Narrows Bridge (built 1959–1964) The bridge between Staten Island and Long Island at the Verrazano Narrows Strait is monumental in every sense of the word. The longest suspension bridge at the time of its 1964 completion, the Verrazano Narrows Bridge also acquired the distinction of becoming the architectural symbol first encountered by maritime traffic entering New York Harbor. Its creative father was an immigrant Swiss engineer, Othmar Hermann Ammann, and the bridge stands as the crowning achievement of Ammann's long career as a bridge builder.

The mile-wide strait was named for the 16th-century Italian explorer Giovanni da Verrazano, who sailed into New York Bay in 1524 on behalf of France. Within 350 years of Verrazano's discovery, New York and its boroughs constituted the most dynamic city in the world, albeit a city of islands that had to be linked by something better than ferry service. One ferryboat operator who made the run between Staten Island and Brooklyn was Cornelius

Vanderbilt, who parlayed his ferry service into a steamship and railroad empire. After the introduction of the automobile in the early 1900s, bridges and tunnels would be much preferred by a nation of drivers.

In 1888, the Baltimore & Ohio Railroad proposed tunneling beneath the Narrows. Another tunnel plan for the Narrows was backed by the New York mayor, John F. Hylan, in 1923; Hylan obtained $500,000 in public funding to finance the partial digging of two access shafts. In the face of rising costs, this plan also evaporated. The New York City Board of Transportation recommended two additional tunnel proposals—in 1929 and 1937—but nothing came of those suggestions.

The initial idea to bridge the Narrows first emerged in detailed form from a bridge designer, David B. Steinman, who, in 1926 sought legislative approval to construct a privately owned toll bridge across the Narrows. The New York congressman Fiorello H. La Guardia successfully opposed the measure, believing such a bridge should not enrich private investors.

During World War II, La Guardia (by then New York City's mayor) began planning the construction of a bridge across the Narrows, but the idea gained little momentum. At almost the same time, one of New York's most famed figures, the Triborough Bridge and Tunnel Authority (TBTA) chairman, Robert M. Moses, began forming his own plan to bridge the Narrows. But wartime demands for materials and money stymied even the redoubtable Moses, at least for a while.

Moses, who was lord of interisland transportation projects, preferred bridges to tunnels. In 1949, he obtained permission from the U.S. Army Corps of Engineers to build a bridge across the Narrows on the condition that the suspended span be tall enough to allow passage of the largest military and civilian ships. To comply, the Verrazano Narrows Bridge would provide clearance of 228 feet.

Still, no action was taken to construct a bridge between Brooklyn and Staten Island until 1954, when the TBTA and the Port Authority of New York issued the Joint Study of Arterial Facilities calling for the construction of the bridge. The study favored a Verrazano Narrows bridge linking Moses's previous road projects in Brooklyn to Staten Island. Othmar Hermann Ammann, who by the 1950s had become the grand old man of bridge design in America, was selected by Moses to design the bridge.

A display of physical strength and an exercise in uncluttered, modernistic design, the Verrazano Narrows Bridge crowns the entrance to New York Harbor known as the Narrows. This view reveals the exceptionally strong, two-level deck of this mighty bridge. *(Historic American Engineering Record, National Park Service)*

Much of this is academic since the finances of Moses's Triborough were stretched so thinly with other projects by 1959 that the Authority was unable to foot the bill. Moses struck a deal with the Port Authority of New York and New Jersey (which did have the money) to initiate and fund the project. The Port Authority was eager to boost auto traffic to Staten Island to feed the impoverished toll booths of its little-used Goethals Bridge and the Outerbridge Crossing. Triborough would buy the bridge from the Port Authority when it had the means. Othmar Ammann, designer of the mighty George Washington Bridge, among others, would design the Narrows crossing.

Not all was smooth sailing. For two years Moses and Triborough conducted complex negotiations with the army to provide easements at two 19th-century forts standing at each side of the Narrows where the bridge reached shore: Fort Hamilton in Brooklyn and Fort Wadsworth on Staten Island. Triborough eventually agreed to pay Uncle Sam $24 million, and today both lie in the shadow of the bridge.

To be effective, the bridge needed high-capacity roadways leading to its approaches, something that did not exist during the mid-1950s. This required construction of the Staten Island Expressway and the Seventh Avenue portion of the Brooklyn-Queens Expressway, which is also known to New Yorkers as the Gowanus Expressway. The Gowanus Expressway extension along Seventh Avenue would provide an eastern approach to the bridge, but the cost was high: the displacement of 7,000 Bay Ridge neighborhood residents and the condemnation of 800 buildings. Construction of the approach roadway left physical and emotional scars through the heart of a closely knit community that have yet to heal.

Unfortunately for Moses, his authority was now subject to challenge, in no small part because of the way his projects laid waste to community cohesion to benefit regional transportation needs. By the late 1950s, cracks were starting to form in his power base, creating a situation that posed a threat to his plans for the Verrazano Narrows Bridge. As the routes of the peripheral roadways leading to the proposed bridge were being laid out and plans made to purchase the necessary right-of-way, a New York gubernatorial candidate, Neslon A. Rockefeller, began to oppose Moses's bridge. Opponents to Moses preferred a span between Bayonne, New Jersey, and northern Brooklyn, which would have caused less disruption.

The plan, not surprisingly, was yet another by Steinman. Disliked by Ammann, with whom he had once worked, Steinman was an inveterate self-promoter still seeking the commission to design the bridge. He was touting the new plan as a neighborhood-friendly alternative to the Verrazano Narrows Bridge amid the growing sentiment against Moses. Despite the challenges to his plan, Moses was able to obtain New York City approval of the road project leading to his Verrazano Narrows Bridge, clearing the way.

The bridge would be the longest, heaviest, and most expensive suspension bridge the world had ever seen. It would have to span 6,690 feet of water with the center span between its towers measuring 4,260 feet—60 feet longer than the center span of the Golden Gate Bridge and 460 feet longer than the center span of Steinman's 1957 Mackinac Bridge in Michigan. The total weight of the deck, towers, foundations, anchorages, and suspension cables—the five basic components of any modern suspension bridge—is 1,265,000 tons.

Although only 60 feet longer than the Golden Gate Bridge, Ammann's design with its double decks and heavier construction weighs 70 percent more and averages 37,000 pounds per linear foot. Building the bridge would cost $324 million. The foundations for its two cable suspension towers would be sunk by using unmanned caissons excavated through steel tubes. With its 6,500 tons of reinforcing steel, the deeper Brooklyn pier foundation weighed 243,425 tons. The Staten Island foundation weighed in at 164,675 tons, including 4,700 tons of reinforcing steel embedded in its concrete.

The towers that would hold the suspension cables aloft would climb 693 feet above the average high-water mark of the Narrows. The internal structure of the towers would be a honeycomb of steel boxes. A total of 26,000 tons of steel plate would be required to create the modular segments for each tower, which would contain 10,000 internal steel boxes yielding phenomenal strength. Each tower would have its steel plates bonded by 3 million rivets.

Because the foundation of the Brooklyn anchorage had to be dug deeper than that at Staten Island, it took 207,000 cubic yards of concrete to complete. The relatively "lightweight" anchorage at Staten Island contained a mere 171,000 cubic yards of concrete. The concrete footings for both measured 230 feet by 345 feet. This broad footing helped to distribute the weight of the anchorages over a wider area to minimize the chance of the structures' sinking.

The $57 million cable-spinning effort was begun after the completion of the cable anchorages and the towers. Four cables weighing 9,798 tons apiece would suspend the live and dead loads of the road deck. Each cable would be spun to a 35 1/2-inch diameter and measure 7,205 feet. Tied to steel eyebar links within the anchorages, the cables could sustain far more loading than the bridge would ever experience.

Of the three worker deaths during the construction of the bridge, one occurred when Gerard McKee, 19, fell through a gap where a portion of the chicken wire sidewall had been removed from a catwalk. The handsome, well-liked son of a veteran "boomer" (a bridge construction worker), he had followed his father into the risky trade. McKee slipped off the catwalk and was hanging by one hand when spotted by another worker, Edward Iannelli. Tragically and ironically, the dangerous nature of heavy construction and bridge-building had sealed McKee's fate: Iannelli's left hand had been permanently mauled in a previous bridge construction accident, and Iannelli was unable to drag McKee back to safety. When McKee finally lost his grip, the wind tore the shirt from his back as he plummeted nearly 300 feet to the water below.

This tragedy gained the young man a chapter in Gay Talese's *The Bridge,* perhaps the only book written about bridgeworkers. Poignantly, Talese revealed that McKee's father had been permanently disabled while working on a bridge and deeply regretted introducing his son to the trade. After the young McKee's death and the installation of safety netting, four men were caught by the net, rescued from falling to their own deaths.

From a distance, the Verrazano Narrows Bridge would be slender, graceful, and beautifully proportioned. Up close, the bridge is a brute. Its road deck contains two levels, each carrying six lanes of traffic. Its double deck was truss-reinforced and measured 27 feet deep and 115 feet in width. The lower deck—which was not opened for use on the completion of the bridge—was to be held in reserve until 1975. However, traffic grew so heavy on the bridge that the lower roadway was opened in 1969, six years ahead of schedule.

See also AMMANN, OTHMAR HERMANN; BROOKLYN BRIDGE; LINDENTHAL, GUSTAVE; MOSES, ROBERT M.; ROEBLING, JOHN AUGUSTUS; STEINMAN, DAVID BARNARD; TACOMA NARROWS BRIDGE; TRIBOROUGH BRIDGE AND TUNNEL AUTHORITY.

viaduct Another name for a bridge, the word *viaduct* literally translated from its Latin roots means "to lead the way," as *via* is Latin for "way" and *ducere* is Latin for "to lead." However, the term *viaduct* in recent years has been reserved for a railroad bridge that crosses a river or gorge using a series of relatively short spans supported by arches. The term *bridge* during the 19th century was usually applied to spans designed to support pedestrian or vehicular traffic.

Fine examples of viaducts are Gustave Eiffel's Garabit Viaduct in France (1884) and his Maria-Pia Viaduct in Portugal (1876). The term is seldom used today, as the word *bridge* has come to designate virtually any span across a waterway.

Confusion over the word has always existed, and no less a group of viaduct experts than the Marx Brothers comedy team discussed the issue in the 1929 movie *Cocoanuts*. In the film, Groucho Marx attempts to provide directions to Chico Marx and tells him of a "viaduct over to the mainland." Chico, confused by the word, keeps asking over and over, "Why a duck? Why not a chicken?" An exasperated Groucho finally announces: "Well, I'm sorry the matter ever came up. All I know is that it's a viaduct."

Before the term *viaduct* was used for 19th-century railroad bridges, it was also applied to Roman Empire arch bridges built of stone and concrete. An example of such a bridge is the Alcántara Bridge in Spain, which is usually referred to as a bridge rather than a viaduct.

Adding to the etymological confusion is the fact that some viaducts simultaneously serve as aqueducts, or bridges designed to "lead" or carry water. Many ordinary bridges are also designed to carry water lines beneath their roadways, making them both a viaduct and an aqueduct. The telltale pipes running beside or beneath a bridge's roadway reveal this double use. The opposite almost occurred to the Pont du Gard aqueduct, a Roman arch bridge structure designed to carry water across a deep gorge in France. During a series of internal wars in the 1960s, the Pont du Gard's water trough was converted into a roadway with destructive effects. Since 1743, however, French officials have repaired and maintained the civil engineering treasure.

See also EIFFEL, GUSTAVE; GARABIT VIADUCT; PONT DU GARD VIADUCT.

Walsh, Homan   When the fiery and brilliant American engineer Charles Ellet, Jr., began construction on a suspension bridge across the Niagara River in the winter of 1847, his first task was to connect both shores with a cable. To do this, Ellet combined practicality with showmanship when he sponsored a contest to see who could fly a kite bearing a line over the Niagara Gorge. The reward offered by Ellet was five dollars, a princely sum at the time, especially for a child.

A 10-year-old boy, Homan Walsh, was an avid kite flyer and built a special kite he named "The Union" for the task. He persuaded his family to let him cross to the Canadian side alone since the prevailing winds blew toward America. On his first attempt, Homan lofted his kite and flew it in a strong wind that carried it over the gorge to the American side. The winds did not subside, and the youngster was forced to fly the kite well past dark.

Spectators built bonfires to mark the far shore as Homan's kite stubbornly hung in the air. Finally the kite, shrouded in darkness, descended into the gorge. Heavy ice on the Niagara River stranded the boy on the Canadian side for eight days, prompting a local family to care for him until he could return to the far shore and retrieve his kite.

Undeterred, Walsh repaired his kite and made a second attempt to fly it across the gorge. This time he met with success and his kite string was attached to progressively heavier lines until an iron cable was finally pulled across the 800-foot gorge. Newspaper accounts of his feat, undoubtedly encouraged by Ellet, appeared all over the country, and Walsh became one of America's youngest heroes.

Walsh, who lived into his 90s, remained proud of his accomplishment for the remainder of his life. He remembered the event in great detail in one of the last interviews he gave during the late 1920s.

Ellet would soon be in conflict with the bridge company that financed the project and leave after having constructed little more than a walkway. Walsh, on the other hand, has remained a well-known American hero, surpassing Ellet in terms of fame, especially in the minds of American schoolchildren who have heard his story.

See also ELLET, CHARLES, JR.; NIAGARA GORGE SUSPENSION BRIDGE.

Watkin, Edward (1819–1901) *English Channel Tunnel promoter and one of the first to dig*   While everyone else talked and sketched, England's Sir Edward Watkin was one of the first promoters actually to begin digging a tunnel beneath the English Channel. No starry-eyed fool, Watkin was a politically connected businessman who was chairman of three railroads, in stark contrast to dreamy tunnel promoters like the Frenchman Aimé Thomé de Gamond.

Although he might at first appear to be a cousin to the American robber barons, Watkin was a progressive thinker who believed in humane working conditions and espoused support for a shorter industrial workweek. The author of a book about America, Watkin also entertained friendships with well-known writers of his time.

As a railroad man, Watkin understood the problems of tunnel construction and had an appreciation of the technology needed to overcome those difficulties. He also dreamed of operating a rail line between Dover, England, and Calais, France. So eager was he to do so that in 1880 he became the first Englishman to mount a serious effort to tunnel beneath the Channel.

Even though Watkin may have been far different from the French dreamer Thomé, it was a disciple of Thomé who proselytized Watkin to back a tunnel effort. This disciple was an English mining engineer, William Low, who, began to work with Thomé in 1867 to develop schemes for tunneling beneath the Channel. Low was a legitimate engineering professional and Watkin enthusiastically became one of Thomé's followers. At one point Watkin announced to a group of investors in his project, "The dream of Thomé de Gamond has come true thanks to our friend and colleague, William Low."

Watkin's initial tunnel-digging tool would be the Brunton tunneling machine, but this device yielded poor results. His second choice was the Beaumont tunneling machine, the first practical device for digging tunnels. The Machine, conceived of by Frederick Beaumont in 1875 and refined by Thomas English in 1880, presented Watkin with the invention he needed for boring beneath the Channel. Powered by compressed air, the machine would claw the chalk strata onto a conveyor belt that carried the spoil to the rear of the machine.

As a businessman, Watkin knew he would need contacts on the far side of the Channel and sought to ally his effort with a French organization also promoting an undersea Channel crossing. Loosely translated, the French name for the group was the Company for an Underwater Railroad Entering England and France. The French concern was well organized and had obtained a concession from its government and cautious approval from the British Parliament to tunnel between the two nations. To partner with the French Watkin too would have to obtain concession legislation from his government.

The French effort was exceptionally well funded: 2 million francs in operating capital and the promise of a 99-year Channel tunnel concession from its own government if it obtained an English partner by 1881. Watkin began a two-pronged attack by both digging beneath the Channel and lobbying the powers that were for permission to become that partner.

Watkin sank access shafts near Shakespeare Cliff in 1880–1881 and began making his way to the Channel. When a substantial tunnel was completed, he hosted picnics and tours for the wealthy and the politically influential.

Watkin formed the Submarine Continental Railway Company and raised money through the sale of stock. The tunnel-boring machine was making good progress carving a tunnel toward the Channel. In 1879, the French began a two-year effort to excavate a 282-foot shaft at Sangatte so they could move horizontally into deeply placed impermeable chalk marl to prevent flooding. The going was tough as the French initially blasted their way through the chalk while dealing with extensive flooding.

Buoyed by his successes, Watkin announced in 1881 that a pilot tunnel seven feet in diameter would cross the Channel within five years. His confidence was based in part on the capabilities of the Beaumont tunneling machine. Even the inventors Beaumont and English were involved in the project, overseeing the excavation using their boring machine on the French and English sides. They were contractually obligated to bore two feet per hour through the chalk and were meeting the terms of their contract. It began to look as if it would be possible to grind a passageway beneath the Channel.

While all this tunneling success excited both investors and the public, the power elite of England was appalled at the prospect of their island's being connected to Europe. A notable opponent was Sir Garnet Wolseley, a fighting general of the British Empire who later become supreme commander of Great Britain's military forces. Wolseley darkly warned that a small force could seize the tunnel and allow an invading force into England.

The tunneling success of Watkin and the French so unnerved the English public that in 1882 a petition opposing the tunnel was published in *The Nineteenth Century*, a prestigious journal of the period. The petition was not the work of a fringe element but contained the signatures of 250 prominent citizens, including military men, the English poet laureate Alfred Tennyson, and a Roman Catholic cardinal.

The French, who never seemed to have worries about a tunnel, ridiculed the fears that seemed to infect the English upper class. They also conducted public relations damage control by offering tours of their own excavations. Panicked by English opposition, the French even offered to route a portion of the rail line outside the cliffs so it would be vulnerable to the Royal Navy's guns.

Unfortunately for Watkin's tunneling effort the damage was done. The influential people who so feared a tunnel had the ear of the government. Watkin's tunneling was craftily halted by an order preventing him from drilling beneath land belonging to the Crown. The railroad man persevered for a while in both his digging and his lobbying, knowing full well that the fears over the tunnel were ludicrous. The government stood firm, and on August 12, 1882, Watson shut down his tunnel effort, which had penetrated 6,177 feet.

Across the Channel the French were at first amused and then confused by the controversy. They decided to continue their tunneling effort and by March 18, 1883, had penetrated 5,475 feet into the chalk strata before halting their excavation. By 1886, the French company was defunct, and Watkin's Submarine Continental Railway Company purchased its assets and concession, forming a new Channel Tunnel Company.

Watkin tried in subsequent years to restart his Channel tunnel project but was continually rebuffed by opposition to the plan.

The Channel Tunnel Company would live on into the 20th century and would review subsequent projects. The company would hold a 25 percent stake in a 1957–1963 effort energized by a group of Americans to build a Channel tunnel. Although the effort could conceivably have built a tunnel nearly 30 years before the Eurotunnel that now exists, the project narrowly missed becoming a reality as a result of an unrelated political matter between Great Britain and France.

See also BEAUMONT TUNNELING MACHINE; BRUNTON TUNNELING MACHINE; EUROTUNNEL; THOMÉ DE GAMOND, AIMÉ.

wet-drilling The problem of dust in hard rock tunnels cannot be overstated since rock dust, in particular dust from quartz rock, can lead to the deadly lung disease silicosis. An incredibly simple way to control what for centuries was a curse on miners is to employ a water spray during drilling to turn the dust into harmless mud.

Although modern tunneling and drilling operations in developed nations have stringent ventilation requirements, a means of keeping dust under control has been in the hands of miners for more than 100 years in the form of drills that could spray a mist of water into the drill hole.

Although wet-drilling was often ignored since it was slower than dry-drilling, the injection of water into a drill hole has long been known as a solution to the problem. The rock drill inventor J. G. Lyner's 1897 invention of a drill that could drive vertical dynamite holes into the ceiling or floor of a tunnel was returned to his doorstop by miners, who said it kicked up too much dust. Leyner, wanting to please the miners, installed a water spray system that turned the dust to mud during drilling.

Wet-drilling is also being advised for above-ground work where quartz dust, otherwise known as silica, could be a problem. The Mine Safety Health Administration has warned that older drills lacking the ability to do wet-drilling are being purchased by contractors who perform their work on the surface. It is surmised that many surface contractors and workers mistakenly believe that they are not at risk from silicosis, once commonly known as miner's consumption.

See also HAWK'S NEST TUNNEL.

Wheeling Bridge (built 1847–1849) *America's first great suspension bridge* The suspension span at Wheeling, West Virginia, is not only the oldest wire cable vehicular bridge still in full use anywhere but an engineering landmark leavened with the engineering expertise of one of America's most colorful civil engineers.

The bridge is the brainchild of an iconoclastic American-born engineer, Charles D. Ellet, Jr., who was instrumental in successfully promoting the much distrusted suspension bridge concept in America. Ellet's design was accepted instead of one submitted by his great rival, the German-American engineer John Augustus Roebling, who had proposed a more heavily reinforced design.

Badly damaged by wind-induced movement five years after its completion, the bridge was immediately repaired by Ellet, despite a common misconception that his rival Roebling did the work. It is true that during subsequent work to refurbish the bridge, Roebling's engineer son, Washington Roebling, added elements to the bridge originally proposed by his father, including diagonal cable stays. Hence, Ellet's magnificent bridge over the Ohio river is the sum of the expertise of not only Ellet but also the elder Roebling.

The majestic Wheeling Bridge across the Ohio River is the oldest vehicular suspension bridge still in use in America. Designed by the brilliant but mercurial Charles Ellet, Jr., the bridge was rebuilt by Ellet after a wind-induced collapse of its deck in 1854. *(Historic American Engineering Record, National Parks Service)*

The Wheeling Bridge was spawned from a desire by Wheeling to make itself a hub of transportation and commerce in competition with the city of Pittsburgh 35 miles to the northeast. This move was prompted by the construction of a National Road, the first federal highway effort, begun in 1813 and destined to reach Wheeling in 1818. In an effort to foster trade and business activity the Wheeling and Belmont Bridge Company was established in 1816 to bridge the Ohio River. Wheeling's leaders knew full well that the river would block not only the National Road but also development on the Ohio River's eastern shore.

After nearly 30 years the bridge company had arrived at the point at which it was ready to raise money for a bridge, and two of the contenders were Ellet and Roebling, who submitted proposals in 1845. Roebling's proposal sought to place piers in the river to support a central span of only 600 feet. Ellet provided a less expensive proposal that would place piers adjacent to each shore of the Ohio River to support a suspension span measuring a then-astonishing 1,010 feet.

Ellet proposed a road deck 22 feet wide, and the more conservative Roebling recommended a heavier deck, 26 feet wide. Roebling, deeply respectful of the little understood phenomenon of aerodynamics, also desired heavier catenary suspension cables. Cables stays radiating from the suspension towers to the roadway would also reinforce the suspenders running from the cables to the deck. Because of the difficulty of comparing the two disparate designs, both engineers were asked to submit estimates for bridges of similar dimensions.

Ellet, who was craftier than Roebling in politicking with bridge companies, offered to accept 200 shares in the Wheeling Bridge in lieu of his $5,000 fee. Additionally, his 1842 completion of a wire suspension bridge over the Schuylkill River near Pittsburgh could not have hurt Ellet's cause. Adding to

Ellet's cachet is the general perception, ably promoted by Ellet, that he was America's premier suspension bridge theorist. It was little surprise that the bridge company's directors in a seven-to-one vote selected Ellet's design.

Ellet began construction of the piers adjacent to the banks of the Ohio, using only cofferdams because it was not necessary to descend far underwater to seat the footings of the piers on bedrock. As work progressed, a legal challenge to the right of Wheeling to build the bridge threw its completion into doubt.

The state of Pennsylvania, dependent on river and canal commerce, sought to have Ellet's bridge declared an obstruction to boats traversing the Ohio River. This legal fight ebbed and flowed for years, as the steamboat interests worked the levers behind the scenes to do away with the bridge. It became commonplace for steamboat companies to mount, or have mounted, a legal fight to thwart bridge-building over major rivers to eliminate competition. The usual specious claim was that the piers and road decks of the bridges were navigation hazards.

During one of the many court hearings in the matter, Roebling was asked to testify on behalf of those opposed to the bridge, in the mistaken belief that the engineer harbored a grudge against Ellet as well as the bridge company that had rejected his design. Roebling, dedicated to his principles, took the side of Ellet and the bridge. Roebling reversed the tables on the steamboat interests by calling for elimination of the tall smokestacks as unnecessary hazards to bridges. This legal challenge, which wound its way to the U.S. Supreme Court, was ended temporarily by a rider attached to the U.S. Postal Bill in 1852 establishing the bridge as a postal route, thereby protecting it from demolition.

The elements of the bridge included a road deck that was 97 feet above the river's surface. Cables secured to masonry anchorages suspended the roadway. The eastern suspension tower rose 153.5 feet above the water and the western tower stood 132.75 feet tall. Three pairs of cables, each consisting of 550 strands of wire and measuring four inches, would be suspended over the towers on each side of the bridge deck. Ellet estimated the bridge capable of sustaining a traffic load, or live load, of 593,400 pounds.

The bridge was completed on October 30, 1849, to great fanfare, allowing Wheeling to thumb its nose at the state of Pennsylvania and the steam-boat interests that had so much sway with both the politicians and courts.

With the legal battle temporarily won, the real disaster awaiting the bridge was yet to come, and Ellet himself brought it upon the bridge. Designed to be far lighter than Roebling's proposed bridge, Ellet's bridge lacked the diagonal stay cables that Roebling would have employed. The cables would have run at an angle from the towers to the bridge deck to provide additional stability. Ellet's bridge also lacked any type of truss reinforcement, a stiffening element that would have also added dead weight to the span, above or below the bridge deck.

On May 17, 1854, a strong prevailing wind excited the Wheeling Bridge deck in much the same way that a 1940 gust caused the Tacoma Narrows Bridge to twist and undulate until collapse. The road deck of the Wheeling Bridge, built too light and lacking adequate reinforcement, crashed into the Ohio River. Ellet immediately obtained permission from the bridge company's directors to spend $37,000 rebuilding the damaged deck, an effort that was opposed once again in court by the steamboat interests.

Although an injunction was obtained to prevent the repairs, Ellet and the bridge company ignored the ruling and performed the work nonetheless. During this effort, much of the bridging material was dredged from the river for reuse. In 1860, after Ellet had turned away from bridge construction, the bridge received additional refurbishment in the form of guy wires running from the bridge to the shore to control wind-induced movement, inspired by a study of John Roebling's Niagara Gorge Suspension Bridge, which employed the same stabilizing method.

Roebling's engineer son, Washington Roebling, would be hired in 1871 to design changes in the bridge to strengthen it and to widen it from one lane to two. Deeply involved with construction of the Brooklyn Bridge, Roebling nonetheless took the time to recommend ways of moving the suspension cables farther apart to allow a pair of walkways to run inside the cables. Roebling also recommended another innovation developed by his father, the installation of diagonal cable stays running perpendicular to the vertical suspenders from the towers to control wind-induced movement. Altogether, the series of changes to Ellet's original design are often referred to as the Roeblingization of Ellet's bridge.

The bridge continues to carry auto traffic across the Ohio River and remains the oldest fully operational suspension bridge in the United States. Ellet would unsuccessfully propose another bridge, one across the Potomac, in 1852, but his bridge-building days were over. Ellet subsequently became more famous for his development of a fleet of steam-powered ram boats used during the Civil War and would die as a result of wounds received during his first and last river battle in 1862.

Ellet's bridge, with its repairs and Roeblingization, was selected as a National Engineering Landmark in 1969 by the American Society of Civil Engineers and in 1976 became a National Historic Landmark. In 1941, the bridge was purchased from the bridge company for $2.15 million, and in the following year it was ceded to the state of West Virginia, which continues to oversee its repair and operation.

See also BRIDGE AERODYNAMICS; ELLET, CHARLES, JR.; ROEBLING, JOHN AUGUSTUS; ROEBLING, WASHINGTON AUGUSTUS; TACOMA NARROWS BRIDGE.

White, Josiah *See* Schuylkill Bridge.

width to span ratio The ideal bridge is one wide enough to accommodate ample traffic flow, thin enough to be attractive, and rigid enough to resist its own weight, the weight of traffic, and the destructive effects of wind and vibration. Determination of the proper ratio of the width of a bridge compared to its length is crucial to flexible suspension bridge designs.

Unfortunately, bridge builders must make trade-offs as they design a span, and one of the most important ratios to consider is the width to span ratio. This ratio considers the width of the deck in relationship to its length and must be carefully considered because of the effects of aerodynamic forces.

A bridge deck, because it is a long, flat surface, can serve as an airfoil, much as the wing of an airplane does. A wide, thin horizontal deck that is sufficiently light can be lifted by the wind, buffeted, or excited into lifting and then dropping destructively against its suspension cabling. A deck that is wider imparts greater stiffness, weight, and lateral rigidity to a bridge. A narrower deck is more susceptible to lateral deflection or sideways movement.

The classic example of a bridge that was built too light and too thin to counteract the lifting effect of its deck is the Tacoma Narrows Bridge, which had a width to span ratio of 1/72, meaning the bridge had 72 feet of length for every foot of width. The Golden Gate Bridge, thought to be shockingly narrow when built, had a width to span ratio of 1/47.

The road deck of the Tacoma Narrows Bridge was twisted, undulated, and finally lifted and dropped by a relatively mild wind because the deck lacked adequate stiffening and weight to counteract its narrowness. The lightness of the bridge was amply revealed by its depth to span ratio, which showed how shallow the depth of the bridge deck was in relation to its length. With only one foot of deck depth for every 350 feet of length, the bridge was both dangerously light and remarkably thin. Added to the narrowness of the deck, these factors spelled doom for the bridge, which collapsed in 1940.

See also DEPTH TO SPAN RATIO; TACOMA NARROWS BRIDGE.

Williamsburg Bridge (built 1896–1903) An exercise in heavy construction, the Williamsburg Bridge was built to relieve the congestion on John Roebling's Brooklyn Bridge, which was becoming overwhelmed by rail, carriage, and pedestrian traffic between Manhattan and Brooklyn.

By the early 1890s, after Roebling's span made it apparent that long-span bridges could be built, people were already clamoring for another bridge over the East River. One such proponent of a new bridge was Frederick Uhlmann, who in 1892 obtained a charter from the state of New York to build one upriver from the Brooklyn Bridge.

Uhlmann, a businessman who owned elevated railways in Brooklyn, wanted to expand his service to Manhattan and obtained official permission to build a bridge between the two locations. Though Uhlmann's proposal came to naught, a new bridge was still needed, and the East River Bridge Commission purchased Uhlmann's charter in 1895 for $200,000. The commission named Leffert Lefferts Buck, a highly competent civil engineer and bridge builder, chief engineer on the project. An engineer, O. F. Nichols, would assist him in his design of what would be named the Williamsburg Bridge.

The design by Buck and Nichols could be built for a modest $7 million, an amount far less than the $15 million it cost to build the Brooklyn Bridge, completed 12 years earlier. As a rule, modern projects tend to cost more than older ones because of inflation, but it was hoped that modern materials

Squat and ugly to many, New York City's Williamsburg Bridge has endured as an example of a suspension bridge built far heavier than was necessary. This view shows the bridge under construction shortly after 1900 after workers prepared a catwalk to allow the spinning of the suspension cables. *(Author's Collection)*

and techniques would reduce the cost of the new bridge. Politicians and the public were still gnashing their teeth over the spiraling cost and the 14-year construction time of the Brooklyn Bridge.

Buck would also gain economy by the nature of the design. There would be no monolithic stone work on the opposing towers of the Williamsburg Bridge; nor would it utilize stone and masonry work on the piers supporting the two side spans. Skeletal steelwork would be employed for all piers and towers above the water.

Whereas suspension spans traditionally employ load-bearing cables that curve downward from the onshore anchorages over the side spans for side span deck support, the Williamsburg Bridge would instead cut cabling costs by employing a massive deck truss, a stiffening truss, on the underside of the side spans. Cables stretched in a straight line from the towers to the onshore anchorages would replace the gracefully curved suspension cables over the side spans that would ordinarily support those spans with connecting suspender cables.

Although competent as a bridge designer, Buck was essentially a railroad bridge engineer whose career had been mainly committed to spans that could resist massive loading and the jarring vibration of engines and heavy rail cars. His design of the Williamsburg Bridge seemed to be a hodgepodge of concepts. It was to carry both rail and horse-drawn wagon traffic, as did the Brooklyn Bridge, but it would resemble a huge truss-reinforced bridge with a suspension cable system added on.

One way Buck intended to keep the costs of the bridge down was to use copious amounts of steel on the superstructure of the span. By the latter half of the 1800s the structural material of choice was steel, as opposed to masonry and iron, two traditional building materials that reigned until the mid-1800s. Buck's design would incorporate steel by the thousands of tons. Unfortunately for Buck's creation, the slender, graceful lines of the nearby Brooklyn Bridge were akin to a slender and strikingly beautiful woman compared with her homely, rawboned younger sister, the Williamsburg Bridge.

Steel, seemingly heavier than stone or masonry, is actually a far lighter building material since it can be fabricated into self-supporting and open shapes such as trusses. The use of steel for the suspension towers meant they could be more slender and therefore have a smaller footprint. The benefit of this was that smaller foundations were needed. Stone and masonry towers, such as those used on the Brooklyn Bridge, would have required foundations with twice the area. Foundation work is devastatingly expensive, and abbreviating it in any way has tremendous cost benefits.

Before construction even began on the bridge, critics who saw drawings of it published in 1896 identified it as an ugly duckling of a structure. If the Brooklyn Bridge was a successful mixture of aesthetics and beauty, Buck's design was merely a late 19th century expression of the heavy hand of industrialism. It would have the distinction of being a paltry four feet longer than the Brooklyn Bridge, but its chief engineer promised it would take no more than

four years to build, as opposed to the 14 years the Brooklyn Bridge had required. Although the cost of the Williamsburg Bridge also spiraled out of control, it was completed relatively quickly, requiring seven years of work. Buck had been off by only three years.

As the bridge began to take shape, it attracted more criticism since it never seemed to be finished to the eye of those passing by. This was more of an issue to the 19th century observer than to present-day passers-by, who have grown accustomed to the industrial minimalism of modern architecture. Large bridges, which were imposing and solid structures made of stone before the 19th century, had been slowly giving way to iron, and later steel, structures. The Williamsburg Bridge seemed to flaunt this difference with temporary-looking pylons that appeared from a distance to be made of nothing more substantial than gray wicker.

Exacerbating this mechanistic and unfinished look were the trusses that added stiffness and weight to the central span and the two side spans. The upper part of the 1,600-foot center span was reinforced with a through-truss design. This type of truss provided a reinforced tunnel of steel composed of truss walls and roof running the length of the bridge's centerline. A deck truss that was attached below the roadway for stiffness and support reinforced both 300-foot side spans.

Suspension bridges are often judged by their girder to length or width to span ratio, which is essentially the thickness of their road decks and accompanying reinforcement compared to their length. Whereas the Golden Gate Bridge had a slender design that allowed a stunning 1 to 164 width-to-span ratio, the Williamsburg Bridge had a plodding 1 to 40 ratio meaning that for every 40 feet of length, the bridge possessed a foot of reinforcing truss.

Its four suspension cables were mounted on an equal number of saddles and were paired off to curve down both sides of the bridge on the outside of the through-truss. These cables contained 17,500 miles of wire and weighed a total of 4,344 tons. Each cable measured 18 3/4 inches in diameter and 2,985 feet in length. The decking, towers and piers were all steel.

The bridge had the appearance of a large piece of industrial machinery but despite its massive appearance, it was not as strong as it appeared. Designed for rail cars and engines of the late 1800s, it was subjected to greater-than-expected stress by far heavier trains when it entered service in 1905. This caused its side spans, which were supported by steel piers alone and not the cables, to sag noticeably. Pairs of additional steel piers were placed under each side span and heavier steel was added to the rail deck following a 1911 report indicating the bridge had to adapt to heavier loading.

During the 1920s, as automobiles filled the roadways, the bridge was altered so it would contain eight lanes of vehicular traffic. These lanes have an unusual configuration since two lanes run down either side of the bridge not encased in the through-truss on an open air roadway. An additional two lanes run along both sides of the subway rail lines inside the through-truss.

The only enemy the steel bridge had was corrosion, as pointed out by the New York bridge commissioner Gustave Lindenthal during the official opening of the span on December 19, 1903. Said Lindenthal of this ungainly bridge, "This colossal structure, if protected against corrosion, its only deadly enemy, will stand hundreds of years in unimpaired strength."

Unfortunately for the Williamsburg Bridge, this was not to be the case. By the 1960s New York, as had virtually every other government entity in the nation, had been generally ignoring the maintenance of its bridges for decades. The Williamsburg Bridge was becoming a rusting hulk by the 1960s, and it

was not uncommon for pedestrians on or below the bridge to be pelted by falling rust. New York's mid-1970s financial crisis and the attendant mismanagement of the city added to its neglect. By 1988, the Williamsburg Bridge was a structure in crisis. Corrosion was discovered in virtually all of its major components, including the suspension cabling, the towers, and its deck trusses. The bridge was closed for two months that year to allow a detailed engineering assessment.

It was then determined that the bridge needed either complete replacement or a massive overhaul to repair corrosion damage. After determining that a replacement bridge could cost as much as three quarters of a billion dollars, city officials and engineering consultants it was decided to spend $250 million to renovate and repair the bridge during a 14-year period. While some vehicular lanes were closed during this work, traffic was routed to the lanes that remained to prevent the bridge from being completely closed. The repairs, originally estimated at $250 million by the New York City Department of Transportation, rose to $600 million by the end of the 1990s. The renovation and repair work on the bridge were expected to continue through 2002.

See also BUCK, LEFFERT LEFFERTS; DEPTH TO SPAN RATIO; GOLDEN GATE BRIDGE; LINDENTHAL, GUSTAVE.

working face   *See* heading.

Yerba Buena Island Tunnel *See* San Francisco–
Oakland Bay Bridge.

APPENDIXES

Appendix I
World's Longest Bridges and Tunnels by Type

Appendix II
Bridge and Tunnel Websites

APPENDIX I

World's Longest Bridges and Tunnels
by Type

(Lengths are in Feet)

Suspension Bridges

Name	Location	Completed	Length
Akashi Kaikyo	Kobe-Naruto, Japan	1998	6,532
Great Belt	Halsskov-Sprogoe, Denmark	1998	5,328
Humber Hull	Humber, England	1981	4,625
Jiangyin	Jiangsu Province, China	1998	4,543
Tsing Ma	Hong Kong, China	1997	4,517
Verrazano Narrows	New York, New York	1964	4,258
Golden Gate	San Francisco, California	1937	4,199
Höga Kusten	Veda, Sweden	1997	3,969
Mackinac	Mackinaw City, Michigan	1957	3,799
Minami Bisan-Seto	Kojima-Sakaide, Japan	1988	3,608
Fatih Sultan Mehmet	Istanbul, Turkey	1988	3,576
Bosporus I	Istanbul, Turkey	1973	3,523
George Washington	New York, New York	1931	3,500
Kurushima-3	Onomichi-Imabari, Japan	1999	3,379
Kurushima-2	Onomichi-Imabari, Japan	1999	3,346
Ponte 25 de Abril	Tejo, Lisbon, Portugal	1966	3,323
Forth Road	Edinburgh, Great Britain	1964	3,300
Kita Bisan-Seto	Kojima-Sakaide, Japan	1988	3,248
Severn	Bristol, Great Britain	1966	3,241
Shimotsui-Seto	Kojima-Sakaide, Japan	1988	3,083

Cable-Stayed Bridges

Name	Location	Completed	Length
Tatara	Onomichi-Imabari, Japan	1999	2,919
Pont de Normandie	Le Havre, France	1995	2,808
Oingzhou Minjiang	Fuzhou, China	1996	1,984
Yangpu	Shanghai, China	1993	1,975
Xupu	Shanghai, China	1997	1,935
Meiko-Chuo	Nagoya, Japan	1997	1,935
Skarnsundet	Trondheim Fjord, Norway	1991	1,738
Tsurumi Tsubasa	Yokohama, Japan	1994	1,673
Ikuchi	Onomichi-Imabari, Japan	1991	1,607
Öresund	Copenhagen, Denmark-Malmö, Sweden	2000	1,607
Higashi-Kobe	Kobe, Japan	1992	1,591
Ting Kau	Hong Kong, China	1997	1,558
Seohae	Dangjin, South Korea	1997	1,541
Annacis Island	Vancouver, Canada	1986	1,525
Yokohama Bay	Yokohama, Japan	1989	1,509
Second Hoogly	Calcutta, India	1992	1,499
Second Severn	Bristol, Great Britain	1996	1,496
Rama IX	Bangkok, Thailand	1987	1,476
Queen Elizabeth II	Dartford, Great Britain	1991	1,476
Chongging-2	Sichuan Province, China	1996	1,456
Carlos Casado	Barrios de Luna, Spain	1983	1,443

Concrete Arch Bridges

Name	Location	Completed	Length
Wanxiang	Yangzi River, China	1996	1,377
Krk-1 (east span)	Krk Island, Croatia	1980	1,279
Jiangjiehe	Wu River, China	1995	1,082
Yongjiang	Guangxi, China	1996	1,023
Gladesville	Sydney, Australia	1964	1,000
Ponte da Amizade	Paraná River, Brazil/Paraguay	1964	951
Bloukrans	Cape Province, South Africa	1983	892
Arrábida	Oporto, Portugal	1963	885
Sandö	Kramfors, Sweden	1943	866
Pont Chateaubriand	La Rance, France	1991	856
Shibenik	Shibenic Bay, Croatia	1966	807
Barelang	Sumatra, Indonesia	1998	803
Krk-2 (west span)	Krk Island, Croatia	1980	800
Jinshajiang-Yibin	Sichuan, China	1990	787
Beppo-Myouban	Oita, Japan	1989	770
Fiumarella	Catenzaro, Italy	1961	757
Zaporoze	Dnepr River, Ukraine	1952	748
El Rincón Viaduct	Las Palmas, Spain	1994	744
Novi Sad	Danube River, Yugoslavia	1961	692
Lingenau	Bregentz, Austria	1968	688
Usagawa	Yamaguchi, Japan	1983	669

Steel Arch Bridges

Name	Location	Completed	Length
New River Gorge	Fayetteville, West Virginia	1977	1,699
Bayonne	New York, New York	1931	1,653
Sydney Harbour	Sydney, Australia	1932	1,650
Fremont	Portland, Oregon	1973	1,256
Port Mann	Vancouver, Canada	1964	1,200
Thatcher	Balboa, Panama	1962	1,128
Laviolette	Trois Rivières, Canada	1967	1,099
Runcorn-Widnes	Mersey River, Great Britain	1961	1,082
Zdákov Lake	Orlik, Czech Republic	1967	1,082
Birchenough	Sabi River, Zimbabwe	1935	1,079
Roosevelt Lake	Roosevelt, Arizona	1990	1,079

Prestressed Concrete Girder Bridges

(Main Span Length)

Name	Location	Completed	Length
Stolmasundet	Austevoll, Norway	1998	987
Raftsundet	Lofoten, Norway	1998	977
Humen	Pearl River, China	1998	915
Varodd	Kristiansand, Norway	1994	853
Gateway	Brisbane, Australia	1986	853
Skye	Skye Island, Great Britain	1995	820
Schottwien	Semmering, Austria	1989	820
Ponte de S. João	Oporto, Portugal	1991	820
Northumberland	New Brunswick, Canada	1997	820
Huangshi	Hubei, China	1996	803
Koror-Babelthuap	Koror, Palau (Collapsed 1996)	1977	790
Hamana	Imagiri-Guchi, Japan	1976	787
Hikoshima	Shimonoseki, Japan	1975	774
Norddalsfjord	Sogn-Fjordane, Norway	1987	757
Urato	Kochi, Japan	1972	754
Ship Channel	Houston, Texas	1982	751
Puente International	Uruguay/Argentina	1976	721
Ponte Tancredo	Brazil/Argentina	1985	721
Mooney Creek	Mount White, Australia	1986	721
Agi-Gawa	Gifu, Japan	1985	721

Courtesy of Professor Hakan Sundquist, head of the Structural Design and Bridges Research Group,
Department of Structural Engineering, Royal Institute of Technology, Sweden.

World's Longest Tunnels

(Lengths are in Miles)

Road Tunnels

Name	Location	Completed	Length
Laerdal	Norway	2000	15.2
St. Gotthard	Switzerland	1980	10.5
Arlberg	Austria	1978	8.68
Fréjus	France-Italy	1980	8
Mont Blanc	France-Italy	1965	7.24
Gudvanga	Norway	1991	7.1
Folgefonn	Norway	2001	6.9
Kan-etsu (South)	Japan	1991	6.86
Kan-Etsu (North)	Japan	1985	6.78
Gran Sasso (East)	Italy	1984	6.32
Gran Sasso (West)	Italy	1995	6.28

Railway Tunnels

Name	Location	Completed	Length
Seikan	Japan	1988	33.46
Eurotunnel	England-France	1994	31.34
Daishimizu	Japan	1982	13.8
Simplon II	Italy-Switzerland	1922	12.31
Simplon I	Italy-Switzerland	1906	12.3
Vereina	Switzerland	1999	11.84
Shinkanmon	Japan	1975	11.63
Appeninno	Italy	1934	11.5
Rokkou	Japan	1972	10
Furka Base	Switzerland	1982	9.6
Haruna	Japan	1982	9.53
Severomuyskly	Russia	1984	9.5
Gorigamine	Japan	1997	9.43
Mone Santomarco	Italy	1987	9.34
St. Gotthard	Switzerland	1882	9.32
Nakayama	Japan	1982	9.23
Lötschberg	Switzerland	1913	9.1

Courtesy of Gunnar Lotsberg, Senior Engineer, Norwegian Public Roads Administration, and editor of the *World's Longest Tunnel Page*.
This website with dozens of additional entries can be located at http://home.no.net/lotsberg/index.html.

APPENDIX II

Bridge and Tunnel Websites

Bailey Bridge

These websites provide historical and technical information on the prefabricated Bailey Bridge developed during World War II.

http://www.mabey.com/Bailey_bridge.htm

http://www.baileybridge.com/index.html

Bat Use of Bridges

An effort that began in Austin, Texas, to make the public aware that certain bridges provide a haven for endangered bats has flapped its way across the nation. These websites address this fascinating and unexpected use of bridges.

http://www.batcon.org/bridge/babintro.html

http://www.fhwa.dot.gov/environment/BATS_TX.htm

http://www.or.blm.gov/coosbay/Bat/bci_rprt.htm

Bridge Builder Magazine

This link will take the viewer to *Bridge Builder Magazine,* an online professional trade magazine containing articles about the various facets of bridge construction and maintenance.

http://www.bridgebuildermagazine.com/3a.html

BridgePros.Com

Geared for the engineering professional, the site's links to contractors and engineering firms will also be of interest to bridge enthusiasts. There are also illustrated sections on past and future bridge projects.

http://bridgepros.com/

Bridges and Tunnels of Allegheny County and Pittsburgh, Pennsylvania

This privately operated website at first appears to be an official agency effort but is the excellent work of the bridge enthusiast Bruce C. Cridlebaugh. The site contains a comprehensive glossary of bridge and tunnel terms and the historical background on some of America's most famous bridges, including Gustav Lindenthal's Smithfield Street Bridge.

http://www.buildingtechnology.com/bcba/bridges/index.htm

Bridges of Metropolitan Cleveland: Past and Present

This website is an online version of an excellent book published in 1981 by Sara Ruth Watson and John R. Wolfs. The site is illustrated with numerous images and contains links to other digitized books about the history of bridges in the Cleveland area. An account of the 1876 Ashtabula Bridge disaster is included.

http://web.ulib.csuohio.edu/SpecColl/bmc/index.html

BridgeSite.Com

This commercial site is designed for professional engineers, bridge enthusiasts, and students. It also has abundant links to bridge image galleries and to educational sites.

http://www.bridgesite.com/index.html

Brunel Engine House and Tunnel Exhibition

This British site is posted by the museum based in the engine house built by Marc Brunel to power water pumps used in the effort to build the first subaqueous tunnel in the world, the subterranean crossing beneath London's Thames River.

http://www.museumweb.freeserve.co.uk/brunel.htm

Calatrava, Santiago

This fascinating website has been posted by the firm of the engineer-artist Santiago Calatrava to showcase his dramatic designs of bridges and other structures.

http://www.calatrava.com/indesflash.html

Central Artery/Tunnel Project

Boston's tunnel and bridge transportation project commonly known by the nickname Big Dig has dug itself into a financial mess, but this official website of the Massachusetts Turnpike Authority provides links to everything you ever wanted to know about America's most expensive infrastructure project.

http://www.bigdig.com/

Crossings of Metro New York, The

This fantastic website provides numerous pages dedicated to the bridge and tunnel crossings of the New York City area. A concise history of each project is provided along with photographs and statistics.

http://www.nycroads.com/crossings/

Detroit-Windsor Tunnel

America's first immersed tube tunnel project still carries traffic between the United States and Canada, and this official site for the tunnel provides information about its history, construction, and present-day operations.

http://www.dwtunnel.com/home.html

Seismic Research Organizations

Bridges, like other structures, are prone to devastating earthquake damage. The following sites are those of university earthquake research programs. The last site provides unsettling information about the New Madrid quakes.

http://eerc.berkeley.edu/

http://mae.ce.uiuc.edu/

http://mceer.buffalo.edu

http://www.hsv.com/genlintr/newmadrd/index.htm

Eurotunnel (English Channel Tunnel)

This official website of the Eurotunnel contains a history page of the historic project, which was dreamed of for more than 200 years. The main pages also provide information on the tunnel's present-day activities.

http://www.eurotunnel.co.uk/corp/history.asp

Federal Highways Administration Office of Bridge Technology

For those interested in the nuts and bolts of bridge technology and a glimpse of state-of-the-art bridge engineering techniques, this site offers both images and text.

http://www.fhwa.dot.gov/bridge/

Golden Gate Bridge and San Francisco–Oakland Bay Bridge

This Museum of the City of San Francisco website provides an engaging look at the history and construction of the two most famous bridges in California.

http://www.sfmuseum.org/assoc/bridge00.html

Helmet Men, Early 19th-Century Mine Rescue Experts

The Mine Safety Health Administration has posted a historical site recounting the development and use of "helmet men," or mine rescue teams that employed early self-contained breathing apparatus. This site contains fascinating period photographs.

http://www.msha.gov/century/rescue/rstart.htm

Historic American Building Survey/Historic American Engineering Record

This Library of Congress site contains the catalog of detailed histories, photographs, and engineering drawings of many of America's most famous structures, including bridges and tunnels. A sizable percentage of the text and photographic data has been digitized and is available for downloading.

http://lcweb2.loc.gov/ammem/hhhtml/hhhome.html

Honshu-Shikoku Bridge Authority
This official website of the Honshu-Shikoku Bridge Authority contains images and engineering information about two of the world's most impressive bridges: the Akashi Kaikyo Bridge and the Tatara Bridge.

http://www.hsba.go.jp/e-index.htm

Hoover Dam Diversion Tunnels
These websites provide a concise history of the Hoover Dam diversion tunnels project with links to the story of the other elements of the Hoover Dam construction effort.

http://www.usbr.gov/history/hoover.htm

http://www.hooverdam.com/History/essays/tunnels.htm

Lake Pontchartrain Causeway
The longest bridge in the world has its own website, which provides information about its history, operation, and toll fees.

http://www.thecauseway.com/default.html

Mackinac Bridge
This official website of the Mackinac Bridge Authority in Michigan contains links to the history and image pages about Michigan's stunning suspension bridge across the Straits of Mackinac.

http://www.mackinacbridge.org/

Maginot Line
This famous series of fortifications built by France along its frontier with Germany was a massive feat of underground construction during the years between the two world wars. These websites provide fascinating images and discussions about this controversial wonder.

http://www.geocities.com/Athens/Acropolis/9173/index.htm#anglais

http://www.ligne-maginot.com/

http://www.lignemaginot.com/ligne/index-en.htm

Metropolitan Transit Authority of New York
This is the official website of the Metropolitan Transit Authority, the subordinate agency of the Triborough Bridge and Tunnel Authority. It features detailed information about bridges and tunnels linking New York City, including the Brooklyn Battery Tunnel and the Verrazano Narrows Bridge.

http://www.mta.nyc.ny.us/bandt/html/btintro.htm

National Association for Trenchless Technology
This professional organization's website for firms and organizations involved in microtunneling contains a link to a glossary of terms related to robotic tunneling methods.

http://www.nastt.org/index.html

National Information Service for Earthquake Engineering
Known as NISEE, this University of California, Berkeley, organization works with the Earthquake Engineering Research Center (EERC) and the Pacific Earthquake Engineering Research Center (PEER) to disseminate information about the effects of earthquakes on artificial structures. The site also contains a link to the affiliated Godden Structural Engineering Slide Library and the EQIIS Image Database, which total more than 12,000 images of structures of all types.

http://nisee.berkeley.edu/

Norwegian Submerged Floating Tunnel
Although not a new idea, the concept of a submerged floating tunnel—a cross between a tunnel and a pontoon bridge—is being touted by the company that established this website complete with images of this daring plan.

http://www.nsft.no/

Oregon Department of Transportation Historic Bridges
This website's "Coastal Bridge" section provides images of bridges designed by the famed Oregon bridge engineer Conde B. McCullough as well as other interesting spans.

http://www.odot.state.or.us/eshtm/br.htm

Port Authority of New York and New Jersey
This official website of the Port Authority of New York and New Jersey contains a page dedicated to the bridges and tunnels under its jurisdiction, including the magnificent George Washington Bridge and the Holland Tunnel.

http://www.panynj.gov/tbt/tbtframe.HTM

Prevention and Control of Highway Tunnel Fires
This Federal Highway Administration site provides an electronic version of a government publication on the causes of tunnel fires around the world and an analysis of how to prevent such calamities.

http://www.fhwa.dot.gov/bridge/prevent1.htm

John A. Roebling's Sons Co.
An extremely detailed website, it explains and illustrates the extraordinary history of the Brooklyn Bridge designer and wire rope inventor John A. Roebling and the wire cable firm he established.

http://www.inventionfactory.com/history/main.html

San Francisco–Oakland Bay Bridge: Bancroft Library, University of California, Berkeley

This extraordinary website has compiled medium- and high-resolution photographs chronicling the construction of the San Francisco–Oakland Bay Bridge during the 1930s to provide a rare behind-the-scenes view of bridge-building. The second site provides additional background on the bridge and five other major crossings of San Francisco Bay.

http://www.oac.cdlib.org/dynaweb/ead/calher/baybridge/

http://www.lib.berkeley.edu/Exhibits/Bridge/index.html

Structurae: International Database and Gallery of Structures
This amazing website features links to more than 1,500 structures, including bridges. Bridges can be searched for by the name of their designer or their type of structure.

http://www.structurae.de/en/index.html?http://www.structurae.de/en/structures/data/st r00162.html

TunnelingOnline.Com
This webzine provides up-to-date information on the tunneling and underground construction industry and is sponsored by *Tunnel Business Magazine*. A photo gallery provides a glimpse of various projects around the world.

http://www.tunnelingonline.com/

Turner-Fairbanks Highway Research Center
This website, with an internal search engine, contains information about research into bridge and highway technology conducted for the Federal Highway Administration. It also contains links to a trio of federally published magazines, including *Public Roads*.

http://www.tfhrc.gov/

Wheeling Bridge
The Ohio County Public Library has posted this excellent website discussing the history of the first great suspension bridge built in America by the maverick civil engineer Charles Ellet.

http://wheeling.weirton.lib.wv.us/landmark/bridges/susp/bridgdex.htm

World's Longest Tunnel Page
This personal website contains a listing of dozens of the world's most interesting and longest tunnels with their statistics.

http://home.no.net/lotsberg/index.html

BIBLIOGRAPHY

American Society of Civil Engineers, Committee of Civil Engineers. *A Biographical Dictionary of American Civil Engineers*. New York: American Society of Civil Engineers, 1972.

American Society of Mechanical Engineers. *Mechanical Engineers in America Born Prior to 1861: A Biographical Dictionary*. New York: American Society of Mechanical Engineers, 1980.

Beaver, Patrick. *A History of Tunnels*. Secaucus, N.J.: The Citadel Press, 1973.

Beckett, Derrick. *Stephensons' Britain*. Newton Abbot, Devon; North Pomfret, Vt.: David & Charles, 1984.

Bell, S. Peter. *A Biographical Index of British Engineers in the 19th Century*. New York: Garland Publishing, 1975.

Bieniawski, T. Z. *Rock Mechanics Design in Mining and Tunneling*. Rotterdam and Boston: A.A. Balkmea, 1984.

Billington, David P. *Robert Maillart's Bridges: The Art of Engineering*. Princeton, N.J.: University Press, 1979.

————. *The Tower and the Bridge: The New Art of Structural Engineering*. New York: Basic Books, 1983.

————. *Robert Maillart: Builder, Designer and Artist*. Cambridge; New York: Cambridge University Press, 1997.

Black, Archibald. *The Story of Bridges*. New York; London: Whittlesey House, McGraw-Hill, 1936.

————. *The Story of Tunnels*. New York; London: Whittlesey House, 1937.

Boardman, Fon W. Jr. *Tunnels*. New York: Henry Z. Walk, 1960.

Bobrick, Benson. *Labyrinths of Iron: A History of the World's Subways*. New York: Newsweek Books, 1981.

Boscardin, Marco D., ed. *Jacked Tunnel Design and Construction: Proceedings of Sessions of GeoCongress 98*. Reston, Va.: American Society of Civil Engineers, 1999.

Brown, David J. *Bridges*. New York: Macmillan, 1993.

Browne, Lionel. *Bridges: Masterpieces of Architecture*. New York: Smithmark Publishers, 1996.

Burr, S. *Tunneling Under the Hudson River*. New York: John Wiley & Sons, 1885.

Byron, Carl R. *A Pinprick of Light: The Troy and Greenfield Railroad and Its Hoosac Tunnel*. Shelburne, Vt.: The New England Press, 1995.

Caro, Robert A. *The Power Broker: Robert Moses and the Fall of New York*. New York: Alfred A. Knopf, 1974.

Cudahy, Brian J. *Under the Sidewalks of New York: The Story of the Greatest Subway System in the World*. New York: Fordham University Press, 1995.

DeLony, Eric. *Landmark American Bridges*. New York: American Society of Civil Engineers; Boston: Little, Brown, 1993.

Derucher, Kenneth N., and Heins, Conrad P., Jr. *Bridge and Pier Protective Systems and Devices.* New York and Basel: Marcel Dekker, 1979.

Dorsey, Florence L. *Road to the Sea: The Story of James B. Eads and the Mississippi River.* New York: Rinehart, 1947.

Doyle, Barry R. *Hazardous Gases Underground: Applications to Tunnel Engineering.* New York: Marcel Dekker, 2001.

Dupré, Judith. *Bridges: A History of the World's Most Famous and Important Spans.* New York: Black Dog & Leventhal Publishers, 1997.

Epstein, Sam, and Epstein, Beryl. *Tunnels.* Boston and Toronto. Little, Brown, 1985.

Fant, Kenne. *Alfred Nobel: A Biography.* New York: Arcade, 1993.

Farraday, R. V., and Charlton, F. G. *Hydraulic Factors in Bridge Design.* Wallingford, Eng.: Hydraulics Research Station, 1983.

Fetherstone, Drew. *The Chunnel: The Amazing Story of the Undersea Crossing of the English Channel.* New York: Times Books, 1997.

Gies, Joseph. *Adventure Underground: The Story of the World's Great Tunnels.* Garden City, N.Y.: Doubleday, 1962.

Gies, Joseph. *Bridges and Men.* Garden City, N.Y.: Doubleday, 1963.

Halliday, Stephen. *The Great Stink of London: Sir Joseph Bazalgette and the Cleansing of the Victorian Capital.* Stroud: Sutton, 1999.

Harrod, Kathryn E. *Master Bridge Builders: The Story of the Roeblings.* New York: Julian Messner, 1958.

Hechler, Ken. *The Bridge at Remagen.* New York: Ballantine Books, 1957.

Hewett, Bertram, and Henry Majendie. *Shield and Compressed Air Tunneling.* New York: McGraw-Hill, 1922.

Hood, Clifton. *722 Miles: The Building of the Subways and How They Transformed New York.* Baltimore and London: The Johns Hopkins University Press, 1995.

Jacobs, David, and Anthony E. Neville. *Bridges, Canals & Tunnels: The Engineering Conquest of America.* New York: American Heritage Publishing, 1968.

Keane, John. *Tom Paine: A Political Life.* Boston, New York, Toronto and London: Little, Brown, 1995.

Lewis, Gene D. *Charles Ellet Jr.: The Engineer as Individualist, 1810–1862.* Urbana, Chicago, and London: University of Illinois Press, 1968.

Low, K. W., ed. *Tunneling in Soil and Rock: Proceedings of Two Sessions at Geotech, 1984.* New York: American Society of Civil Engineers, 1984.

Loyrette, Henri. *Gustave Eiffel.* New York: Rizzoli, 1985.

Mangold, Tom, and John Penycate. *The Tunnels of Cu Chi.* New York: Berkley Books, 1986.

McCullough, David G. *The Great Bridge: The Epic Story of the Building of the Brooklyn Bridge.* New York: Simon & Schuster, 1972.

Miller, Howard S. *The Eads Bridge.* Columbia: University of Missouri Press, 1979.

Murray, Don. *Man Against Earth: The Story of Tunnels and Tunnel Builders.* Philadelphia and New York: J. B. Lippincott, 1961.

Mysak, Joe, and Judith Schiffer. *Perpetual Motion: The Illustrated History of the Port Authority of New York & New Jersey.* Los Angeles: General Publishing Group, 1997.

Organisation for Economic Co-operation and Development. *Road Research: Bridge Maintenance.* Paris: Organisation for Economic Co-operation and Development, 1981.

———. *Road Transport Research: Bridge Rehabilitation and Strengthening.* Paris: Organisation for Economic Co-operation and Development, 1983.

Petroski, Henry. *Engineers of Dreams: Great Bridge Builders and the Spanning of America.* New York: Vintage Books, 1996.

Plowden, David. *Bridges: The Spans of North America.* New York and London: W. W. Norton, 1974.

Podolny, Walter, Jr., and John B. Scalzi. *Construction and Design of Cable-Stayed Bridges.* New York, Chichester, Brisbane, Toronto, and Singapore: John Wiley & Sons, 1986.

Powell, Fred Wilbur. *The Bureau of Mines: It's History, Activities and Organization.* New York and London: D. Appleton, 1922.

Prebble, John. *Disaster at Dundee: The Tay Bridge Disaster.* New York: Harcourt, Brace, 1957.

Pugsley, Alfred, ed. *The Works of Isambard Kingdom Brunel: An Engineering Appreciation.* Cambridge, Eng. and New York: Cambridge University Press, 1980.

Rastorfer, Darl. *Six Bridges: The Legacy of Othmar H. Ammann.* New York and London: Yale University Press, 2000.

Roe, Vivian. *The Great Wall of France: The Triumph of the Maginot Line.* New York: G. P. Putnam's Sons, 1961.

Rolt, Lionel Thomas Caswell. *Isambard Kingdom Brunel: A Biography.* New York: St. Martin's Press, 1959.

Ryall, M. J. "Britannia Bridge: From Concept to Construction," *Proceedings of the Institution of Civil Engineers, Civil Engineering,* May/August, p. 132.

Sandström, Gösta E. *The History of Tunneling: Underground Workings Through the Ages.* London: Barrie and Rockliff, 1963.

———. *Tunnels: A History of Man's Quest for Passage Through the Earth from Ancient Egyptian Rock Temples to the Tunnel Under the English Channel.* New York, Chicago, and San Francisco: Holt, Rinehart & Winston, 1963.

Scott, Quinta. *The Eads Bridge: A Photographic Essay.* Columbia: University of Missouri Press, 1979.

Slater, Humphrey, and Correlli Barnett. *The Channel Tunnel.* London: Allan Wingate, 1957.

Smiles, Samuel. *The Life of George Stephenson and of His Son Robert Stephenson; Comprising Also a History of the Invention and Introduction of the Railway Locomotive.* New York: Harper, 1868.

———. *Lives of the Engineers with an Account of Their Principal Works; Comprising Also a History of Inland Communication in Britain.* London: J. Murray, 1904.

Smith, H. Shirley. *The World's Great Bridges.* New York: Harper & Row, 1965.

Steinman, David B. *The Builders of the Bridge: The Story of John Roebling and His Son.* New York: Harcourt, Brace, 1945.

———. *Famous Bridges of the World.* New York: Random House, 1953.

Steinman, David B., and John T Neville. *Miracle Bridge at Mackinac.* Grand Rapids, Mich.: Wm. B. Erdmans, 1957.

Steinman, David B., and Sara Ruth Watson. *Bridges and Their Builders.* New York: G. P. Putnam's Sons, 1941.

Szechy, Karoly. *The Art of Tunneling.* Budapest: Akademiai Kiado, 1966.

Van der Zee, John. *The Gate: The True Story of the Design and Construction of the Golden Gate Bridge.* New York: Simon & Schuster, 1986.

Vreden, Werner, and John Scott. *Curved Continuous Beams for Highway Bridges.* New York: Frederick Ungar, 1969.

Wahlstrom, Ernest Eugene. *Tunneling in Rock.* Amsterdam, N.Y.: Elsevier Scientific, 1973.

Watson, Wilber J., and Sara Ruth Watson. *Bridges in History and Legend.* Cleveland: J. H. Jansen, 1937.

West, Graham. *Innovation and the Rise of the Tunneling Industry.* Cambridge and New York: Cambridge University Press, 1988.

White, Kenneth R., et al. *Bridge Maintenance Inspection and Evaluation.* New York and Basel: Marcel Dekker, 1981.

Whitney, Charles S. *Bridges: Their Art, Science and Evolution.* New York: Greenwich House, 1983.

Young, Edward M. *The Great Bridge: The Verrazano Narrows Bridge.* New York: Ariel, 1965.

INDEX